REINFORCED CONCRETE

Analysis and Design

REINFORCED CONCRETE
Analysis and Design

S.S. RAY
BE (Cal), CEng, FICE, MBGS

b

Blackwell
Science

© 1995 by
Blackwell Science Ltd
Editorial Offices:
Osney Mead, Oxford OX2 0EL
25 John Street, London WC1N 2BL
23 Ainslie Place, Edinburgh EH3 6AJ
238 Main Street, Cambridge,
   Massachusetts 02142, USA
54 University Street, Carlton,
   Victoria 3053, Australia

Other Editorial Offices:
Arnette Blackwell SA
1, rue de Lille
75007 Paris
France

Blackwell Wissenschafts-Verlag GmbH
Kurfürstendamm 57
10707 Berlin
Germany

Blackwell MZV
Feldgasse 13
A-1238 Wien
Austria

First published 1995

Set by Setrite Typesetters, Hong Kong
Printed and bound in Great Britain by
Bell and Bain Ltd., Glasgow

DISTRIBUTORS

Marston Book Services Ltd
PO Box 87
Oxford OX2 0DT
(*Orders*: Tel: 01865 791155
          Fax: 01865 791927
          Telex: 837515)

USA
Blackwell Science, Inc.
238 Main Street
Cambridge, MA 02142
(*Orders*: Tel: 800 215-1000
                617 876-7000
          Fax: 617 492-5263)

Canada
Oxford University Press
70 Wynford Drive
Don Mills
Ontario M3C 1J9
(*Orders*: Tel: 416 441-2941)

Australia
Blackwell Science Pty Ltd
54 University Street
Carlton, Victoria 3053
(*Orders*: Tel: 03 347-5552)

A catalogue record for this book is available
from the British Library

ISBN 0-632-03724-5

Library of Congress
Cataloging in Publication Data

Ray, S.S.
   Reinforced concrete: analysis and
   design/S.S. Ray.
       p.   cm.
   Includes bibliographical references and
   index.
   ISBN 0-632-03724-5
   1. Reinforced concrete construction.
   I. Title.
TA683.R334   1994
624.1′8341−dc20   94-13306
                                        CIP

Dedicated to my father Professor K.C. Ray

Contents

Preface

I believe that the contents of this book will prove to be extremely valuable to practising engineers, students and teachers in the field of reinforced concrete design. There are many excellent books available dealing with the design of reinforced concrete elements but, in my opinion, they lack completeness in certain ways. The design of a reinforced concrete member requires many checks in a systematic structured manner and the step-by-step approach adopted in this book is intended to ensure that the design process is complete in all respects. It is my view that the member itself, when fully designed, does not constitute a complete design because it ignores the connections to other members and to the foundation that are needed to provide true completeness of design for the structure. I have attempted here to elucidate the necessary global analysis. Also, most books on reinforced concrete design do not deal with the aspects of soil structure interaction problems and are hence incomplete.

The highly structured step-by-step methodology I have used makes the book fully comprehensive and user-friendly. Accordingly, the task of quality assurance becomes less arduous and the product or output of a design office becomes fully standardised if this approach is strictly followed. For students, the book should prove to be invaluable because the essential elements of the theory of reinforced concrete are discussed, followed by a structured approach to the design of all elements in a building, including foundations and the connections of the reinforced concrete members to each other to create a complete building. The numerous worked examples should be very useful to students and practitioners alike. The book also presents practical advice on designing reinforced concrete elements and the student should benefit from learning the methods adopted in a design consultancy.

My intention has been to illustrate the design principles at each stage by using a profusion of sketches. The book includes many more illustrations than a standard textbook on reinforced concrete because it was felt necessary to clear all ambiguities in the codes of practice by the use of diagrams, an approach which should appeal to both practising engineers and students.

The book includes a lot more new design aids than are usually found in the available books. For instance, the tables and charts included in this book for the design of solid slabs and flat slabs cannot be found in other published textbooks on the subject. References to many published books on the subject of reinforced concrete are also given.

I would like to thank the British Standards Institution for their kind permission to reproduce some of the essential tables from the codes of practice. I also wish to thank the US Army Armament Research and Development Centre, Picatinny Arsenal, NJ and Amman and Whitney, Consulting Engineers, New York for granting permission to reproduce the extremely useful charts on the yield-line design of slabs in Chapter 3.

Finally this undertaking could not have been successfully achieved without the active encouragement of my wife.

S.S. Ray
Great Bookham
Surrey

References

1. British Standards Institution (1985) *Structural use of concrete*. Parts 1, 2 and 3. BSI, London, BS 8110.

2. British Standards Institution (1986) *Foundations*. BSI, London, BS 8004.

3. British Standards Institution (1979) *Code of practice for design of composite bridges*. Part 5. BSI, London, BS 5400.

4. American Concrete Institute (1983) *Building code requirements for reinforced concrete*. M83. ACI, Detroit, Michigan, USA, ACI 318.

5. Tomlinson, M.J. (1982) *Foundation Design and Construction*, 3rd edn. Pitman Publishing, London.

6. Bowles, J.E. (1982) *Foundation Analysis and Design*, 3rd edn. McGraw-Hill International, Tokyo.

7. Tomlinson, M.J. (1987) *Pile Design and Construction Practice*, 3rd edn. E. & F.N. Spon, London.

8. US Army Armament Research and Development Centre (1987) *Structures to resist the effects of accidental explosions, Volume IV: Reinforced concrete design*. US Army ARDEC, N.J., USA, US Army Standard ARLCD, SP 84001.

9. Moody, W.T. *Moments and Reactions for Rectangular Plates*. US Department of the Interior, Engineering Monograph No. 27, Denver, Colorado, USA.

10. Reynolds, C.E. & Steedman, J.C. (1988) *Reinforced Concrete Designer's Handbook*, 10th edn. E. & F.N. Spon, London.

11. Wood, R.H. (1968) The reinforcement of slabs in accordance with a predetermined field of moments. *Concrete*, **2**, No. 2, Feb. pp. 69–76.

12. Armer, G.S.T. (1968) Discussion of Reference 16. *Concrete*, **2**, No. 8, Aug. pp. 319–20.

13. Cheng-Tzu Thomas Hsu (1986) Reinforced concrete members subject to combined biaxial bending and tension. *ACI Journal*, Jan./Feb. American Concrete Institution, Detroit, Michigan, USA.

14. British Standards Institution (1972) *Wind loads*. CP3: Chapter V: Part 2. BSI, London.

General references

Allen, A.H. (1983) *Reinforced Concrete Design to BS8110 − Simply Explained*. E. & F.N. Spon, London.

Batchelor & Beeby (1983) *Charts for the design of circular columns to BS8110*. British Cement Association, Slough, UK.

British Standards Institution (1987) *Design of concrete structures to retain aqueous liquids*. BSI, London, BS8007.

Park, R. & Paulay, T. (1975) *Reinforced Concrete Structures*. John Wiley & Sons, New York.

Pucher, A. (1977) *Influence Surfaces of Elastic Plates*. Springer Verlag, Vienna, Austria.

Roark, R.J. & Young, W.C. (1975) *Formulae for Stress and Strain*, 5th edn. McGraw-Hill International, Tokyo.

Chapter 1

Theory of Reinforced Concrete

1.0 NOTATION

a_u	Deflection of column due to slenderness
A_c	Net area of concrete in a column cross-section
A_s	Area of steel in tension in a beam
A_s'	Area of steel in compression in a beam
A_{sb}	Area of bent shear reinforcement
A_{sc}	Area of steel in column
A_{sv}	Area of steel in vertical links
b	Width of reinforced concrete section
b_w	Width of web in a beam
c_0	Effective crack height at 'no slip' at steel
C	Internal compressive force in reinforced concrete section
d	Effective depth of tensile reinforcement
d'	Effective depth of compressive reinforcement
E_c	Modulus of elasticity of concrete
E_s	Modulus of elasticity of steel
f_c	Service stress in concrete
f_k	Characteristic strength of material
f_m	Mean strength of material from test results
f_s	Service stress in steel
f_y	Characteristic yield strength of steel
f_{cu}	Characteristic cube strength of concrete at 28 days
h	Overall depth of a concrete section
h_f	Thickness of flange in a T-beam
h_o	Initial crack height in reinforced concrete member
h_{max}	Maximum overall dimension of a rectangular concrete section
h_{min}	Minimum overall dimension of a rectangular concrete section
I	Moment of inertia
M	Applied bending moment
M'	Maximum moment of resistance of concrete section
M_f	Moment of resistance of concrete in flange
M_w	Moment of resistance of concrete in web
N	Ultimate axial load on column
p	Percentage of tensile reinforcement in a beam $= 100A_s/bd$
p	Percentage of total reinforcement in a column $= 100A_s/bh$
q	Shear flow (kN/m)
Q	First moment of area above plane of interest

r_b	Curvature of a member in bending
s	Standard deviation
S	Spacing of shear reinforcement
T	Internal tensile force in steel reinforcement
v	Shear stress in concrete (N/mm^2)
v_c	Design concrete shear stress (N/mm^2)
v_t	Shear stress in concrete due to torsion (N/mm^2)
V	Shear force in concrete section
V_c	Design concrete shear capacity
V_s	Design shear capacity of shear reinforcement
x	Depth of neutral axis from compression face
y	Distance from neutral axis
z	Depth of lever arm
α	Angle of inclination to horizontal of shear reinforcement
β	Angle of inclination to horizontal of concrete strut in truss analogy
β_a	Empirical factor governing deflection of slender columns
β_b	Ratio of redistributed moment over elastic analysis moment
β_f	Factor governing moment of resistance of concrete T-section
γ_m	Material factor
δ	Deflection of beam
ε_y	Strain at yield of steel reinforcement

1.1 INTRODUCTION

The criteria which govern the design of a structure for a particular purpose may be summarised as follows:

Fitness for purpose
Safety and reliability
Durability
Good value for money
External appearance
User comforts
Robustness.

Fitness for purpose is generally covered by the overall geometry of the structure and its components. It should be possible to have unrestricted and unhindered use of the structure for the purpose for which it is built.

Safety and reliability are assured by following the Codes of Practice for loading, materials, design, construction and fire-resistance.

Durability is taken care of by the choice of the right material for the purpose and also by bearing in mind during the design process, the requirements for proper maintenance.

Good value for money is perhaps the most important criterion. The designer should take into account not only the cost of materials but also the buildability, the time required to build, the cost of temporary structures, the cost of maintenance over a period of time and in some cases the cost of demolition/decommissioning.

External appearance of structures changes over a period of time. The designer should be aware of the effects of cracking, leaking, staining, spalling, flaking, etc. of the materials in use. The designer should make appropriate allowances to avoid the degradation of appearance.

User comforts are influenced by the vibration of the structure due to wind, road/rail traffic or vibrating machinery. Large deflections under load also cause alarm to the users. The designer should pay adequate attention to alleviation of these anticipated discomforts.

Robustness comes with the chosen structural form and is determined by the additional inherent strength of the structure as a whole to withstand accidental loadings. Collapse of one key member in the structure must not initiate global collapse. The design must foresee the 'domino effect' in the structure and avoid it by careful planning.

1.2   CHARACTERISTIC STRENGTH OF MATERIALS

The characteristic strength of a material is defined as the strength below which 1 in 20 test results are likely to fall.

The value of the characteristic strength is defined statistically by the following formula

$$f_k = f_m - 1.64s$$

where f_k = characteristic strength of material
f_m = mean strength of material from test results
1.64 is a factor which defines the 1 in 20 test results falling below f_k
s is the standard deviation.

The characteristic strength of concrete, f_{cu}, is the cube strength of concrete at 28 days.
The characteristic strength of reinforcing steel, f_y is the strength at yield.

1.3   MATERIAL FACTORS

To obtain the design strength of materials a further factor called the material factor γ_m is applied. The material factor takes into account the tolerances associated with the geometry, the variability of materials on site, the inconsistency in the manufacture and curing on site and the effects of long-term degradation.

The values of γ_m for the ultimate limit state are as follows:

reinforcement   1.15
concrete in flexure or axial load   1.50
concrete in shear   1.25
bond strength in concrete   1.40
bearing stress   1.50

For exceptional loads and for localised damage, γ_m may be taken equal to 1.3 for concrete and 1.0 for reinforcement.

1.4 MATERIAL STRESS–STRAIN RELATIONSHIP

1.4.1 Short-term design stress–strain curve for normal weight concrete

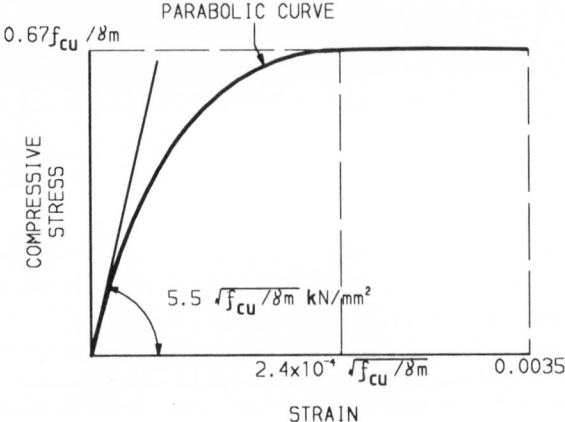

SK 1/1 Short-term design stress–strain curve for normal weight concrete.

The features of this design curve are as follows:

- The *initial elastic modulus* of concrete may be defined as the initial tangent to the parabolic curve which is given by:

$$E_c = 5.5\left(\frac{f_{cu}}{\gamma_m}\right)^{\frac{1}{2}} \text{kN/mm}^2$$

- The *ultimate stress* in concrete for design purposes is defined as:

$$\sigma_u = 0.67\frac{f_{cu}}{\gamma_m} \text{N/mm}^2$$

- The *ultimate strain* in concrete for design purposes is taken as 0.0035. Beyond that strain level the concrete loses its compressive stiffness.

The strain in concrete when the parabolic stress–strain relationship reaches the ultimate stress level is given by:

$$\varepsilon_y = 2.4 \times 10^{-4}\left(\frac{f_{cu}}{\gamma_m}\right)^{\frac{1}{2}}$$

Note: Concrete can withstand compressive stresses only. The tensile stress in concrete is ignored in the design.

1.4.2   Short-term design stress–strain curve for reinforcement

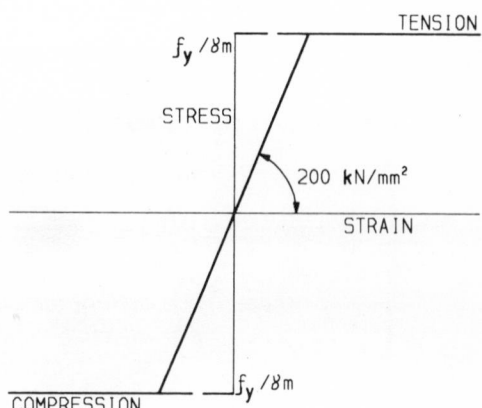

SK 1/2 Short-term design stress–strain curve for reinforcement.

The features of this design curve are as follows:

- The *elastic modulus* of steel reinforcement regardless of grade of steel may be assumed as $200 \, kN/mm^2$, which is the slope of the curve up to yield.
- The *yield stress* of steel reinforcement is f_y, but for design purposes it will be taken as f_y/γ_m.
- The *stress* after yield remains constant and is represented by a constant stress line.

The stress–strain relationship is identical in tension and compression.

$$\varepsilon_y = 0.87 \frac{f_y}{200}$$

$$= 2.0 \times 10^{-3} \qquad \text{for } f_y = 460 \, N/mm^2$$

1.5 DESIGN FORMULAE FOR REINFORCED CONCRETE SECTIONS

1.5.1 Singly reinforced rectangular section

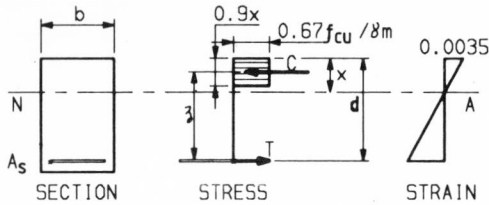

SK 1/3 Stress–strain diagrams of a reinforced concrete section subject to bending moment.

Plane section remains plane.
Applied moment on the section $= M$; $\gamma_m = 1.5$ for concrete, 1.15 for steel.

$\quad C =$ compressive force in section

$$= \left(\frac{0.67}{1.5}\right)f_{cu}b(0.9x) = 0.402f_{cu}bx$$

$\quad T =$ force in steel reinforcement

$$= \left(\frac{f_y}{\gamma_m}\right)A_s = 0.87f_yA_s$$

where $A_s =$ area of tensile steel in section
$\qquad\quad d =$ effective depth from outer compressive fibre to centroid of steel reinforcement.

By internal force equilibrium,

$$C = T$$

or $\quad 0.402f_{cu}bx = 0.87f_yA_s$

or $\quad x = 2.164\left(\dfrac{f_yA_s}{f_{cu}b}\right)$

$$z = d - 0.45x = d\left(1 - 0.97\left(\frac{f_yA_s}{f_{cu}bd}\right)\right)$$

or $\quad \dfrac{z}{d} = 1 - 0.97\left(\dfrac{f_yA_s}{f_{cu}bd}\right)$

or $\quad A_s = \left(1 - \dfrac{z}{d}\right)\left(\dfrac{f_{cu}bd}{0.97f_y}\right)$

$$M = 0.87f_yA_sz = \left(\frac{0.87}{0.97}\right)\left(1 - \frac{z}{d}\right)f_{cu}bdz$$

$$= 0.90\left(1 - \frac{z}{d}\right)f_{cu}bd^2\left(\frac{z}{d}\right)$$

$$\frac{M}{f_{cu}bd^2} = K = 0.90\left(1 - \frac{z}{d}\right)\left(\frac{z}{d}\right)$$

or $\quad \dfrac{z}{d} = \left[0.5 + \left(0.25 - \dfrac{K}{0.9} \right)^{\frac{1}{2}} \right]$

Maximum moment of resistant of concrete section is obtained for *redistribution not exceeding 10%*, when $x = d/2$.

or $\quad z = d - 0.45x = 0.775d$

Moment of resistance of concrete (maximum), M', is given by

$$M' = 0.402 f_{cu} bxz$$

$$= 0.402 f_{cu} b \left(\dfrac{d}{2} \right) (0.775d)$$

$$= 0.156 f_{cu} bd^2$$

Where redistribution exceeds 10%,

$$x \le (\beta_b - 0.4)d$$

Similarly,

$$M' = 0.402 f_{cu} bxz$$
$$= 0.402 f_{cu} b(\beta_b - 0.4)d[d - 0.45(\beta_b - 0.4)d]$$
$$= [0.402(\beta_b - 0.4) - 0.18(\beta_b - 0.4)^2] f_{cu} bd^2$$

or $\quad K' = 0.402(\beta_b - 0.4) - 0.18(\beta_b - 0.4)^2$

where $\quad \beta_b = \dfrac{\text{(moment after redistribution)}}{\text{(moment before redistribution)}} < 0.9$

1.5.2   The concept of balanced design and redistribution of moments

In a singly reinforced section, if the yield strain in steel $\varepsilon_y = 0.002$ and the ultimate strain in concrete ($= 0.0035$) are simultaneously reached then a balanced failure condition exists.

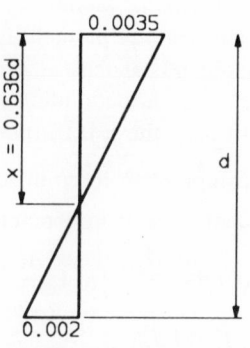

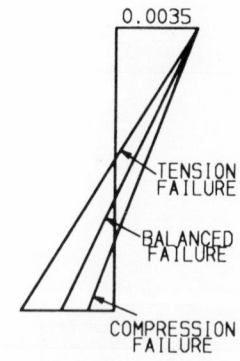

SK 1/4 Strain diagrams of reinforced concrete section.

STRAIN DIAGRAM FOR
BALANCED FAILURE

COMPARATIVE
STRAIN DIAGRAM

From strain diagram,

$$\frac{x}{(d-x)} = \frac{0.0035}{0.002} = 1.75 \quad \text{for } f_y = 460\,\text{N/mm}^2$$

or $x = 0.636d$

The Code does not allow x to be larger than $0.5d$ ensuring that the steel reaches its yield strain before the concrete reaches the ultimate strain. This is designed to allow sufficient rotational capacity in the section.

The more redistribution of moment is allowed, the more rotational capacity is needed from the section. The amount of rotation is dependent on how under-reinforced the section is, or in other words, how quickly the steel in the section reaches the yield strain before the concrete reaches the ultimate strain. To make sure that the rotational capacity exists in the section to allow redistribution, the depth of neutral axis for the design is fixed corresponding to the ratio β_b of redistribution. The compression failure is extremely brittle and must be avoided.

On the other hand, the Code has also put a limit to the minimum value of x. It has done so by limiting z to a maximum value of $0.95d$, which limits x to $0.11d$. This limitation is to avoid a very thin stress block at the ultimate state.

1.5.3 Doubly reinforced rectangular section

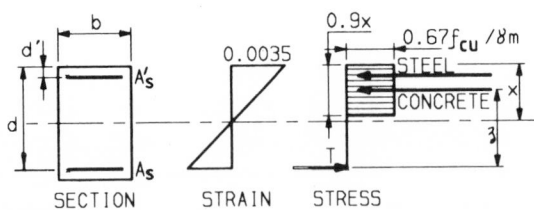

SK 1/5 Stress–strain diagram for doubly reinforced section.

Plane section remains plane.

The design bending moment is greater than $K'f_{cu}bd^2$, which means the concrete moment capacity is exceeded. The neutral axis is fixed by the Code depending on the amount of redistribution or $x = (\beta_b - 0.4)d \leq 0.5d$. This in turn fixes the lever arm z to concrete compression.

C = compressive force in section

   = compression in concrete and compression in steel

$$= 0.9x\left(\frac{0.67f_{cu}}{\gamma_m}\right)b + \frac{A_s' f_y}{\gamma_m}$$

$$= \left(\frac{K'f_{cu}bd^2}{z}\right) + 0.87f_y A_s'$$

T = tensile force = $0.87f_y A_s$

Equating $C = T$,

$$A_s = \left(\frac{K'f_{cu}bd^2}{0.87f_{yz}}\right) + A_s'$$

Applied moment M is equal to the moment of the internal forces. Taking moment about the centre of steel in tension,

$$M = K'f_{cu}bd^2 + 0.87f_y A_s'(d - d')$$

$$A_s' = \frac{M - K'f_{cu}bd^2}{0.87f_y(d - d')}$$

$$= \frac{(K - K')f_{cu}bd^2}{0.87f_y(d - d')}$$

In the above formula it is assumed that the compressive steel will attain yield. This is true provided d' is less than or equal to $0.43x$ or the strain in the steel is at least 0.002 for $f_y = 460\,\text{N/mm}^2$. If d'/x is greater than $0.43x$, the steel stress f_s' will be proportionately modified to account for the reduced strain. Use f_s' in the equation for A_s' instead of $0.87f_y$.

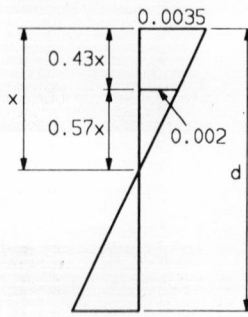

SK 1/6 Doubly reinforced beam strain diagram.

$$A_s' = \frac{(K - K')f_{cu}bd^2}{f_s'(d - d')}$$

where $\quad f_s' = \left(\frac{x - d'}{0.57x}\right)\varepsilon_y E_s$

$$\varepsilon_y = \frac{f_y}{\gamma_m E_s}$$

1.5.4   Singly reinforced flanged beams

The formulation is exactly the same as in a rectangular beam with b equal to the width of the flange provided 0.9 times the depth of the neutral axis x is less than or equal to the depth of the flange.

When $0.9x$ is greater than the depth of flange, then the following analysis will apply.

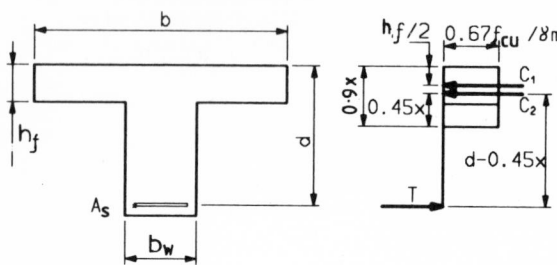

C_1 = compressive force in flange only without web

$$= \left(\frac{0.67 f_{cu}}{\gamma_m}\right)(b - b_w)h_f$$

$$= 0.45 f_{cu}(b - b_w)h_f$$

C_2 = compressive force in web as in a singly reinforced beam

$$= 0.45 f_{cu}b_w(0.9x) = 0.402 f_{cu}b_w x$$

T = tension in steel

$$= \left(\frac{f_y}{\gamma_m}\right)A_s = 0.87 f_y A_s$$

The maximum allowable value of x equals $0.5d$ when the concrete moment of resistance reaches its maximum value.

Assume $x = d/2$.

Taking moment about the centre of tensile steel,

$$M' = C_1\left(d - \frac{h_f}{2}\right) + C_2(d - 0.45x)$$

$$= 0.45 f_{cu}\,(b - b_w)h_f\left(d - \frac{h_f}{2}\right) + 0.201 f_{cu}b_w d(d - 0.225d)$$

$$= f_{cu}bd^2\left[\left(0.45\frac{h_f}{d}\right)\left(1 - \frac{b_w}{b}\right)\left(1 - \frac{h_f}{2d}\right) + 0.157\frac{b_w}{b}\right]$$

$$= \beta_f f_{cu}bd^2$$

Values of β_f for different ratios of b/b_w and d/h_f are found in Fig. 2.1 (Chapter 2).

If the applied moment exceeds $\beta_f f_{cu}bd^2$, then compressive steel in the flange will be required.

To find the tensile steel take moment about C_1 assuming $x = d/2$.

$$M = T\left(d - \frac{h_f}{2}\right) - C_2\left(0.45x - \frac{h_f}{2}\right)$$

$$= 0.87 f_y A_s\left(d - \frac{h_f}{2}\right) - 0.1 f_{cu}b_w d(0.45d - h_f)$$

$$A_s = \frac{M + 0.1 f_{cu}b_w d(0.45d - h_f)}{0.87 f_y(d - 0.5h_f)}$$

Another approach to the design of flanged beams is presented below. When $x = h_f/0.9$, the stress block is situated entirely in the flange.

$$C = \text{compressive force} = bh_f\left(0.67\frac{f_{cu}}{\gamma_m}\right)$$

$$\text{lever arm} = d - \frac{h_f}{2}$$

$$M_f = 0.45f_{cu}bh_f\left(d - \frac{h_f}{2}\right)$$

This is the flange resistance and if the applied moment exceeds this value, then the web comes into compression.

The moment to be carried by the web is M_w, when M is the applied moment.

$$M_w = M - (\text{compression in flange only outside web}) \times (\text{lever arm of flange})$$

$$= M - 0.45f_{cu}(b - b_w)h_f\left(d - \frac{hf}{2}\right)$$

$$= M - M_f\frac{(b - b_w)}{b}$$

$$= M - M_f\left(1 - \frac{b_w}{b}\right)$$

Find M_f, and if M_f is less than M, then find M_w by the above formula. Design for M_w as for a rectangular beam with width equal to b_w. Find A_{s1} for M_f and A_{s2} for M_w.

Total $A_s = A_{s1} + A_{s2}$

$$A_{s1} = \frac{0.45f_{cu}(b - b_w)h_f}{0.87f_y}$$

$$A_{s2} = \frac{M_w}{0.87f_yz}$$

when $K = \dfrac{M_w}{f_{cu}b_wd^2}$

and $z = d\left[0.5 + \left(0.25 - \dfrac{K}{0.9}\right)^{\frac{1}{2}}\right]$

Note: The design against M_w may follow Section 1.5.3, which means that the flanged section may be doubly reinforced, if required.

1.6 ULTIMATE LIMIT STATE – SHEAR

The horizontal shear stress in a homogeneous, isotropic, uncracked beam is given by the classical expression:

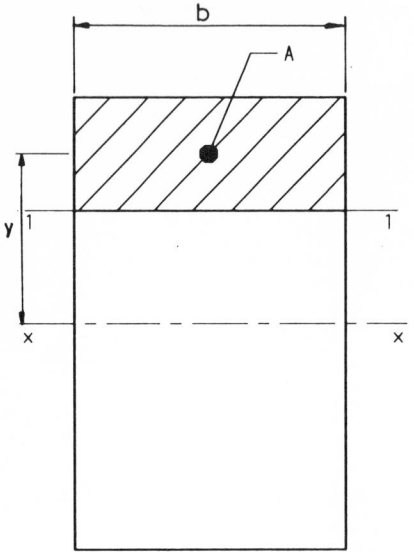

SK 1/8 Calculation of shear stress
in a homogeneous section.

$$v = \frac{VQ}{Ib} \, \text{N/mm}^2$$

where Q = first moment of area above line 1−1 = Ay
 I = second moment of area of the section about x−x
 b = width of section on line 1−1
 v = shear stress at line 1−1
 V = shear at the section
 A = area of section above line 1−1
 y = distance of the centroid of area A from the neutral axis.

Shear flow, $q = \dfrac{VQ}{I}$

In a concrete beam where concrete is ignored, the horizontal shear stress can be included in the expression below for horizontal equilibrium under the neutral axis.

$$\text{d}T = vb_w\text{d}x$$

The shear flow in the tension zone of concrete will be constant because the concrete is ignored.

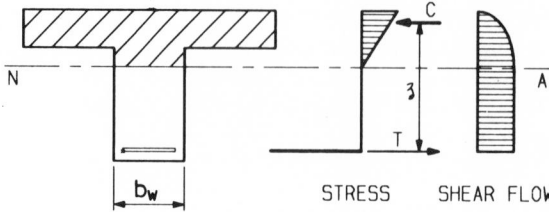

STRESS   SHEAR FLOW   SK 1/9 Calculation of shear stress.

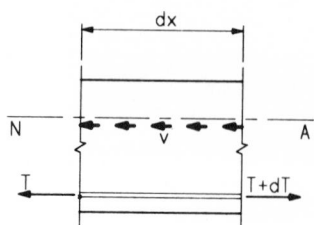

SK 1/10 Shear stress in a
reinforced concrete section.

From the above expression,

$$v = \left(\frac{1}{b_w}\right)\left(\frac{dT}{dx}\right)$$

$$\frac{M}{z} = T$$

or $\quad\dfrac{dM}{z} = dT$

$$\therefore \quad v = \left(\frac{1}{b_w}\right)\left(\frac{1}{z}\right)\left(\frac{dM}{dx}\right)$$

$$V = \frac{dM}{dx}$$

$$\therefore \quad v = \frac{V}{b_w z}$$

For convenience in the ultimate limit state the Code shear stress index is
taken as:

$$v = \frac{V}{b_w d}$$

Effective shear in haunched beams

$$V_{eff} = V - C \sin \theta'$$
$$= V - C' \tan \theta'$$
$$= V - \frac{M}{z} \tan \theta'$$

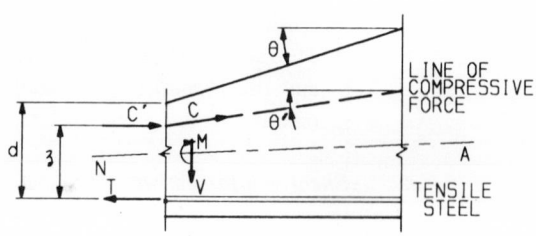

SK 1/11 Effective shear force in a
beam with variable depth.

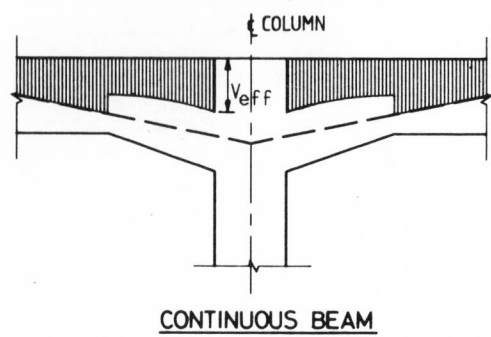

CONTINUOUS BEAM

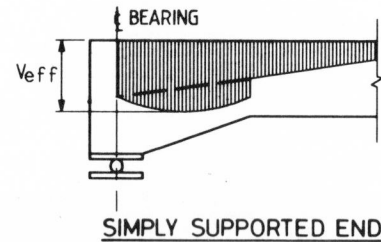

SIMPLY SUPPORTED END

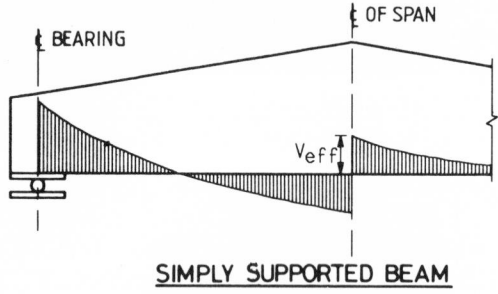

SIMPLY SUPPORTED BEAM

SK 1/12 Effective shear force diagrams for beams with variable depth.

1.6.1 Principle of 'design concrete shear stress'

Shear is resisted in concrete beams by the combined action of the following:

- Shear resistance of concrete in compression zone. Dowel force in tension bars across a crack. Aggregate interlocking across the inclined crack in tension zone.
- The Code formula takes into account the dowel force of the tensile steel and the formula is essentially of empirical nature arrived at from test results.

Shear reinforcement – truss analogy

V_s = shear force to be resisted by reinforcement

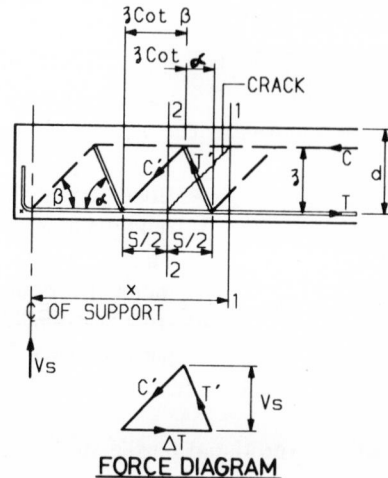

SK 1/13 Truss analogy of shear reinforcement.

From the force diagram,

$$V_s = C' \sin\beta = T' \sin\alpha$$

where C' = concrete strut force
T' = tensile force in shear reinforcement
= resultant of all forces in shear reinforcement within the spacing S.

From geometry:

$$S = z(\cot\alpha + \cot\beta)$$

$$\frac{T'}{S}\sin\alpha = \frac{V_s}{S} = \frac{V_s}{z(\cot\alpha + \cot\beta)}$$

or $$\frac{T'}{S} = \frac{V_s}{z\sin\alpha(\cot\alpha + \cot\beta)}$$

or $$\frac{0.87f_y A_{sb}}{S} = \frac{V_s}{z\sin\alpha(\cot\alpha + \cot\beta)}$$

or $$V_s = 0.87f_y A_{sb}(\cos\alpha + \sin\alpha \cot\beta)\left(\frac{z}{S}\right)$$

The Code uses $(d - d')$ in place of z in the formula.

When vertical stirrups are used and the concrete struts are assumed inclined at 45° to horizontal then $\alpha = 90°$ and $\beta = 45°$.

$$\therefore\ V_s = 0.87f_{yv}A_v\left(\frac{z}{S}\right)$$

$$v_s = \frac{V_s}{bz} = \frac{0.87f_{yv}A_v}{bS}$$

$$v_s \geq (v - v_c)$$

or $\quad A_{sv} \geq \dfrac{bS(v - v_c)}{0.87 f_{yv}}$

Note: The assumptions in the above truss analogy are:

- Bond forces are sustained along the length of the beam where shear reinforcement is required.
- The lever arm z is assumed constant over the section with variable moment producing the shear to be resisted. The diagonal compressive stress in concrete struts calculated from the analogy is equal to $v_s/[\sin^2\beta(\cot\alpha + \cot\beta)]$ is sustainable.

1.6.2 Additional tensile steel in conjunction with shear reinforcement

Referring to SK 1/13, assume that a diagonal crack forms in concrete when the shear force exceeds the concrete shear capacity, V_c. Assume that the ultimate shear force is $V_u = V_c + V_s$, where V_s is resisted by shear reinforcement.

Sections 1–1 and 2–2 are taken at two ends of a diagonal crack.

Consider the free body diagram of internal forces.

Assume that the tensile force requirement at section 2 is also divided in two parts. When the shear becomes V_c, the tensile force required is T_c, and for V_s it is T_s.

Assume that the moment at section 2 is M_{2c} corresponding to V_c and M_{2s} corresponding to V_s.

$$\therefore \ V_u = V_c + V_s$$
$$T_u = T_c + T_s$$
$$M_{2u} = M_{2c} + M_{2s}$$

T' is the tensile force in the shear reinforcement.

Initially assume a shear of V_c on section.
Taking moment about the concrete compressive force C at section 1–1,

$$M_{1c} = V_c x = M_{2c} + V_c z \cot\beta = T_c z$$

or $\quad T_c = \dfrac{M_{2c}}{z} + V_c \cot\beta$

Note: At this stage $T' = 0$.

Next assume a shear of V_s on section.
Taking moment about the concrete compressive force C at section 1–1,

$$M_{1s} = V_s x = M_{2s} + V_s z \cot\beta = T_s z + \left(\dfrac{S}{2}\right) T' \sin\alpha$$

or $\quad T_s = \dfrac{M_{2s}}{z} + V_s \cot\beta - \left(\dfrac{S}{2z}\right) T' \sin\alpha$

Substituting:

$$T' \sin \alpha = V_s \quad \text{and} \quad S = z(\cot \alpha + \cot \beta)$$

we get:

$$T_s = \frac{M_{2s}}{z} + \left(\frac{V_s}{2}\right)(\cot \beta - \cot \alpha)$$

$$T_u = T_c + T_s = \left(\frac{M_{2c} + M_{2s}}{z}\right) + V_c \cot \beta + \left(\frac{V_s}{2}\right)(\cot \beta - \cot \alpha)$$

$$= \frac{M_u}{z} + V_c \cot \beta + \left(\frac{V_s}{2}\right)(\cot \beta - \cot \alpha)$$

This demonstrates quite clearly that when diagonal cracks form in concrete due to shear exceeding V_c, additional tensile steel will be required over and above M_u/z.

This requirement is not explicitly covered in the Code. The rules of curtailment of reinforcement are deemed to satisfy this requirement.

Note: At locations of very high shear this additional requirement should be checked.

1.7   SERVICEABILITY LIMIT STATE – CRACK WIDTH

The basic assumptions to find crack width for flexure are summarised below:

(1) Plane section remains plane before and after bending.
(2) Concrete compressive stress diagram is linear and triangular. The stress is directly proportional to strain.
(3) The short-term Young's modulus of concrete may be used.
(4) The steel reinforcement does not go beyond yield.
(5) The loading is at serviceability limit state.
(6) Effective crack height at 'no slip' is C_o, which is the minimum cover to reinforcement.
(7) Mean crack spacing is $1.5\ C_o$.
(8) Initial crack height h_o is up to the neutral axis and the maximum crack width is a function of the ratio C_o/h_o to take into account slip and internal cracks in concrete.
(9) Stiffening effect of tension in concrete is allowed for by an empirical term.
(10) The equation for the determination of maximum crack width is empirical and is a close fit of test results. It is anticipated that 1 in 5 results will exceed the prediction by the formula.

For the formula and its application, see Chapter 2.

1.8   SERVICEABILITY LIMIT STATE – DEFLECTION

Calculate moments at service load without redistribution.

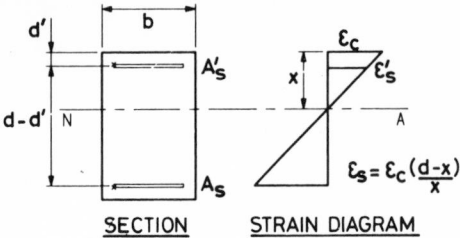

SK 1/14 Doubly reinforced
section.

The assumptions for the analysis of the section are similar to assumptions
(1) to (5) in Section 1.7.

Find the depth of neutral axis, x, and find the stresses in concrete, f_c,
and the stress in steel, f_s, by following Step 25 of worked example in
Chapter 2. This method ignores the concrete under the neutral axis.

$$\text{Curvature} = \frac{1}{r_b} = \frac{f_c}{xE_c} = \frac{f_s}{(d - x)E_s}$$

Alternatively, calculate x as previously and then allow a tensile stress in
concrete up to $1\,\text{N/mm}^2$ short-term and $0.55\,\text{N/mm}^2$ long-term.

$$\text{Deflection, } \delta = Kl^2\left(\frac{1}{r_b}\right)$$

where K depends on the shape of the bending moment diagram.

BS 8110: Part 2: 1985 [1] in Table 3.1 gives different values of K for
various loadings and support conditions. The principle of superposition
may be used to combine different types of loading.

1.9   ULTIMATE LIMIT STATE – TORSION

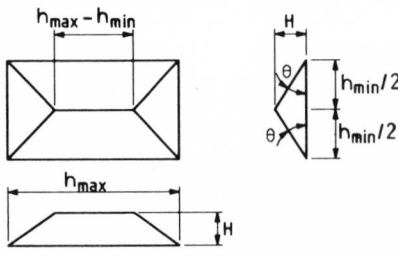

SK 1/15 Membrane analogy for
torsion.

By principles of membrane analogy it is known that 2 times the volume
included between the surface of a deflected membrane and the plane of its
outline is equal to the torque in a twisted member.

Applying this analogy to a rectangular section gives a pyramidal deflected membrane.

$$\text{Volume of pyramid} = \tfrac{1}{3} h_{min}^2 H + \tfrac{1}{2} h_{min}(h_{max} - h_{min})H$$

$$= \left(\frac{h_{min}H}{2}\right)\left(h_{max} - \frac{h_{min}}{3}\right)$$

By membrane theory it is known that the torsional shear stress is the slope of the angle of the deflected membrane.

$$\tan\theta = \frac{H}{(h_{min}/2)} = v_t$$

$$T = 2 \times \text{volume of pyramid} = h_{min}H\left(h_{max} - \frac{h_{min}}{3}\right)$$

Substituting $H = v_t(h_{min}/2)$,

$$T = \left(\frac{v_t h_{min}^2}{2}\right)\left(h_{max} - \frac{h_{min}}{3}\right)$$

$$\text{or} \quad v_t = \frac{2T}{h_{min}^2\left(h_{max} - \dfrac{h_{min}}{3}\right)}$$

1.10   ULTIMATE LIMIT STATE – COLUMNS

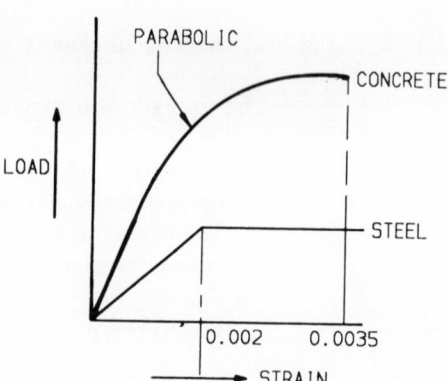

SK 1/16 Typical load–strain curve of a column.

1.10.1   Axial load capacity of columns

Taking creep and shrinkage of concrete into consideration it is difficult to predict the actual stresses in a short concrete column subjected to service axial load in the elastic range, because initial compressive stress from concrete gets transferred to the steel due to creep of concrete. But at the ultimate load stage it is easier to predict the ultimate load-carrying capacity because the concrete ultimate strain of 0.0035 is much higher than steel yield strain. Hence, the steel reaches its *ultimate* load-carrying capacity long before concrete gets there.

The ultimate load-carrying capacity of a short reinforced column may be written as,

$$N_{uz} = \left(\frac{0.67 f_{cu}}{\gamma_m}\right) A_c + \left(\frac{f_y}{\gamma_m}\right) A_{sc}$$

$$= 0.45 f_{cu} A_c + 0.87 f_y A_{sc}$$

where A_c = net concrete cross-sectional area
A_{sc} = area of compressive steel reinforcement.

The Code equations allow for a nominal eccentricity and the formulae are changed to one of the following depending on application:

$$N = 0.4 f_{cu} A_c + 0.75 f_y A_{sc}$$

for a column with nominal eccentricity of load − meaning a column with no design moments and eccentric loads. The eccentricity is allowed for the constructional tolerances.

$$N = 0.35 f_{cu} A_c + 0.67 f_y A_{sc}$$

for a column supporting an approximately symmetrical arrangement of beams. The spans of the beams on either side of the column should not differ by more than 15%. To allow for a certain eccentricity of loading due to the variations in spans and the location and disposition of live loadings on spans, the equation has been modified.

1.10.2   Axial load capacity of slender columns

The strength of a slender column depends on:

(1) Effective height-to-width ratio, where the effective height depends on the rotational end restraints and the lateral restraints by bracing.
(2) The flexural rigidity of the column section which determines the Euler critical buckling load.
(3) The duration of loading which influences the strength and deflections due to creep.

The Code uses the 'Moment Magnifier Method', whereby the effect of slenderness is transferred into an equivalent deflection and an additional moment given by the product of this deflection and the applied direct load.

$$M_{add} = N a_u$$

where $a_u = \beta_a K h$

$$\beta_a = \frac{1}{2000} \left(\frac{l_e}{b'}\right)^2$$

$$K = \left(\frac{N_{uz} - N}{N_{uz} - N_{bal}}\right) \leq 1$$

$N_{bal} = 0.25 f_{cu} bd$ approximately (see Section 1.10.4.1).

h = dimension of column in the plane of bending considered

b' = shorter dimension of column for uniaxial bending, or

= dimension of column in the plane of bending considered for significant biaxial bending

l_e = effective length of column in the plane of bending considered.

(See Chapter 4 for further explanation.)

1.10.3   Axial load and moment on column

The assumptions for the analysis of the section are exactly the same as in the case of beams and the analysis depends on strain compatibility. The design is usually carried out by the use of published charts which have been derived using the following assumptions:

(1) Plane section remains plane or the strain compatibility is assumed.
(2) The concrete stress block is assumed rectangular-parabolic.
(3) The stress–strain curve for steel is bilinear.

To use the charts to find the total area of steel required, the following parameters are required: f_{cu}, f_y, N/bh, M/bh^2 and the d/h ratio. (See Chapter 4 for further explanation.)

1.10.4   Column interaction diagrams

1.10.4.1   *Rectangular section*

If column charts are not available and a hand calculation is required or, where the column size and reinforcement are known and the column load-carrying capacity with variable eccentricity is required for assessment purposes, the following design procedure may be followed. The interaction diagram of a column with known areas of steel will illustrate the ultimate

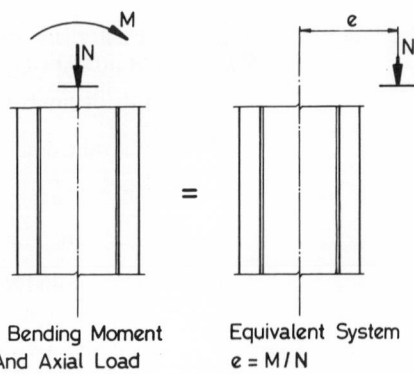

Bending Moment
And Axial Load

Equivalent System
$e = M/N$

SK 1/17 Elevation of a column.

load-carrying capacity of the column section subjected to uniaxial bending moment.

N = applied ultimate direct load
M = applied ultimate coacting bending moment
$e = M/N$ = eccentricity of direct load
C_c = resistance of concrete in compression
C_s = resistance of steel in compression
T = resistance of steel in tension

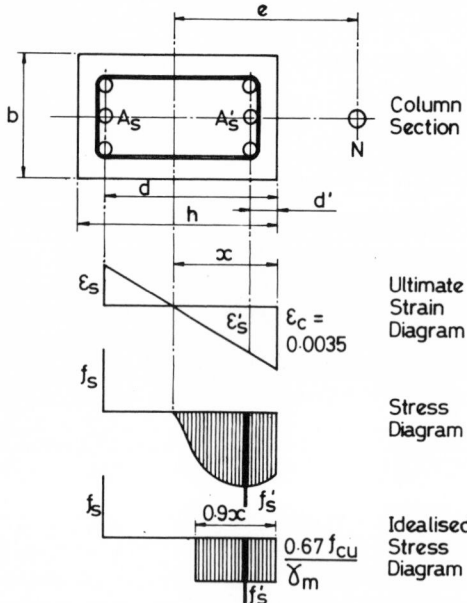

Column Section

Ultimate Strain Diagram

Stress Diagram

Idealised Stress Diagram

SK 1/18 Stress and strain diagrams.

Balanced failure

A *balanced failure* occurs when the tension steel just reaches yield at the same time as the extreme compression fibre in concrete reaches the ultimate strain of 0.0035.

f_y = characteristic yield strength of steel
E_s = modulus of elasticity of steel = $200\,\text{kN/mm}^2$
x_b = depth of neutral axis at 'balanced failure'

From the strain diagram:

$$\frac{0.0035}{x_b} = \frac{f_y/E_s\gamma_m}{(d - x_b)}$$

assuming $f_y = 460\,\text{N/mm}^2$
$\gamma_m = 1.15$ for steel
$E_s = 200\,\text{kN/mm}^2$

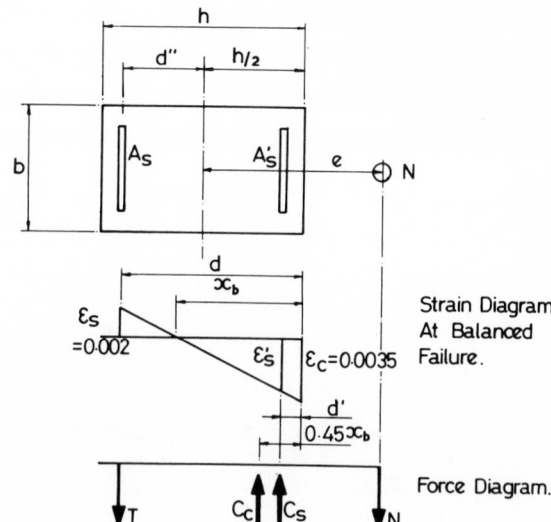

SK 1/19 Strain and force diagram
at balanced failure.

the above equation reduces to:

$$\frac{0.0035}{x_b} = \frac{0.002}{(d - x_b)}$$

or $x_b = 0.636d$

N_b = ultimate load at balanced failure
= $C_s + C_c - T$

$$= A'_s (0.87f_y) + \left(\frac{0.67f_{cu}}{\gamma_m}\right) 0.9x_b b - A_s(0.87f_y)$$

For a symmetrically reinforced section where $A_s = A'_s$, and the compression
reinforcement reaches yield strain, the terms for C_s and T cancel each
other out.

∴ $N_b = 0.256f_{cu}bd$

The strain in compression steel is governed by the value of d'.
The yield strain in compression steel is $f_y/E_s\gamma_m$.
$E_s = 200\,\text{kN/mm}^2$ and $\gamma_m = 1.15$.
For Grade $460\,\text{N/mm}^2$ steel, this yield strain $= 0.002$.
At 'balanced failure' condition:

$$\frac{0.0035}{x_b} = \frac{\varepsilon'_s}{(x_b - d')}$$

Assuming $\varepsilon'_s = 0.002$ and $x_b = 0.636d$, the maximum value of d' to produce
yield strain in compression steel is given by:

$d' = 0.273d$

or $\dfrac{d}{h} = 0.78$ minimum to produce yield strain in compression reinforcement at 'balanced failure'.

For a symmetrically reinforced section,

M_b = moment to produce balanced failure

$d/h = k$

$p = 100A_s/bh$ = percentage of reinforcement with respect to tension reinforcement only

$f_y = 460\,\text{N/mm}^2$

$d'/h = (1 - k)$

$d''/h = (k - 0.5)$

Taking moment about the centroid of the section,

for $k \geq 0.78$

$$M_b = 0.402f_{cu}x_b b(0.5h - 0.45x_b) + 0.87A_s f_y h(k - 0.5) + 0.87A_s f_y h(k - 0.5)$$

for $k < 0.78$

$$M_b = 0.402f_{cu}x_b b(0.5h - 0.45x_b) + 0.87A_s f_y h(k - 0.5) + A_s f_s h(k - 0.5)$$

where $f_s = \varepsilon_s E_s$ and $\varepsilon_s = \dfrac{0.0035(x_b - d')}{x_b}$

for $k \geq 0.78$

$$\frac{M_b}{bh^2} = 0.256f_{cu}k(0.5 - 0.286k) + 8p(k - 0.5)$$

for $k < 0.78$

$$\frac{M_b}{bh^2} = 0.256f_{cu}k(0.5 - 0.286k) + 4p(k - 0.5)$$

$$+ 11p(1.636k - 1)\frac{(k - 0.5)}{k}$$

Note: In the above equations use $0.5p$ instead of p if p is the percentage of total reinforcement in the column.

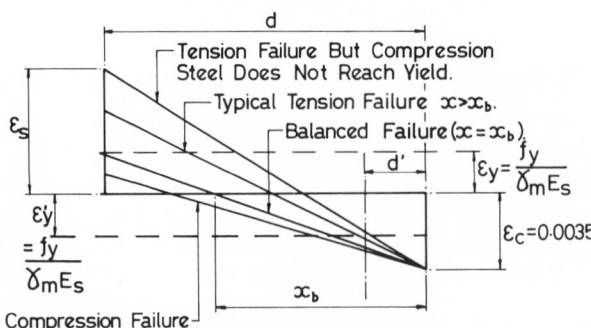

Compression Failure—
Tensile Steel Does Not Reach Yield $(x > x_b)$.

SK 1/20 Strain diagram of column for various types of failure.

Tension failure

When $N < N_b$ and $f_s = f_y/\gamma_m$, a tension failure condition will apply. The column behaves more like a beam in this condition.

$$N = C_c + C_s - T$$

Assuming symmetrical reinforcement and yield strain in both tension and compression steel,

$$N = C_c = 0.402 f_{cu}bx$$

or $$x = \frac{N}{0.402 f_{cu}b}$$

Check that

$$\varepsilon'_s = \frac{0.0035(x - d')}{x} \geq 0.002 \qquad \text{for } f_y = 460\,\text{N/mm}^2$$

or $$x \geq 2.331h(1 - k)$$

This ensures that the compression steel has reached yield.
For a symmetrically reinforced rectangular section,

$$\frac{x}{h} = 2.4875\left(\frac{N}{f_{cu}bh}\right) \qquad \text{for } \frac{x}{h} \geq 2.331(1 - k)$$

$$= 0.5[(B^2 + 4C)^{\frac{1}{2}} - B] \qquad \text{for } \frac{x}{h} < 2.331(1 - k)$$

where $$B = 7.463\left(\frac{p}{f_{cu}}\right) - 2.4875\left(\frac{N}{f_{cu}bh}\right)$$

$$C = 17.413(1 - k)\left(\frac{p}{f_{cu}}\right)$$

Taking moment about the centroid of the section,

for $\dfrac{x}{h} \geq 2.331(1 - k)$

$$M = 0.402 f_{cu}xb(0.5h - 0.45x) + 0.87A_sf_yh(k - 0.5) \\ + 0.87A_sf_yh(k - 0.5)$$

for $\dfrac{x}{h} < 2.331(1 - k)$

$$M = 0.402 f_{cu}xb(0.5h - 0.45x) + 0.87A_sf_yh(k - 0.5) \\ + A_sf_sh(k - 0.5)$$

for $\dfrac{x}{h} \geq 2.331(1 - k)$

$$\frac{M}{f_{cu}bh^2} = 0.402\left(\frac{x}{h}\right)\left(0.5 - 0.45\left(\frac{x}{h}\right)\right) + \frac{8p(k - 0.5)}{f_{cu}}$$

for $\dfrac{x}{h} < 2.331(1 - k)$

$$\frac{M}{f_{cu}bh^2} = 0.402\left(\frac{x}{h}\right)\left[0.5 - 0.45\left(\frac{x}{h}\right)\right] + \frac{11p(k - 0.5)}{f_{cu}}$$
$$- \frac{7p[(1 - k)/(x/h)](k - 0.5)}{f_{cu}}$$

Compression failure

When $N > N_b$, a compression failure condition applies.

$$N = C_c + C_s - T$$

Assuming symmetrical reinforcement and yield strain in both tension and compression steel,

$$N = C_c = 0.402 f_{cu}bx$$

or

$$\frac{x}{h} = \frac{N}{0.402 f_{cu}bh}$$

$$= 2.4875\left(\frac{N}{f_{cu}bh}\right)$$

For the tensile steel to be at yield,

$$\varepsilon_s \geq 0.002 \qquad \text{for } f_y = 460\,\text{N/mm}^2$$

or

$$0.0035\frac{(d - x)}{x} \geq 0.002$$

or

$$\frac{x}{h} \leq 0.636k$$

For both tensile steel and compression steel to be at yield,

$$0.636k \geq \frac{x}{h} \geq 2.331(1 - k)$$

When compression steel does not yield,

for $0.636k \geq \dfrac{x}{h} < 2.331(1 - k)$,

$$\frac{x}{h} = 0.5[(B^2 + 4C)^{\frac{1}{2}} - B]$$

where $B = 7.463\left(\dfrac{p}{f_{cu}}\right) - 2.4875\left(\dfrac{N}{f_{cu}bh}\right)$

$$C = 17.413(1 - k)\left(\frac{p}{f_{cu}}\right)$$

When tension steel does not yield,

for $0.636k < \dfrac{x}{h} \geq 2.331(1 - k)$,

$$\frac{x}{h} = 0.5[(B^2 + 4C)^{\frac{1}{2}} - B]$$

where $B = 27.363\left(\dfrac{p}{f_{cu}}\right) - 2.4875\left(\dfrac{N}{f_{cu}bh}\right)$

$$C = 17.413k\left(\frac{p}{f_{cu}}\right)$$

When tension steel and compression steel do not yield,

for $0.636k < \dfrac{x}{h} < 2.331(1-k)$,

$$\frac{x}{h} = 0.5[(B^2 + 4C)^{\frac{1}{2}} - B]$$

where $B = 34.826\left(\dfrac{p}{f_{cu}}\right) - 2.4875\left(\dfrac{N}{f_{cu}bh}\right)$

$$C = 17.413\left(\frac{p}{f_{cu}}\right)$$

Taking moment about the centroid of section.
For both tension and compression steel going into yield,

$$M = 0.402f_{cu}xb(0.5h - 0.45x) + 0.87A_sf_yh(k - 0.5) \\ + 0.87A_sf_yh(k - 0.5)$$

For tension steel only going into yield,

$$M = 0.402f_{cu}xb(0.5h - 0.45x) + 0.87A_sf_yh(k - 0.5) \\ + A_sf_s'h(k - 0.5)$$

For compression steel only going into yield,

$$M = 0.402f_{cu}xb(0.5h - 0.45x) + A_sf_sh(k - 0.5) \\ + 0.87A_sf_yh(k - 0.5)$$

For both tension and compression steel not going into yield,

$$M = 0.402f_{cu}xb(0.5h - 0.45x) + A_sf_sh(k - 0.5) + A_sf_s'h(k - 0.5)$$

for $0.636k \geq \dfrac{x}{h} \geq 2.331(1-k)$

$$\frac{M}{f_{cu}bh^2} = 0.402\left(\frac{x}{h}\right)\left[0.5 - 0.45\left(\frac{x}{h}\right)\right] + \frac{8p(k - 0.5)}{f_{cu}}$$

for $0.636k \geq \dfrac{x}{h} < 2.331(1-k)$

$$\frac{M}{f_{cu}bh^2} = 0.402\left(\frac{x}{h}\right)\left[0.5 - 0.45\left(\frac{x}{h}\right)\right] + \left[\frac{11p(k - 0.5)}{f_{cu}}\right]$$
$$- \left\{\frac{7p[(1 - k)/(x/h)] (k - 0.5)}{f_{cu}}\right\}$$

for $0.636k < \dfrac{x}{h} \geq 2.331(1-k)$

$$\frac{M}{f_{cu}bh^2} = 0.402\left(\frac{x}{h}\right)\left[0.5 - 0.45\left(\frac{x}{h}\right)\right] + \left[\frac{7pk(k-0.5)}{(x/h)f_{cu}}\right] - \left[\frac{3p(k-0.5)}{f_{cu}}\right]$$

for $0.636k < \dfrac{x}{h} < 2.331\,(1-k)$.

$$\frac{M}{f_{cu}bh^2} = 0.402\left(\frac{x}{h}\right)\left[0.5 - 0.45\left(\frac{x}{h}\right)\right] + \frac{7p(2k-1)(k-0.5)}{(x/h)f_{cu}}$$

Note: In all above equations use $0.5p$ instead of p if p is the percentage of total reinforcement in the column.

Symmetrical rectangular column – interaction tables for columns subject to uniaxial bending and direct load
Tables 11.8 to 11.17 (Chapter 11) have been prepared by solving by iteration the following equations. Assuming equal reinforcement in each face of the rectangular section and assuming that all steel reinforcement is in tension,

b = width of rectangular section
d_1 = depth of the top layer of reinforcement near the compression face from the compression face
d = depth of the bottom layer of reinforcement near the tension face i.e. the effective depth
h = overall depth of section
e = M/N = eccentricity of load from the centre of the rectangular section, assuming that the concentrated load N always acts at the centre of the rectangular section
f_{s1} = stress in steel at depth d_1
f_s = stress in steel at depth d
A_{s1} = $A_s/2$, i.e. equal reinforcement on each face
p = percentage of reinforcement in column = $100A_s/bh$
x = depth of neutral axis from compression face
k = d/h
$d_1/h = 1 - k$
$0.87f_y = 0.87 \times 460 = 400\,\text{N/mm}^2$
$f_{s1} = E_s\varepsilon_{s1} = 200 \times 10^3 \times 0.0035 \times (d_1 - x)/x = 700\,(d_1 - x)/x \leq |400|\,\text{N/mm}^2$
$f_s = 700\,(d - x)/x \leq |400|\,\text{N/mm}^2$
A_c = area of concrete in compression = $0.9bx = 0.9bh(x/h)$
$N = (0.67f_{cu}/\gamma_m)A_c - (A_s/2)(f_{s1} + f_s)$

$$\frac{N}{bh} = 0.402f_{cu}\left(\frac{x}{h}\right) - \left(\frac{p}{200}\right)(f_{s1} + f_s)$$

Taking moment about the centre of the rectangular section,

$$Ne = \left(\frac{0.67 f_{cu}}{\gamma_m}\right) A_c \left(\frac{h}{2} - 0.45x\right) + \left(\frac{A_s}{2}\right)\left(\frac{h}{2} - (1-k)h\right)(f_s - f_{s1})$$

$$= 0.402 f_{cu}\left(\frac{x}{h}\right) bh^2 \left(0.5 - 0.45\left(\frac{x}{h}\right)\right) + bh^2\left(\frac{p}{200}\right)(k - 0.5)(f_s - f_{s1})$$

Dividing Ne/bh^2 by N/bh we get ε/h.

$$\therefore \quad \frac{e}{h} = \frac{0.402 f_{cu}\left(\frac{x}{h}\right)\left(0.5 - 0.45\left(\frac{x}{h}\right)\right) + \left(\frac{p}{200}\right)(k - 0.5)(f_s - f_{s1})}{0.402 f_{cu}\left(\frac{x}{h}\right) - \left(\frac{p}{200}\right)(f_{s1} + f_s)}$$

$$f_{s1} = \frac{700\left(\dfrac{d_1}{h} - \dfrac{x}{h}\right)}{\dfrac{x}{h}}$$

$$\text{or} \quad f_{s1} = \frac{700\left(1 - k - \dfrac{x}{h}\right)}{\dfrac{x}{h}} \leq |400|\,\text{N/mm}^2$$

$$f_s = \frac{700\left(k - \dfrac{x}{h}\right)}{\dfrac{x}{h}} \leq |400|\,\text{N/mm}^2$$

For a range of values of f_{cu}, k and p, the above equations can be solved for different values of e/h. Tables 11.8 to 11.17 give N/bh for different values of e/h.

Note: The above equations are valid up to $x = 1.111h$.

1.10.4.2   Circular section

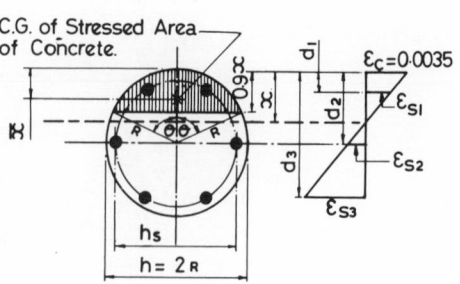

SK 1/21 Circular column – strain diagram.

R = radius of circular section
A_c = area of equivalent uniform stressed section of concrete bounded by a line at $0.9x$ from compression face
$\bar{x}$ = centroid of stressed section of concrete

x = depth of neutral axis from compression face
e = eccentricity of applied load = M/N
N = applied direct load at centre of section
M = applied equivalent uniaxial moment
p = percentage of reinforcement = $100A_s/\pi R^2$
k = h_s/h
2θ = angle to the corner of equivalent uniformly stressed area subtended at the centre of section, or the angle subtended to the line at $0.9x$ from compression face
A_s = total area of steel in six bars
A_c = $R^2(\theta - \sin\theta\cos\theta)$

$$\bar{x} = R\left[1 - \frac{2\sin^3\theta}{3(\theta - \sin\theta\cos\theta)}\right]$$

First layer of steel is at d_1, second layer of steel is at d_2, and third layer is at d_3.

Note: If x is greater than d_1, d_2, or d_3 then the corresponding steel is in compression.

$$N = \left(\frac{0.67f_{cu}}{\gamma_m}\right)A_c - \left(\frac{A_s}{3}\right)(f_{s1} + f_{s2} + f_{s3})$$

$$= 0.45R^2f_{cu}(\theta - \sin\theta\cos\theta) - \left(\frac{p\pi R^2}{300}\right)(f_{s1} + f_{s2} + f_{s3})$$

$$\frac{N}{R^2} = 0.45f_{cu}(\theta - \sin\theta\cos\theta) - \left(\frac{p\pi}{300}\right)(f_{s1} + f_{s2} + f_{s3})$$

$$f_{s1} = 0.0035E_s\left(\frac{d_1 - x}{x}\right) \leq |400| \, \text{N/mm}^2$$

$$\text{or} \quad f_{s1} = \frac{700\left(\dfrac{d_1}{R} - \dfrac{x}{R}\right)}{\dfrac{x}{R}} \leq |400| \, \text{N/mm}^2$$

$$f_{s2} = \frac{700\left(\dfrac{d_2}{R} - \dfrac{x}{R}\right)}{\dfrac{x}{R}} \leq |400| \, \text{N/mm}^2$$

$$f_{s3} = \frac{700\left(\dfrac{d_3}{R} - \dfrac{x}{R}\right)}{\dfrac{x}{R}} \leq |400| \, \text{N/mm}^2$$

$$\frac{d_1}{R} = 1 - k\cos 30° \qquad \frac{d_2}{R} = 1 \qquad \frac{d_3}{R} = 1 + k\cos 30°$$

$$\frac{x}{R} = \left(\frac{1}{0.9}\right)(1 - \cos\theta) = 1.11\,(1 - \cos\theta)$$

$$\frac{\bar{x}}{R} = 1 - \frac{2\sin^3\theta}{3(\theta - \sin\theta\cos\theta)}$$

Taking moment about the centre of section assuming that the applied load N is always at the centre of section,

$$Ne = 0.45R^2 f_{cu}(\theta - \sin\theta\cos\theta)(R - \bar{x}) + \left(\frac{A_s}{3}\right)(kR\sin 60°)(f_{s3} - f_{s1})$$

$$= 0.45R^3 f_{cu}(\theta - \sin\theta\cos\theta)\left(1 - \frac{\bar{x}}{R}\right) + \left(\frac{p\pi R^3}{300}\right)(k\sin 60°)(f_{s3} - f_{s1})$$

Dividing Ne/R^3 by N/R^2 we get e/R.

$$\frac{e}{R} = \frac{0.45 f_{cu}(\theta - \sin\theta\cos\theta)\left(1 - \dfrac{\bar{x}}{R}\right) + \left(\dfrac{p\pi}{300}\right)(k\sin 60°)(f_{s3} - f_{s1})}{0.45 f_{cu}(\theta - \sin\theta\cos\theta) - \left(\dfrac{p\pi}{300}\right)(f_{s1} + f_{s2} + f_{s3})}$$

$$\frac{d}{R} = 1.0866k$$

$$\frac{z}{R} = \frac{d}{R} - \frac{\bar{x}}{R}$$

For a range of values of f_{cu}, k and p, the above equations can be solved for different values of e/R. Tables 11.18 to 11.27 give N/R^2 and z/R for different values of e/R.

Note: The above equations are valid up to $x = 1.111h = 2.22R$.

1.11 ULTIMATE LIMIT STATE – CORBELS

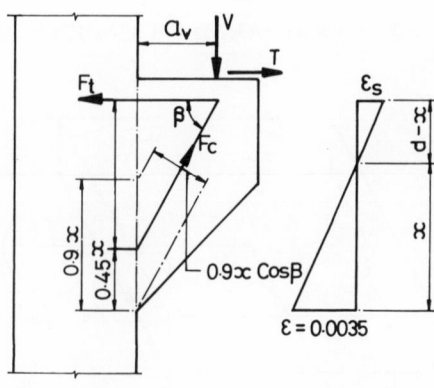

SK 1/22 Concrete corbel.   Strut and Tie Diagram   Strain Diagram

The derivation of Fig. 5.1

$$v = \frac{V}{bd}$$

From the strut and tie diagram:

$$F_t = T + F_c \cos \beta = T + \frac{Va_v}{z}$$

$$F_c = \left(\frac{0.67 f_{cu}}{1.5}\right) b \; 0.9x \cos \beta$$

$$= 0.402 f_{cu} bx \cos \beta$$

$$V = F_c \sin \beta$$

$$= 0.402 f_{cu} bx \cos \beta \sin \beta$$

$$\cos \beta = \frac{a_v}{(a_v^2 + z^2)^{\frac{1}{2}}}$$

$$\sin \beta = \frac{z}{(a_v^2 + z^2)^{\frac{1}{2}}}$$

$$v = \frac{V}{bd} = 0.402 f_{cu} x \; \frac{za_v}{(a_v^2 + z^2)d}$$

$$\frac{v}{f_{cu}} = 0.402 \left(\frac{z}{d}\right)\left(\frac{xa_v}{a_v^2 + z^2}\right)$$

Substituting $x = (d - z)/0.45$,

$$\frac{v}{f_{cu}} = \frac{0.893 \left(\dfrac{z}{d}\right)\left(\dfrac{a_v}{d}\right)\left(1 - \dfrac{z}{d}\right)}{\left(\dfrac{a_v}{d}\right)^2 + \left(\dfrac{z}{d}\right)^2}$$

From the above equation the graphs in Fig. 5.1 have been drawn.

1.12   WOOD–ARMER COMBINATION OF MOMENT TRIADS

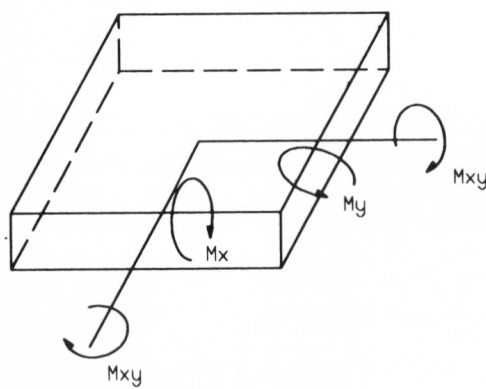

SK 1/23 Moment triad in a slab panel.

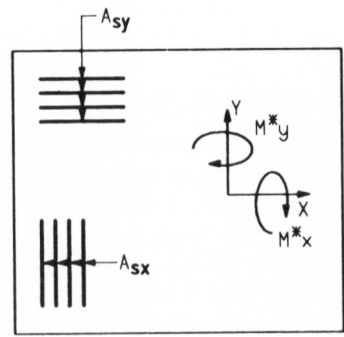

SK 1/24 Direction of orthogonal reinforcement in a slab panel.

Orthogonal reinforcement
Bottom steel

$$M_x^* = M_x + |M_{xy}|$$
$$M_y^* = M_y + |M_{xy}|$$

If $M_x^* < 0$, then $M_x^* = 0$

and $\quad M_y^* = M_y + \left| \dfrac{M_{xy}^2}{M_x} \right|$

If $M_y^* < 0$, then $M_y^* = 0$

and $\quad M_x^* = M_x + \left| \dfrac{M_{xy}^2}{M_y} \right|$

Top steel

$$M_x^* = M_x - |M_{xy}|$$
$$M_y^* = M_y - |M_{xy}|$$

If $M_x^* > 0$, then $M_x^* = 0$

and $\quad M_y^* = M_y - \left| \dfrac{M_{xy}^2}{M_x} \right|$

If $M_y^* > 0$, then $M_y^* = 0$

and $\quad M_x^* = M_x - \left| \dfrac{M_{xy}^2}{M_y} \right|$

Skew reinforcement

SK 1/25 Direction of skew reinforcement in a slab panel.

Bottom steel

$$M_x^* = M_x + 2M_{xy}\cot\alpha + M_y\cot^2\alpha + \left| \frac{M_{xy} + M_y\cot\alpha}{\sin\alpha} \right|$$

$$M_\alpha^* = \frac{M_y}{\sin^2\alpha} + \left| \frac{M_{xy} + M_y\cot\alpha}{\sin\alpha} \right|$$

If $M_x^* < 0$, then $M_x^* = 0$

and $\quad M_\alpha^* = \left(\dfrac{1}{\sin^2\alpha}\right)\left(M_y + \left|\dfrac{(M_{xy} + M_y\cot\alpha)^2}{M_x + 2M_{xy}\cot\alpha + M_y\cot^2\alpha}\right|\right)$

If $M_\alpha^* < 0$, then $M_\alpha^* = 0$

and $\quad M_x^* = M_x + 2M_{xy}\cot\alpha + M_y\cot^2\alpha + \left|\dfrac{(M_{xy} + M_y\cot\alpha)^2}{M_y}\right|$

Top steel

$M_x^* = M_x + 2M_{xy}\cot\alpha + M_y\cot^2\alpha - \left|\dfrac{M_{xy} + M_y\cot\alpha}{\sin\alpha}\right|$

$M_\alpha^* = \dfrac{M_y}{\sin^2\alpha} - \left|\dfrac{M_{xy} + M_y\cot\alpha}{\sin\alpha}\right|$

If $M_x^* > 0$, then $M_x^* = 0$

and $\quad M_\alpha^* = \left(\dfrac{1}{\sin^2\alpha}\right)\left(M_y - \left|\dfrac{(M_{xy} + M_y\cot\alpha)^2}{M_x + 2M_{xy}\cot\alpha + M_y\cot^2\alpha}\right|\right)$

If $M_\alpha^* > 0$, then $M_\alpha^* = 0$

and $\quad M_x^* = M_x + 2M_{xy}\cot\alpha + M_y\cot^2\alpha - \left|\dfrac{(M_{xy} + M_y\cot\alpha)^2}{M_y}\right|$

1.13   SERVICEABILITY LIMIT STATE − BENDING AND DIRECT LOADS

1.13.1   Serviceability limit state: uniaxial bending

C' = compressive force in bars in compression with allowance for area of concrete occupied by bars

C = compressive force in concrete stress block assumed triangular

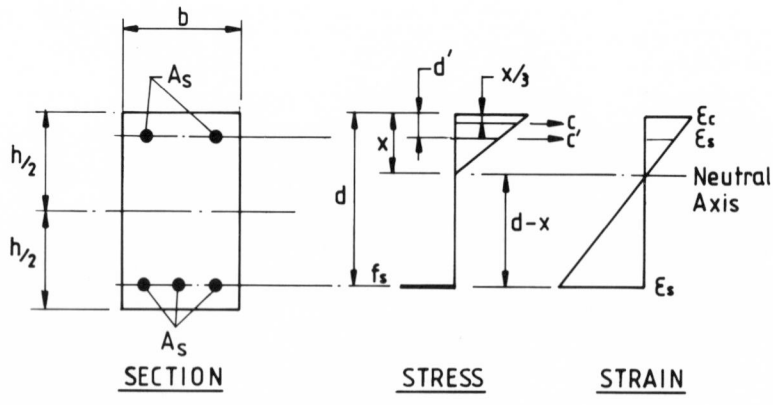

SK 1/26 Serviceability uniaxial bending.

T = tensile force in bars in tension
M = applied bending moment
x = depth of neutral axis from compressive face
ε_c = strain in extreme compressive fibre of concrete
ε_s' = strain in compressive steel
ε_s = strain in tensile steel
m = E_s/E_c

From strain diagram,

$$\frac{\varepsilon_c}{\varepsilon_s'} = \frac{x}{x - d'} \quad \text{or} \quad \frac{f_c}{f_s'} = \frac{E_c}{E_s}\left(\frac{x}{d - d'}\right)$$

$$\frac{\varepsilon_c}{\varepsilon_s} = \frac{x}{x - d} \quad \text{or} \quad \frac{f_c}{f_s} = \frac{E_c}{E_s}\left(\frac{x}{d - x}\right)$$

$$C = 0.5bxf_c$$

$$C' = f_s'A_s' - f_cA_s'\left(\frac{x - d'}{x}\right)$$

$$= (m - 1)f_cA_s'\left(\frac{x - d'}{x}\right)$$

$$T = f_sA_s = mf_cA_s\left(\frac{d - x}{x}\right)$$

Taking moment about the steel in tension,

$$C\left(d - \frac{x}{3}\right) + C'(d - d') = M$$

or $\quad 0.5bxf_c\left(d - \frac{x}{3}\right) + (m - 1)f_cA_s'\left(\frac{x - d'}{x}\right)(d - d') = M$

Equating the loads on the section,

$$C + C' = T$$

or $\quad 0.5bxf_c + (m - 1)f_cA_s'\left(\frac{x - d'}{x}\right) = mf_cA_s\left(\frac{d - x}{x}\right)$

Eliminating f_c and multiplying by $2x/bd^2$,

$$\frac{x^2}{d^2} + \frac{2(m - 1)A_s'(x - d')}{bd^2} - \frac{2mA_s(d - x)}{bd^2} = 0$$

Simplifying, and substituting $p = A_s/bd$ and $p' = A_s'/bd$,

$$\left(\frac{x}{d}\right)^2 + 2[(m - 1)p' + mp]\left(\frac{x}{d}\right) - 2\left[(m - 1)p'\left(\frac{d'}{d}\right) + mp\right] = 0$$

or $\quad \dfrac{x}{d} = \left\{\left[mp + (m - 1)p'\right]^2 + 2\left[mp + (m - 1)\left(\frac{d'}{d}\right)p'\right]\right\}^{\frac{1}{2}}$
$\qquad\qquad - [mp + (m - 1)p']$

Put $p' = 0$, where compressive steel is not present.

Having found x using the above expression, find f_c.

$$f_c = \frac{M}{0.5bx(d - x/3) + (m - 1)A_s'(x - d')(d - d')/x}$$

$$= \frac{M}{k_2bd^2 + k_3A_s'(d - d')}$$

where $k_2 = \left(\frac{x}{2d}\right)\left(1 - \frac{x}{3d}\right)$

$$k_3 = (m - 1)\left(1 - \frac{d'}{x}\right)$$

$$f_s = \text{tensile stress in steel} = mf_c\left(\frac{d}{x} - 1\right)$$

1.13.2 Serviceability limit state: uniaxial bending and compression

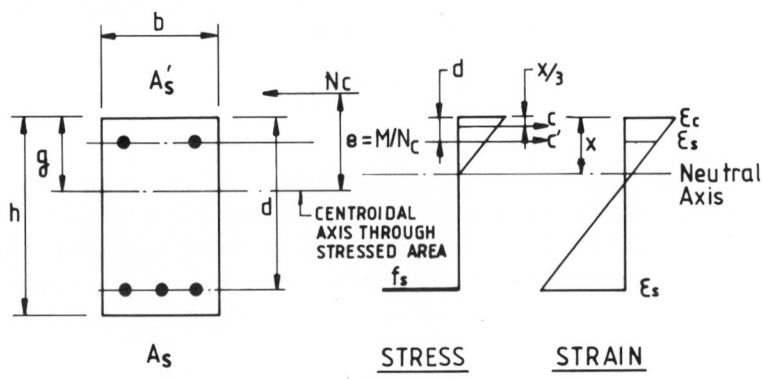

SK 1/27 Serviceability uniaxial bending and thrust.

Using the same symbols as in Section 1.13.1 and N_c is the compressive force, equating the loads on the section:

$$C' + C = N_c + T \quad \text{and} \quad e = \frac{M}{N_c}$$

The output from a computer program will give an axial compression of a member with a coacting moment. This axial compression theoretically for a reinforced concrete section may be considered as acting at the centroid of the stressed section. This will mean finding the centroid of the stressed area ignoring concrete in tension. On the other hand, the line of application of the compressive load may also be assumed to coincide with the centroid of the full concrete section ignoring any reinforcement.

The distance of the centroid of the stressed area from the compressive face of the rectangular section is called g and may be found by taking moments of all transformed areas of steel about the compressive face of the section

$$g = \frac{bx(x/2) + (m-1)A'_s d' + mA_s d}{bx + (m-1)A'_s + mA_s}$$

$g = \dfrac{h}{2}$ where the load is assumed to act at the centroid of the full concrete section ignoring steel.

(See Section 1.13.1 for expressions of C and C'.)

Taking moment about tension steel,

$$-N_c(e + d - g) + C'(d - d') + C\left(d - \frac{x}{3}\right) = 0$$

or $\quad -N_c(e + d - g) + (m - 1)f_c A'_s\left(\dfrac{x - d'}{x}\right)(d - d')$

$$+ 0.5bxf_c\left(d - \frac{x}{3}\right) = 0$$

or $\quad (m - 1)\left(1 - \dfrac{d'}{x}\right)(d - d')\dfrac{f_c A'_s}{d} + 0.5bx\left(1 - \dfrac{x}{3d}\right)f_c$

$$= N_c\left[\frac{(e - g)}{d} + 1\right]$$

$$k_1 = \left[\frac{(e - g)}{d} + 1\right]$$

$$k_2 = \left(\frac{x}{2d}\right)\left(1 - \frac{x}{3d}\right)$$

$$k_3 = (m - 1)\left(1 - \frac{d'}{x}\right)$$

where k_1, k_2 and k_3 are non-dimensional constants.

$$k_3\left(1 - \frac{d'}{d}\right)f_c A'_s + k_2 bdf_c = k_1 N_c$$

or $\quad f_c = \dfrac{k_1 N_c}{k_3\left(1 - \dfrac{d'}{d}\right)A'_s + k_2 bd}$

$$T = C' + C - N_c$$

or $\quad A_s f_s = (m - 1)f_c A'_s\left(\dfrac{x - d'}{x}\right) + 0.5bxf_c - N_c$

$$= k_3 f_c A'_s + 0.5bxf_c - N_c$$

or $\quad f_s = \dfrac{f_c(k_3 A'_s + 0.5bx) - N_c}{A_s}$

From strain diagram, see Section 1.13.1

$$\frac{f_s}{f_c} = \frac{m(d - x)}{x}$$

or $x = \dfrac{d}{\left(1 + \dfrac{f_s}{mf_c}\right)}$

The procedure is to assume x and then calculate f_c and f_s, and then check x. Repeat this process until convergence is reached.

1.13.3   Serviceability limit state: uniaxial bending and tension

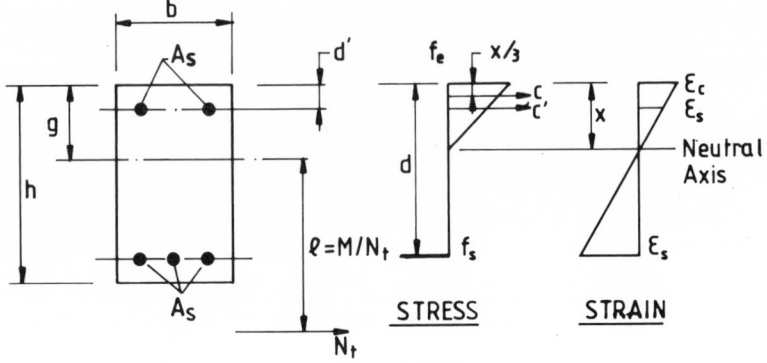

SK 1/28 Serviceability uniaxial bending and tension.

Using the same symbols as in Section 1.13.1 and N_t is the tensile force, equating the loads:

$$C' + C + N_t = T \quad \text{and} \quad e = \frac{M}{N_t}$$

The expressions for e, g, k_2 and k_3 are exactly the same as in Section 1.13.2 and g may be taken equal to $h/2$ where the point of application of the tensile load is at the centroid of the full concrete section ignoring steel. (See Section 1.13.1 for expressions of C and C'.)

Taking moment about tensile steel,

$$-N_t(e + g - d) + C'(d - d') + C\left(d - \frac{x}{3}\right) = 0$$

or $(m - 1)\left(1 - \dfrac{d'}{x}\right)(d - d')\dfrac{f_c A'_s}{d} + 0.5bx\left(1 - \dfrac{x}{3d}\right)f_c$

$= N_t\left[\dfrac{(e + g)}{d} - 1\right]$

$$k_1 = \left[\frac{(e + g)}{d} - 1 \right]$$

$$k_2 = \left(\frac{x}{2d} \right) \left(1 - \frac{x}{3d} \right)$$

$$k_3 = (m - 1) \left(1 - \frac{d'}{x} \right)$$

or $\quad f_c = \dfrac{k_1 N_t}{k_3 \left(1 - \dfrac{d'}{d} \right) A_s' + k_2 bd}$

$$A_s f_s = T = C' + C + N_t$$

or $\quad f_s = \dfrac{f_c(k_3 A_s' + 0.5bx) + N_t}{A_s}$

$$x = \frac{d}{\left(1 + \dfrac{f_s}{mf_c} \right)}$$

as in Section 1.13.2. Check assumed value of x and repeat until convergence is reached.

Chapter 2
Design of Reinforced Concrete Beams

2.0 NOTATION

a'	Compression face to point on surface of concrete where crack width is calculated
a_b	Centre-to-centre distance between bars or groups of bars
a_{cr}	Point on surface of concrete to nearest face of a bar
A_c	Gross area of concrete in a section
A_s	Area of steel in tension
A_s'	Area of steel in compression
A_{sv}	Area of steel in vertical links
b	Width of reinforced concrete section
b_c	Breadth of compression face of beam mid-way between restraints
b_t	Width of section at centroid of tensile steel
b_w	Average web width
c	Coefficient of torsional stiffness
c_{min}	Minimum cover to tensile reinforcement
C	Torsional stiffness
d	Effective depth of tensile reinforcement
d'	Effective depth of compressive reinforcement
d_1	From tension face of concrete section to centre of tensile reinforcement
E_c	Modulus of elasticity of concrete
E_s	Modulus of elasticity of steel
f_s	Service stress in steel reinforcement
f_y	Characteristic yield strength of steel
f_y'	Revised compressive stress in steel taking into account depth of neutral axis
f_{cu}	Characteristic cube strength of concrete at 28 days
f_{yv}	Characteristic yield strength of reinforcement used as links
F	Coefficient for calculation of cracked section moment of inertia
F_{bt}	Tensile force in a bar at start of a bend
G	Shear modulus
h	Overall depth of a concrete section
h_f	Thickness of flange in a T-beam
h_{max}	Maximum overall dimension of a rectangular concrete beam
h_{min}	Minimum overall dimension of a rectangular concrete beam
I	Moment of inertia
l	Clear span or span face-to-face of support

l_e	Effective span
l_o	Centre-to-centre distance between supports
m	modular ratio $= E_s/E_c$
M	Applied bending moment
M_d	Design bending moment modified to account for axial load
M_r	Moment of resistance of concrete in flanged beams
N	Axial load
p	Percentage of tensile reinforcement
p'	Percentage of compressive reinforcement
r	Internal radius of a bend in a bar
S_b	Spacings of bent bars used as shear reinforcement
S_v	Spacing of vertical links
T	Applied torsion
T_s	Proportion of total torsion carried by each rectangle of an I-, T- or L-section
v	Shear stress in concrete (N/mm^2)
v_c	Design concrete shear stress (N/mm^2)
v_t	Shear stress in concrete due to torsion (N/mm^2)
v_{tu}	Ultimate permissible torsional shear stress (N/mm^2)
$v_{t,min}$	Design concrete torsional shear stress (N/mm^2)
V	Shear force in concrete section
V_b	Shear force carried by bent bars
V_c	Shear force capacity of concrete section
V_s	Shear force carried by vertical links
V_{max}	Ultimate maximum shear forces allowed on section
V_{nom}	Shear force capacity of concrete section with minimum vertical links
V_{conc}	Design shear resistance of concrete
W_{max}	Maximum crack width (mm)
x	Depth of neutral axis from compression face
x_1	Centre-to-centre of two external vertical legs of a link
y_1	Centre-to-centre of two external horizontal legs of a link
z	Depth of lever arm
α	Angle of inclination to horizontal of shear reinforcement
β	Angle of inclination to horizontal of concrete strut in truss analogy
β_b	Ratio of redistributed moment over elastic analysis moment
β_f	Factor governing moment of resistance of concrete T-section
γ_m	Material factor
ε_h	Calculated strain in concrete at depth h
ε_m	Strain with stiffening effect corrected
ε_s	Strain at centre of steel reinforcement
ε_y	Yield strain in steel reinforcement
ε_s'	Strain at centre of compressive reinforcement
ε_{mh}	Strain at depth h corrected for stiffening effect
ε_1	Calculated strain in concrete ignoring stiffening effect
μ	Poisson's ratio
ϕ	Diameter of a reinforcing bar or equivalent diameter of a group of bars

2.1   ANALYSIS OF BEAMS

2.1.1   Effective spans

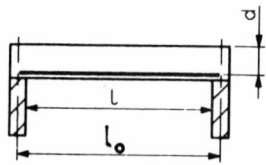

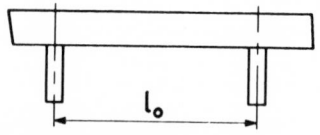

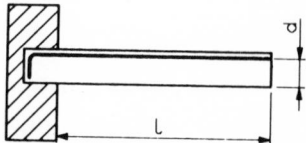

SK 2/2 Continuous beam.

SK 2/1 Simply supported beam.

SK 2/3 Cantilever beam.

Simply supported or encastré l_e = smaller of $(l + d)$ or l_o

Continuous $l_e = l_o$

Cantilever $l_e = l + \dfrac{d}{2}$

where l_o = centre-to-centre distance between supports
l_e = effective span
l = clear span or span to face of support
d = effective depth of tension reinforcement.

2.1.2   Effective width of compression flange

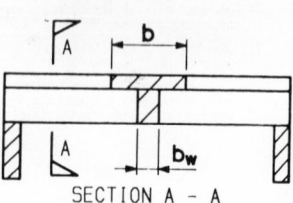

SECTION A - A

SK 2/4 Effective width of compression flange.

Simply supported T-beam $b = \dfrac{l_e}{5} + b_w$

Simply supported L-beam $b = \dfrac{l_e}{10} + b_w$

Continuous or encastré T-beams $b = \dfrac{l_e}{7.14} + b_w$

Continuous or encastré L-beams $b = \dfrac{l_e}{14.29} + b_w$

where b = effective width of compression flange
       b_w = average width of web.

Note: Use actual b if it is less than the calculated b using the above formulae. A typical example may be a precast T-beam.

2.1.3   Moment of inertia

Method 1   Gross concrete section only
Find moment of inertia of gross concrete section − see Table 11.2.

Method 2   Uncracked transformed concrete
If reinforcement quantities are known, find moment of inertia of trans-formed concrete section using Table 11.2.

Method 3   Average of gross concrete section and cracked section

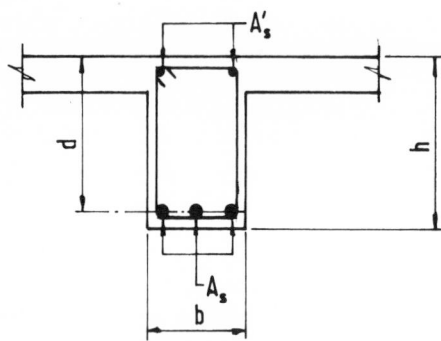

SK 2/5 Doubly reinforced beam.

$$I = 0.5 \left(\frac{1}{12}bh^3 + Fbh^3 \right)$$

where I = moment of inertia of rectangular concrete section
       b = width of rectangular concrete section
       h = overall depth of rectangular concrete section
       F = factor − see Fig. 11.1 for values of F.

$$p = 100 \frac{A_s}{bd}$$

where A_s = area of tensile reinforcement
       d = effective depth to tensile reinforcement.

$$p' = 100 \frac{A_s'}{bd}$$

where A_s' = area of compressive reinforcement.

$$m = \text{modular ratio} = \frac{E_s}{E_c}$$

The graphs in Fig. 11.1 have been drawn for $p' = 0$ and $p = p'$. Intermediate values may be interpolated.

Note: The preferred method is Method 3 for rectangular sections. Where reinforcement quantities are not known, an assumption may be made of the percentage of reinforcement.

T-beams and L-beams in a frame or continuous beam structure should be treated as rectangular beams for the purpose of determining moment of inertia. The width of the beam will be taken equal to b_w.

2.1.4   Modulus of elasticity

Modulus of elasticity of reinforcement steel

$$E_s = 200 \, \text{kN/mm}^2$$

Modulus of elasticity of concrete, E_c, for short-term and long-term loadings is given in Table 2.1.

Table 2.1 Modulus of elasticity of concrete: short-term and long-term loading.

f_{cu} (N/mm^2)	Short-term loading, E_c (kN/mm^2)	Long-term loading, E_c (kN/mm^2)
20	24	12
25	25	12.5
30	26	13
40	28	14
50	30	15
60	32	16

Note: Wind load is short-term loading and dead load is long-term loading.

2.1.5   Torsional stiffness

For a rectangular section the torsional stiffness, C, is given by

$$C = ch_{min}^3 h_{max}$$

where c = coefficient from Table 2.2
     h_{max} = maximum overall dimension of rectangular section
     h_{min} = minimum overall dimension of rectangular section.

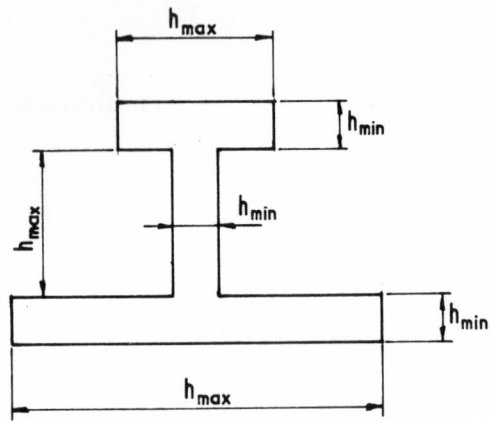

The torsional stiffness of a non-rectangular section may be obtained by dividing the section into a series of rectangles and summing the torsional stiffness of these rectangles.

Table 2.2 Values of coefficient c.

$\dfrac{h_{max}}{h_{min}}$	1	1.5	2	3	5	10
c	0.14	0.20	0.23	0.26	0.29	0.31

The coefficient c is given by the following formula:

$$c = \frac{1}{16}\left[\frac{16}{3} - 3.36k\left(1 - \frac{k^4}{12}\right)\right]$$

where $k = \dfrac{h_{min}}{h_{max}}$.

2.1.6 Shear modulus

Shear modulus, G, is given by

$$G = \frac{E}{2}(1 + \mu) = 0.42E_c \qquad \text{for concrete}$$

where μ = Poisson's ratio.

Note: In normal slab and beam or framed construction, torsional rigidity of RC beams may be ignored in the analysis and the torsional stiffness may be given a very small value in the computer analysis. Torsional rigidity becomes important only where torsion is relied on to carry the load, as in curved beams.

2.1.7 Poisson's ratio

Poisson's ratio for concrete $= 0.2$

2.1.8 Shear area

Shear area of concrete $= 0.8A_c$

where $A_c =$ gross cross-sectional area of concrete.

Note: The shear area of concrete is entered as input to some computer programs when the analysis is required to take into account the deformations due to shear.

2.1.9 Thermal strain

The coefficients of thermal expansion are given in Table 2.3 for different types of aggregate used.

Table 2.3 Coefficient of thermal expansion.

Aggregate type	Coefficient ($\times 10^{-6}$/°c)
Flint, Quartzite	12
Granite, Basalt	10
Limestone	8

Note: Normally for ultimate limit state no specific calculations are necessary for thermal loads. Thermal calculations should be produced for structures in contact with hot gases or liquid.

2.2 LOAD COMBINATIONS

2.2.1 General rules

The following load combinations and partial load factors should be used in carrying out the analysis of beams:

LC_1:   1.4 *DL* + 1.6 *LL* + 1.4 *EP* + 1.4 *WP*
LC_2:   1.0 *DL* + 1.4 *EP* + 1.4 *WP*
LC_3:   1.4 *DL* + 1.4 *WL* + 1.4 *EP* + 1.4 *WP*
LC_4:   1.0 *DL* + 1.4 *WL* + 1.4 *EP* + 1.4 *WP*
LC_5:   1.2 *DL* + 1.2 *LL* + 1.2 *WL* + 1.2 *EP* + 1.2 *WP*

Note: Load combinations LC_2 and LC_4 should be considered when the effects of dead load and live load are beneficial.

where DL = dead load
LL = live load or imposed load
WL = wind load
WP = water pressure
EP = earth pressure.

The general principle of load combination is to leave out the loads which have beneficial effect. If the load is of a permanent nature, like dead load, earth load or water load, use the partial load factor of 1 for that load which produces a beneficial rather than adverse effect. This rule of combination will be used for design as well as for the check of stability of a structure.

2.2.2   Rules of load combination for continuous beams

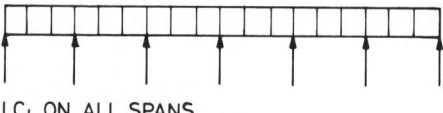

LC₁ ON ALL SPANS

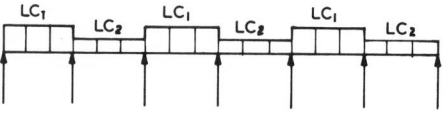

LC₁ ALTERNATE WITH LC₂ FOR MAXIMUM
MIDSPAN MOMENT

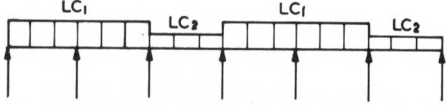

SK 2/7 Continuous beam loading
sequences.

LC₁ ON ADJACENT SPANS ALTERNATE WITH
LC₂ FOR MAXIMUM SUPPORT MOMENT

(1) Load all spans with LC_1.
(2) Load alternate spans with LC_1 and other spans with LC_2.
(3) Load beam in the repeated sequence of two adjacent spans loaded with LC_1 and one span loaded with LC_2. This sequence gives the maximum support moment between adjacent spans. This is not a normal requirement, as per clause 3.2.1.2.2 of BS 8110: Part 1: 1985.[1]

2.2.3 Redistribution of moments

2.2.3.1 Continuous beams

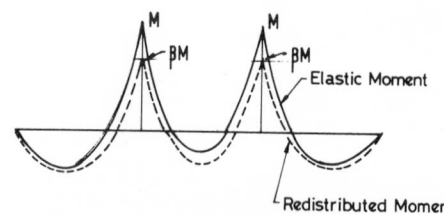

SK 2/8 Continuous beam – typical moment redistribution.

Usually 10% redistribution of moments may be allowed from those obtained by elastic analysis. Redraw bending moment diagram with redistributed moments. Calculate revised shear. Reduction of support moment means a corresponding increase in span moment. For structural frames over four stories high providing lateral stability, the redistribution of moments should not exceed 10%. Resistance moment at any section must be at least 70% of moment at that section obtained by elastic analysis.

2.2.3.2 Frame structures

No reduction or redistribution of moments is allowed from the columns.

2.2.3.3 Continuous one-way spanning slab panels

Usually 10% redistribution of moments may be allowed from those obtained by elastic analysis.

2.2.4 Exceptional loads

Exceptional loads may be any of the following.

(1) Accidental loads of very low probability properly quantified. The definition of low probability may vary from project to project and will be agreed with the client.
(2) Probable misuse and its effect accurately quantified.
(3) Once in a lifetime very short-term loads which are accurately quantified.

Note: With exceptional loads some rectification of local damage after the incident may be necessary.

Load combination to be considered:

$$LC_6 = 1.05DL + 1.05LL_1 + 1.05EL + 1.05WL_1$$

where DL = full expected dead load
LL_1 = full expected live load if this is a storage building, otherwise, one-third of expected maximum live load
EL = exceptional load
WL_1 = one-third of expected maximum wind load.

2.3   STEP-BY-STEP DESIGN PROCEDURE FOR BEAMS

Step 1   Analysis

Carry out analysis — follow Section 2.1.

Step 2   Moment envelope

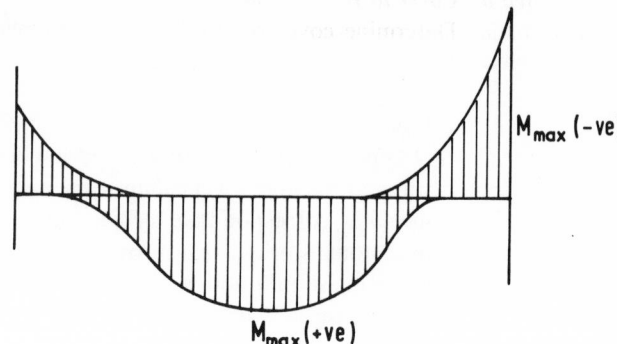

M_{max} (−ve)

SK 2/9 Typical moment envelope
of a continuous beam.

M_{max}(+ve)

Draw maximum−minimum ultimate load bending moment envelope after
redistribution.

Step 3   Shear envelope

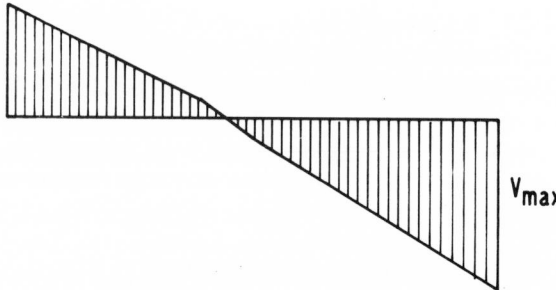

V_{max}

SK 2/10 Typical shear envelope of
a continuous beam.

Draw maximum−minimum ultimate load shear force envelope after
redistribution.

Step 4   Axial loads

Determine coacting axial loads with maximum and minimum bending
moments respectively. Ignore axial load if less than $0.1f_{cu}bh$.

Step 5   Torsions

Determine coacting torsions with maximum and minimum shear forces
respectively.

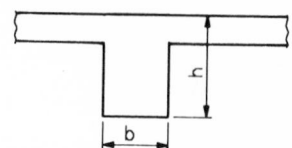

SK 2/11 Dimensions to compute
axial load in beam.

Step 6   Cover to reinforcement
Determine cover required to reinforcement as per Tables 11.6 and 11.7.
Find effective depth d, assuming reinforcement diameter.

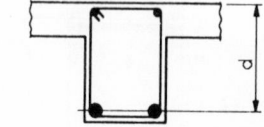

SK 2/12 Rectangular beam –
effective depth.

Step 7   Effective span
Determine effective span – see Section 2.1.1.

Step 8   Effective width of compression flange
Determine effective width of compression flange – see Section 2.1.2.

Step 9   Slenderness ratio

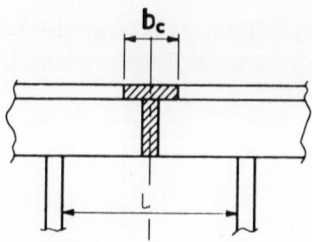

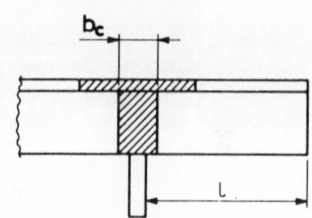

SK 2/14 Cantilever beam.

SK 2/13 Simply supported and continuous
beams.

Check slenderness of beam as per clause 3.4.1.6 of BS 8110: Part 1:
1985.[1]

For simply supported and continuous beams,

$$l < 60b_c \quad \text{or} \quad 250 \frac{b_c^2}{d}$$

For cantilever beams,

$$l < 25b_c \quad \text{or} \quad 100 \frac{b_c^2}{d}$$

where $d = \sqrt{\left(\dfrac{M}{0.156 f_{cu} b}\right)}$ or actual d, whichever is lesser

b = effective width of the compression flange

M = design ultimate moment.

Step 10   Design for moment — rectangular beam

Select critical sections on beam for bending moment. Find the following parameters at all critical sections, for rectangular beams and flanged beams.

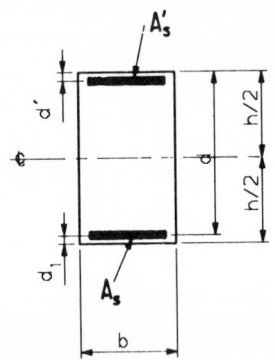

SK 2/15 Rectangular beam — doubly reinforced.

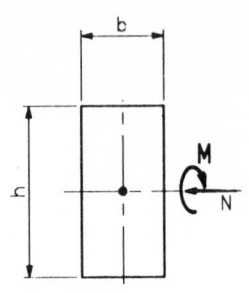

SK 2/16 Rectangular beam — moment and axial load.

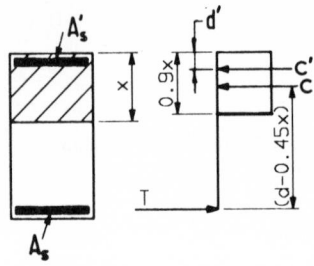

SK 2/17 Rectangular beam — stress diagram.

$$M_d = M + N (0.5h - d_1) \qquad \text{for } N \leq 0.1 f_{cu} bd$$

Note: For $N > 0.1 f_{cu} bd$, design as column (see Chapter 4). M_d may also be taken equal to M where $N \leq 0.1 f_{cu} bd$ and N may be totally ignored. (Sign convention: N is +ve for compression.)

$$K = \frac{M_d}{f_{cu} bd^2}$$

$$z = d \left[0.5 + \sqrt{\left(0.25 - \frac{K}{0.9}\right)} \right] \leq 0.95d$$

$$x = \frac{d - z}{0.45}$$

$$A_s = \frac{M_d}{0.87 f_{yz}} - \frac{N}{0.87 f_y}$$

$K' = 0.156$ when redistribution does not exceed 10%

$K' = 0.402(\beta_b - 0.4) - 0.18(\beta_b - 0.4)^2$ when redistribution exceeds 10%

$$\beta_b = \frac{M^2}{M'} < 0.9$$

where M^2 = moment after redistribution; M' = moment before redistribution

When $K > K'$,

$$z = d\left[0.5 + \sqrt{\left(0.25 - \frac{K'}{0.9}\right)}\right]$$

$$x = \frac{d - z}{0.45} \leq 0.5d$$

$$A'_s = \frac{(K - K')f_{cu}bd^2}{0.87f_y\,(d - d')}$$

$$A_s = \frac{K'f_{cu}bd^2}{0.87f_{y}z} + A'_s - \frac{N}{0.87f_y}$$

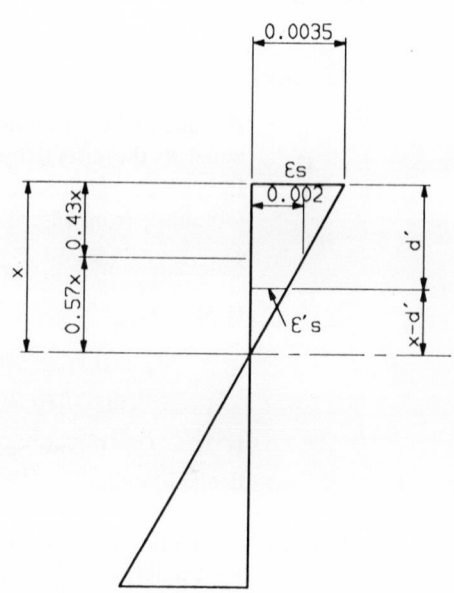

SK 2/18 Strain diagram.

If $d'/x > 0.43x$,

$$A'_s = \frac{(K - K')f_{cu}bd^2}{f'_s\,(d - d')}$$

$$A_s = \frac{K'f_{cu}\,bd^2}{0.87f_{y}z} + A'_s - \frac{N}{0.87f_y}$$

$$f'_s = \left(\frac{x - d'}{0.57x}\right)\frac{f_y}{\gamma_m} \quad \text{because steel strain } \varepsilon'_s = \left(\frac{x - d'}{0.57x}\right)\varepsilon_y$$

where ε_y corresponds to steel stress f_y/γ_m, as in Section 1.4.2.

Note: The flanged beam becomes a rectangular beam if the bending moment produces tension in the flange.

Design charts in BS 8110: Part 3: 1985[1] may be used if design parameters fall within the scope of the charts.

Step 11   ***Design for moment – flanged beam***

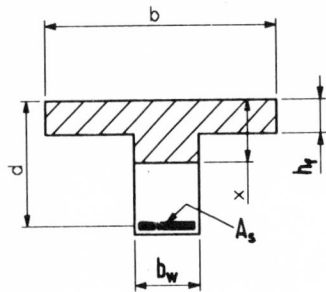

SK 2/19 Flanged beam – singly reinforced.

A flanged beam should be designed as a rectangular beam with width equal to the effective width of flange in compression if $x \leq 1.1h_f$.
If $x > 1.1h_f$, find b/b_w and d/h_f.
Obtain β_f from Fig. 2.1.

Calculate $M_r = \beta_f f_{cu} b d^2$

If $M_r \geq M_d$,

$$A_s = \frac{M_d + 0.1f_{cu}b_w d(0.45d - h_f)}{0.87f_y(d - 0.5h_f)} - \frac{N}{0.87f_y}$$

If $M_r < M_d$, follow Section 1.5.4 using the second approach to design of flanged beams.

Step 12   ***Maximum allowable shear stress***
Find maximum shear in beam from shear envelope at the face of support or under a concentrated load.
Find $v = V/bd$.

Check $v \leq 0.8 \sqrt{f_{cu}} \leq 5 \, \text{N/mm}^2$

Change beam geometry if this condition is not satisfied.

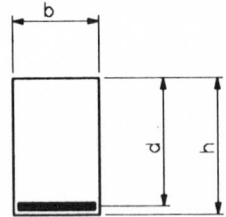

SK 2/20 Rectangular beam.

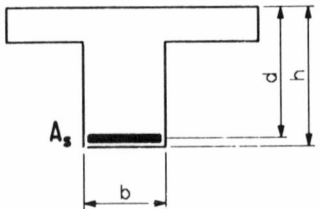

SK 2/21 Flanged beam.

Step 13   Design for shear

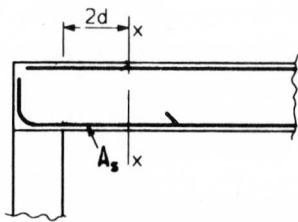

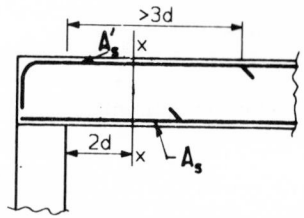

SK 2/22 Critical section for shear. Shear check based on bottom reinforcement adequately anchored.

SK 2/23 Critical section for shear. Shear check based on top reinforcement adequately anchored (A'_s to continue at least $3d$ from face of support).

Select critical sections on beam which are at a distance of $2d$ from the face of support or concentrated load. Find the following parameters for rectangular and flanged beams.

V = design ultimate shear force

$$v = \frac{V}{bd} \le 0.8 \sqrt{f_{cu}} \le 5\,\text{N/mm}^2$$

$$p = \frac{100A_s}{bd}$$

Find v_c from Figs 11.2 to 11.5.
When axial load in compression, N, is present,

$$v'_c = v_c + 0.6 \left(\frac{NVh}{A_cM}\right) \le 0.8 \sqrt{f_{cu}} \le 5\,\text{N/mm}^2$$

Note: N/A_c is average stress in the concrete section. $Vh/M \le 1$ and moment and shear at the section under consideration must be for the same load combination. N is +ve for compression and −ve for tension. To avoid shear cracks at ultimate load, limit shear stress to

$$v'_c = v_c \sqrt{\left(1 + \frac{N}{A_c v_c}\right)}$$

Replace v_c by v'_c where axial load is present.

Find $A_{sv} = \dfrac{0.4bS_v}{0.87f_{yv}}$

$f_{yv} \le 460\,\text{N/mm}^2$ for links.

Provide minimum area of links, A_{sv}, at a spacing of S_v for the zone where shear is less or equal to V_{nom}. From the shear force envelope determine zones where V exceeds $V_{nom} = (v_c + 0.4)bd$

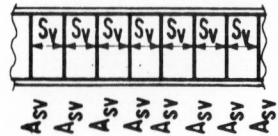

SK 2/24 Elevation of beam showing shear reinforcement.

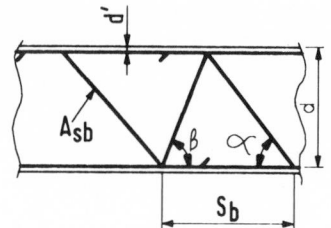

SK 2/25 Single system of bent-up shear reinforcement.

SK 2/26 Multiple system of bent-up shear reinforcement.

Find $A_{sv} = \dfrac{bS_v(v - v_c)}{0.87f_{yv}}$.

Replace v_c by v'_c where axial load is present.
Provide area of links, A_{sv}, at a spacing of S_v at the section of the beam in consideration.
For a mixture of links and bent-up bars,

$$V_b + V_s \geq (v - v_c)bd$$

where $V_b = A_{sb}(0.87f_{yv})(\cos\alpha + \sin\alpha\cot\beta)\dfrac{(d - d')}{S_b}$ α and $\beta \geq 45°$

$$V_s = A_{sv}(0.87f_{yv})\dfrac{d}{S_v} \geq 0.5(v - v_c)bd$$

Replace v_c by v'_c where axial load in compression is present.

Note: Step 14 below may be omitted if at Step 13 the critical section is selected at a distance of d from the face of support or from the concentrated load. No further checks will be necessary at the face of support or at the concentrated load.

Step 14 *Alternative design for shear*
Find $V_{max} = 0.8\sqrt{f_{cu}}\ bd$, or $= 5bd$, whichever is less.
Complete the table below:

Distance from face of support or concentrated load	$V_{nom} = V_{conc} + 0.4bd$	$V_{conc} < V_{max}$
2.00d	$(v_c + 0.4)bd$	$v_c bd$
1.75d	$(1.143v_c + 0.4)bd$	$1.143v_c bd$
1.50d	$(1.333v_c + 0.4)bd$	$1.333v_c bd$
1.25d	$(1.6v_c + 0.4)bd$	$1.6v_c bd$
1.00d	$(2.0v_c + 0.4)bd$	$2.0v_c bd$
0.75d	$(2.67v_c + 0.4)bd$	$2.67v_c bd$
0.50d	$(4v_c + 0.4)bd$	$4.0v_c bd$
0.25d	$(8v_c + 0.4)bd$	$8.0v_c bd$

Distance from face of support or concentrated load	V (actual shear force)	$V - V_{conc}$
2.00d		
1.75d		
1.50d		
1.25d		
1.00d		
0.75d		
0.50d		
0.25d		

Satisfy the following conditions:
when $V \leq V_{nom}$,

$$A_{sv} = \frac{0.4bS_v}{0.87f_{yv}}$$

when $V > V_{nom}$, calculate

$$V_s = A_{sv}(0.87f_{yv})\frac{d}{S_v} \geq V - V_{conc}$$

The shear resistance may be provided by a combination of links and bent-up bars.

Step 15 *Minimum tension reinforcement*

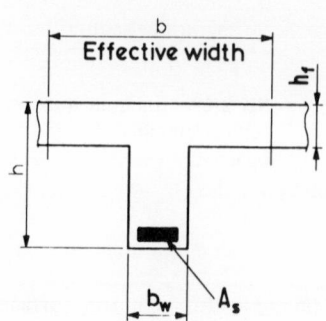

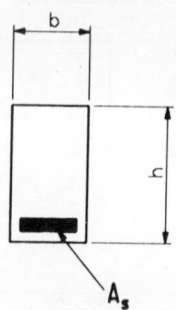

SK 2/27 Flanged beam.   **SK 2/28** Rectangular beam.

For flanged beam web in tension, find b_w/b.
If $b_w/b < 0.4$

$$A_s \geq 0.0018b_wh \qquad \text{for } f_y = 460\,\text{N/mm}^2$$

If $b_w/b \geq 0.4$

$$A_s \geq 0.0013b_wh \qquad \text{for } f_y = 460\,\text{N/mm}^2$$

For flanged beam flange in tension,

for T-beam $A_s \geq 0.0026 b_w h$ for $f_y = 460 \, \text{N/mm}^2$

for L-beam $A_s \geq 0.0020 b_w h$ for $f_y = 460 \, \text{N/mm}^2$

For rectangular beams,

$$A_s \geq 0.0013 bh \quad \text{for } f_y = 460 \, \text{N/mm}^2$$

Step 16 *Minimum compression reinforcement − when designed as doubly reinforced*
For flanged beam flange in compression,

$$A_s' \geq 0.004 bh_f$$

For flanged beam web in compression,

$$A_s' \geq 0.002 b_w h$$

For rectangular beam,

$$A_s' \geq 0.002 bh$$

Note: Minimum compression reinforcement in beams will be used only when compression reinforcement is required.

Step 17 *Minimum transverse reinforcement in flange*

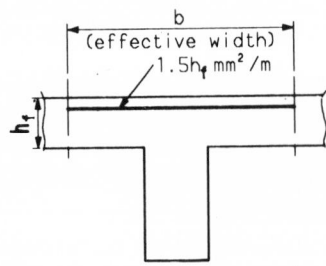

SK 2/29 Minimum transverse reinforcement in flange of flanged beam.

For flanged beams over full effective flange width near top surface, use $1.5 h_f \, \text{mm}^2/\text{m}$ reinforcement for the whole length of the beam. Normally this amount of reinforcement is provided in the slab at the top surface over the beam as part of slab reinforcement when the flanged beam forms part of a beam−slab construction.

Step 18 *Minimum reinforcement in side face of beams*

$$d = \text{dia. of bar} \geq \sqrt{\left(\frac{S_b b}{f_y}\right)}$$

$S_b \leq 250 \, \text{mm}$
$b = $ actual, or 500 mm, whichever is the lesser.

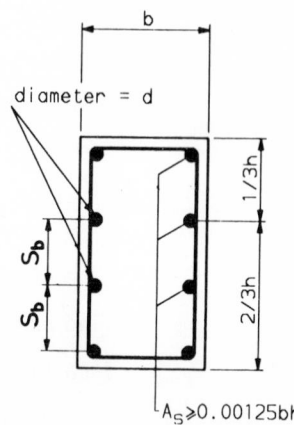

SK 2/30 Minimum reinforcement side face of beam.

$A_s \geqslant 0.00125bh$

Note: To control cracking on the side faces of beams use small diameter bars at close spacings. The distribution of these bars should be over two-thirds of beam's overall depth measured from tension face.

$A_s \geq 0.00125bh$ on each side face as shown.

Step 19 Deflection

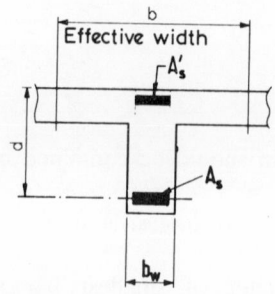

SK 2/31 Doubly reinforced flanged beam.

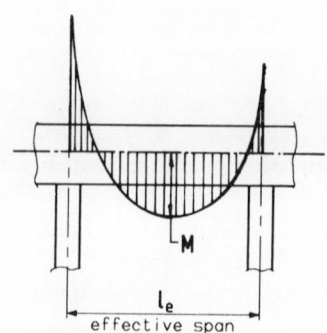

SK 2/32 Simply supported or continuous beam. $M =$ moment at midspan.

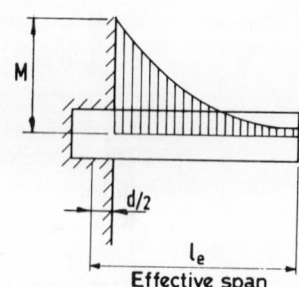

SK 2/33 Cantilever beam. $M =$ moment at support.

Find b_w/b for flanged beams.
Find l_e/d.
Find basic span/effective depth ratio from Table 11.3.

Note: If b_w/b is greater than 0.3, then interpolate between values in Table 11.3 assuming b_w/b equal to 1 for rectangular beams and 0.3 for flanged beams.

Find service stress $f_s = f_y \left(\dfrac{5}{8\beta_b} \right) \left(\dfrac{A_{s\ reqd}}{A_{s\ prov}} \right)$

where $\beta_b = M/M'$
M = moment after redistribution
M' = moment before redistribution
$A_{s\ reqd}$ = area of steel required from calculations
$A_{s\ prov}$ = area of steel actually provided.

Find M/bd^2.

Find modification factor for tension reinforcement from Chart 11.5.

Find $100A'_s/bd$.

Find modification factor for compression reinforcement from Chart 11.4.

Find modified span/depth ratio by multiplying the basic span/depth ratio by the modification factor of tension and compression reinforcement.

Check $l_e/d <$ modified span/depth ratio

Note: Table 11.3 can be used for up to a 10 m span. Beyond a 10 m span multiply these values by 10/span except for cantilevers where deflection should be calculated (see Section 1.8 for calculation of deflection.)

Step 20   Maximum areas of reinforcement
For all beams,

$A_s \leq 0.04b_w h$
$A'_s \leq 0.04b_w h$

Step 21   Containment of compression reinforcement
Designed compression reinforcement in a beam should be contained by links.
Minimum diameter of links = 0.25 times diameter of largest compression bar, or 6 mm, whichever is greater.
Maximum spacing of links = 12 times diameter of smallest bar in compression.

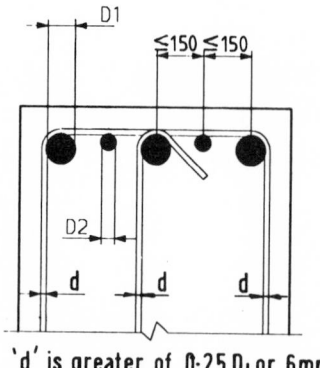

'd' is greater of $0.25 D_1$ or 6mm

SK 2/34 Containment of
compression reinforcement.

Step 22 *Bearing stress inside bend*

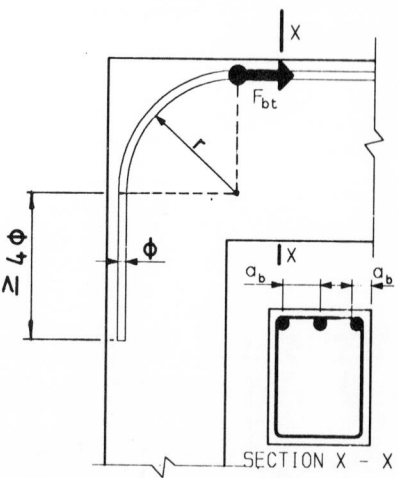

SK 2/35 Bearing stress inside
bend.

SECTION X - X

Check bearing stress inside bend where it is required to extend the bar for
more than 4 × diameter beyond the bend because the anchorage requirement is not otherwise satisfied.
Satisfy that

$$\text{bearing stress} = \frac{F_{bt}}{r\phi} \leq \frac{2f_{cu}}{1 + 2\left(\dfrac{\phi}{a_b}\right)}$$

where F_{bt} = tension in bar at the start of bend
a_b = centre to centre of bar, or, cover plus diameter of bar.

Step 23 **Curtailment of bars**
Follow simplified detailing rules for beams where the load is predominantly
uniformly distributed and spans in a continuous beam are approximately
equal. Follow bending moment diagram for other cases.

Step 24 **Spacing of bars**
Minimum clear spacing horizontally = $MSA + 5 \geq$ diameter of bar

where MSA = maximum size of aggregate.

Minimum clear spacing vertically between layers = $\dfrac{2MSA}{3}$

Maximum clear spacing of bars in tension $\leq \dfrac{47\,000}{f_s} \leq 300$

Service stress in bar $f_s = f_y\left(\dfrac{5}{8\beta_b}\right)\left(\dfrac{A_{s\text{ reqd}}}{A_{s\text{ prov}}}\right)$

(See Step 19 for explanation of β_b.)

The distance between the corner of the beam and the nearest longitudinal bar in tension should not be greater than half the maximum clear spacing.

Note: In normal internal or external condition of exposure where the limitation of crack widths to 0.3 mm is appropriate, Step 24 will deem to satisfy the crack width criteria.

Step 25 Torsional shear stress

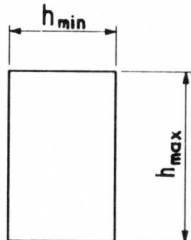

SK 2/36 Rectangular section torsional shear stress.

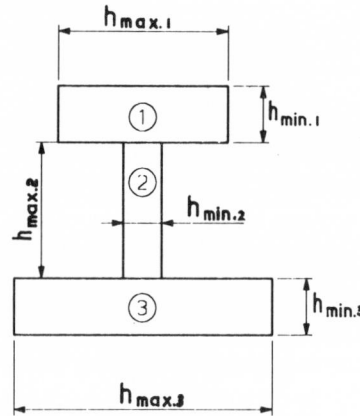

SK 2/37 Division into rectangles of composite section.

Check torsional shear stress.
Find ultimate torsion T from analysis.
For a rectangular section, torsional shear stress, v_t, is given by

$$v_t = \frac{2T}{h_{min}^2\left(h_{max} - \dfrac{h_{min}}{3}\right)}$$

For I,T or L-section, divide each section into component rectangles.

Proportion of total torsion carried by each rectangle $= T_s$

$$\frac{T h_{min}^3 h_{max}}{\Sigma\,(h_{min}^3\,h_{max})} = T_s$$

Torsional shear stress for each section

$$v_t = \frac{2T_s}{h_{min}^2\left(h_{max} - \dfrac{h_{min}}{3}\right)}$$

For hollow and other box sections, follow the method in Chapter 8.
If wall thickness in a rectangular hollow section exceeds one quarter of the
dimension in that direction, treat the hollow section as a solid rectangle.
Calculate

$$v_{t,min} = 0.067 \sqrt{f_{cu}} < 0.4\,\text{N/mm}^2$$

$$v_{tu} = 0.8 \sqrt{f_{cu}} \leq 5\,\text{N/mm}^2$$

If $v_t < v_{t,min}$, no torsional reinforcement is required.

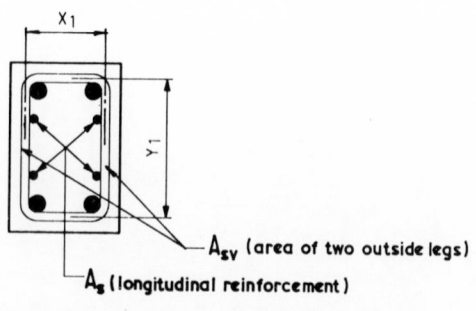

A_{sv} (area of two outside legs)

A_s (longitudinal reinforcement)

$S_v \leq x_1, \ Y_1/2 \text{ and } 200$

SK 2/38 Torsional reinforcement in beam.

SK 2/39 Elevation of torsional reinforcement in
beam.

If $v_{t,min} < v_t < v_{tu}$, provide torsional shear reinforcement by closed links
and longitudinal bars.

Check $(v + v_t) < v_{tu}$

where $v =$ flexural shear stress.

Check $v_t < v_{tu}\left(\dfrac{y_1}{550}\right)$

$$\frac{A_{sv}}{S_v} \geq \frac{T}{0.8x_1y_1(0.87f_{yv})}$$

$$A_s \geq \frac{A_{sv}f_{yv}(x_1 + y_1)}{S_v f_y}$$

Note: Add torsional reinforcement to already calculated shear reinforcement.

$$S_v < x_1, \text{ or } \frac{y_1}{2}, \text{ or } 200\,\text{mm, whichever is the least}$$

Step 26   Crack width in flexure

Serviceability limit state

Load combination $LC_7 = 1.0DL + 1.0LL + 1.0EP + 1.0WP + 1.0WL$

Note: Omit loadings from LC_7 which produce beneficial rather than adverse
effect.

64   Reinforced Concrete

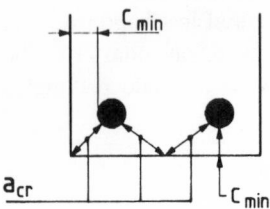

SK 2/40 Critical dimensions for crack width calculations.

$$\frac{b_t\,(h-x)(a'-x)}{3E_s\,A_s\,(d-x)}$$

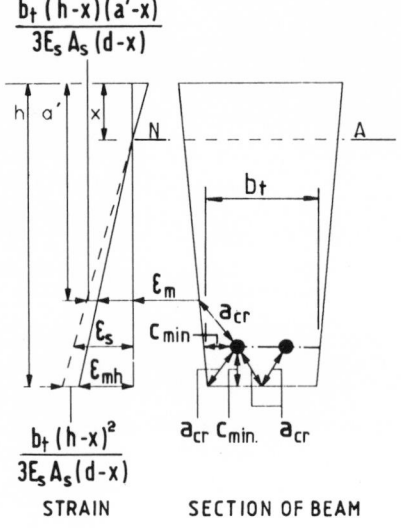

$$\frac{b_t\,(h-x)^2}{3E_s\,A_s(d-x)}$$

STRAIN   SECTION OF BEAM

SK 2/41 Strain diagram for crack width calculations.

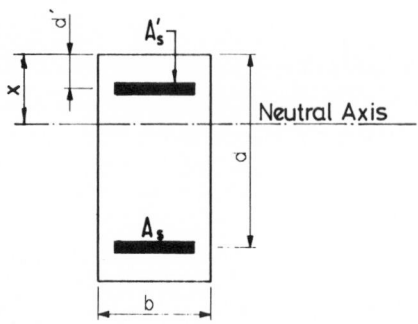

SK 2/42 Doubly reinforced rectangular beam.

$$W_{max} = \frac{3a_{cr}\,\varepsilon_m}{1 + \dfrac{2(a_{cr} - c_{min})}{(h - x)}}$$

$$\varepsilon_m = \varepsilon_1 - b_t(h - x)\frac{(a' - x)}{3E_sA_s(d - x)}$$

Note: ε_1 is the strain due to load combination LC_7.

where b_t is the width of the section at the centroid of tensile reinforcement.
For a rectangular section,

Note: A flanged beam is a rectangular section if $x \leq 1.1h_f$.

$$m = \frac{E_s}{E_c}$$

$$x = d\left\{\left[(mp + (m - 1)p')^2 + 2(mp + (m - 1)\left(\frac{d'}{d}\right)p')\right]^{\frac{1}{2}}\right.$$
$$\left. - (mp + (m - 1)p')\right\}$$

(See Section 1.13.1.)

$$p = \frac{A_s}{bd} \qquad p' = \frac{A'_s}{bd}$$

$$f_c = \frac{M}{k_2bd^2 + k_3A'_s(d - d')}$$

$$k_2 = \left(\frac{x}{2d}\right)\left(1 - \frac{x}{3d}\right)$$

$$k_3 = (m - 1)\left(1 - \frac{d'}{x}\right)$$

$$f_s = mf_c\left(\frac{d}{x} - 1\right)$$

$$\varepsilon_s = \frac{f_s}{E_s}$$

$$\varepsilon_h = \left(\frac{h - x}{d - x}\right)\varepsilon_s$$

$$\varepsilon_{mh} = \varepsilon_h - \frac{b(h - x)^2}{3\,E_sA_s(d - x)}$$

Note: In normal internal or external condition of exposure where the limitation of crack widths to 0.3 mm is appropriate, Step 24 will deem to satisfy the crack width criteria.

Step 27 *Design of connections to other components*
Follow Chapter 10.

2.4 WORKED EXAMPLES

Example 2.1 *Simply supported rectangular beam*
Clear span = 6.0 m

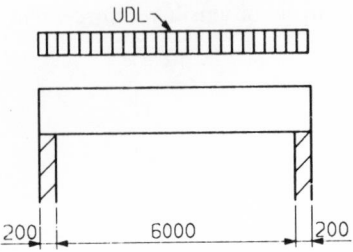

200 | 6000 | 200

SK 2/43 Simply supported beam.

Overall depth = 500 mm
Width = 300 mm
Width of supporting walls = 200 mm
All reinforcement to be used is high yield steel with $f_y = 460 \, \text{N/mm}^2$.

Note: Steps 1–5 form part of the analysis and are excluded from the worked
example. For a typical analysis see Example 2.3.

Step 6   *Determination of cover*
Maximum size of aggregate = 20 mm
Maximum bar size assumed = 32 mm
Maximum size of link assumed = 10 mm
Exposure condition = severe
Fire resistance required = 2 hours.

Refer to the following tables in Chapter 11:

Table 11.6   grade of concrete = C40 for severe exposure
Table 11.6   minimum cement content = 325 kg/m^3
Table 11.6   maximum free water/cement ratio = 0.55
Table 11.6   nominal cover = 40 mm
Table 11.7   nominal cover to beams for 2 hours fire resistance = 40 mm

For 2 hours fire resistance, minimum width of beam = 200 mm, from Figure
3.2 of BS 8110: Part 1: 1985.[1]

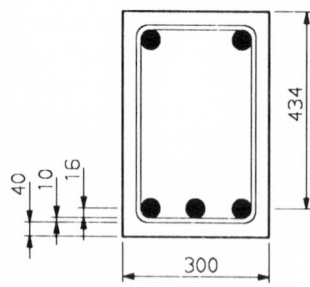

SK 2/44 Section of rectangular
beam.

Effective depth, d, is given by:

d = overall depth − nominal cover − dia. of link − half dia. of bar
  = 500 − 40 − 10 − 16 = 434 mm

Step 7 *Determination of effective span*
$l + d = 6.0 + 0.434 = 6.434$ m

$l_o = 6.2$ m

Therefore $l_e = l_o = 6.2$ m

Step 8 *Determination of effective width*
Not required.

Step 9 *Check slenderness of beam*
$l = 6.0$ m

$60b_c = 60 \times 300$ mm $= 18.0$ m

$$\frac{250\,b_c^2}{d} = \frac{250 \times 300^2}{434} = 51.84 \text{ m}$$

Satisfied $l < 60bc < \dfrac{250b_c^2}{d}$

Step 10 *Design for moment − rectangular beam*

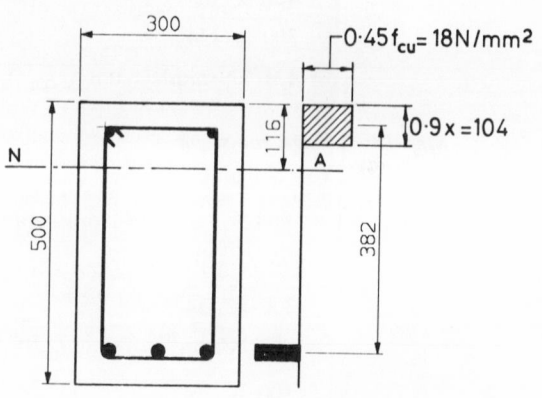

SK 2/45 Stress diagram of
rectangular beam.

Maximum ultimate bending moment = 216 kNm
Maximum shear at face of support = 140 kN
Shear at $2d$ from face of support = 96 kN
Shear at d from face of support = 116 kN
Direct load, $N = 0$ kN
$M_d = M = 216$ kNm
$f_{cu} = 40$ N/mm^2

$$K = \frac{M_d}{f_{cu}bd^2} = \frac{216 \times 10^6}{40 \times 300 \times 434^2} = 0.0956 < 0.156$$

$$\begin{aligned} z &= d\left[0.5 + \sqrt{\left(0.25 - \frac{K}{0.9}\right)}\right] \\ &= d\left[0.5 + \sqrt{\left(0.25 - \frac{0.0956}{0.9}\right)}\right] \\ &= 0.88d = 382\,\text{mm} \end{aligned}$$

$$x = \frac{d - z}{0.45} = \frac{434 - 382}{0.45} = 116\,\text{mm}$$

$$\begin{aligned} A_s &= \frac{M_d}{0.87f_yz} \\ &= \frac{216 \times 10^6}{0.87 \times 460 \times 382} = 1413\,\text{mm}^2 \end{aligned}$$

Use 3 no. 25 mm dia. Grade $460 = 1472\,\text{mm}^2$

Step 11 **Design for moment − flanged beam**
Not required.

Step 12 **Check maximum allowable shear**
$$v = \frac{V}{bd} \quad \text{at face of support}$$
$$= \frac{140 \times 10^3}{300 \times 434}$$
$$= 1.075\,\text{N/mm}^2 < 0.8\,\sqrt{f_{cu}} = 5\,\text{N/mm}^2$$

Step 13 **Design for shear**
$$2d = 870\,\text{mm}$$

$$V = 96\,\text{kN} \quad \text{at } 2d \text{ from support face}$$

$$v = \frac{96 \times 10^3}{300 \times 434}$$
$$= 0.74\,\text{N/mm}^2 < 0.8\,\sqrt{f_{cu}} = 5\,\text{N/mm}^2$$

$$p = \frac{100\,A_s}{bd}$$
$$= \frac{100 \times 1472}{300 \times 434}$$
$$= 1.13\%$$

$$\begin{aligned} v_c &= 0.65 \times 1.17\,\text{N/mm}^2 \quad \text{for Grade 40 concrete} \\ &= 0.76\,\text{N/mm}^2 \quad \text{from Fig. 11.5} \end{aligned}$$

$$V_{nom} = (v_c + 0.4)\,bd$$
$$= 151\,\text{kN} > 140\,\text{kN} \qquad \text{at face of support}$$
$$V < V_{nom} \text{ at all points in the beam.}$$

Nominal links $A_{sv} = \dfrac{0.4 b_{sv}}{0.87 f_{yv}}$

Assume $S_v = 300\,\text{mm}$

$$A_{sv} = \dfrac{0.4 \times 300 \times 300}{0.87 \times 460}$$

$$= 90\,\text{mm}^2$$

Use 8 mm dia. single closed link $= A_{sv} = 100\,\text{mm}^2$ $(f_y = 460\,\text{N/mm}^2)$ at 300 mm centre to centre.

Step 14 *Alternative design for shear*
$V_{nom} > V$ at face of support so Step 14 is superfluous — use nominal links everywhere on the beam.

Step 15 *Minimum tensile reinforcement*
Minimum tensile reinforcement $= 0.0013 bh$
$$= 0.0013 \times 300 \times 500\,\text{mm}^2$$
$$= 195\,\text{mm}^2 < 1472\,\text{mm}^2 \text{ provided}$$

2 no. 12 diameter $(= 226\,\text{mm}^2)$ provided at top of beam.

Step 16 *Minimum compression reinforcement*
Not required.

Step 17 *Minimum transverse reinforcement in flange*
Not required.

Step 18 *Minimum reinforcement in side face of beams*

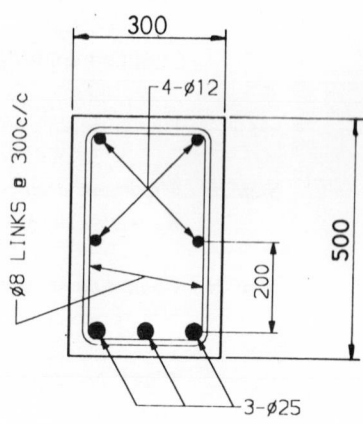

SK 2/46 Section through beam.

ALL REINFORCEMENT HIGH-YIELD

b = actual, or 500 mm, whichever is the lesser.

Minimum dia. of bar in side face of beam $= \sqrt{\left(\dfrac{S_b b}{f_y}\right)}$

(assume S_b = 200 mm)

$$= \sqrt{\left(\frac{200 \times 300}{460}\right)} = 11.4\,\text{mm}$$

Use 12 dia. Grade 460 bars at approximately 200 centres on the side face of beam.

Reinforcement on each side face of beam = 2 no. 12 dia. + 1 no. 25 dia.
$$= 716\,\text{mm}^2$$

$A_s = 0.00125bh$
$\quad = 0.00125 \times 300 \times 500$
$\quad = 188\,\text{mm}^2 < 716\,\text{mm}^2$ OK

Note: Strictly speaking these bars on the side face are not required for beams less than 750 mm overall depth but it is good practice to use them in order to avoid shrinkage cracks.

Step 19   **Check deflection**
$$\frac{l_e}{d} = \frac{6200}{434} = 14.3$$

Basic span/depth ratio = 20   from Table 11.3

$\beta_b = 1,$

$$f_s = f_y\left(\frac{5}{8}\right)\left(\frac{A_{s\ \text{reqd}}}{A_{s\ \text{prov}}}\right) = 460\left(\frac{5}{8}\right)\left(\frac{1413}{1472}\right) = 275\,\text{N/mm}^2$$

$$\frac{M}{bd^2} = \frac{216 \times 10^6}{300 \times 434^2} = 3.8$$

Modification factor for tension reinforcement = 0.90   from Table 11.5

Modified span/depth ratio = $20 \times 0.90 = 18 > \dfrac{l_e}{d} = 14.3$

Hence deflection is OK.

Step 20   **Maximum areas of reinforcement**
A_s is less than 4%.

Step 21   **Containment of compression reinforcement**
Not required.

Step 22   **Check bearing stress inside bend**
Not required.

Step 23 Curtailment of bars
$0.08l = 0.08 \times 6000 = 480\,\text{mm}$

The central 25 mm dia. bar will be stopped 250 mm from the face of the support.

Step 24 Spacing of bars

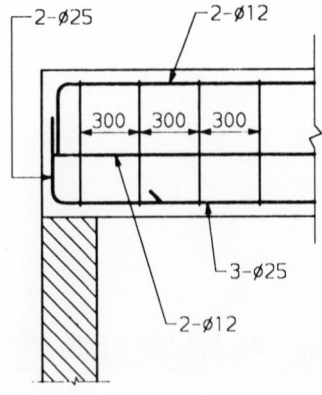

SK 2/47 Elevation of beam near support.

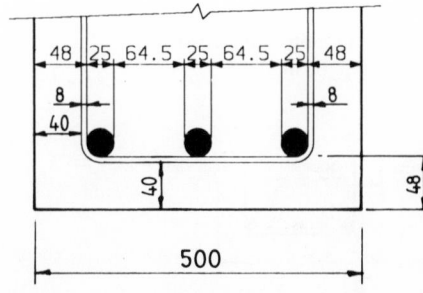

SK 2/48 Arrangement of bars at the bottom of beam.

Clear spacing between bars in tension $= 64.5\,\text{mm}$

Minimum required spacing $= 20 + 5 = 25\,\text{mm}$

$$\text{Maximum spacing} = \frac{47\,000}{f_s}$$

$$= \frac{47\,000}{275} = 170\,\text{mm}$$

where $f_s = 275\,\text{N/mm}^2$ (see Step 19)

Spacing of bars is OK.

Step 25 Check torsional shear stress
Not required.

Step 26 Crack width calculations
Service maximum moment $= 144\,\text{kNm}$
$A_s = 1472\,\text{mm}^2$ $d' = 54\,\text{mm}$
$A_s' = 226\,\text{mm}^2$ $d = 439\,\text{mm}$

$$m = \frac{E_s}{E_c} = \frac{200}{20} = 10$$

E_c assumed halfway between long and short-term.

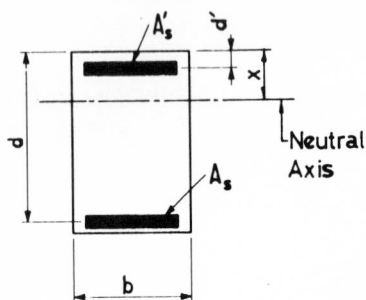

SK 2/49 Doubly reinforced rectangular beam.

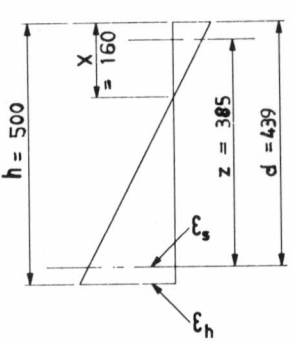

SK 2/50 Strain diagram.

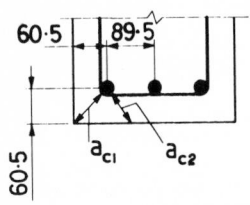

SK 2/51 Crack width calculations.

$$p = \frac{A_s}{bd} = \frac{1472}{300 \times 439} = 0.0112 \qquad p' = \frac{A'_s}{bd} = 0.0017$$

$$x = d\left\{\left[(mp + (m-1)p')^2 + 2(mp + (m-1)\left(\frac{d'}{d}\right)p')\right]^{\frac{1}{2}} - (mp + (m-1)p'\right\} = 160 \text{ mm}$$

$$k_2 = \left(\frac{x}{2d}\right)\left(1 - \frac{x}{3d}\right)$$

$$= \left(\frac{160}{2 \times 439}\right)\left(1 - \frac{160}{3 \times 439}\right)$$

$$= 0.16$$

$$k_3 = (m-1)\left(1 - \frac{d'}{x}\right) = (10-1)\left(1 - \frac{54}{160}\right) = 5.96$$

$$f_c = \frac{M}{k_2 bd^2 + k_3 A'_s(d - d')}$$

$$= \frac{144 \times 10^6}{0.16 \times 300 \times 439^2 + 5.96 \times 226 \times (439 - 54)}$$

$$= 14.74 \text{ N/mm}^2$$

$$f_s = mf_c\left(\frac{d}{x} - 1\right)$$

$$= 10 \times 14.74 \times \left(\frac{439}{160} - 1\right)$$

$$= 257\,\text{N/mm}^2$$

$$\varepsilon_s = \frac{f_s}{E_s} = \frac{257}{200 \times 10^3} = 1.285 \times 10^{-3}$$

$$\varepsilon_h = \left(\frac{h-x}{d-x}\right)\varepsilon_s = \left(\frac{340}{279}\right)\varepsilon_s = 1.566 \times 10^{-3}$$

$$\varepsilon_{mh} = \varepsilon_h - \frac{b(h-x)^2}{3E_s A_s(d-x)}$$

$$= 1.566 \times 10^{-3} - \frac{300 \times 340^2}{3 \times 200 \times 10^3 \times 1472 \times 279}$$

$$= 1.425 \times 10^{-3}$$

$$a_{c1} = 1.414 \times 60.5 - 12.5 = 73.0\,\text{mm}$$

$$a_{c2} = \sqrt{(60.5^2 + 45^2)} - 12.5 = 62.9\,\text{mm}$$

$$a_{c1} > a_{c2}$$

$$W_{cr} = \frac{3 a_{cr}\, \varepsilon_m}{1 + \left[\dfrac{2(a_{cr} - C_{min})}{h - x}\right]}$$

$$= \frac{3 \times 73.0 \times 1.425 \times 10^{-3}}{1 + \left[\dfrac{2(73 - 48)}{340}\right]}$$

$$= 0.27\,\text{mm} < 0.3\,\text{mm} \qquad \text{OK}$$

Example 2.2 *Three span continuous beam*

SK 2/52 Three-span continuous beam.

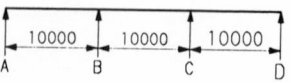

10000   10000   10000
A   B   C   D

Three equal spans of 10 m centre-to-centre of columns.

Width of column = 0.4 m
clear span = 9.6 m
slab depth = 150 mm
beam spacing = 4.0 m
beam overall depth = 550 mm
beam width = 300 mm

Redistribution of moments = 10%

Note: Steps 1−5 form part of the analysis and have been excluded. For a typical analysis see Example 2.3.

All reinforcement to be used will be high yield steel with $f_y = 460 \, \text{N/mm}^2$. It is expected that the analysis will be carried out using a computer program with the load combination shown in Section 2.2.

From moment and shear envelope,

$M_A = 0$ $V_{AB} = 300 \, \text{kN}$ $V_{A'B} = 250 \, \text{kN}$

$M_{AB} = +600 \, \text{kNm}$ $V_{AB} = \text{negligible}$

$M_B = -650 \, \text{kNm}$ $V_{BA} = 370 \, \text{kN}$ $V_{B'A} = 320 \, \text{kN}$

$V_{BC} = 325 \, \text{kN}$ $V_{B'C} = 275 \, \text{kN}$

$M_{BC} = +370 \, \text{kNm}$ or $-150 \, \text{kNm}$

where $V_{A'B} = \text{shear at a distance of } d \text{ from face of support.}$

Step 6   *Determination of cover*

Maximum size of aggregate = 20 mm

maximum bar size = 32 mm

maximum size of link = 8 mm

exposure condition = severe

fire resistance required = 2 hours

grade of concrete = C40

maximum cement content = 325 kg/m^3

maximum free water cement ratio = 0.55

nominal cover = 40 mm   from Tables 11.6 and 11.7

effective depth, $d = 550 - 40 - 8 - 16 = 486 \, \text{mm}$

Step 7   *Effective span*

$l_e = l_o = 10.0 \, \text{m}$

Step 8   *Effective width of compression flange*

Actual $b = 4.0 \, \text{m}$ (centre-to-centre of beams)

Calculated $b = \dfrac{l_e}{7.14} + b_w$

$= \dfrac{10\,000}{7.14} + 300$

$= 1700 \, \text{mm}$

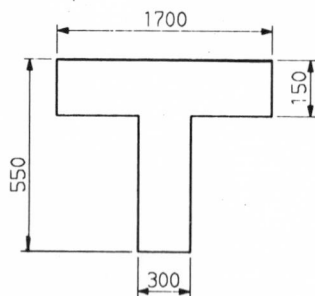

SK 2/53 Effective width of compression flange.

Step 9 **Slenderness check**
May be ignored.

Step 10 **Design for moment**
$M_{AB} = 600 \, \text{kNm}$

Flanged beam
$M_d = M_{AB} = 600 \, \text{kNm}$

$$K = \frac{M_d}{f_{cu}bd^2}$$

$$= \frac{600 \times 10^6}{40 \times 1700 \times 486^2}$$

$$= 0.0373$$

$$z = d\left[0.5 + \sqrt{\left(0.25 - \frac{K}{0.9}\right)}\right]$$

$$= d\left[0.5 + \sqrt{\left(0.25 - \frac{0.0373}{0.9}\right)}\right]$$

$$= 0.95d = 462 \, \text{mm}$$

$$x = \frac{d - z}{0.45}$$

$$= \frac{486 - 462}{0.45}$$

$$= 53 \, \text{mm} < h_f = 150 \, \text{mm}$$

Neutral axis in the slab

$$A_s = \frac{M_d}{0.87 f_y z}$$

$$= \frac{600 \times 10^6}{0.87 \times 460 \times 462} = 3245 \, \text{mm}^2$$

Use 3 no. 32 dia. bars in bottom layer plus 2 no. 25 dia. bars in top layer.

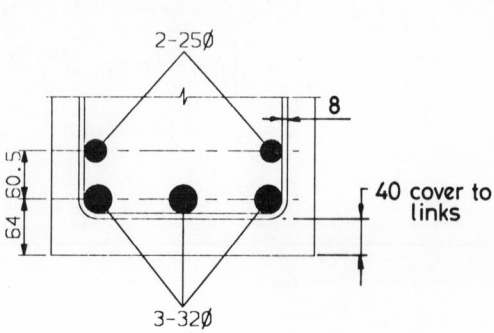

SK 2/54 Arrangement of reinforcement at bottom of beam at midspan.

Total area of steel provided $= 3394\,\text{mm}^2$

Check effective depth.

Centre of gravity of group of 5 bars

$$x = \frac{3 \times 804 \times 64 + 2 \times 491 \times 124.5}{3394}$$

$$= 81.5\,\text{mm}$$

$$d = 550 - 81.5 = 468.5\,\text{mm}$$

Recheck reinforcement requirement with revised effective depth:

$$K = 0.040$$

$$z = 0.95 \times 468.5 = 445\,\text{mm}$$

$$A_\text{s} = 3369\,\text{mm}^2 \qquad \text{(required)}$$

$$A_\text{s}\ \text{provided} = 3394\,\text{mm}^2 \qquad \text{OK}$$

$$M_\text{B} = -650\,\text{kNm}$$

Rectangular beam

$$M_\text{d} = 650\,\text{kNm}$$

Effective depth, $d = 550 - 40 - 32 - 16 - 8 = 454\,\text{mm}$

(assuming two layers of 32 dia. bars)

$$K = \frac{650 \times 10^6}{40 \times 300 \times 454^2}$$

$$= 0.263 > 0.156$$

Compression reinforcement required.

Redistribution is 10%

$$A'_\text{s} = \frac{(K - 0.156)f_\text{cu}bd^2}{0.87f_\text{y}(d - d')}$$

$$= \frac{(0.263 - 0.156) \times 40 \times 300 \times 454^2}{0.87 \times 460 \times (454 - 64)}$$

$$= 1696\,\text{mm}^2$$

Use 3 no. 32 dia. bars ($2412\,\text{mm}^2$) − bottom of beam.

$$z = d\left[0.5 \sqrt{\left(0.25 - \frac{0.156}{0.9}\right)}\right]$$

$$= 0.775d = 352\,\text{mm}$$

$$x = 0.5d = 227\,\text{mm} \quad \text{and} \quad \frac{d'}{x} = \frac{64}{227} = 0.28 < 0.43$$

$$A_\text{s} = \frac{0.156f_\text{cu}bd^2}{0.87f_\text{y}z} + A'_\text{s}$$

$$= \frac{0.156 \times 40 \times 300 \times 454^2}{0.87 \times 460 \times 352} + 1696 = 4435 \, \text{mm}^2$$

Use 6 no. 32 dia. bars (4824 mm^2) – top of beam in two layers.

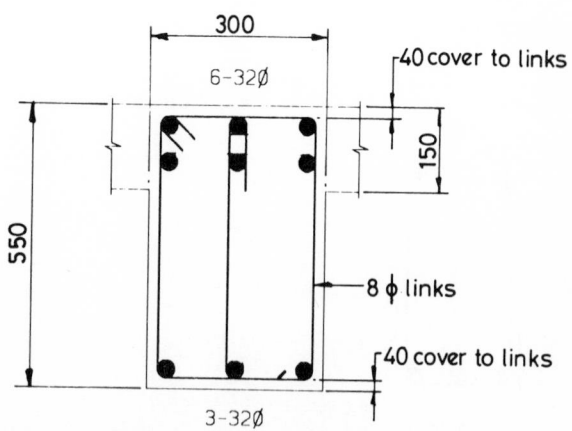

300

6-32∅

40 cover to links

150

550

8 ∅ links

40 cover to links

SK 2/55 Arrangement of reinforcement at top of beam over support.

3-32∅

$M_{BC} = +370 \, \text{kNm}$

Flanged beam

$b = 1700 \, \text{mm}$

$d = 550 - 40 - 8 - 16 = 486 \, \text{mm}$

$K = \dfrac{M_d}{f_{cu}bd^2} = 0.023$

$z = 0.95d = 426 \, \text{mm}$

$A_s = \dfrac{M_d}{0.87f_{y}z}$

$$= \frac{370 \times 10^6}{0.87 \times 460 \times 462} = 2001 \, \text{mm}^2$$

Use 3 no. 32 dia. bar (2412 mm^2) – bottom of beam

$M_{BC} = -150 \, \text{kNm}$

Rectangular beam

$b = 300 \, \text{mm}$

$d = 486 \, \text{mm}$

$K = \dfrac{150 \times 10^6}{300 \times 486^2 \times 40} = 0.053$

$$z = d\left[0.5 + \sqrt{\left(0.25 - \frac{K}{0.9}\right)}\right]$$

$$= 0.94d = 456\,\text{mm}$$

$$A_s = \frac{150 \times 10^6}{0.87 \times 460 \times 456} = 822\,\text{mm}^2$$

Use 2 no. 32 dia. bar (1608 mm^2) − top of beam.

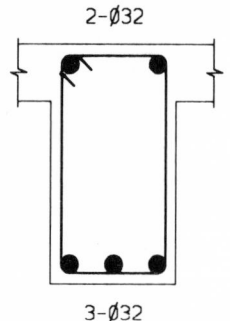

2-∅32

3-∅32

SK 2/56 Section through midspan BC.

Step 11 Design for moment − flanged beam
Not required.

Step 12 Maximum shear stress

$$v = \frac{V_{\text{max}}}{bd} = 2.716\,\text{N/mm}^2 < 5\,\text{N/mm}^2$$

Step 13 Design for shear
Maximum shear = 370 kN = V_{BA}

$$v = \frac{V}{bd} = \frac{370 \times 10^3}{300 \times 454}$$

$$= 2.716\,\text{N/mm}^2 < 0.8\,\sqrt{f_{cu}} = 5\,\text{N/mm}^2 \quad \text{OK}$$

Check shear stress at d from face of column.

$$V_{A'B} = 250\,\text{kN}$$

$$d = 468.5\,\text{mm} \qquad \text{for span AB}$$

$$v = \frac{250 \times 10^3}{300 \times 468.5}$$

$$= 1.78\,\text{N/mm}^2$$

$$p = \frac{100A_s}{bd} = \frac{100 \times 3394}{300 \times 468.5}$$

$$= 2.41$$

$$v_c = 0.85 \times 1.17 \qquad \text{from Fig. 11.5}$$
$$= 0.99 \,\text{N/mm}^2$$

$$V_{\text{nom}} = (v_c + 0.4)bd$$
$$= 195 \,\text{kN}$$

$$v > v_c + 0.4 = 1.39 \,\text{N/mm}^2$$

$$A_{sv} = \frac{bS_v(v - v_c)}{0.87f_y}$$

(assume $S_v = 150 \,\text{mm}$)

$$= \frac{300 \times 150 \times (1.78 - 0.99)}{0.87 \times 460}$$

$$= 89 \,\text{mm}^2$$

Use 8 mm dia. links = 100 mm^2 (two legs) at 150 centre-to-centre up to the point where shear falls to 195 kN. High yield reinforcement ($f_y = 460 \,\text{N/mm}^2$).

$$\text{Nominal } A_{sv} = \frac{0.4bS_v}{0.87f_y}$$

$$= \frac{0.4 \times 300 \times 300}{0.87 \times 460} = 90 \,\text{mm}^2$$

Use 8 mm dia. links = 100 mm^2 (two legs) at 300 centre-to-centre as nominal links ($f_y = 460 \,\text{N/mm}^2$).

$$V_{B'A} = 320 \,\text{kN}$$

$$v = \frac{320 \times 10^3}{300 \times 454} \qquad (d = 454 \,\text{mm at B})$$

$$= 2.35 \,\text{N/mm}^2$$

$$p = \frac{100A_s}{bd} = \frac{100 \times 4435}{300 \times 454} = 3.25$$

$$v_c = 0.91 \times 1.17 = 1.065 \,\text{N/mm}^2 \text{ from Fig. 11.5}$$

$$A_{sv} = \frac{bS_v(v - v_c)}{0.87f_y}$$

$$= \frac{300 \times 150 \times (2.35 - 1.063)}{0.87 \times 460}$$

$$= 144.5 \,\text{mm}^2$$

Use 8 mm dia. links = 150 mm^2 (3 legs) at 150 centre to centre up to the point where shear falls to 195 kN.

Step 14 Alternative design for shear
Omitted.

Step 15 Minimum tension reinforcement

Flanged beam

$$\frac{b_w}{b} = \frac{300}{1700} = 0.176 < 0.4$$

For web in tension

$$A_s > 0.0018 b_w h = 297\,\text{mm}^2$$

For flange in tension

$$A_s > 0.0026 b_w h = 429\,\text{mm}^2$$
Both conditions satisfied.

Step 16 ***Minimum compression reinforcement***
$$A_s' > 0.002 b_w h = 330\,\text{mm}^2$$

Provided $A_s' = 2412\,\text{mm}^2$

Condition satisfied.

Step 17 ***Transverse reinforcement in flange***
Minimum transverse reinforcement in flange $= 1.5 h_f\,\text{mm}^2/\text{m}$
$$= 1.5 \times 150\,\text{mm}^2/\text{m}$$
$$= 225\,\text{mm}^2/\text{m}$$

Reinforcement in the slab over the beam will be a lot more than this quantity.

Step 18 ***Reinforcement in side face of beam***
For a 550 mm overall depth of beam with 150 mm slab, side reinforcement will not be required.

Step 19 ***Check deflection***
$$\frac{l_e}{d} = \frac{10\,000}{468.5} = 21.3$$

$d = 468.5\,\text{mm}$ for span AB

$$\frac{b_w}{b} = 0.176 < 0.3$$

Basic span/effective depth ratio from Table $11.3 = 20.8$

Since the ultimate moment at midspan is greater after redistribution than the ultimate elastic moment, the service elastic stress may be taken as $(5/8) f_y$.

Service stress, $f_s = \dfrac{5}{8} f_y$ (assumed)

$$= \frac{5}{8} \times 460$$

$$= 288\,\text{N/mm}^2$$

$$\frac{A_{s\ reqd}}{A_{s\ prov}} = \frac{3369}{3394} = 1.0$$

$$\frac{M}{bd^2} = \frac{600 \times 10^6}{1700 \times 468.5^2} = 1.6$$

Modification factor $= 1.19$ from Chart 11.5

Modified span/effective depth ratio $= 20.8 \times 1.19$
$$= 24.75 > 21.3 \qquad \text{OK}$$

Step 20 *Maximum areas of reinforcement*
$A_s \leq 0.04b_w h = 6600\,\text{mm}^2$

Maximum tensile reinforcement used $= 4824\,\text{mm}^2$ OK

Step 21 *Containment of compression reinforcement*
Minimum dia. of links $= 0.25 \times$ max. dia. of bar
$$= 0.25 \times 32 = 8\,\text{mm} \qquad \text{OK}$$

Maximum spacing of links $= 12 \times$ dia. of bar
$$= 12 \times 32\,\text{mm} = 384\,\text{mm} \qquad \text{OK}$$

Note: At least one link at the centre of columns B and C will be required for containment.

Step 22 *Check bearing stress inside bend*
Not required.

Step 23 *Curtailment of bars*

$0.15l = 1500\,\text{mm}$
$0.10l = 1000\,\text{mm}$
$0.25l = 2500\,\text{mm}$

Span AB
Continue 3 no. 32 dia. + 2 no. 32 dia. up to 1000 mm from A (end support).

Stop 1 no. 32 dia. and 2 no. 32 dia. at 1000 mm from A.

(See Step 26: reinforcement in span AB increased.)

Over support B (top bars)

Continue 6 no. 32 dia. bar top up to 1500 mm on either side of B.

Stop 2 no. 32 dia. bar at 1500 from B.

Stop 2 no. 32 dia. bar at 2500 from B.

Continue 2 no. 32 dia. through span.

Step 24 *Spacing of bars*
Minimum clear spacing $= MSA + 5 = 25\,\text{mm}$

Clear spacing of bars in tension $= 54\,\text{mm} > 25\,\text{mm}$

$$\text{Maximum clear spacing} = \frac{47\,000}{f_s} = \frac{47\,000}{288} = 163\,\text{mm}$$

(See Step 19 for f_s)

At span BC top tension reinforcement

Clear spacing of bars (2 no. 32 dia.) $= 140\,\text{mm}$ OK

Note: Under normal circumstances this step will deem to satisfy the 0.3 crack width limitation criteria, but, as Step 26 will prove, when crack width calculations are actually carried out this may not be the case. In span AB the maximum clear spacing criterion is satisfied but the calculations show that the crack widths may be exceeded.

Step 25   ***Check torsional shear stress***
Not required.

Step 26   ***Crack width calculations***

Span AB
Maximum service moment $= 400\,\text{kNm}$

$d = 468.5\,\text{mm}$ $C_{min} = 48\,\text{mm}$

$b = 1700\,\text{mm}$

$A_s = 3394\,\text{mm}^2$

$A'_s = 1608\,\text{mm}^2$ (ignored in the computation)

$$m = \frac{E_s}{E_c} = 10$$

$$x = \sqrt{\left[\left(\frac{mA_s}{b}\right)^2 + \frac{2mA_sd}{b}\right]} - \frac{mA_s}{b}$$

$$= 118\,\text{mm} < h_f(= 150\,\text{mm})$$

$$z = d - \frac{x}{3} = 429\,\text{mm}$$

$$f_s = \frac{400 \times 10^6}{429 \times 3394} = 274\,\text{N/mm}^2$$

$$\varepsilon_s = \frac{f_s}{E_s} = \frac{274}{200 \times 10^3} = 1.37 \times 10^{-3}$$

$$\varepsilon_h = \left(\frac{h - d}{d - x}\right)\varepsilon_s = \left(\frac{550 - 118}{468.5 - 118}\right) \times 1.37 \times 10^{-3}$$

$$= 1.69 \times 10^{-3}$$

$$\varepsilon_{mh} = \varepsilon_h - \frac{b(h - x)^2}{3E_s A_s(d - x)}$$

$$= 1.69 \times 10^{-3} - \frac{300 \times 432^2}{3 \times 200 \times 10^3 \times 350.5 \times 3394}$$

$$= 1.61 \times 10^{-3}$$

$$a_{c1} = \sqrt{(64^2 + 64^2)} - 16 = 74.5 \, \text{mm}$$

$$a_{c2} = \sqrt{(64^2 + 43^2)} - 16 = 61.1 \, \text{mm}$$

$$a_{cr} = 74.5 \, \text{mm} \qquad \text{at the corner of the beam}$$

$$W_{cr} = \frac{3a_{cr} \, \varepsilon_m}{1 + \dfrac{2(a_{cr} - c_{min})}{(h - x)}} = \frac{3 \times 74.5 \times 1.61 \times 10^{-3}}{1 + \dfrac{2(74.5 - 48)}{(550 - 118)}}$$

$$= 0.32 \, \text{mm} > 0.3 \, \text{mm}$$

The calculated crack width is greater than allowable. Increase reinforcement to 5 no. 32 dia. bar instead of 3 no. 32 dia. plus 2 no. 25 dia. No more checks are necessary.

Over support B

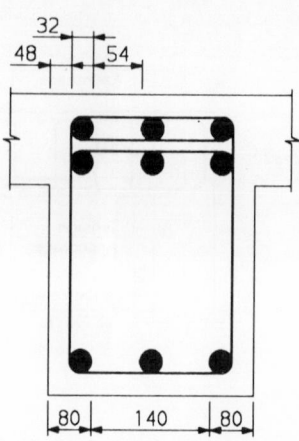

SK 2/57 Arrangement of bars over support.

At face of column,

maximum service moment $= 390 \, \text{kNm}$

$d = 454 \, \text{mm} \qquad d' = 64 \, \text{mm}$

$b = 300 \, \text{mm}$

$A_s = 4824 \, \text{mm}^2$

$A_s' = 2412 \, \text{mm}^2$

$C_{min} = 48\,mm$

$m = 10$

See Step 26 of Example 2.1 for explanation of symbols and the equations.

$x = 225\,mm$

$K_2 = 0.2068$

$K_3 = 6.44$

$$f_c = \frac{M}{K_2bd^2 + K_3A_s'(d - d')} = 20.69\,N/mm^2$$

$f_s = 211.6\,N/mm^2$

$\varepsilon_s = 1.058 \times 10^{-3}$

$\varepsilon_h = 1.502 \times 10^{-3}$

$ac_1 = 74.5\,mm$ at the top corner

$W_{cr} = 0.297\,mm < 0.3\,mm$ OK

Step 27 *Design of connections to other elements*
See Chapter 10.

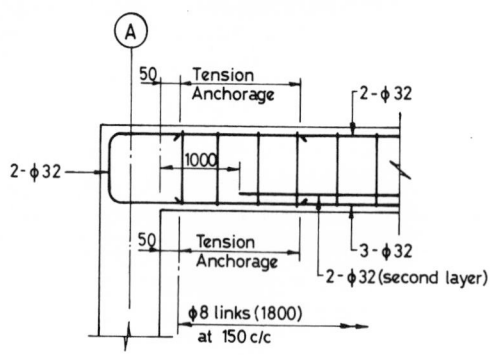

SK 2/58A Detail of beam at A.

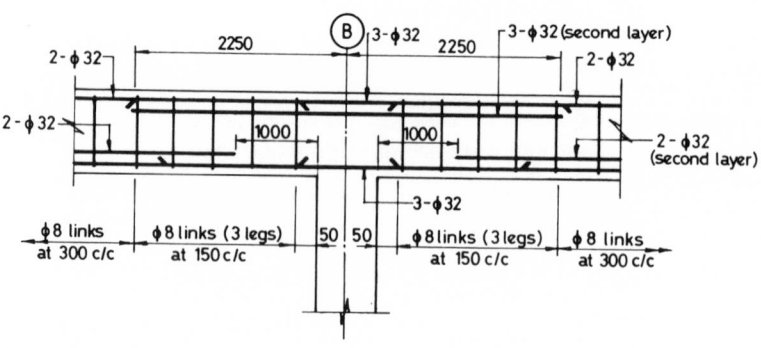

SK 2/58B Detail of beam at B.

Example 2.3 **Design of beam with torsion**

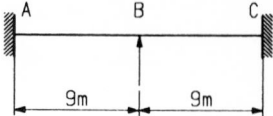

SK 2/59 Two-span edge beam with nib.

Edge beam to carry precast floor slabs on nibs.

Clear gap between beams = 4.5 m
Effective span of beam = 9.0 m

See Example 5.2 for details of precast floor slabs and nib geometry computations.

Two-span beam is fully restrained at the rigid supports.

Step 1 **Analysis of beam**

Properties of section

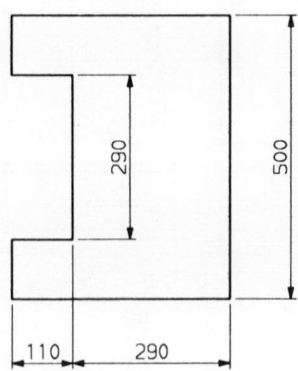

SK 2/60 Section of beam with nibs.

Area of section = $500 \times 290 + 2 \times 110 \times 105$
$$= 168\,100\,\text{mm}^2$$

Self-weight of beam = $0.1681 \times 24\,\text{kN/m}^3 = 4\,\text{kN/m}$

$$\bar{x} = \frac{500 \times 290 \times 145 + 2 \times 110 \times 105 \times (290 + 55)}{168\,100}$$

$$= 172.5\,\text{mm}$$

$$I_{xx} = \frac{1}{12} \times 400 \times 500^3 - \frac{1}{12} \times 110 \times 290^3$$

$$= 3.943 \times 10^9\,\text{mm}^4 \quad (\textit{gross section})$$

Assume $\dfrac{p'}{P} = 0$

Assume $p = 1\%$

Assume $m = \dfrac{E_s}{E_c} = 10$

From Fig. 11.1,

$F = 6 \times 10^{-2}$

Cracked moment of inertia $= Fbd^3$
$$= 6 \times 10^{-2} \times (400 \times 500^3 - 110 \times 290^3)$$
$$= 2.839 \times 10^9 \, \text{mm}^4$$

Average moment of inertia, $I_{xx} = 0.5(3.943 + 2.839) \times 10^9 \, \text{mm}^4$
$$= 3.391 \times 10^9 \, \text{mm}^4$$

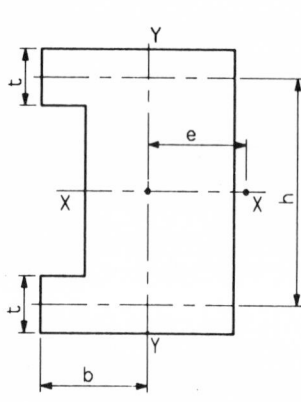

SK 2/61 Beam geometry to find shear centre e.

Shear centre, $e = \dfrac{b^2 h^2 t}{4 I_{xx}}$

$b = 400 - 145 = 255 \qquad h = 500 - 105 = 395 \qquad t = 105$

$e = \dfrac{255^2 \times 395^2 \times 105}{4 \times 3.391 \times 10^9} = 78.5 \, \text{mm}$

Loading
Dead load from slab $= 5 \, \text{kN/m}^2 \times 2.25 \, \text{m} = 11.25 \, \text{kN/m}$.

Self-weight of beam $= 0.1681 \times 24 \, \text{kN/m}^3 = 4.0 \, \text{kN/m}$

Total dead load on beam including self-weight $= 15.25 \, \text{kN/m}$

Live load from slab @ $5 \, \text{kN/m}^2 = 5 \times 2.25 = 11.25 \, \text{kN/m}$

Ultimate limit state,

$LC_1 = 1.4DL + 1.6LL = 1.4 \times 15.25 + 1.6 \times 11.25 = 22\,kN/m + 18\,kN/m$
$LC_2 = 1.0DL = 15.25\,kN/m$

Load both spans with LC_1 to get maximum support moment at B.

Load span AB with LC_1 and span BC with LC_2 to get maximum support moment A and maximum span moment at AB.

Steps 2 and 3 *Draw moment and shear envelope*
Non-linear analysis with 10% redistribution.

Boundary condition	Loading	Force	Support A	Span AB	Support B	Span BC	Support C	
A and C fully restrained	LC_1 on both spans	BM	−270	+135	+270	+135	−270	kNm
		Shear	180	—	180 180	—	180	kN
	1.4DL on AB	BM	−159.8	+79.9	−125.9	+46.0	−92.0	kNm
	1.0DL on BC	Shear	102.8	—	95.2 72.6	—	65.1	kN
	LC_1 on AB	BM	−311.7	+155.9	−186.7	+28.0	−70.6	kNm
	LC_2 on BC	Shear	193.9	—	166.1	—	—	kN
	1.0 kN/m LL	BM	−8.44	+4.22	−3.38	−1.0	+1.19	kNm
	on AB	Shear	5.06	—	3.94	—	—	kN
Plastic hinge at A, C fully restrained	1.0 kN/m LL on AB	BM	0	+7.23	−5.78	−1.45	+2.89	kNm
		Shear	3.86	—	5.14	—	—	kN

Assume 10% redistribution. Support moment at A is fixed at $0.9 \times 311.7 = 280\,kNm$. The support moment at A reaches 280 kNm elastically with live load on span AB equal to $(280 − 159.8)/8.44 = 14.24\,kNm$. At that point a plastic hinge forms at A and the boundary condition of the structure changes. The remaining live load to go on the span with changed boundary condition is $(18\,kN/m − 14.24\,kN/m) = 3.76\,kN/m$.

Design bending moment at support $A = 280\,kNm$

Design bending moment at midspan AB at centre of span $= 79.9 + 14.24 \times 4.22 + 3.76 \times 7.23 = 167.2\,kNm$

A conservative design span moment $= 175\,kNm$

allowing for the maximum span moment to occur away from the centre of span.

Design bending moment at support $B = 270\,kNm$ ⠀⠀⠀from elastic analysis (LC_1 on both spans)

Design shear at support $A = 102.8 + 5.06 \times 14.24 + 3.86 \times 3.76$
⠀⠀⠀⠀⠀⠀⠀⠀⠀⠀⠀⠀$= 189.5\,kN$ ⠀⠀say 190 kN

Design shear at support B = 180 kN   (LC_1 on both spans)

Step 4 Determine axial loads
Not required.

Step 5 Determine torsion
Ultimate load from slab = 1.4 × 11.25 + 1.6 × 11.25
= 34 kN/m

Load assumed to act on edge of nib.

Eccentricity of load from shear centre of beam = 110 − 15(chamfer)

$+ \dfrac{290}{2} + 78.5$

(e = 78.5 = shear centre)

= 318.5 mm

Torsion per unit length
= 34 × 0.3185
= 10.83 kNm/m

Ultimate self-weight of beam
= 5.6 kN/m

Eccentricity of self-weight from shear centre
$= \bar{x} - \dfrac{290}{2} + e$
= 172.5 − 145 + 78.5
= 106 mm

Torsion per unit length
= 5.6 × 0.106
= 0.59 kNm/m

Total ultimate torsion in beam
= (10.83 + 0.59) × 4.5
= 51.4 kNm   at the supports restraining rotation

Step 6 Cover to reinforcement
Maximum size of aggregate = 20 mm
Maximum size of bar = 25 mm assumed
Maximum size of link = 10 mm
Exposure condition = mild
Fire resistance required = 1 hour
Grade of concrete = C40
Minimum cement content = 325 kg/m³
Maximum free water/cement ratio = 0.55
Nominal cover = 20 mm
Effective depth, d = 500 − 20 − 10 − 12.5 = 457.5 mm

Step 7 Effective span
Effective span = 9.0 m

Step 8 Effective width of flange
Not required.

Step 9 ***Slenderness ratio***

$l = 8.5\,\text{m} = \text{clear span}$

$b_c = 400\,\text{mm} \qquad 60b_c = 60 \times 400 = 24\,000\,\text{mm} > 8500\,\text{mm}$

$$d = \left(\frac{M}{0.156 f_{cu} b}\right)^{\frac{1}{2}} = \left(\frac{280 \times 10^6}{0.156 \times 40 \times 400}\right)^{\frac{1}{2}} = 335\,\text{mm}$$

$$\frac{250 b_c^2}{d} = \frac{250 \times 400^2}{335} = 119\,402\,\text{mm} > 8500\,\text{mm}$$

Slenderness check is satisfied.

Step 10 ***Design for flexure***

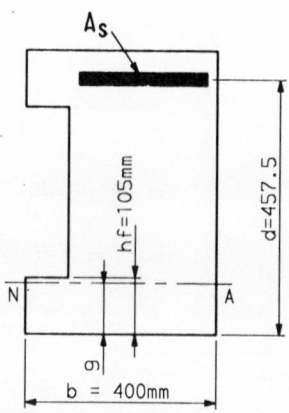

SK 2/62 Calculation of tensile steel at support.

Support bending moments at A or C = 280 kNm

$$K = \frac{M}{f_{cu} b d^2}$$

$$= \frac{280 \times 10^6}{40 \times 400 \times 457.5^2}$$

$$= 0.0836 < 0.156$$

No compressive reinforcement required.

$$z = d\left[0.5 + \sqrt{\left(0.25 - \frac{K}{0.9}\right)}\right]$$

$$= 457.5\left[0.5 + \sqrt{\left(0.25 - \frac{0.0836}{0.9}\right)}\right]$$

$$= 410\,\text{mm}$$

$$x = \frac{d - z}{0.45}$$

$$= \frac{457.5 - 410}{0.45}$$

$$= 105\,\text{mm} = h_{\text{f}}$$

∴ Neutral axis is in the flange.

$$A_{\text{s}} = \frac{M}{0.87 f_{\text{y}} z}$$

$$= \frac{280 \times 10^6}{0.87 \times 460 \times 410} = 1706\,\text{mm}^2$$

Use 4 no. 25 mm dia. bars (1964 mm²).

Midspan bending moment = 175 kNm

$$K = \frac{M}{f_{\text{cu}} b d^2}$$

$$= \frac{175 \times 10^6}{40 \times 400 \times 457.5^2}$$

$$= 0.052$$

$$z = 0.94d$$

$$= 430\,\text{mm}$$

$$x = \frac{d - z}{0.45}$$

$$= 61\,\text{mm} < 105\,\text{mm} = h_{\text{f}}$$

$$A_{\text{s}} = \frac{M}{0.87 f_{\text{y}} z}$$

$$= \frac{175 \times 10^6}{0.87 \times 460 \times 430}$$

$$= 1017\,\text{mm}^2$$

Use 2 no. 25 mm dia. bars (982 mm²) + 1 no. 12 mm dia. bar (113 mm²).

Step 11 **Flanged beam**
Not required.

Step 12 **Check maximum shear stress at support**

$$v = \frac{V}{bd}$$

$$= \frac{190 \times 10^3}{290 \times 457.5}$$

$$= 1.43\,\text{N/mm}^2$$

$$0.8\sqrt{f_{\text{cu}}} = 0.8 \times \sqrt{40} = 5\,\text{N/mm}^2$$

Step 13 **Check flexural shear stress**
$$d = 457.5\,\text{mm}$$

$$V_{\text{A}} = 190 - 40 \times 0.457$$

$$= 172\,\text{kN} \quad \text{at effective depth away from support}$$

$$v = \frac{V}{bd}$$

$$= \frac{172 \times 10^3}{290 \times 457.5}$$

$$= 1.30 \,\text{N/mm}^2$$

$$p = \frac{100A_\text{s}}{bd}$$

$$= \frac{100 \times 1964}{290 \times 457.5}$$

$$= 1.48$$

$$v_\text{c} = 0.72 \times 1.7 = 0.84 \,\text{N/mm}^2 \qquad \text{From Fig. 11.5}$$

$$V_\text{nom} = (v_\text{c} + 0.4)bd$$
$$= (0.84 + 0.4) \times 290 \times 457.5 \times 10^{-3}$$
$$= 164.5 \,\text{kN}$$

$$v > v_\text{c} + 0.4$$

$$A_\text{sv} = \frac{bS_\text{v}(v - v_\text{c})}{0.87f_\text{y}}$$

$$= \frac{290 \times 200 \times (1.30 - 0.84)}{0.87 \times 460}$$

$$= 66.7 \,\text{mm}^2 \qquad \text{at 200 mm c/c (2 legs)}$$

$$\frac{A_\text{sv}}{S_\text{v}} = \frac{66.7}{200 \times 2} = 0.17 \text{ for each leg}$$

$$\text{Nominal } \frac{A_\text{sv}}{S_\text{v}} = \frac{0.4b}{0.87f_\text{y}}$$

$$= \frac{0.4 \times 290}{0.87 \times 460}$$

$$= 0.29 \quad \text{(2 legs)}$$
$$= 0.145 \quad \text{(for each leg)}$$

Area of tension reinforcement required to carry weight of slab on the nib

$$= \frac{34 \,\text{kN/m}}{0.87 \times 460}$$

$$= 85 \,\text{mm}^2/\text{m}$$

$$\frac{A_\text{s}}{S} = \frac{85}{1000} = 0.085 \quad \text{for each leg}$$

Step 14 *Alternative design for shear*
Not required since design shear is calculated at d from support.

Step 15 *Minimum tension reinforcement*
Assume channel section as L-beam.

$$A_s > 0.0020b_w h = 0.0020 \times 290 \times 500$$
$$= 290 \,\text{mm}^2 < 1964 \,\text{mm}^2 \text{ provided}$$

Step 16   Minimum compression reinforcement
Not required.

Step 17   Transverse reinforcement in flange
$$A_s = 1.5 h_f \,\text{mm}/^2\text{m}$$
$$= 1.5 \times 105 = 158 \,\text{mm}^2/\text{m minimum}$$

(See Example 5.2.)
Reinforcement in nib $= 201 \,\text{mm}^2/\text{m}$ provided.

Step 18   Minimum reinforcement in side face of beams
Not required.

Step 19   Check deflection
$$\frac{b_w}{b} = \frac{290}{400} = 0.725 > 0.3$$
From Table 11.3,
Basic span/effective depth ratio

for rectangular section $= 26$ for $b_w/b = 1.0$
         for flanged beams $= 20.8$ for $b_w/b = 0.3$

Interpolated basic ratio $= 20.8 + \left(\dfrac{26 - 20.8}{0.7}\right) \times (0.725 - 0.3) = 24$

$$\beta_b = \frac{M}{M'} = \frac{167.2}{155.9}$$

Midspan service stress $= \left(\dfrac{5}{8\beta_b}\right) f_y \left(\dfrac{A_{s\ \text{reqd}}}{A_{s\ \text{prov}}}\right)$
$$= \left(\frac{5}{8}\right) \times \left(\frac{155.9}{167.2}\right) \times 460 \times \left(\frac{1017}{1095}\right)$$
$$= 249 \,\text{N/mm}^2$$

$$\frac{M}{bd^2} = \frac{175 \times 10^6}{400 \times 457.5^2} = 2.09$$

Modification factor for tension reinforcement from Table 11.5 $= 1.20$

Modified span/depth ratio $= 24 \times 1.20 = 28.8$

$$\frac{l_e}{d} = \frac{9000}{457.5}$$
$$= 19.67 < 28.80 \quad \text{OK}$$

Step 20   Maximum areas of reinforcement
$$A_s < 0.04 b_w h = 5800 \,\text{mm}^2$$

Satisfied.

Step 21   **Containment of compression reinforcement**
Not required.

Step 22   **Check bearing stress inside bend**
Not required.

Step 23   **Curtailment of bars**
$45 \times$ bar dia. $= 45 \times 25 = 1125$ mm

$0.15l = 0.15 \times 9000 = 1350$ mm
$0.25l = 0.25 \times 9000 = 2250$ mm

2 no. 25 mm dia. top and bottom throughout.
2 no. 25 m dia. extra top at A, B and C − 5000 long at B, 2500 mm into

span at A and C and properly anchored at A and C.
1 no. 12 mm dia. bottom in spans AB and BC.
Follow simplified detailing rules for beams as in Fig. 2.2.

Step 24   **Spacing of bars**
Minimum clear spacing $= MSA + 5 = 20 + 5 = 25$ mm

Actual minimum clear spacing used $= 43$ mm   (support)

Actual maximum clear spacing used $= 84$ mm   (midspan)

Maximum clear spacing allowed $= \dfrac{47\,000}{f_s} = \dfrac{47\,000}{249} = 189$ mm > 84 mm

where $f_s = 249\,\text{N/mm}^2$ (see Step 19.)

Step 25   **Check torsional shear stress**
Ultimate torsion $= 51.4$ kNm   (see Step 5)

Divide section into 3 rectangles of maximum total torsional stiffness.
First choice

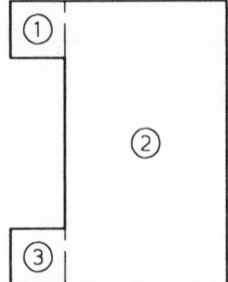

FIRST CHOICE

SK 2/63 Calculation of torsional
shear stress.

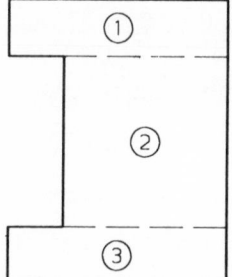

SECOND CHOICE

SK 2/64 Calculation of torsional
shear stress.

$$500 \times 290 - \text{stiffness} = h_{min}^3 h_{max} = 290^3 \times 500 = 1.22 \times 10^{10}$$
$$2 \times 110 \times 105 - \text{stiffness} = 2 \times 105^3 \times 110 = 0.025 \times 10^{10}$$
$$\text{TOTAL} = 1.245 \times 10^{10}$$

Second choice

$$290 \times 290 - \text{stiffness} = 290^3 \times 290 = 0.707 \times 10^{10}$$
$$2 \times 400 \times 105 - \text{stiffness} = 2 \times 105^3 \times 400 = 0.0926 \times 10^{10}$$
$$\text{TOTAL} = 0.7996 \times 10^{10}$$

Hence the first choice is critical.

$$\text{Proportion of torsional moment carried by the web} = \frac{T \times 1.22 \times 10^{10}}{1.245 \times 10^{10}}$$

$$= \frac{51.4 \times 1.22}{1.245}$$

$$= 50.4\,\text{kNm}$$

Torsion carried by flanges $= 0.5(51.4 - 50.4) = 0.5\,\text{kNm}$

$$\text{Torsional shear stress } v_t \text{ in web} = \frac{2T}{h_{min}^2\left(h_{max} - \dfrac{h_{min}}{3}\right)}$$

$$= \frac{2 \times 50.4 \times 10^6}{290^2\left(500 - \dfrac{290}{3}\right)}$$

$$= 2.97\,\text{N/mm}^2$$

$$\text{Torsional shear stress } v_t \text{ in flange} = \frac{2 \times 0.5 \times 10^6}{105^2\left(110 - \dfrac{105}{3}\right)}$$

$$= 1.21\,\text{N/mm}^2$$

$$v_{t,min} = 0.067\sqrt{f_{cu}}$$
$$= 0.067\sqrt{40}$$
$$= 0.4\,\text{N/mm}^2$$

$$v_{tu} = 0.8\sqrt{f_{cu}} = 5\,\text{N/mm}^2$$

$$\frac{v_{tu}y_1}{550} = \frac{5 \times 450}{550}$$

$$= 4.1\,\text{N/mm}^2$$

$v_t > v_{t,\,min}$, torsional reinforcement required.

Torsional shear stress + flexural shear stress $= 2.97 + 1.30$ (see Step 13)
$$= 4.27\,\text{N/mm}^2$$
$$< 5\,\text{N/mm}^2 \quad \text{OK}$$

Torsional reinforcement in web (vertical)

$$\frac{A_{sv}}{S_v} = \frac{T}{0.8x_1y_10.87f_y}$$

$$= \frac{50.4 \times 10^6}{0.8 \times 238 \times 448 \times 0.87 \times 460}$$

$$= 1.48 \quad \text{(for 2 legs)}$$

$$= 0.74 \quad \text{(for each leg)}$$

Longitudinal reinforcement for torsion

$$A_s = \left(\frac{A_{sv}}{S_v}\right)\left(\frac{f_{yv}}{f_y}\right)(x_1 + y_1)$$

$$= 1.48 \times 1 \times (238 + 448)$$

$$= 1012\,\text{mm}^2$$

Use 10 no. bars at $101\,\text{mm}^2$ each in the longitudinal direction evenly placed on the perimeter of web cross-section ($f_y = 460\,\text{N/mm}^2$).

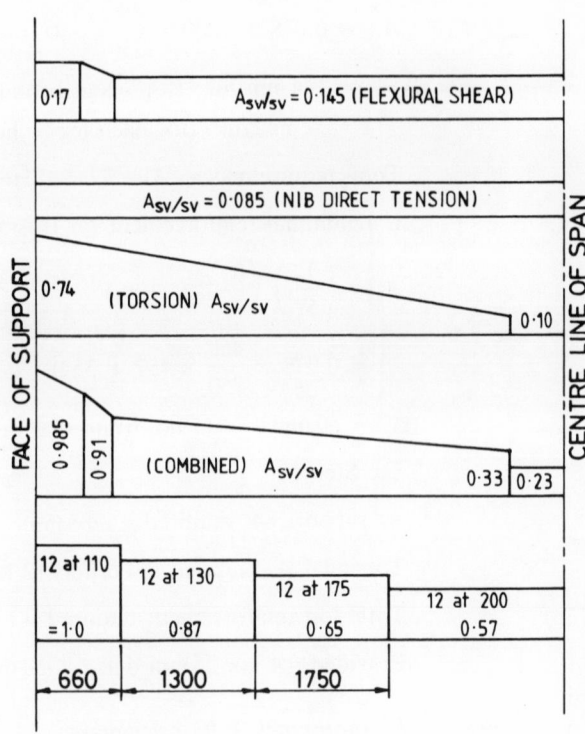

SK 2/65 A_{sv}/S_v diagram for Example 2.3.

Torsional reinforcement in flange

$$\frac{A_{sv}}{S_v} = \frac{T}{0.8x_1y_10.87f_y}$$

$$= \frac{0.5 \times 10^6}{0.8 \times (105 - 48) \times (400 - 48) \times 0.87 \times 460}$$

$$= 0.078$$

Maximum spacing $= x_1$, or $\frac{y_1}{2}$, or 200 mm

$$= 57\,\text{mm}$$

Use 8 mm dia. links at 50 mm centres ($1006\,\text{mm}^2/\text{m}$) ($f_y = 460\,\text{N/mm}^2$) Could also use 6 mm dia. mild steel links at 50 mm centres ($566\,\text{mm}^2/\text{m}$) ($f_y = 250\,\text{N/mm}^2$).

See Example 5.2, Step 4.

$$A_{s\,\text{reqd}} = 131\,\text{mm}^2/\text{m} \quad (460\ \text{grade steel})$$
$$(\text{for flexure}) = 241\,\text{mm}^2/\text{m} \quad (\text{mild steel Grade 250})$$

$$\frac{A_{sv}}{S_v} = 0.078 \quad \text{for torsion (Grade 460)}$$

$$A_{sv} = 0.078 \times 1000 \times \frac{460}{250} \quad (\text{Grade 250})$$

$$= 144\,\text{mm}^2/\text{m} \quad (\text{for 2 legs of mild steel})$$
$$= 72\,\text{mm}^2/\text{m} \quad (\text{for each leg} - \text{horizontal})$$

Total requirement $= 241 + 72 = 313\,\text{mm}^2/\text{m} < 566\,\text{mm}^2/\text{m}$

Longitudinal reinforcement for torsion in flange

$$A_s = \left(\frac{A_{sv}}{S_v}\right)\left(\frac{f_{yv}}{f_y}\right)(x_1 + y_1)$$

$$= 0.078 \times \frac{460}{250} \times (57 + 352)$$

$$= 59\,\text{mm}^2 \quad (\text{4 no. 6 mm dia. mild steel: } f_y = 250\,\text{N/mm}^2)$$

See Step 10.

At support, A_s required $= 1706\,\text{mm}^2$

Torsional A_s required at corners (2 bars) $= 202\,\text{mm}^2$ \quad (Step 25)

Total top reinforcement required $= 1706 + 202 = 1908\,\text{mm}^2$

Provided $= 4$ no. 25 mm dia. $= 1964\,\text{mm}^2$ \quad OK

Step 26 *Flexural crack width calculations*
By elastic analysis: no redistribution.

Maximum support moment at A or C $= 201\,\text{kNm}$ \quad (serviceability limit state)

$$d = 500 - 20 - 12 - 12.5 = 455.5$$
$$b = 400\,\text{mm}$$
$$A_s = 1964\,\text{mm}^2 \quad p = 0.0108$$

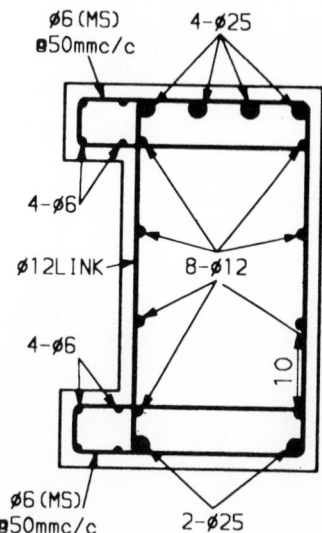

SK 2/66 Typical section at
support.

$$A'_s = 982\,\text{mm}^2 \qquad p' = 0.0054$$
$$m = 10$$
$$h_f = 105\,\text{mm}$$
$$h = 500\,\text{mm}$$
$$b_t = 290\,\text{mm}$$
$$d' = 42.5\,\text{mm}$$

$$x = d\left\{\left[(mp + (m-1)p')^2 + 2\left(mp + (m-1)\left(\frac{d'}{d}\right)p'\right)\right]^{\frac{1}{2}}\right.$$
$$\left. - (mp + (m-1)p')\right\}$$

$$= 156.3\,\text{mm} > h_f = 105\,\text{mm}$$

Using Reference 10, Table 117,

$$x = \frac{mdA_s + 0.5bh_f^2}{mA_s + bh_f}$$

$$= \frac{10 \times 455.5 \times 1964 + 0.5 \times 400 \times 105^2}{10 \times 1964 + 400 \times 105}$$

$$= 181\,\text{mm}$$

$$z = d - \frac{h_f(3x - 2h_f)}{3(2x - h_f)}$$

$$= 455.5 - \frac{105(3 \times 181 - 2 \times 105)}{3(2 \times 181 - 105)}$$

$$= 410\,\text{mm}$$

$$f_s = \frac{M}{A_s z}$$

$$= \frac{201 \times 10^6}{1964 \times 410}$$

$$= 250 \,\text{N/mm}^2$$

$$\varepsilon_s = \frac{f_s}{E_s} = \frac{250}{200 \times 10^3} = 1.25 \times 10^{-3}$$

$$\varepsilon_h = \left(\frac{h - x}{d - x}\right)\varepsilon_s$$

$$= \left(\frac{500 - 181}{455.5 - 181}\right) \times 1.25 \times 10^{-3}$$

$$= 1.45 \times 10^{-3}$$

$$\varepsilon_{mh} = \varepsilon_h - \frac{b_t(h - x)^2}{3E_s A_s(d - x)}$$

$$= 1.45 \times 10^{-3} - \frac{290 \times (500 - 181)^2}{3 \times 200 \times 10^3 \times 1964 \times (455.5 - 181)}$$

$$= 1.36 \times 10^{-3}$$

$$a_{cr} = \sqrt{(44.5^2 + 44.5^2)} - 12.5 = 50.4 \,\text{mm}$$

$$W_{cr} = \frac{3a_{cr}\,\varepsilon_{mh}}{1 + \dfrac{2(a_{cr} - c_{min})}{(h - x)}}$$

$$= \frac{3 \times 50.4 \times 1.36 \times 10^{-3}}{1 + \dfrac{2(50.4 - 32)}{(500 - 181)}}$$

$$= 0.18 \,\text{mm} < 0.3 \,\text{mm}$$

Step 27   Design of connections to other components
Follow Chapter 10.

2.5 FIGURES FOR CHAPTER 2

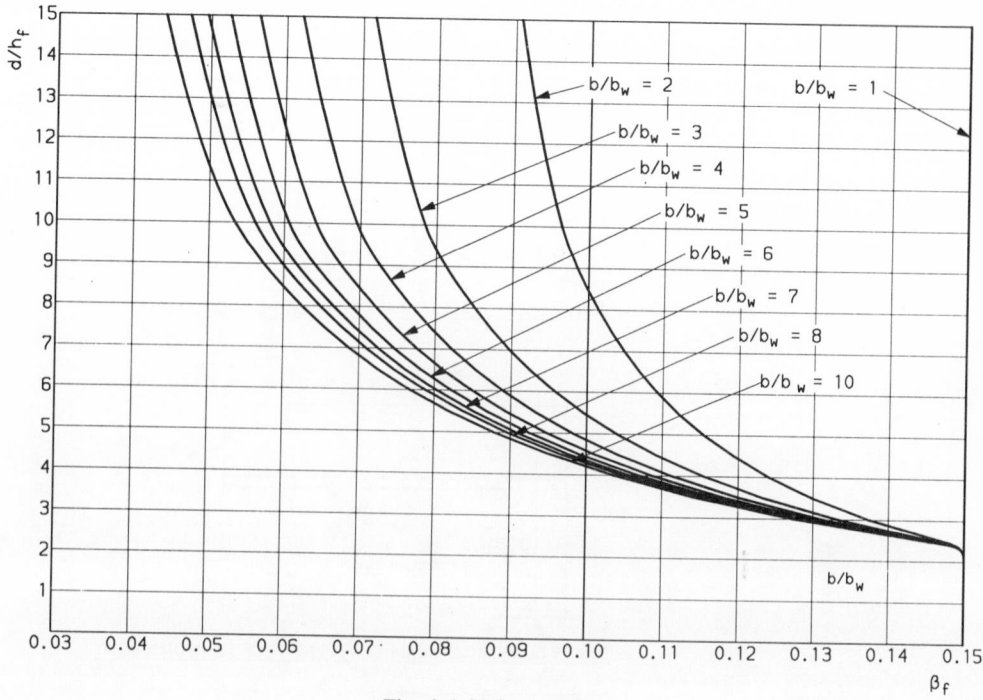

Fig. 2.1 Values of β_f.

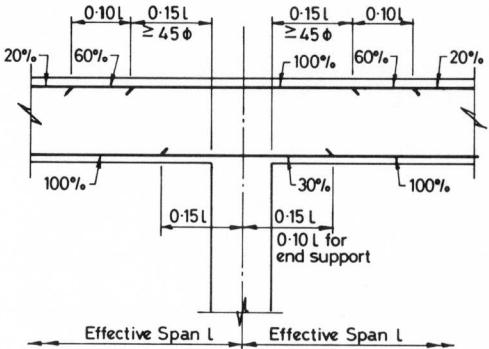

CONTINUOUS BEAM : CURTAILMENT OF REINFORCEMENT

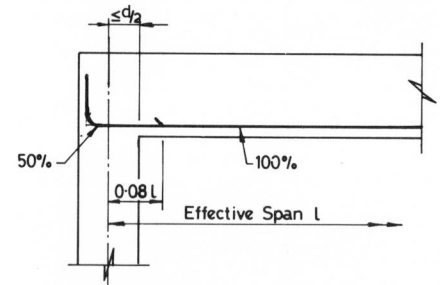

SIMPLY SUPPORTED BEAM : CURTAILMENT OF REINFORCEMENT

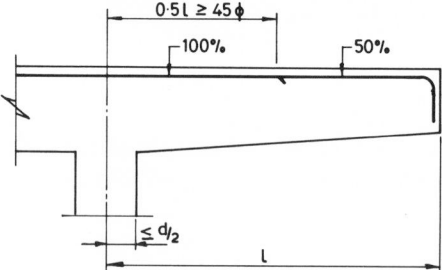

CANTILEVER BEAM : CURTAILMENT OF REINFORCEMENT

Fig. 2.2 Simplified detailing rules for beams.

Chapter 3

Design of Reinforced Concrete Slabs

3.0 NOTATION

a'	Compression face to point on surface of concrete where crack width is calculated
a_b	Centre-to-centre distance between bars or groups of bars
a_{cr}	Point on surface of concrete to nearest face of a bar
A_c	Gross area of concrete in a section
A_s	Area of steel in tension
A'_s	Area of steel in compression
A_{sb}	Minimum area of reinforcement at bottom of slab
A_{st}	Minimum area of reinforcement at top of slab
A_{sx}	Reinforcement in y-direction to resist M_x about x-axis
A_{sy}	Reinforcement in x-direction to resist M_y about Y-axis
A_{sbx}	Area of inclined shear reinforcement to resist V_x
A_{sby}	Area of inclined shear reinforcement to resist V_y
A_{sv}	Area of vertical shear reinforcement
A_{svx}	Area of vertical shear reinforcement to resist V_x
A_{svy}	Area of vertical shear reinforcement to resist V_y
b	Width of reinforced concrete section
b_t	Width of section at centroid of tensile steel
c_{min}	Minimum cover to tensile reinforcement
d	Effective depth of tensile reinforcement
d'	Effective depth of compressive reinforcement
d_1	Distance from tension face of concrete section to centre of tensile reinforcement
E_c	Modulus of elasticity of concrete
E_s	Modulus of elasticity of steel
f_y	Characteristic yield strength of steel
f'_y	Revised compressive stress in steel taking into account depth of neutral axis
f_{cu}	Characteristic cube strength of concrete at 28 days
f_{yv}	Characteristic yield strength of reinforcement used as links
F	Coefficient for calculation of cracked section moment of inertia
G	Shear modulus
h	Overall depth of slab
H	Shorter dimension of a rectangular panel of slab for use of yield-line charts
I	Moment of inertia using b as unit width for slab
l	Clear span or span face-to-face of support

l_e	Effective span
l_o	Centre-to-centre distance between supports
L	Longer dimension of a rectangular panel of slab for yield line calculations
m	Modular ratio $= E_s/E_c$
M_d	Design bending moment per unit width of slab modified to account for axial load
M_x	Moment per unit width about x-axis
M_y	Moment per unit width about y-axis
M_{xy}	Torsional moment per unit width
M_{xt}	Wood–Armer design moment for top reinforcement in y-direction
M_{xb}	Wood–Armer design moment for bottom reinforcement in y-direction
M_{yt}	Wood–Armer design moment for top reinforcement in x-direction
M_yb	Wood–Armer design moment for bottom reinforcement in x-direction
M_{HN}	Ultimate negative moment capacity of slab per unit width about an axis parallel to H
M_{HP}	Ultimate positive moment capacity of slab per unit width about an axis parallel to H
M_{VN}	Ultimate negative moment capacity of slab per unit width about an axis parallel to L
M_{VP}	Ultimate positive moment capacity of slab per unit width about an axis parallel to L
N_x	Axial load per unit width of slab in x-direction to be combined with M_y
N_y	Axial load per unit width of slab in y-direction to be combined with M_x
p	Percentage of tensile reinforcement
p'	Percentage of compressive reinforcement
p_x	Percentage of tensile steel to resist M_x about x-axis
p_y	Percentage of tensile steel to resist M_y about y-axis
r	Loading per unit area used in yield-line analysis (kN/m^2)
r_u	Ultimate loading per unit area
R	Restraint factor for computation of early thermal cracking
R_u	Ultimate total load on panel of slab
S_v	Spacing of vertical links
S_{bx}	Spacing of inclined shear reinforcement to resist V_x per unit width
S_{by}	Spacing of inclined shear reinforcement to resist V_y per unit width
S_{vx}	Spacing of vertical shear reinforcement to resist V_x per unit width
S_{vy}	Spacing of vertical shear reinforcement to resist V_y per unit width
T_1	Differential temperature in a concrete pour for calculation of early thermal cracking
U_n	Perimeter of concentrated load on slab at prescribed multiples of effective depth
U_o	Perimeter of concentrated load footprint on slab
v_c	Design concrete shear strength
v_n	Calculated punching shear stress at perimeter U_n
v_x	Calculated shear stress in concrete due to V_x
v_y	Calculated shear stress in concrete due to V_y
v_{cx}	Design concrete shear stress to compare with V_x for bending about x-axis
v_{cy}	Design concrete shear stress to compare with V_y for bending about y-axis

v_1	Calculated punching shear stress at perimeter U_1
V_x	Shear force per unit width for bending about x-axis
V_y	Shear force per unit width for bending about y-axis
W_{max}	Maximum crack width (mm)
x	Depth of neutral axis from compression face
x	Distance from edge in L-direction to start of a yield line
y	Distance from edge in H direction to start of a yield line
z	Depth of lever arm
α	Angle of inclination to horizontal of shear reinforcement
α	Coefficient of thermal expansion of concrete
β	Angle of inclination to horizontal of concrete strut in truss analogy
β_b	Ratio of redistributed moment over elastic analysis moment
γ_m	Material factor
ε_h	Calculated strain in concrete at depth h
ε_m	Strain with stiffening effect corrected
ε_r	Tensile strain in concrete due to temperature differential causing early thermal cracking
ε_s	Strain at centre of steel reinforcement
ε_s'	Strain at centre of compressive reinforcement
ε_{mh}	Strain at depth h corrected for stiffening effect
ε_l	Calculated strain in concrete ignoring stiffening effect
ρ_{crit}	Critical percentage of steel required to distribute early thermal cracking

3.1   ANALYSIS OF SLABS

3.1.1   Slabs: properties

3.1.1.1   Effective spans

Simply supported or encastré l_e = smaller of $(l+d)$ or l_o

Continuous $l_e = l_o$

Cantilever $l_e = l + \dfrac{d}{2}$

where l_o = centre-to-centre distance between supports
l_e = effective span
l = clear span or span to face of support
d = effective depth of tension reinforcement.

3.1.1.2   Moment of inertia

Method 1   Gross concrete section only
See Section 2.1.3 − use Table 11.2 with b equal to unity.

Method 2   Uncracked transformed concrete
See Section 2.1.3 − use Table 11.2 with b equal to unity and A_s and A_s'

are for unit width. Convert A_s and A_s' into equivalent concrete areas by multiplying by $m = E_s/E_c$. Moment of inertia increment due to steel $= mA_s(x')^2$ where x' is the distance of the steel from the centroidal axis of the section. The shift of the centroidal axis due to the presence of reinforcing steel may be neglected.

Method 3 *Average of gross concrete section and cracked section*

$$I = 0.5 \left(\frac{1}{12} bh^3 + Fbh^3 \right)$$

where I = moment of inertia of rectangular concrete section
  b = unit width of slab
  h = overall depth of slab
  F = factor − see Fig. 11.1 for values of F

$$p = \frac{100A_s}{bd}$$

where A_s = area of tensile reinforcement per unit width of slab

$$p' = \frac{100A_s'}{bd}$$

where A_s' = area of compressive reinforcement per unit width of slab

$$m = \text{modular ratio} = \frac{E_s}{E_c}$$

Note: For slabs, b is taken equal to unity.
  The preferred method is Method 3 for rectangular sections. Where reinforcement quantities are not known, an assumption may be made of the percentage of reinforcement.

3.1.1.3 *Modulus of elasticity*
See Section 2.1.4.

3.1.1.4 *Shear modulus*
Shear modulus $G = 0.42E_c$ for concrete.

3.1.1.5 *Poisson's ratio*
Poisson's ratio for concrete is 0.2

3.1.1.6 *Thermal strain*
See Section 2.1.9.

3.1.2 Analysis of slabs

The objective is to find the following internal forces by analysis:

(1) Moments M_x, M_y and M_{xy}

(2) Shears V_x and V_y
(3) Wood−Armer moments M_{xt}, M_{xb}, M_{yt} and M_{yb}
(4) In-plane loads N_x and N_y

Method 1
BS 8110: Part 1: 1985, clauses 3.5.2 and 3.5.3, Table 3.15.[1]

Method 2
Yield-line method: non-linear − use Figs 3.18 to 3.33.

Method 3
Finite difference: linear elastic − Moody's table.[9]

Method 4
Finite element analysis: linear elastic − use general purpose computer program or Figs 3.1 to 3.17.

Commentary
Method 1 is a non-conservative approach. If cracking has to be avoided, an elastic method of analysis, i.e. finite element or finite difference, will be more appropriate. For complicated loadings and complex layout of slab panels and supporting arrangements, it is always recommended to use finite element analysis. Finite element analysis will give Wood-Armer design moments for top and bottom reinforcement in a panel of slab. Method 2 (yield-lines) may be successfully used for uniformly loaded slab panels with different boundary conditions. Method 2 gives a better representation of internal forces in a slab panel than Method 1.

Recommendations
Use Method 2 or Method 3 generally. Use Method 4 (finite element analysis) only where complicated loadings and geometry render the other methods unusable. Use elastic analysis charts if boundary conditions and loadings are appropriate.

3.1.3 Distribution of loads on beams

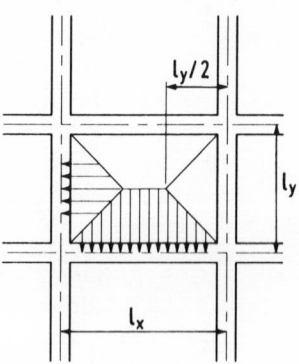

SK 3/1 Distribution of load on
beams (Method 2).

Method 1
BS 8110: Part 1: 1985, clause 3.5.3.7.[1]

Method 2
Triangular and trapezoidal distribution of uniform load.

Method 3
Finite difference – Moody's Table.[9] Use the coefficients R_x and R_y to calculate distribution of loads on the edge beams.

Method 4
Finite element analysis. Use the support reactions as loading on the beam.

Recommendations
Method 2 may be used for all applications. Method 3 and Method 4 may be used when similar methods are used for the analysis of the slab panels.

3.1.4   Concentrated load on slab

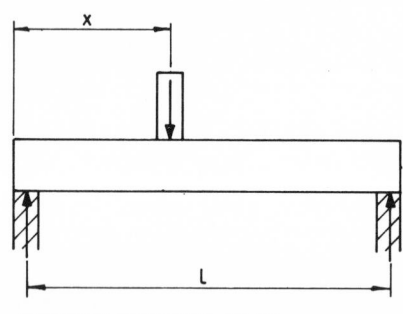

SECTION THROUGH SLAB

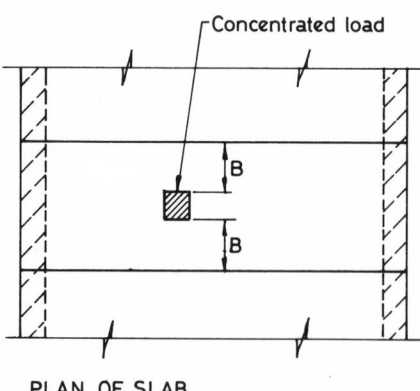

PLAN OF SLAB

SK 3/2 Effective width of slab to be considered for spread of a concentrated load on a simply supported one-way slab.

Simply supported slabs spanning in one direction only the width B on each side of load over which the load may be assumed to spread is given by:

$$B = 1.2x\left(1 - \frac{x}{l}\right)$$

where x = distance of load from support closest to load
          l = effective span.

For slabs spanning in both directions published tables and charts should be used to find bending moment and shear per unit width of slab. A finite element model may be created to analyse a complicated loading arrangement.

3.2 LOAD COMBINATIONS

3.2.1 General rules   See Section 2.2.1.

3.2.2 Rules of load combination for continuous one-way spanning slab panels   See Section 2.2.2.

3.2.3 Redistribution of moments   See Section 2.2.3.

3.2.3.1 Two-way spanning slab panels

No redistribution is allowed when Method 1 or Method 2 of analysis in Section 3.1.2 is followed. Redistribution of 10% may be allowed when Method 3 or Method 4 is adopted. Note that reduction of support moments means a corresponding increase in span moment.

3.2.4 Exceptional loads   See Section 2.2.4.

3.3 STEP-BY-STEP DESIGN PROCEDURE FOR SLABS

Step 1 Analysis
Carry out analysis (follow Section 3.1.2).

Note: One-way spanning slabs should be treated as beams of unit width and Chapter 2 should be followed except for minimum shear reinforcement.

Step 2 Design forces
Draw panel of slab and indicate maximum design moments, shears and in-plane loads, if any, per unit width of slab.

Step 3 Cover to reinforcement
Determine cover required to reinforcement as per Tables 11.6 and 11.7. Find effective depth d, assuming reinforcement diameter. Use actual effective depth in each direction.

Step 4   Design of slab for flexure

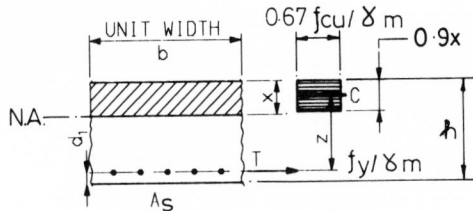

SK 3/3 Section through slab showing stress due to moment.

Find the following parameters for design moments in Step 2 per unit width of slab.

$$M_d = M + N\left(\frac{h}{2} - d_1\right) \qquad \text{for } N \le 0.1 f_{cu} bd$$

Note: For $N > 0.1 f_{cu} bd$, design as wall (see Chapter 8). M_d may also be taken equal to M where $N \le 0.1 f_{cu} bd$ and N may be ignored. (Sign convention: N is +ve for compression.)

$$K = \frac{M_d}{f_{cu} bd^2}$$

$$z = d\left[0.5 + \sqrt{\left(0.25 - \frac{K}{0.9}\right)}\right] \le 0.95d$$

$$x = \frac{d - z}{0.45}$$

$$A_s = \frac{M_d}{0.87 f_{yz}} - \frac{N}{0.87 f_y}$$

$K' = 0.156$ when redistribution does not exceed 10%
$K' = 0.402(\beta_b - 0.4) - 0.18(\beta_b - 0.4)^2$ when redistribution exceeds 10%

$$\beta_b = \frac{M}{M'} < 0.9$$

where M = moment after redistribution

         M' = moment before redistribution.

Note: If K is greater than K', increase depth of slab and start from Step 1 unless links are provided in the zone where steel in compression is used. The links are required to provide lateral restraint to bars in compression. Links in slab should normally be avoided.

When $K > K'$,

$$z = d\left[0.5 + \sqrt{\left(0.25 - \frac{K'}{0.9}\right)}\right]$$

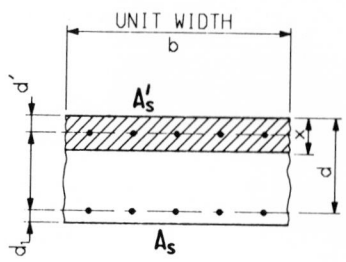

SK 3/4 Section through doubly
reinforced slab.

$$x = \frac{d - z}{0.45} \leq 0.5d$$

$$A'_s = \frac{(K - K')f_{cu}bd^2}{f'_y(d - d')}$$

$$A_s = \frac{K'f_{cu}bd^2}{0.87f_{yz}} + A'_s - \frac{N}{0.87f_y}$$

If $d'/x > 0.43x$,

$$A'_s = \frac{(K - K')f_{cu}bd^2}{f'_s(d - d')}$$

$$A_s = \frac{K'f_{cu}bd^2}{0.87f_{yz}} + A'_s - \frac{N}{0.87f_y}$$

$$f'_s = \left(\frac{x - d'}{0.57x}\right)\varepsilon_y E_s \qquad \text{because steel strain } \varepsilon'_s = \left(\frac{x - d'}{0.57x}\right)\varepsilon_y$$

$$\varepsilon_y = \left(\frac{f_y}{\gamma_m}\right)E_s$$

where ε_y corresponds to steel stress f_y/γ_m, as in Section 1.4.2.

Note: Follow detailing rules in clause 3.5.3.5 of BS 8110: Part 1: 1985[1] if
analysis has been carried out using Table 3.15 of BS 8110. Design charts in
BS 8110: Part 3: 1985 may be used.

Step 5 *Detailing*

Convert areas of steel per unit width found in Step 4 to diameter and
spacing of bars.

Step 6 *Check shear*

SK 3/5 Plan of a panel of slab
showing direction of
reinforcement.

Find the following parameters at critical sections for shear.

$$v_x = \frac{V_x}{bd} \leq 0.8 \sqrt{f_{cu}} \leq 5\,\text{N/mm}^2$$

$$v_y = \frac{V_y}{bd} \leq 0.8 \sqrt{f_{cu}} \leq 5\,\text{N/mm}^2$$

$$p_x = \frac{100A_{sx}}{bd}$$

$$p_y = \frac{100A_{sy}}{bd}$$

Find v_{cx} and v_{cy} from Figs 11.2 to 11.5, depending on strength of concrete.

If $v_x < v_{cx}$ and $v_y < v_{cy}$, no shear reinforcement is required.

If $v_{cx} < v_x \leq (v_{cx} + 0.4)$ or $v_{cy} < v_y \leq (v_{cy} + 0.4)$, nominal links are required in the zone where v_x or v_y is greater than v_{cx} or v_{cy} respectively.

Find nominal links:

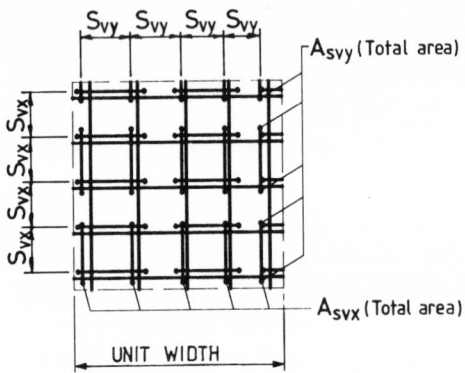

SK 3/6 Plan of unit area of slab showing shear reinforcement by links.

$$A_{svx} \geq \frac{0.4bS_{vx}}{0.87f_y} \quad \text{or} \quad A_{svy} \geq \frac{0.4bS_{vy}}{0.87f_y}$$

Note: Single vertical bars may be used instead of closed links provided proper anchorage bond length is available.

If $\quad v_{cx} < v_x < (v_{cx} + 0.4)$

and $\quad v_{cy} < v_y < (v_{cy} + 0.4)$

nominal links in both directions are required.

Assume $S_{vx} = S_{vy} = S_v$,

$$A_{sv} = \frac{0.8b\,S_v}{0.87f_y}$$

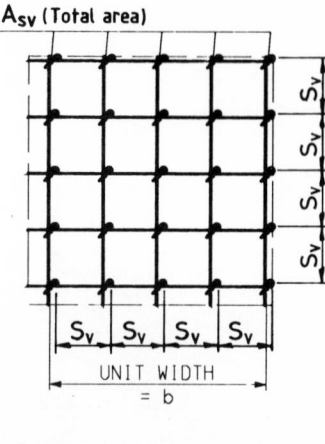

A_{sv} (Total area)

UNIT WIDTH = b

SK 3/7 Plan of unit area and
section showing shear
reinforcement by single vertical
bars.

Provide single vertical bars with proper anchorage over the whole zone at
a grid spacing of S_v.

If $(v_{cx} + 0.4) < v_x \le 0.8 \sqrt{f_{cu}} \le 5\,\text{N/mm}^2$
or $(v_{cy} + 0.4) < v_y \le 0.8 \sqrt{f_{cu}} \le 5\,\text{N/mm}^2$

use links or bent-up bars.

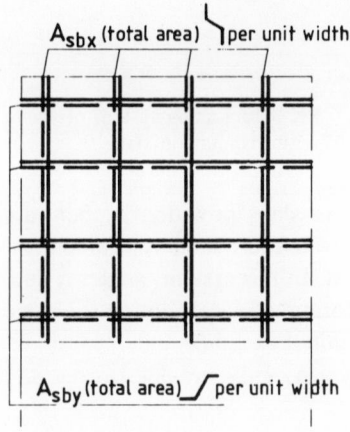

A_{sbx} (total area) ⌐per unit width

A_{sby} (total area) ⌐per unit width

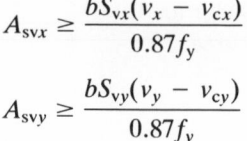

SK 3/8 Plan of slab showing
bent-up bars as shear
reinforcement.

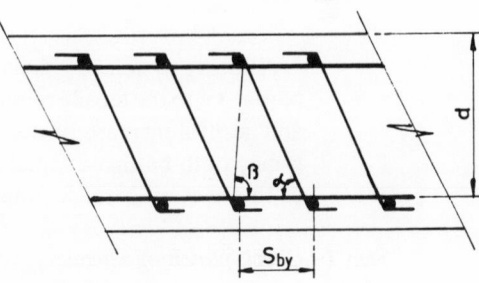

SK 3/9 Section through slab showing bent-up
shear reinforcement.

$$A_{svx} \ge \frac{bS_{vx}(v_x - v_{cx})}{0.87f_y}$$ when using links for V_x, or

$$A_{svy} \ge \frac{bS_{vy}(v_y - v_{cy})}{0.87f_y}$$ when using links for V_y, or

$$A_{sbx} \geq \frac{bdS_{bx}(v_x - v_{cx})}{0.87f_y(\cos\alpha + \sin\alpha\cot\beta)(d - d')}$$

when using bent-up bars for V_x, or

$$A_{sby} \geq \frac{bdS_{by}(v_y - v_{cy})}{0.87f_y(\cos\alpha + \sin\alpha\cot\beta)(d - d')}$$

when using bent-up bars for V_y

If $(v_{cx} + 0.4) < v_x \leq 0.8\sqrt{f_{cu}} \leq 5\,\text{N/mm}^2$
and $(v_{cy} + 0.4) < v_y \leq 0.8\sqrt{f_{cu}} \leq 5\,\text{N/mm}^2$

use bent-up bars in two orthogonal directions.

$$A_{sbx} \geq \frac{bdS_{bx}(v_x - v_{cx})}{0.87f_y\,(\cos\alpha + \sin\alpha\cot\beta)\,(d - d')}$$

and $$A_{sby} \geq \frac{bdS_{by}(v_y - v_{cy})}{0.87f_y\,(\cos\alpha + \sin\alpha\cot\beta)\,(d - d')}$$

Note: A_{sbx} and A_{sby} are the areas of bent-up bar required per unit width of slab equal to b.

Recommendation

Avoid using links or bent-up bars in slabs to resist shear. No shear reinforcement should be used in slabs up to 200 mm thick.

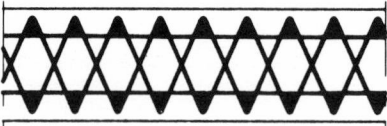

SK 3/10 Lacing system of shear reinforcement in slab.

A lacing system of shear reinforcement in slabs provided by bent-up bars at 45° to the tensile reinforcement works well where shear reinforcement and general increase of ductility are required. In this system, angles α and β may both be taken equal to 45°. In the formula for calculating the area of the bent-up bars, S_{bx} and S_{by} may be limited to $1.5d$.

Step 7   **Check punching shear**

Check punching shear stress.

$$v_{max} = \frac{V}{U_{od}} \leq 0.8\,\sqrt{f_{cu}} \leq 5\,\text{N/mm}^2$$

where $U_o = 2(a + b)$ for rectangular load, or
        = perimeter of loaded area.

$$v_1 = \frac{V}{U_1d}$$

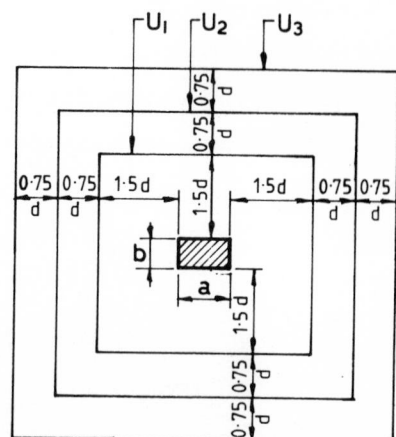

SK 3/11 Plan of slab around a concentrated load showing successive perimeters for punching shear check.

where $U_1 = 2(a + b + 6d)$ for rectangular loaded area, or
$\quad\quad = $ perimeter at 1.5d from face of loaded area.

$\quad v_c = $ design concrete shear stress $\quad$ from Figs 11.2 to 11.5.

$\quad V = $ concentrated load on slab

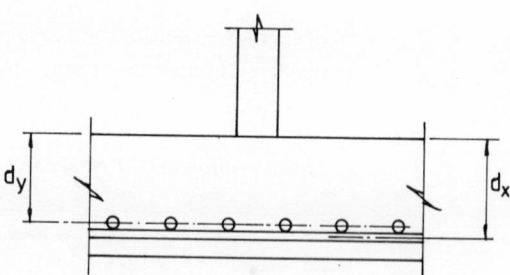

SK 3/12 Section through slab showing effective depths.

Calculate $p = 100\ A_s/bd$ under concentrated load to find v_c.

Note: Take p as the average of p_x and p_y where $p_x = 100A_{sx}/bd_x$ and $p_y = 100A_{sy}/bd_y$.

Shear reinforcement in first failure zone
If $v_1 \leq v_c$, no shear reinforcement is required and no further checks are necessary.

If $v_1 \leq 1.6\ v_c$,

$$A_{sv} \sin\alpha \geq \frac{(v_1 - v_c)\ U_1 d}{0.87 f_y} \geq \frac{0.4 U_1 d}{0.87 f_y}$$

If $1.6 v_c < v_1 \leq 2\ v_c$,

$$A_{sv} \sin\alpha \geq \frac{5(0.7 v_1 - v_c)\ U_1 d}{0.87 f_y} \geq \frac{0.4 U_1 d}{0.87 f_y}$$

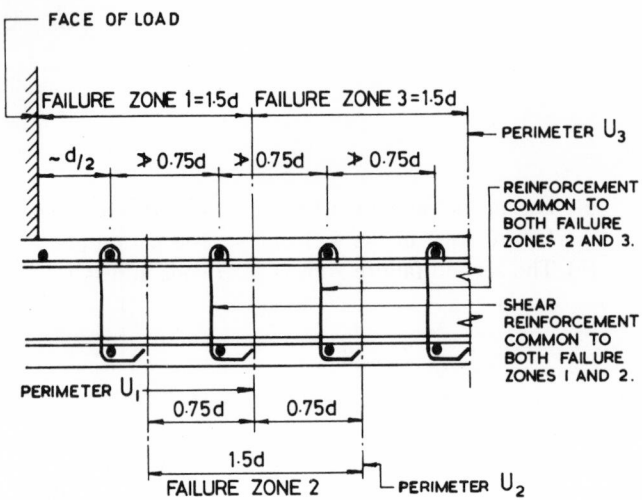

SK 3/13 Typical shear reinforcement for concentrated load on slab.

where A_{sv} is summation of areas of all shear reinforcement in a failure zone and α is the angle between the shear reinforcement and the plane of the slab. If v is greater than $2v_c$ then redesign slab with increased thickness or increased tensile steel or a combination of these parameters. It has been observed in tests that shear reinforcement in slabs does not work effectively if $v > 2v_c$.

Shear reinforcement in second failure zone

$$v_2 = \frac{V}{U_2 d}$$

where $U_2 = 2 (a + b + 9d)$ for rectangular loaded area, or
= perimeter at $2.25d$ from face of loaded area.

If $v_2 \leq v_c$, no shear reinforcement is required and no further checks are necessary.

If $v_2 \leq 1.6 \, v_c$,

$$A_{sv} \sin \alpha \geq \frac{(v_2 - v_c)U_2 d}{0.87 f_y} \geq \frac{0.4 U_2 d}{0.87 f_y}$$

If $1.6 \, v_c < v_2 \leq 2 \, v_c$,

$$A_{sv} \sin \alpha \geq \frac{5(0.7 v_2 - v_c)U_2 d}{0.87 f_y} \geq \frac{0.4 U_2 d}{0.87 f_y}$$

Similarly check successive failure zones $0.75d$ apart till $v \leq v_c$ is satisfied. Reinforcement to resist shear will be provided on at least two perimeters within a failure zone. Spacing of shear reinforcement on the perimeter should not exceed $1.5d$.

Steps to be followed for the determination of punching shear reinforcement in slabs

(1) The first failure zone is from the face of the loaded area to the perimeter 1.5*d* away.

(2) The first perimeter of shear reinforcement should be placed at *d*/2 from the face of the loaded area.

(3) The second perimeter of shear reinforcement should be placed at 0.75*d* from the first perimeter of shear reinforcement.

(4) A_{sv} is the sum of areas of all the legs of shear reinforcement in a failure zone in the first and second perimeter.

(5) The second failure zone is 1.5*d* wide and starts at 0.75*d* from the face of the loaded area.

(6) The successive failure zones are 1.5*d* wide and are 0.75*d* apart.

(7) The first perimeter reinforcement in the second failure zone is the same as the second perimeter reinforcement in the first failure zone.

Step 8 *Modification due to holes*

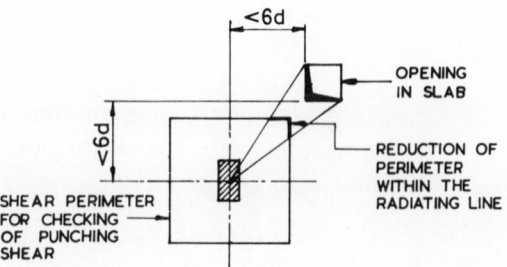

SK 3/14 Modification of shear perimeter due to presence of holes.

Carry out modification of *U* in Step 7 to allow for holes and proximity to edge.

The perimeter under consideration, *U*, in Step 7 will be reduced.

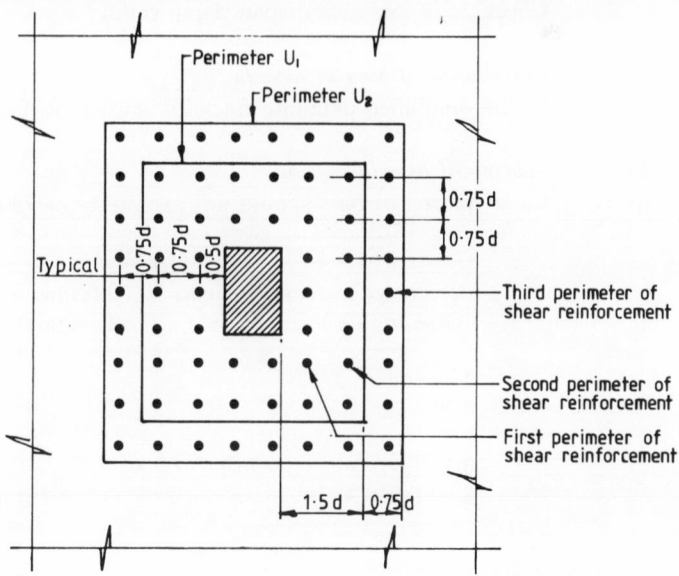

SK 3/15 Plan of slab near a concentrated load showing distribution of shear reinforcement.

Step 9 **Minimum tension reinforcement**

$A_s \geq 0.0013bh$ in both directions

At end support of slabs where simple support has been assumed, provide in the top of slab half the area of bottom steel at midspan or $0.0013bh$, whichever is greater.

Step 10 **Torsional reinforcement**

Special torsional reinforcement will be required at the corners of slab panels when the method of analysis follows clause 3.5.3.4 of BS 8110: Part 1: 1985. Follow clause 3.5.3.5 to determine the amount of torsional reinforcement.

Step 11 **Check span/effective depth**

Find l_e/d, where l_e is the effective span in the shorter direction. Find basic span/effective depth ratio from Table 11.3.

Find service stress, $f_s = \left(\dfrac{5}{8\beta_b}\right)\left(\dfrac{A_{s\ reqd}}{A_{s\ prov}}\right)f_y$

where $\beta_b = M/M'$
       M = moment after redistribution
       M' = moment before redistribution.

Find M/bd^2.

Find modification factor for tension reinforcement from Chart 11.5 and modification factor for compression reinforcement from Chart 11.4.

Find modified span/depth ratio by multiplying the basic span/depth ratio with the modification factor for tensile reinforcement and compression reinforcement, if used.

Check $l_e/d <$ modified span/depth ratio.

Step 12 **Curtailment of bars in tension**

Follow simplified detailing rules for slabs as in Fig. 3.34.

Step 13 **Spacing of bars in tension**

Clear spacing of bars should not exceed $3d$ or 750 mm.

Percentage of reinforcement, $100\ A_s/bd$ (%)	Maximum clear spacing of bars in slabs (mm)
1 or over	160
0.75	210
0.5	320
0.3	530
less than 0.3	$3d$ or 750, whichever is less

A_s is the area required at the ultimate limit state. The clear spacings as given above may be multiplied by β_b to account for redistribution of moments. β_b is the ratio of moment after redistribution to moment before redistribution. These clear spacings deem to satisfy 0.3 mm crack width at serviceability limit state.

Step 14   Check early thermal cracking

Early thermal cracking should be checked for the following pour configurations:

(1) Thin wall cast on massive base: $R = 0.6$ to 0.8 at base, $R = 0.1$ to 0.2 at top.
(2) Massive pour cast on blinding: $R = 0.1$ to 0.2.
(3) Massive pour cast on existing mass concrete: $R = 0.3$ to 0.4 at base, $R = 0.1$ to 0.2 at top.
(4) Suspended slabs: $R = 0.2$ to 0.4.
(5) Infill panels i.e. rigid restraint: $R = 0.8$ to 1.0.

where $R =$ restraint factor

Typical values of T_1 for Ordinary Portland Cement (OPC) concrete are:

Section thickness (mm)	Steel formwork	Plywood formwork	Cast on ground
300	13°C	25°C	17°C
500	22°C	35°C	28°C
700	32°C	42°C	28°C
1000	42°C	47°C	28°C

These figures are based on average cement content of $350 \, \text{kg/m}^3$.
Calculate:

$$\varepsilon_r = 0.8 \, T_1 \, \alpha \, R$$

obtain α (coefficient of thermal expansion) from Table 2.3 in Section 2.1.9.

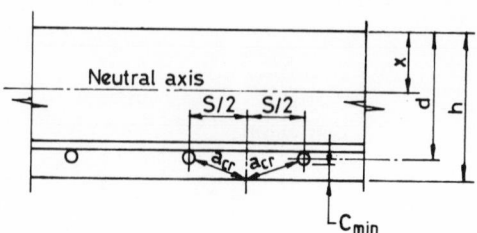

SK 3/16 Section of slab for crack width calculation.

$$W_{max} = \frac{3a_{cr} \, \varepsilon_r}{1 + \dfrac{2(a_{cr} - C_{min})}{h - x}}$$

Assume $x = h/2$

Note: If W_{max} is greater than design crack width, which is normally taken equal to 0.3 mm, then suggest means for reducing T_1.

Step 15   ***Check minimum reinforcement to distribute early thermal cracking***

$\rho_{crit} = 0.0035$ for Grade 460 steel reinforcement

$A_s = A_c \rho_{crit}$

For suspended slabs and walls,

$A_c = \dfrac{bh}{2}$ or $250b$ whichever is smaller

$A_s = 0.0035\, A_c$ near each face in each direction of slab and wall

For ground slabs and foundation bases,
up to 300 mm thickness:

$A_{st} = 0.00175bh$ near top surface in each direction

from 300 mm to 500 mm thickness:

$A_{st} = 0.00175bh$ near top surface in each direction

$A_{sb} = 0.35b$ near bottom surface in each direction

over 500 mm thickness:

$A_{st} = 0.875b$ near top surface in each direction

$A_{sb} = 0.35b$ near bottom surface in each direction

Step 16   ***Check flexural crack width***

Serviceability limit state

$LC_7 = 1.0DL + 1.0LL + 1.0EP + 1.0WP + 1.0WL$

Note: Omit loadings from LC_7 which produce beneficial rather than adverse effect.

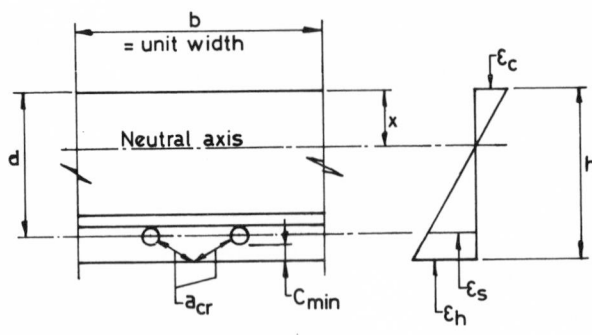

SK 3/17 Section through slab for the calculation of flexural crack width.

$$W_{max} = \frac{3a_{cr}\,\varepsilon_m}{1 + \dfrac{2(a_{cr} - C_{min})}{h - x}}$$

$$\varepsilon_{mh} = \varepsilon_h - \frac{b(h - x)^2}{3E_sA_s(d - x)}$$

Note: ε_h is the strain due to load combination LC_7 at depth h from compression face, b is the unit width of slab, and A_s is the area of tensile steel per unit width of slab.

For slab, b is taken equal to unit width.

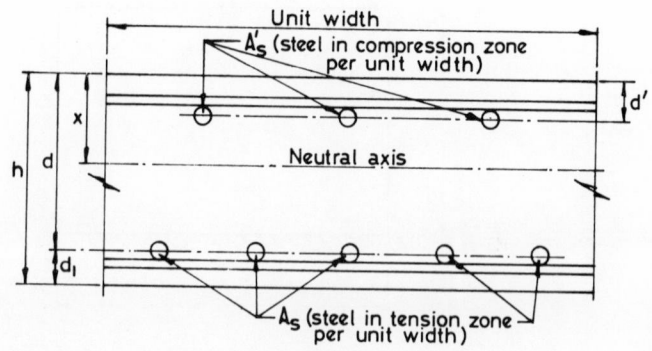

SK 3/18 Section of slab with steel in compression zone.

$$m = \frac{E_s}{E_c} \qquad p = \frac{A_s}{bd} \qquad p' = \frac{A'_s}{bd}$$

$$x = d\left\{\left[(mp + (m - 1)p')^2 + 2\left(mp + (m - 1)\left(\frac{d'}{d}\right)p'\right)\right]^{\frac{1}{2}} - (mp + (m - 1)p')\right\}$$

$$k_2 = \left(\frac{x}{2d}\right)\left(1 - \frac{x}{3d}\right)$$

$$k_3 = (m - 1)\left(1 - \frac{d'}{x}\right)$$

$$f_c = \frac{M}{k_2bd^2 + k_3A'_s(d - d')}$$

$$f_s = mf_c\left(\frac{d}{x} - 1\right)$$

$$\varepsilon_s = \frac{f_s}{E_s}$$

$$\varepsilon_h = \left(\frac{h - x}{d - x}\right)\varepsilon_s$$

$$\varepsilon_{mh} = \varepsilon_h - \frac{b(h - x)^2}{3E_s A_s(d - x)}$$

Note: In normal internal or external condition of exposure where the limitation of crack widths to 0.3 mm is appropriate, Step 13 will deem to satisfy the crack width criteria.

3.4 WORKED EXAMPLE

Example 3.1 Design of a two-way slab panel

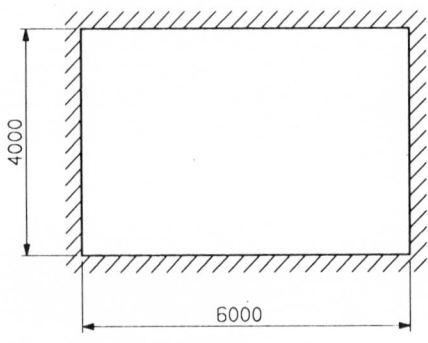

SK 3/19 Plan of a panel of slab continuous on all sides.

Clear panel size is 6 m × 4 m
Thickness of slab = 150 mm
Imposed loading = 20 kN/m²
Finishes = 2 kN/m²
Panel of slab continuous on all four sides
Width of beam = 300 mm

Step 1 **Analysis of slab panel**
Effective span, $l_e = l_o$

$l_x = 4.3$ m
$l_y = 6.3$ m

$l_x/l_y = 0.68$

Elastic analysis

Read coefficients from Fig. 3.12:

$m_{x1} = 0.035$
$m_{y1} = 0.021$
$m'_{x2} = 0.075$
$m'_{y2} = 0.060$

Characteristic dead load = 0.15 m × 25 kN/m³ × 1.4 + 2 × 1.4
$$= 8.0 \, \text{kN/m}^2$$

Characteristic imposed load $= 1.6 \times 20 = 32\,\text{kN/m}^2$

n = ultimate load on slab $= 8 + 32 = 40\,\text{kN/m}^2$

$$M_{x1} = m_{x1}nl_x^2$$
$$= 0.035 \times 40 \times 4.3^2$$
$$= 25.9\,\text{kNm/m}$$

$$M_{y1} = 0.021 \times 40 \times 4.3^2$$
$$= 15.5\,\text{kNm/m}$$

$$M'_{x3} = 0.075 \times 40 \times 4.3^2$$
$$= -55.5\,\text{kNm/m}$$

$$M'_{y2} = 0.060 \times 40 \times 4.3^2$$
$$= -44.4\,\text{kNm/m}$$

Allowing for 10% redistribution of moments,
Design moments:

$M_{x1} = 31.4\,\text{kNm/m}$
$M_{y1} = 19.9\,\text{kNm/m}$
$M'_{x3} = -50.0\,\text{kNm/m}$
$M'_{y2} = -40.0\,\text{kNm/m}$

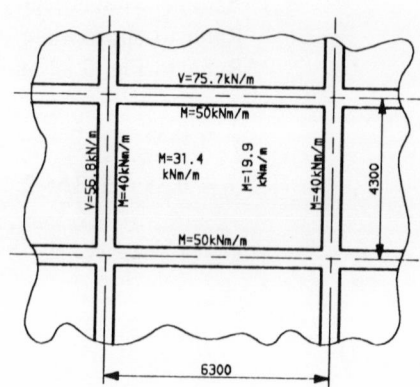

SK 3/20 Plan of panel of slab
showing bending moments and
shears.

Note: These moments do not take into account the Wood−Armer effect due to
the presence of M_{xy} and may be unconservative locally. In ultimate load
design local plastic hinge formation may be tolerated when there is a
possibility of redistribution of loads.

Analysis following BS 8110: Part 1: 1985[1]
Coefficients from Table 3.15.

Interior panel $l_y/l_x = 1.46$

$m_{sx1} = 0.039$	$M_{x1} = 28.8\,\text{kNm/m}$
$m_{sy1} = 0.024$	$M_{y1} = 17.8\,\text{kNm/m}$
$m'_{sx3} = 0.052$	$M'_{x3} = 38.5\,\text{kNm/m}$
$m'_{sx2} = 0.032$	$M'_{y2} = 23.7\,\text{kNm/m}$

Note: These moments are considerably less than the redistributed design moments found from elastic analysis. Elastic analysis gives peak values, whereas the BS 8110 coefficients tend to smear them across a long stretch of slab.

It is desirable and practical to use the elastic analysis results and allow 10% redistribution with a view to minimising the appearance of unsightly cracks in the slab. This is a conservative approach.

Check by yield-lines analysis

Assume that the elastic analysis moments are ultimate capacity moments in the panel of slab.

$$M_{VN} = 50\,\text{kNm/m} \quad \text{(Vertical Negative)}$$
$$M_{VP} = 31.4\,\text{kNm/m} \quad \text{(Vertical Positive)}$$
$$M_{HN} = 40\,\text{kNm/m} \quad \text{(Horizontal Negative)}$$
$$M_{HP} = 19.9\,\text{kNm/m} \quad \text{(Horizontal Positive)}$$

Assume that the elastic analysis results will be the maximum plastic moments in the panel of slab.

$$\frac{L}{H}\left(\frac{M_{VN} + M_{VP}}{M_{HN} + M_{HP}}\right)^{\frac{1}{2}} = \frac{6.3}{4.3}\left(\frac{50 + 31.4}{40 + 19.9}\right)^{\frac{1}{2}}$$
$$= 1.70$$

Assume symmetrical yield-lines − see Table 3.2.
Refer to appropriate diagram from Figs 3.18 to 3.33.
Refer to Fig. 3.22 and find x/L

$$\frac{x}{L} = 0.35$$
$$x = 0.35 \times 6.3 = 2.20\,\text{m}$$

$$\text{Unit resistance, } r = \frac{5(M_{HN} + M_{HP})}{x^2} \quad \text{from Table 3.2}$$

$$= \frac{5 \times 59.9}{2.2^2} = 61.9\,\text{kN/m}^2 > 40\,\text{kN/m}^2$$

Alternatively,

$$r = \frac{8(M_{VN} + M_{VP})(3L - x)}{H^2(3L - 4x)} \quad \text{from Table 3.2}$$

$$= \frac{8(50 + 31.4)(3 \times 6.3 - 2.2)}{4.3^2(3 \times 6.3 - 4 \times 2.2)}$$

$$= 58.23\,\text{kN/m}^2 > 40\,\text{kN/m}^2$$

Note: The values of M_{VN}, M_{VP}, M_{HN} and M_{HP} could be readjusted to arrive at r as close to $40\,\text{kN/m}^2$ as possible.

Designed by the results of elastic analysis the slab panel has a large reserve of strength because the failure loading is $58.23\,\text{kN/m}^2$ against design ultimate loading of $40\,\text{kN/m}^2$. Similarly, designed by the results of the BS 8110 method of analysis, the panel of slab has a small reserve of strength because the calculated collapse loading is 46.3 kN/m².

To check crack widths and deflection due to service load the BS 8110 coefficients may not be used. Always use the elastic analysis results.

Determination of shear at supports
Use BS 8110: Part 1: 1985, Table 3.16.[1]

Shear coefficients 0.44 and 0.33

$V_x = 0.44 \times 40 \times 4.3$
$\quad = 75.7\,\text{kN/m}$

$V_y = 0.33 \times 40 \times 4.3$
$\quad = 56.8\,\text{kN/m}$

Refer to Table 3.4.

By yield-line principle: assuming $r = 40\,\text{kN/m}^2$,

$$V_x = \frac{3\,r\,H\,(1 - x/L)}{2(3 - x/L)}$$

$$\quad = \frac{3 \times 40 \times 4.3 \times (1 - 0.35)}{2(3 - 0.35)}$$

$$\quad = 63.3\,\text{kN/m}$$

$$V_y = \frac{3\,r\,x}{5} = \frac{3 \times 40 \times 2.2}{5} = 52.8\,\text{kN/m}$$

Step 2 *Draw diagram of panel of slab*
See diagram with moments and shears marked on the panel (in Step 1).

Step 3 *Determination of cover*
Assume diameter of main reinforcement = 12 mm
Maximum size of aggregate = 20 mm
Condition of exposure = mild
Grade of concrete = C40
Minimum cement content = 325 kg/m³
Maximum free water/cement ratio = 0.55
Fire resistance required = 1 hour
Nominal cover, as per Tables 11.6 and 11.7 = 20 mm
Effective depth, $d_x = 150 - 20 - 6 = 124\,\text{mm}$
Effective depth, $d_y = 150 - 20 - 12 - 6 = 112\,\text{mm}$

SK 3/21 Section through slab showing effective depths.

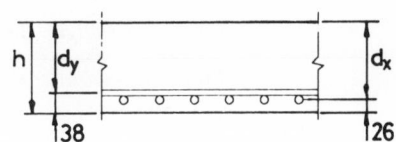

Step 4 Design of slab

Over continuous long edge, M = 50 kNm/m

$$K = \frac{M}{f_{cu}bd_x^2} = \frac{50 \times 10^6}{40 \times 1000 \times 124^2} = 0.081$$

$$z = d\left[0.5 + \sqrt{\left(0.25 - \frac{K}{0.9}\right)}\right] = 0.9d = 111.6 \,\text{mm}$$

$$x = \frac{d - z}{0.45} = 27.5 \,\text{mm}$$

$$A_s = \frac{M}{0.87 f_y z} = \frac{50 \times 10^6}{0.87 \times 460 \times 111.6} = 1120 \,\text{mm}^2/\text{m}$$

Over continuous short edge, M = 40 kNm/m

$$K = \frac{M}{f_{cu}bd_y^2} = \frac{40 \times 10^6}{40 \times 1000 \times 112^2} = 0.08$$

$$z = 0.9d = 100.8 \,\text{mm}$$

$$x = 24.9 \,\text{mm}$$

$$A_s = 992 \,\text{mm}^2/\text{m}$$

Positive midspan moment in short direction

$$M = 31.4 \,\text{kNm/m}$$

$$K = 0.051$$

$$z = 116.5 \,\text{mm}$$

$$A_s = 673 \,\text{mm}^2/\text{m}$$

Positive midspan moment in long direction

$$M = 19.9 \,\text{kNm/m}$$

$$K = 0.04$$

$$z = 0.95d = 106.4 \,\text{mm}$$

$$A_s = 467 \,\text{mm}^2/\text{m}$$

Step 5 Diameter and spacing of bars
Use:

Over long edge 12 dia. at 100 centre-to-centre (top) $(1131 \,\text{mm}^2/\text{m})$

Over short edge 12 dia. at 100 centre-to-centre (top) $(1131 \,\text{mm}^2/\text{m})$

Short direction at midspan 12 dia. at 150 centre-to-centre (bottom) $(754 \,\text{mm}^2/\text{m})$

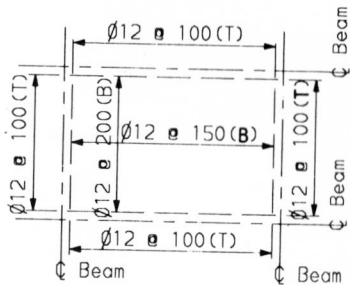

SK 3/22 Plan of panel of slab
showing design steel requirement.

Long direction at midspan 12 dia. at 200 centre-to-centre (bottom)
(565 mm²/m)

Step 6   Check shear stress

$$v_x = \frac{V_x}{bd_x} = \frac{75.7 \times 10^3}{1000 \times 124} = 0.61 \, \text{N/mm}^2$$

$$v_y = \frac{V_y}{bd_y} = \frac{56.8 \times 10^3}{1000 \times 112} = 0.51 \, \text{N/mm}^2$$

$$p_x = \frac{100A_{sx}}{bd_x} = \frac{100 \times 1131}{1000 \times 124} = 0.91\%$$

$$p_y = \frac{100A_{sy}}{bd_y} = \frac{100 \times 1131}{1000 \times 112} = 1.0\%$$

From Fig. 11.5,

$$v_{cx} = 0.97 \, \text{N/mm}^2 > v_x = 0.61 \, \text{N/mm}^2$$

No shear reinforcement required.

Step 7   Check punching shear stress
Not required.

Step 8   Modification due to holes
Not required.

Step 9   Minimum tension reinforcement

$$A_s = 0.0013bh$$
$$= 0.0013 \times 1000 \times 150$$
$$= 195 \, \text{mm}^2/\text{m} \quad \text{satisfied}$$

Step 10   Torsional reinforcement
Not required.

Step 11   Check span/effective depth

$$\frac{l_{ex}}{d_x} = \frac{4.3 \times 10^3}{124}$$

$$= 34.7$$

Basic span/effective depth ratio $= 26$ from Table 11.3

$$\beta_b = \frac{M'}{M} = \frac{31.4}{25.9} = 1.21$$

where $M' =$ moment after redistribution; $M =$ moment before redistribution

$$f_s = \frac{5}{8} f_y \left(\frac{A_{s\ reqd}}{A_{s\ prov}}\right)\left(\frac{1}{\beta_b}\right)$$

$$= \frac{5}{8} \times 460 \times \frac{673}{754} \times \frac{1}{1.21}$$

$$= 212\,\text{N/mm}^2$$

$$\frac{M}{bd^2} = \frac{31.4 \times 10^6}{1000 \times 124^2} = 2.0$$

From Chart 11.5,

modification factor $= 1.33$

Modified span/effective depth ratio $= 26 \times 1.33 = 34.58 < 34.7$

Code deflection limits have been exceeded slightly.
May be ignored.

Step 12   Curtailment of bars

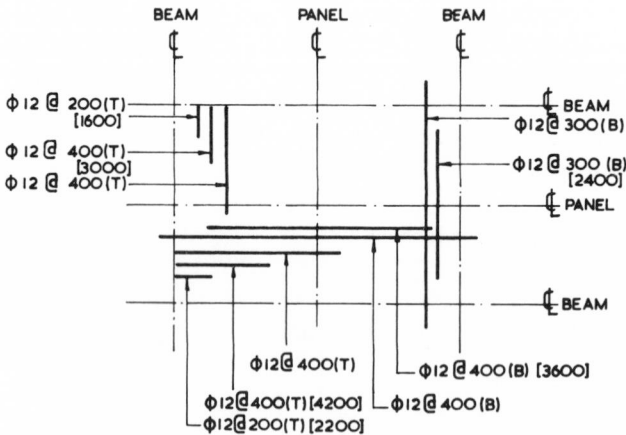

SK 3/23 Plan of panel of slab showing arrangement of reinforcement.

$45 \times$ dia. of bars $= 45 \times 12 = 540$ mm

$0.15\ l_{ex} = 0.15 \times 4.3 = 645$ mm
$0.30\ l_{ex} = 0.30 \times 4.3 = 1290$ mm
$0.20\ l_{ex} = 0.20 \times 4.3 = 860$ mm
$0.15\ l_{ey} = 0.15 \times 6.3 = 945$ mm
$0.30\ l_{ey} = 0.30 \times 6.3 = 1890$ mm
$0.20\ l_{ey} = 0.20 \times 6.3 = 1260$ mm

Direction l_x − top reinforcement

12 dia. @ 100 c/c to 800 mm from centre of beam (top)
12 dia. @ 200 c/c to 1500 mm from centre of beam (top)

Direction l_y − top reinforcement

12 dia. @ 100 c/c to 1100 mm from centre of beam (top)
12 dia. @ 200 c/c to 2100 mm from centre of beam (top)
Elsewhere use 12 dia. @ 400 c/c (top) both directions (282 mm^2)

Direction l_x − bottom reinforcement

12 dia. @ 150 c/c up to 800 mm from centre of beam (bottom)
12 dia. @ 300 c/c over beam (bottom)

Direction l_y − bottom reinforcement

12 dia. @ 200 c/c up to 1200 mm from centre of beam (bottom)
12 dia. @ 400 c/c over beam (bottom)

Step 13   **Spacing of bars**
Percentage of reinforcement in slab $= 1\%$

Maximum clear spacing allowed $= 160$ mm

Actual spacing used $= 100$ mm   OK

Maximum spacing of bars in tension $= 3d = 3 \times 112 = 336$ mm

Maximum spacing used for designed bars in tension $= 200$ mm   OK

Maximum spacing of nominal reinforcement to control early thermal cracking $= 400$ mm

Step 14   **Check thermal cracking**
For suspended slab, $R = 0.3$ assumed

$T_1 = 12°C$ assumed for 150 mm thick slab

$\alpha = 12 \times 10^{-6}$ per degree C

$\varepsilon_r = 0.8\ T_1\ \alpha\ R$
$= 0.8 \times 12 \times 12 \times 10^{-6} \times 0.3$
$= 34.56 \times 10^{-6}$

$$C_{min} = 20\,mm + 12\,mm \quad \text{(dia. of bar)}$$
$$= 32\,mm \quad \text{(direction } l_y)$$

$$x = d/2 \text{ assumed} = 112/2 = 56\,mm \text{ (direction } l_y)$$

$$a_{cr} = \sqrt{(200^2 + 38^2)} - 6 = 197.6\,mm$$

$$W_{max} = \cfrac{3a_{cr}\,\varepsilon_r}{1 + \cfrac{2(a_{cr} - C_{min})}{(h - x)}}$$

$$= \cfrac{3 \times 197.6 \times 34.56 \times 10^6}{1 + \cfrac{2(197.6 - 32)}{(150 - 56)}}$$

$$= 0.0045\,mm < 0.3\,mm \qquad \text{OK}$$

Step 15   Check minimum reinforcement to distribute cracking

$$A_c = \frac{bh}{2} = \frac{1000 \times 150}{2} = 75000\,mm^2$$

$$A_s = 0.0035\,A_c = 262.5\,mm^2/m$$
$$A_s \text{ provided} = 12\,dia.\ @\ 400\ c/c\ (282\,mm^2/m)$$

Step 16   Assessment of crack width in flexure

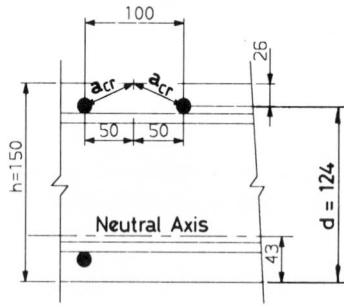

SK 3/24 Section through slab over beam for crack width calculations.

Service load on slab $= 25.75\,kN/m^2$

By elastic analysis,

maximum bending moment over long support
$$= 0.075 \times 25.75 \times 4.3^2$$
$$= 35.7\,kNm/m$$

$$A_s = 1131\,mm^2/m \qquad A_s/bd = 9.12 \times 10^{-3}$$

$$b = 1000\,mm \qquad d = 124\,mm$$

$$m = 10 = E_s/E_c$$

$$A_s' = \text{neglected}$$

$$x = d[((mp)^2 + 2\,mp)^{\frac{1}{2}} - mp] = 43\,mm$$

$$z = d - \frac{x}{3} = 124 - \frac{43}{3} = 109.7 \, \text{mm}$$

$$f_s = \frac{M}{A_s z} = \frac{35.7 \times 10^6}{1131 \times 109.7} = 288 \, \text{N/mm}^2$$

$$\varepsilon_s = \frac{f_s}{E_s} = \frac{288}{200 \times 10^3} = 1.44 \times 10^{-3}$$

$$\varepsilon_h = \left(\frac{h - x}{d - x}\right) \varepsilon_s = \left(\frac{150 - 43}{124 - 43}\right) \times 1.44 \times 10^{-3} = 1.90 \times 10^{-3}$$

$$\varepsilon_{mh} = \varepsilon_h - \frac{b(h - x)^2}{3 E_s A_s (d - x)} = 1.90 \times 10^{-3}$$

$$- \frac{1000(150 - 43)^2}{3 \times 200 \times 10^3 \times 1131 \times (124 - 43)}$$

$$= 1.69 \times 10^{-3}$$

$$C_{min} = 20 \, \text{mm}$$

$$a_{cr} = \sqrt{(26^2 + 50^2)} - 6 = 50 \, \text{mm}$$

$$W_{cr} = \frac{3 a_{cr} \, \varepsilon_m}{1 + \dfrac{2(a_{cr} - C_{min})}{(h - x)}} = 0.16 \, \text{mm} < 0.3 \, \text{mm} \qquad \text{OK}$$

3.5 FIGURES AND TABLES FOR CHAPTER 3

Edge conditions and loading diagrams	Elastic resistance, r_e	Elasto-plastic resistance, r_{ep}
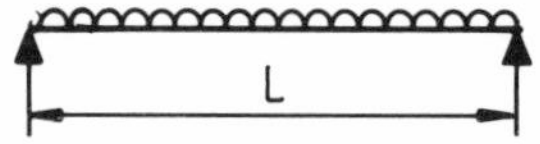 Simply supported beam, uniform load over span L	r_u	—
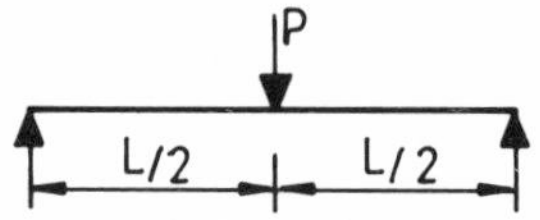 Simply supported beam, point load P at midspan ($L/2$, $L/2$)	R_u	—
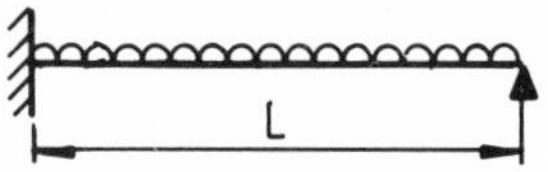 Fixed-simple beam, uniform load over span L	$\dfrac{8M_N}{L^2}$	r_u
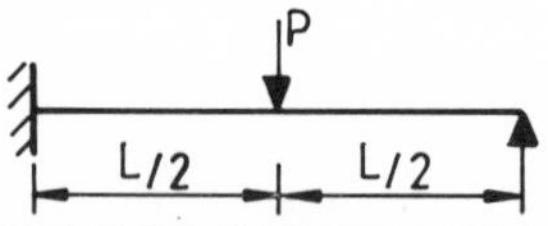 Fixed-simple beam, point load P at midspan ($L/2$, $L/2$)	$\dfrac{16M_N}{3L}$	R_u
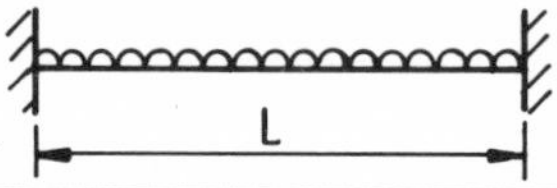 Simple-fixed beam, uniform load over span L	$\dfrac{12M_N}{L^2}$	r_u
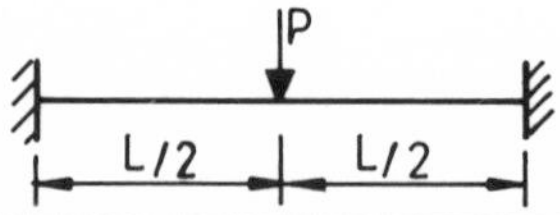 Fixed-fixed beam, point load P at midspan ($L/2$, $L/2$)	$\dfrac{8M_N}{L}$	R_u
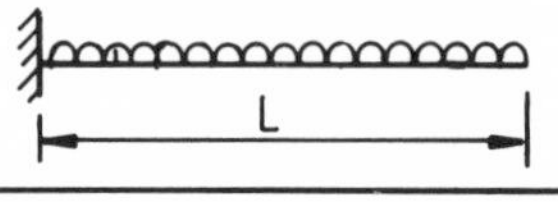 Fixed cantilever, uniform load over span L	r_u	—
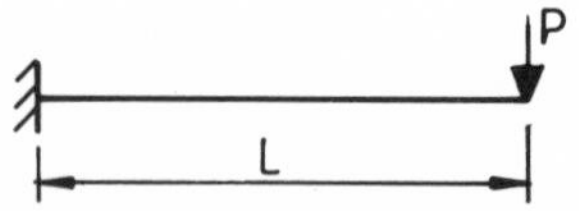 Fixed cantilever, point load P at free end, span L	R_u	—
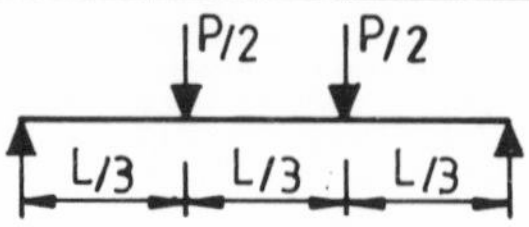 Simply supported beam, two point loads $P/2$ at third points ($L/3$, $L/3$, $L/3$)	R_u	—

Fig. 3.1 Elastic and elasto-plastic unit resistances for one-way elements.

Edge conditions and loading diagrams	Support reactions, V_s
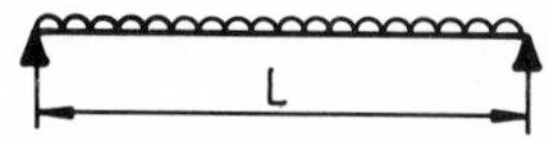 L	$\dfrac{r_u\,L}{2}$
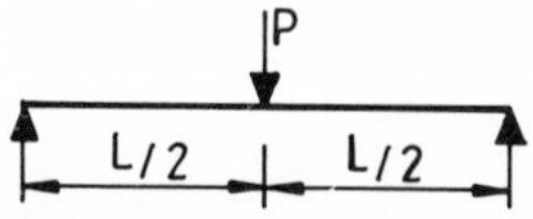 P $L/2$ $L/2$	$\dfrac{R_u}{2}$
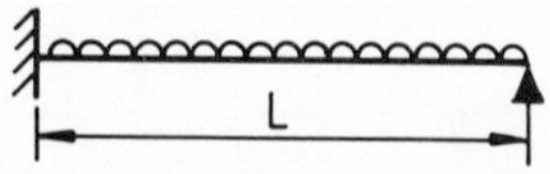 L	L. reaction $\dfrac{5r_u L}{8}$ R. reaction $\dfrac{3r_u L}{8}$
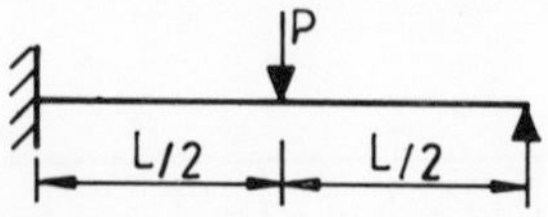 P $L/2$ $L/2$	L. reaction $\dfrac{11R_u}{16}$ R. reaction $\dfrac{5R_u}{16}$
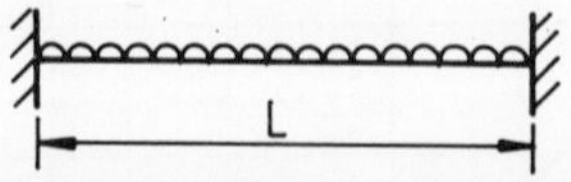 L	$\dfrac{r_u L}{2}$
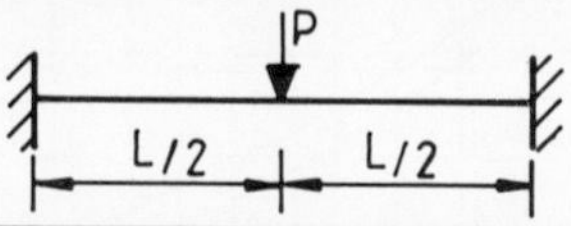 P $L/2$ $L/2$	$\dfrac{R_u}{2}$
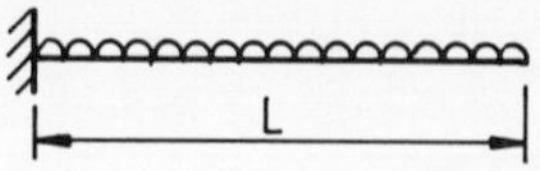 L	$r_u L$
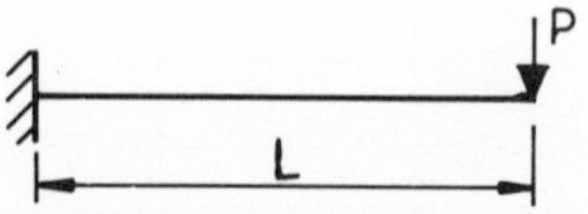 P L	R_u
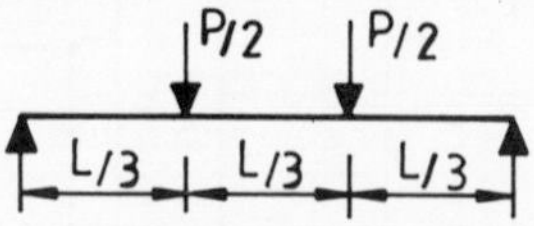 $P/2$ $P/2$ $L/3$ $L/3$ $L/3$	$\dfrac{R_u}{2}$

Fig. 3.2 Support shears for one-way elements (to be read in conjunction with Fig. 3.1).

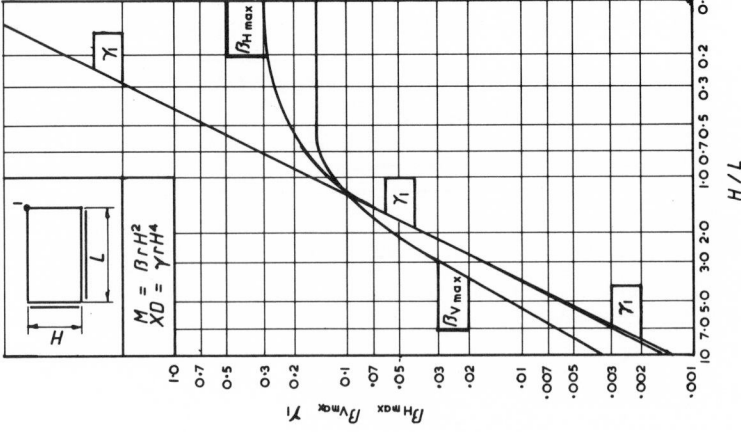

Fig. 3.5 Moment and deflection coefficients for uniformly loaded two-way element with two adjacent edges simply supported and two edges free.[8]

Fig. 3.4 Moment and deflection coefficients for uniformly loaded two-way element with one edge fixed, an adjacent edge simply supported and two edges free.[8]

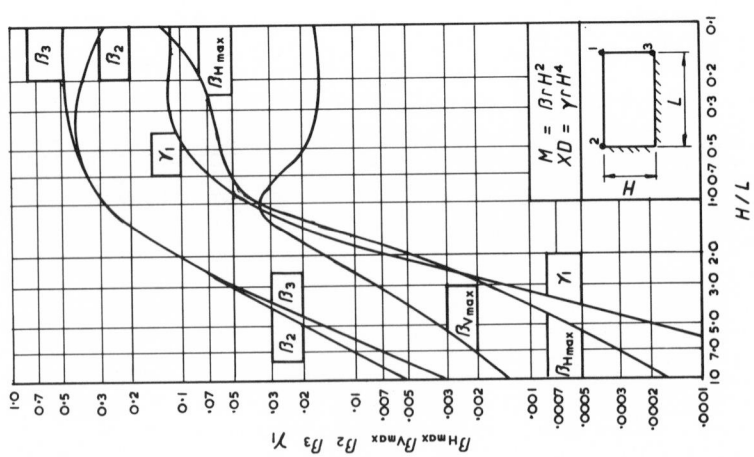

Fig. 3.3 Moment and deflection coefficients for uniformly loaded two-way element with two adjacent edges fixed and two edges free.[8]

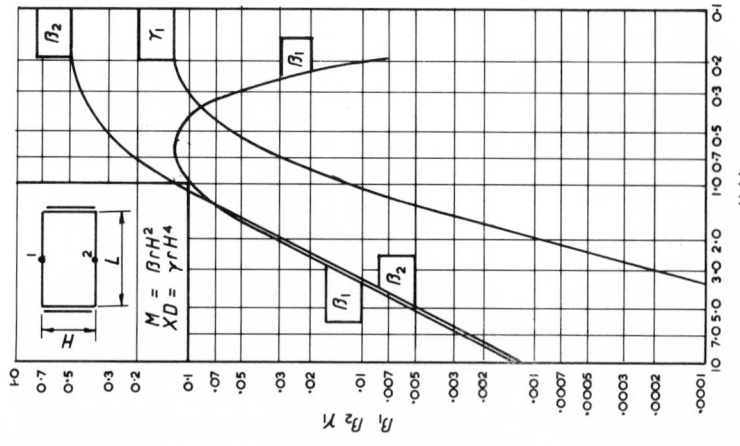

Fig. 3.8 Moment and deflection coefficients for uniformly loaded two-way element with two opposite edges simply supported, one edge fixed and one edge free.[8]

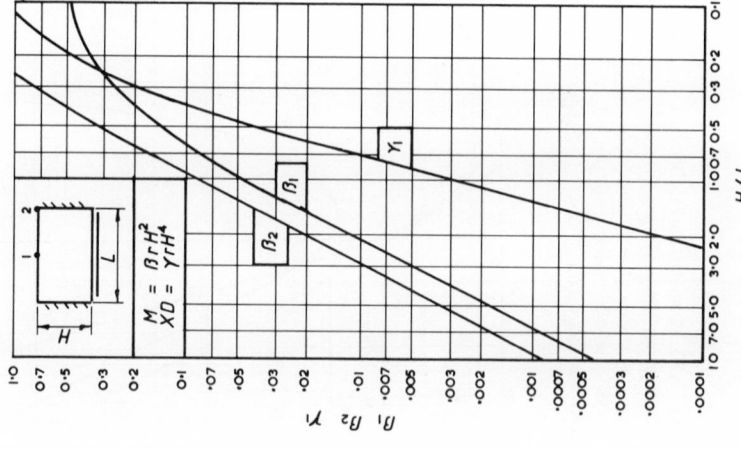

Fig. 3.7 Moment and deflection coefficients for uniformly loaded two-way element with two opposite edges fixed, one edge simply supported and one edge free.[8]

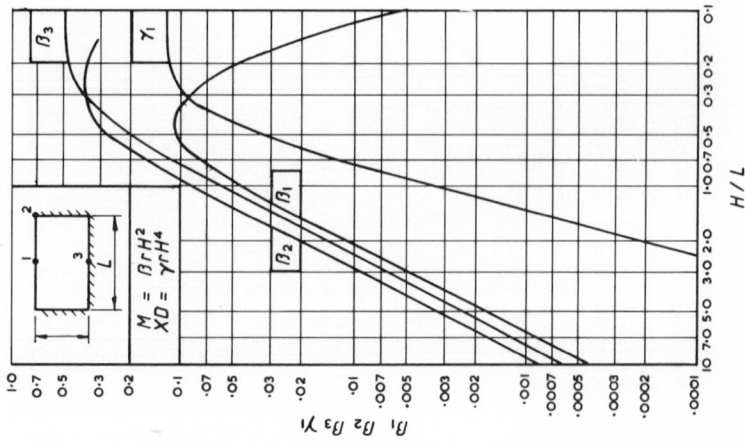

Fig. 3.6 Moment and deflection coefficients for uniformly loaded two-way element with three edges fixed and one edge free.[8]

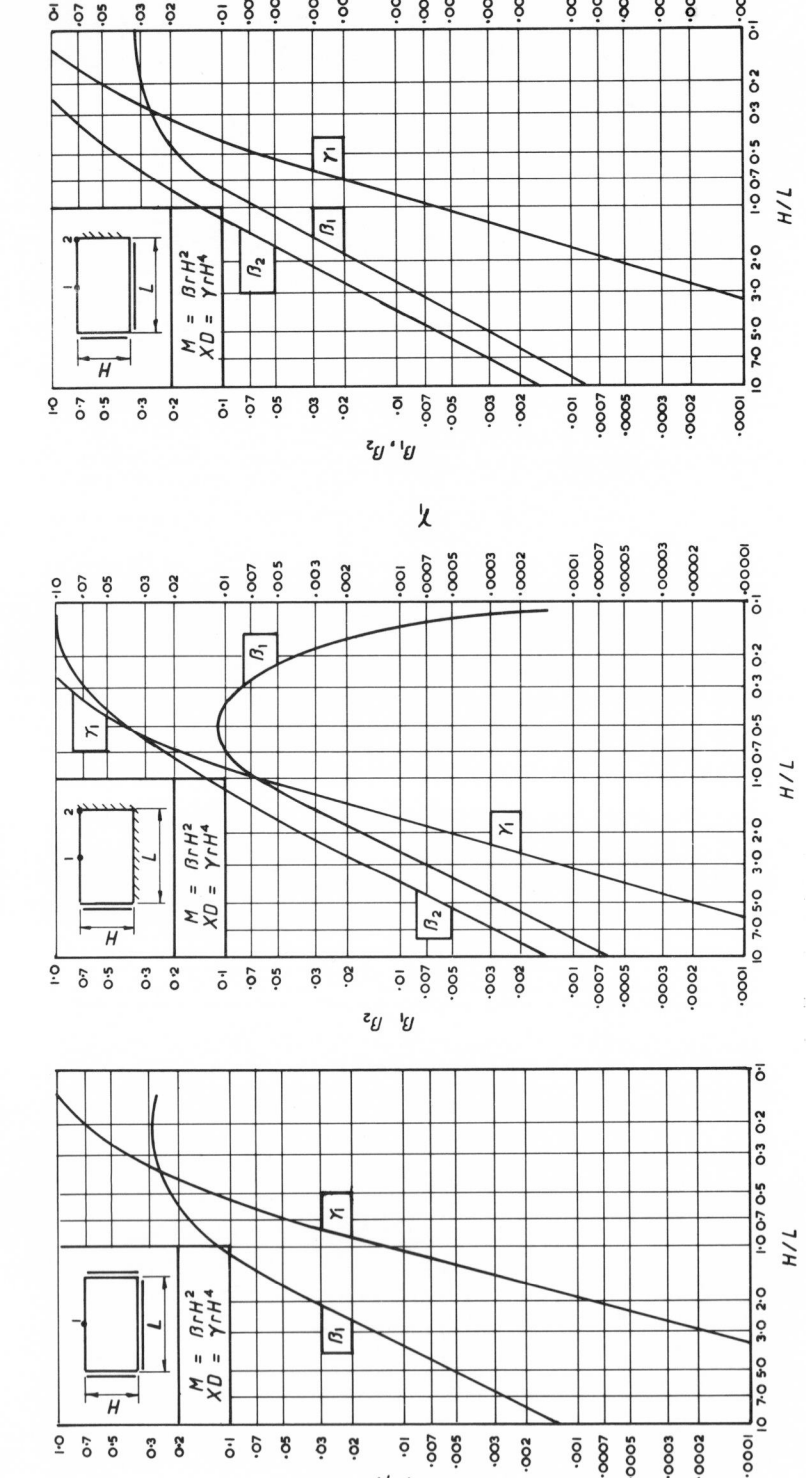

Fig. 3.11 Moment and deflection coefficients for uniformly loaded two-way element with two adjacent edges simply supported, one edge fixed and one edge free.[8]

Fig. 3.10 Moment and deflection coefficients for uniformly loaded two-way element with two edges fixed, one edge simply supported and one edge free.[8]

Fig. 3.9 Moment and deflection coefficients for uniformly loaded two-way element with three edges simply supported and one edge free.[8]

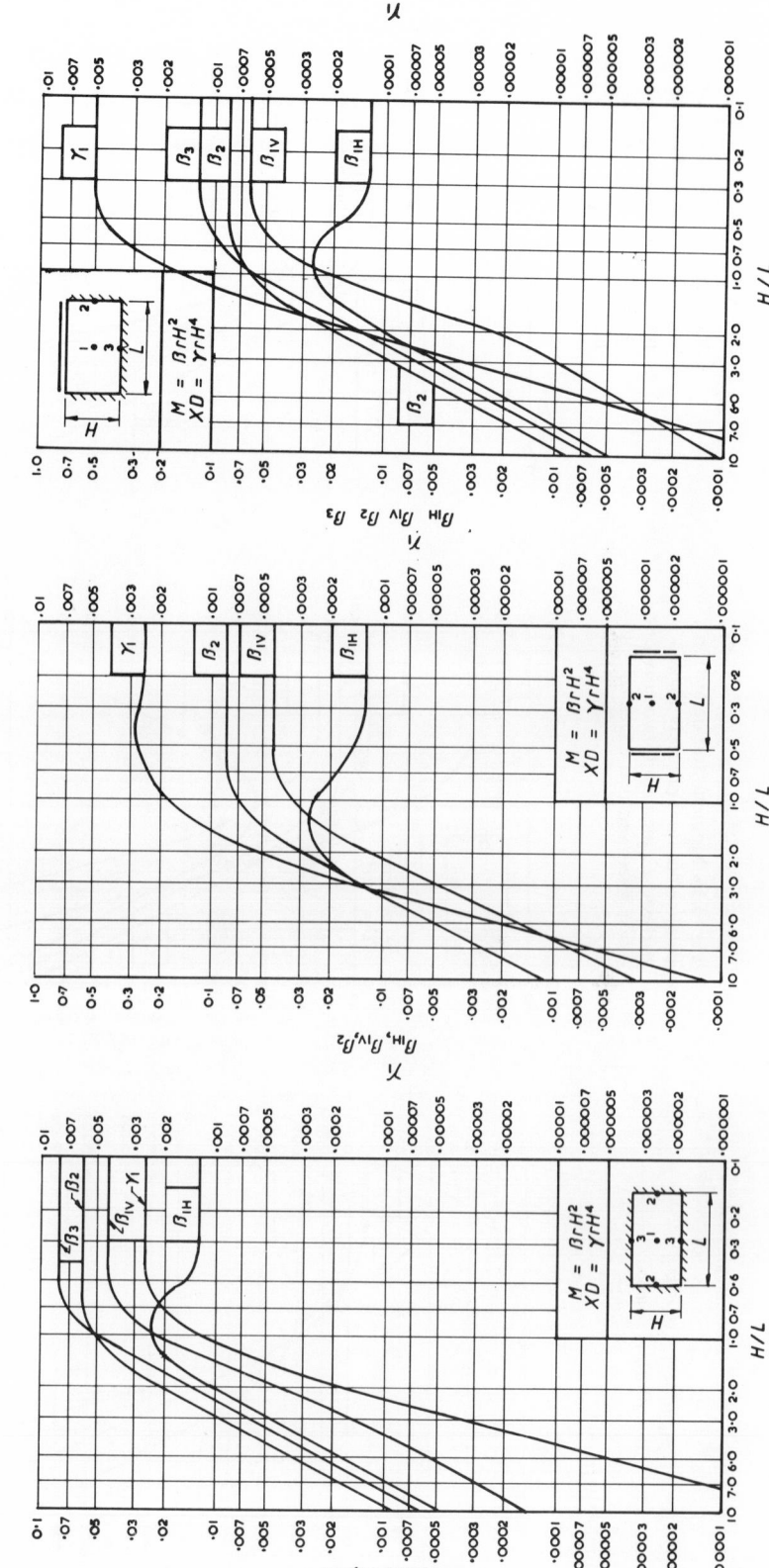

Fig. 3.14 Moment and deflection coefficients for uniformly loaded two-way element with three edges fixed and one edge simply supported.[8]

Fig. 3.13 Moment and deflection coefficients for uniformly loaded two-way element with two opposite edges fixed and two edges simply supported.[8]

Fig. 3.12 Moment and deflection coefficients for uniformly loaded two-way element with all edges fixed.[8]

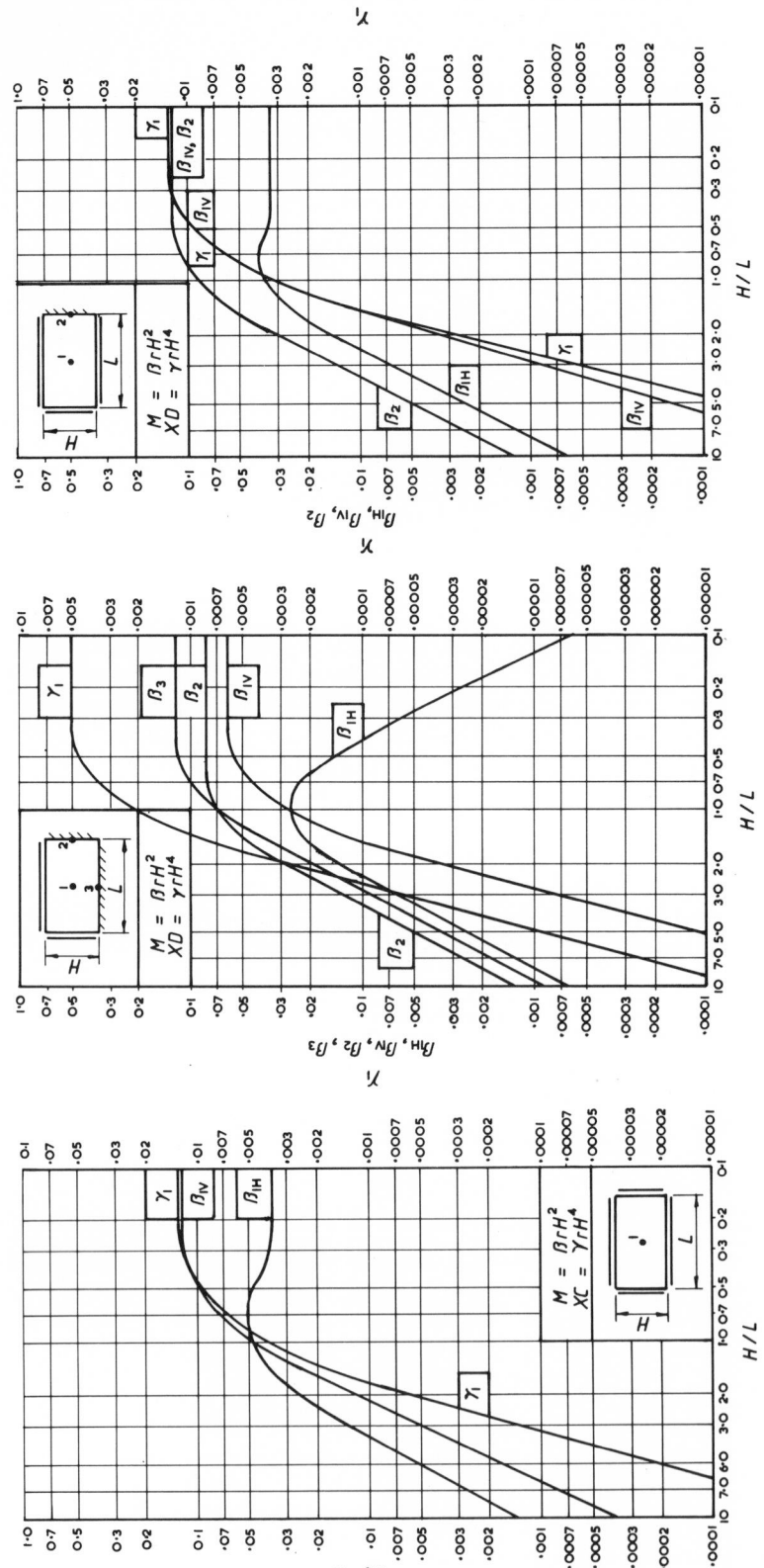

Fig. 3.17 Moment and deflection coefficients for uniformly loaded two-way element with three edges simply supported and one edge fixed.[8]

Fig. 3.16 Moment and deflection coefficients for uniformly loaded two-way element with two adjacent edges fixed and two edges simply supported.[8]

Fig. 3.15 Moment and deflection coefficients for uniformly loaded two-way element with all edges simply supported.[8]

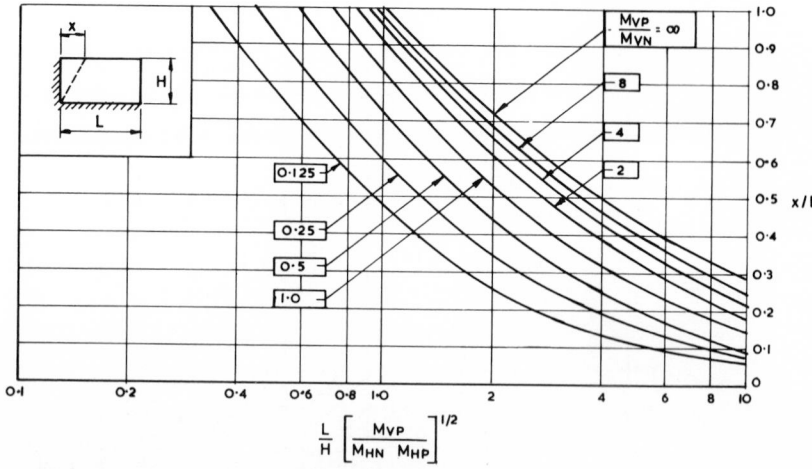

Fig. 3.18 Location of yield lines for two-way element with two adjacent edges supported and two edges free (values of x).[8]

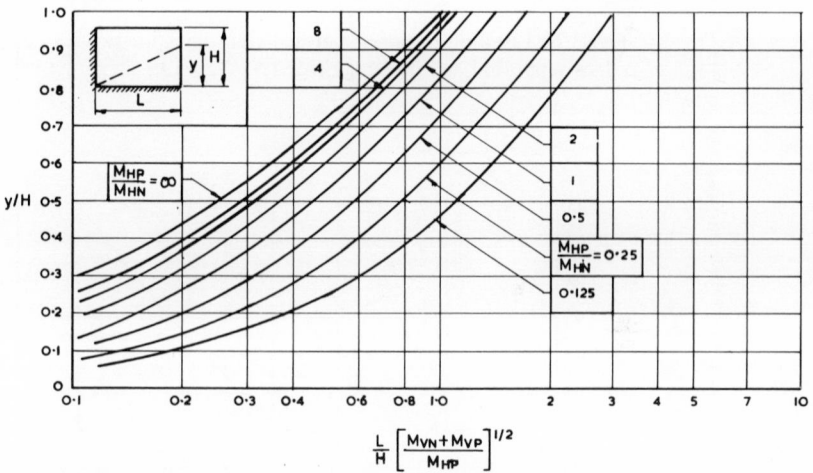

Fig. 3.19 Location of yield lines for two-way element with two adjacent edges supported and two edges free (values of y).[8]

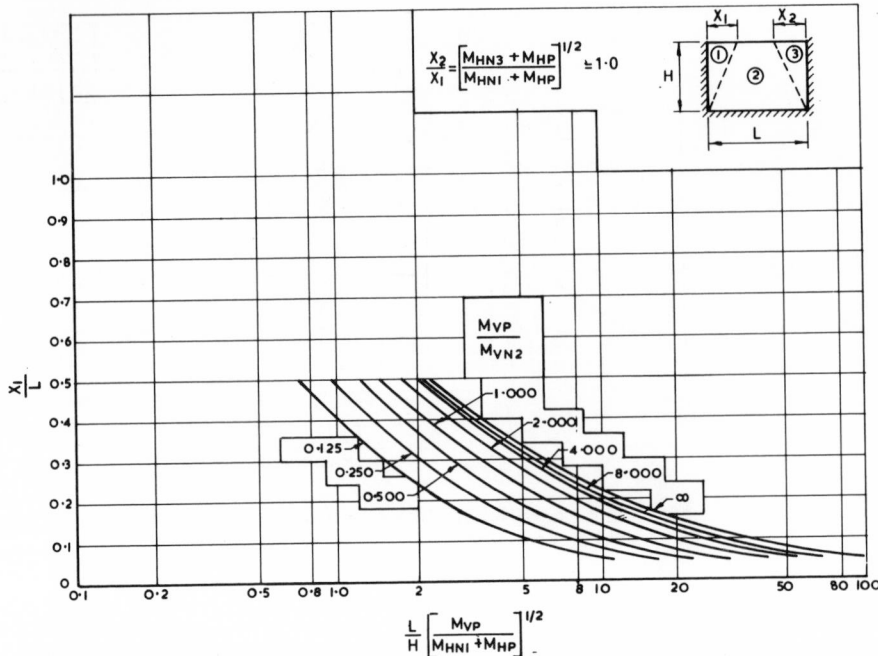

Fig. 3.20 Location of unsymmetrical yield lines for two-way element with three edges supported and one edge free $(X_2/X_1 = 1.0)$.[8]

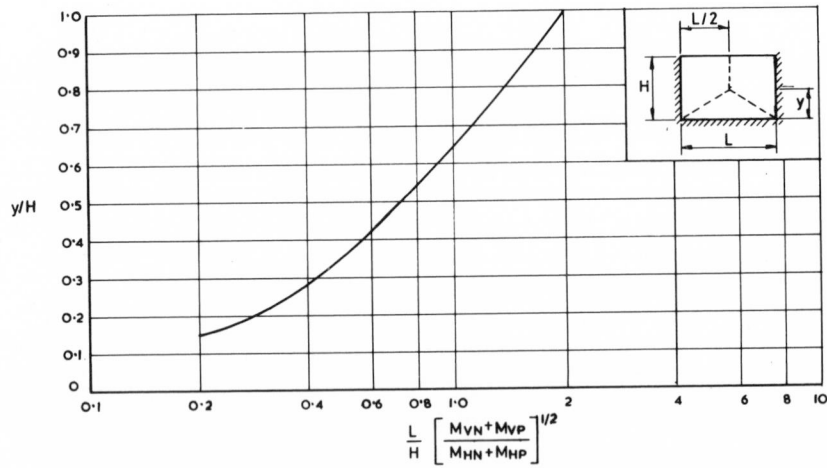

Fig. 3.21 Location of symmetrical yield lines for two-way element with three edges supported and one edge free (value of v).[8]

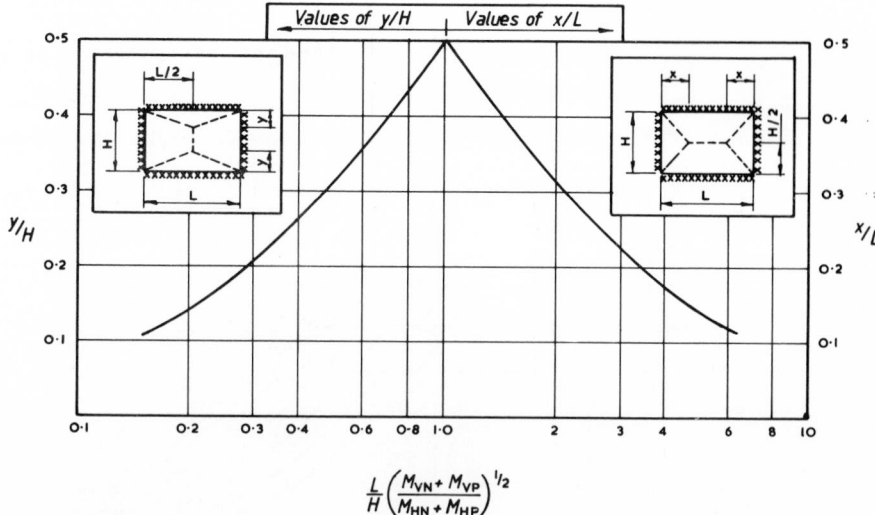

Fig. 3.22 Location of symmetrical yield lines for two-way element with four edges supported.[8]

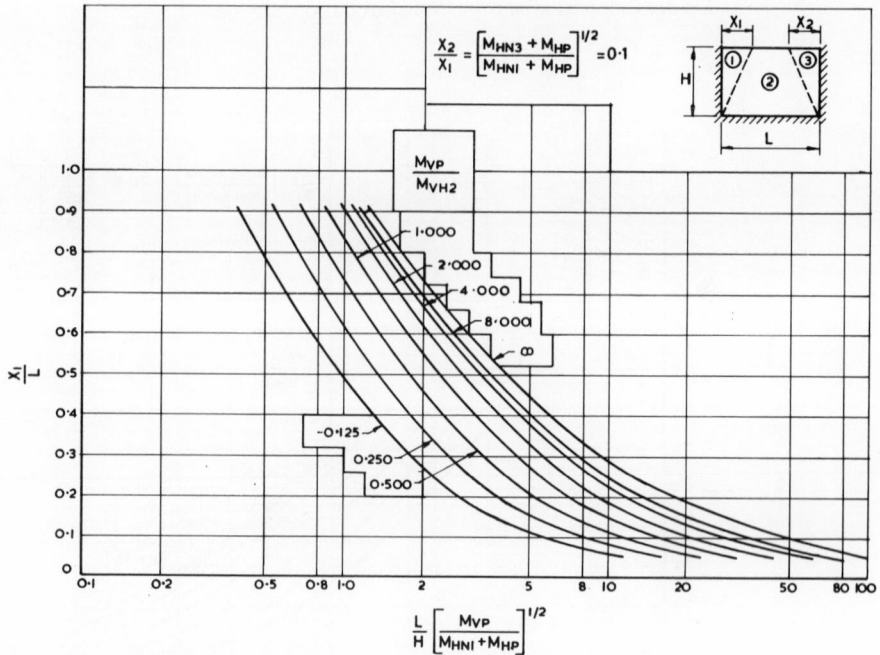

Fig. 3.23 Location of unsymmetrical yield lines for two-way element with three edges supported and one edge free $(X_2/X_1 = 0.1)$.[8]

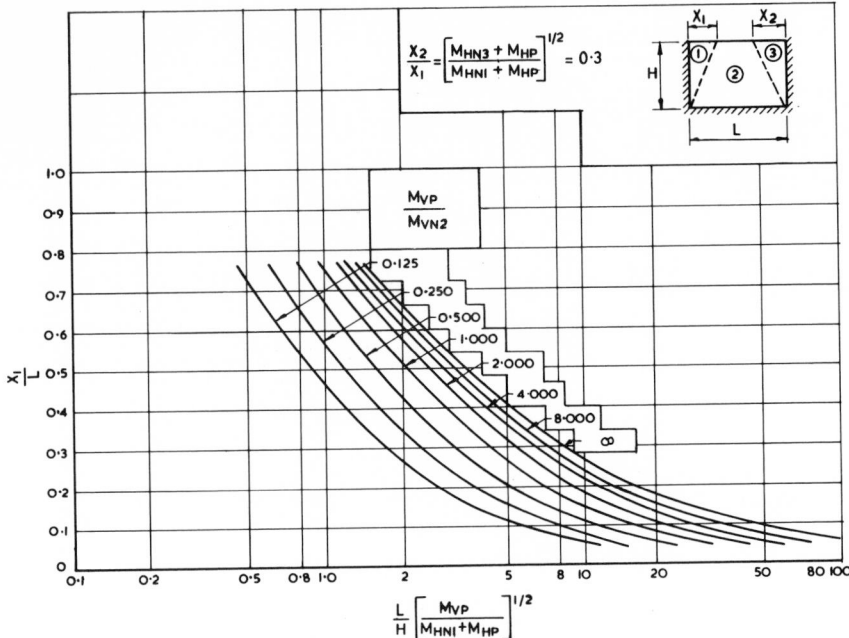

Fig. 3.24 Location of unsymmetrical yield lines for two-way element with three edges supported and one edge free $(X_2/X_1 = 0.3)$.[8]

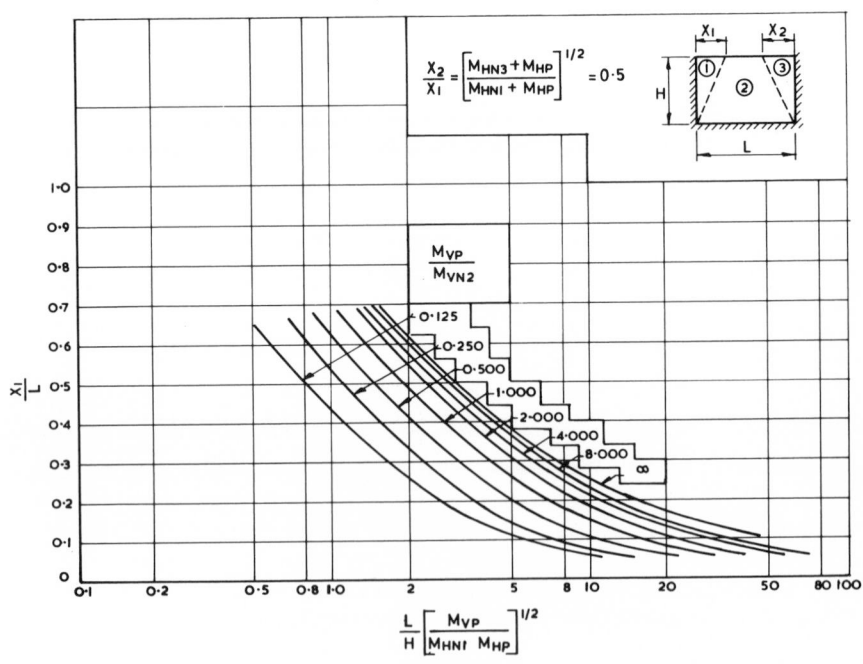

Fig. 3.25 Location of unsymmetrical yield lines for two-way element with three edges supported and one edge free $(X_2/X_1 = 0.5)$.[8]

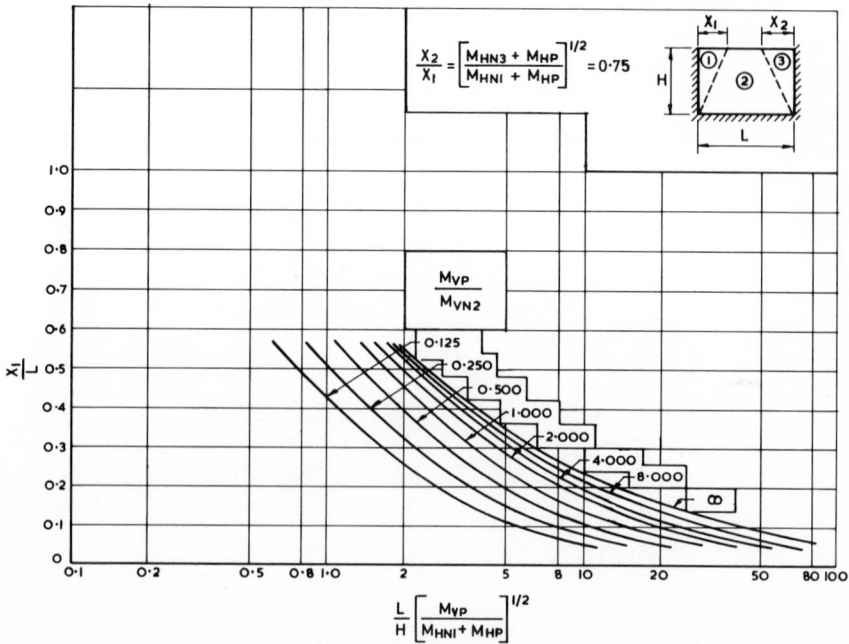

Fig. 3.26 Location of unsymmetrical yield lines for two-way element with three edges supported and one edge free ($X_2/X_1 = 0.75$).[8]

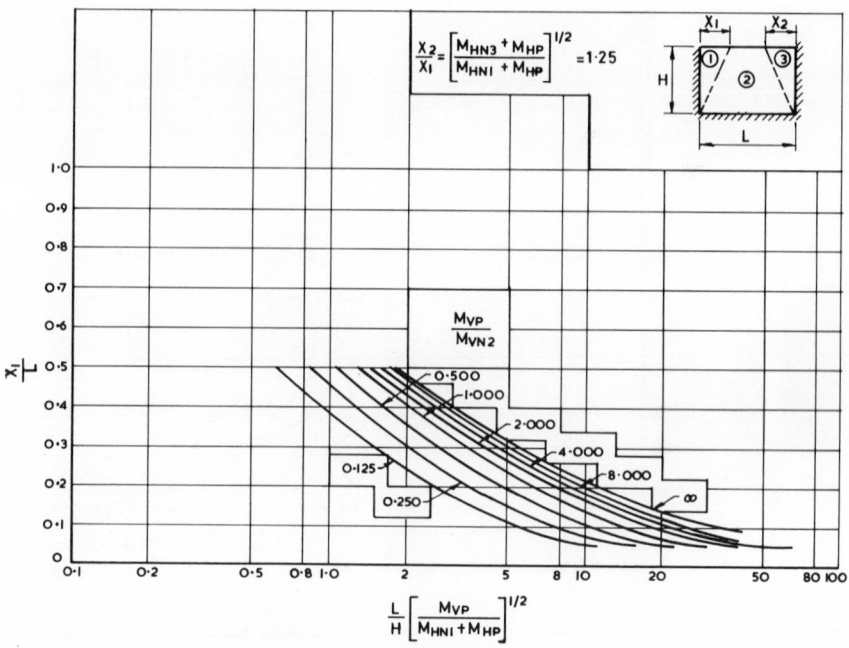

Fig. 3.27 Location of unsymmetrical yield lines for two-way element with three edges supported and one edge free ($X_2/X_1 = 1.25$).[8]

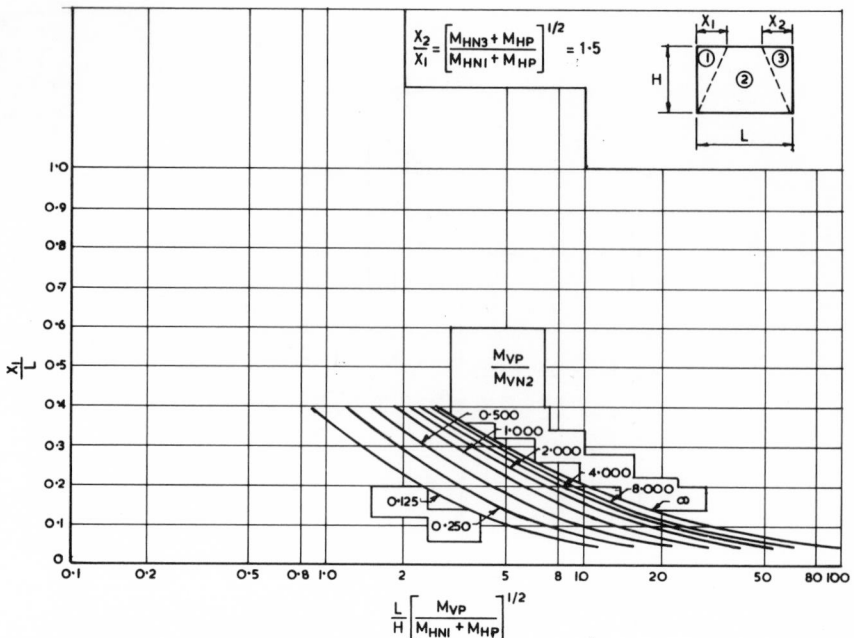

Fig. 3.28 Location of unsymmetrical yield lines for two-way element with three edges supported and one edge free ($X_2/X_1 = 1.5$).[8]

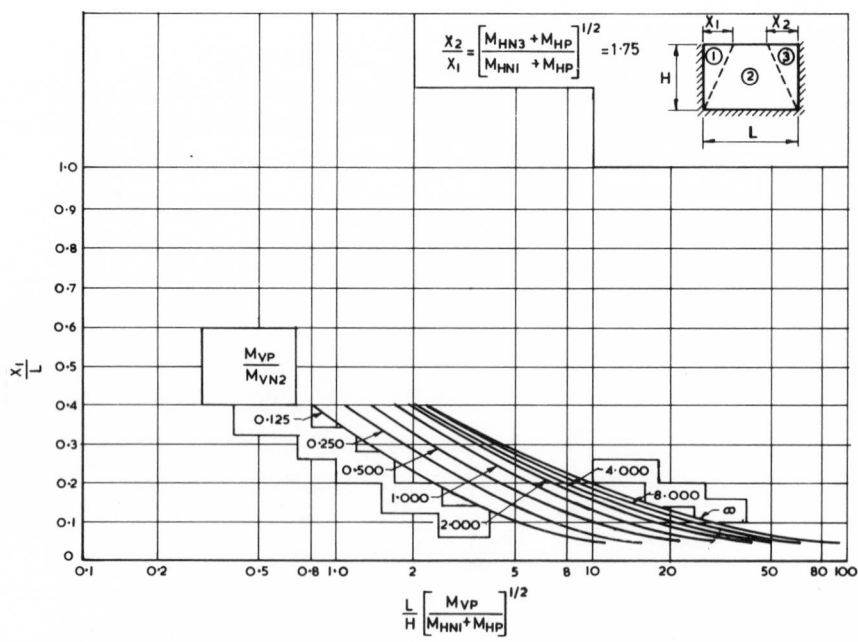

Fig. 3.29 Location of unsymmetrical yield lines for two-way element with three edges supported and one edge free ($X_2/X_1 = 1.75$).[8]

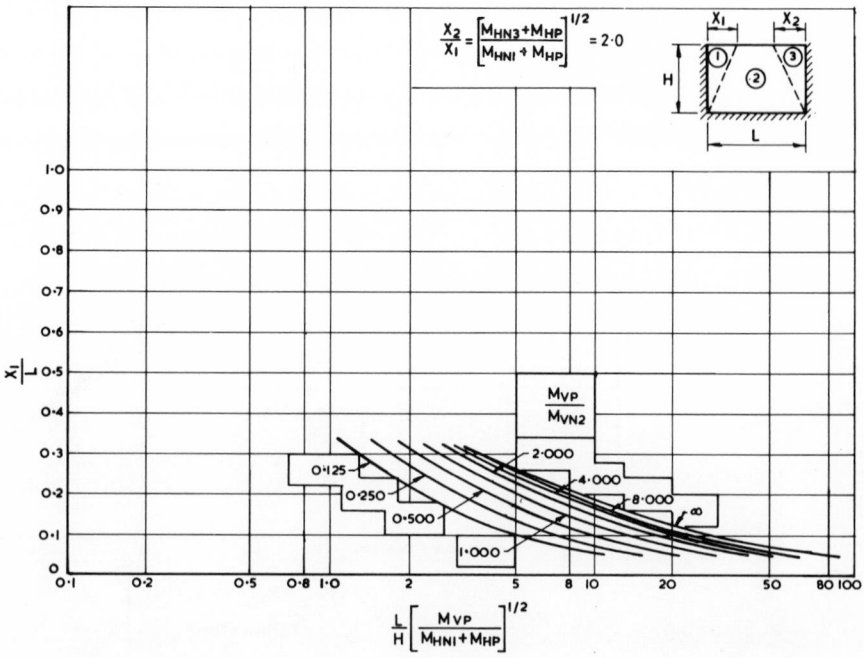

Fig. 3.30 Location of unsymmetrical yield lines for two-way element with three edges supported and one edge free ($X_2/X_1 = 2.0$).[8]

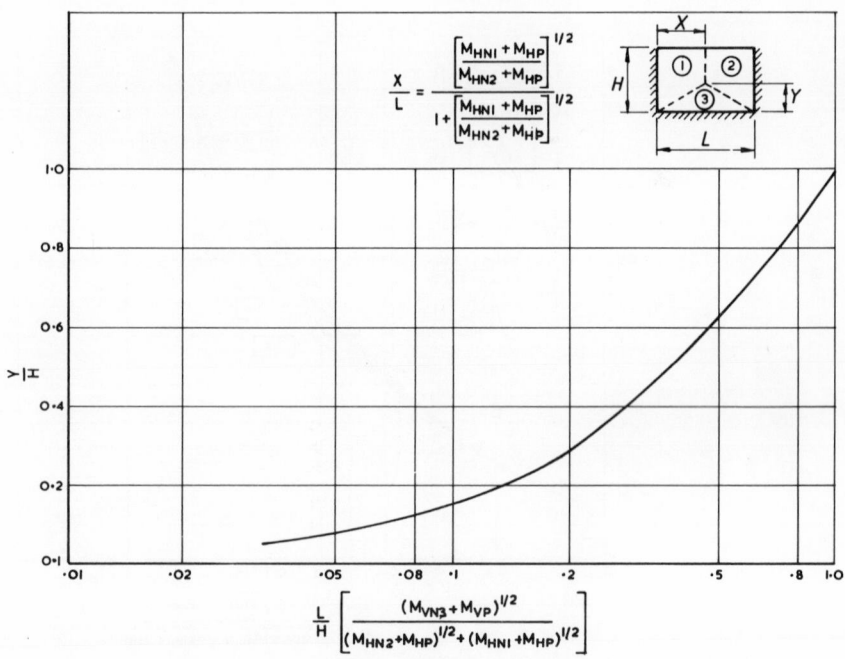

Fig. 3.31 Location of unsymmetrical yield lines for two-way element with three edges supported and one edge free (values of y).[8]

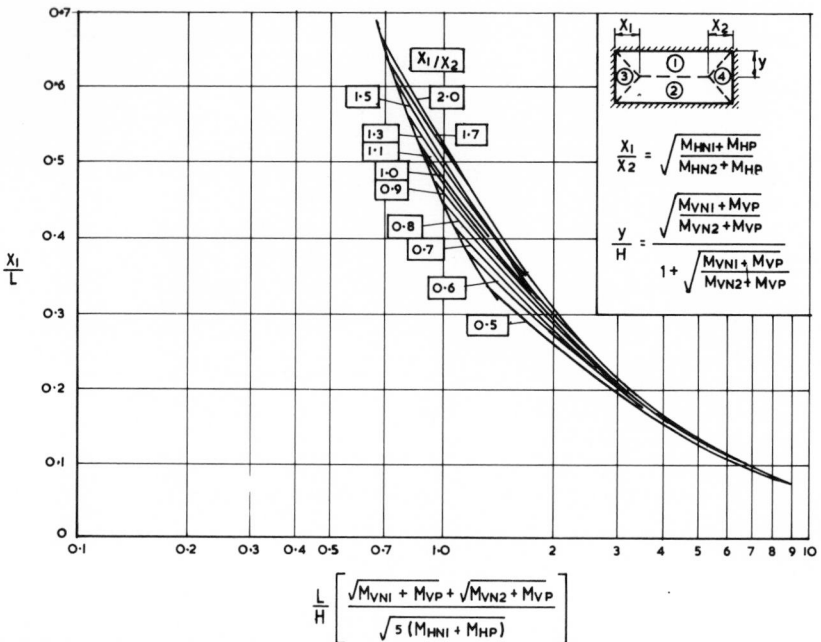

Fig. 3.32 Location of unsymmetrical yield lines for two-way element with four edges supported (values of X_1).[8]

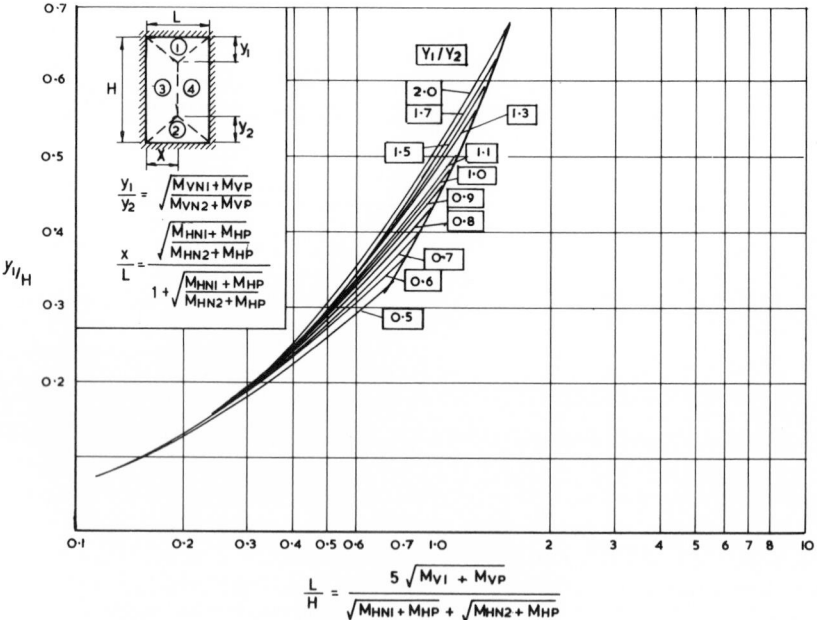

Fig. 3.33 Location of unsymmetrical yield lines for two-way element with four edges supported (values of Y_1).[8]

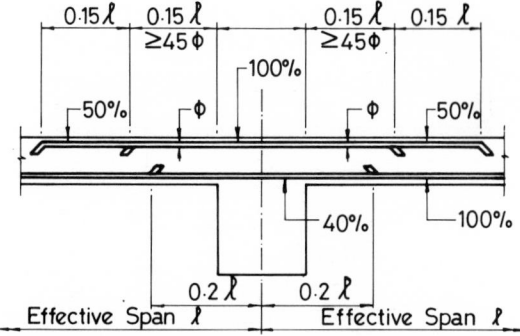

Continuous Slab : Approximate equal spans

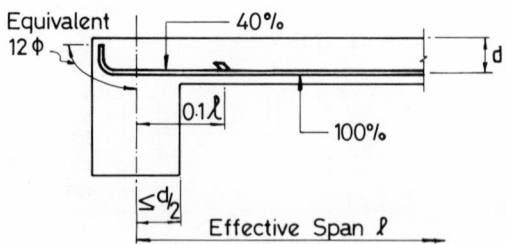

Simply Supported Slab

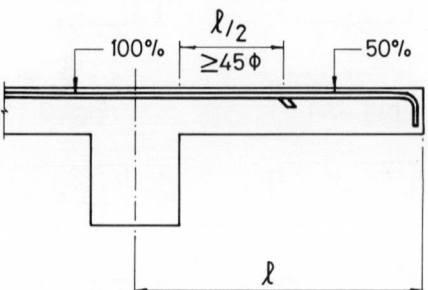

Cantilever Slab

Fig. 3.34 Simplified detailing rules for slabs.

Table 3.1 Graphical summary of two-way elements to be used in conjunction with Figures 3.3 to 3.17

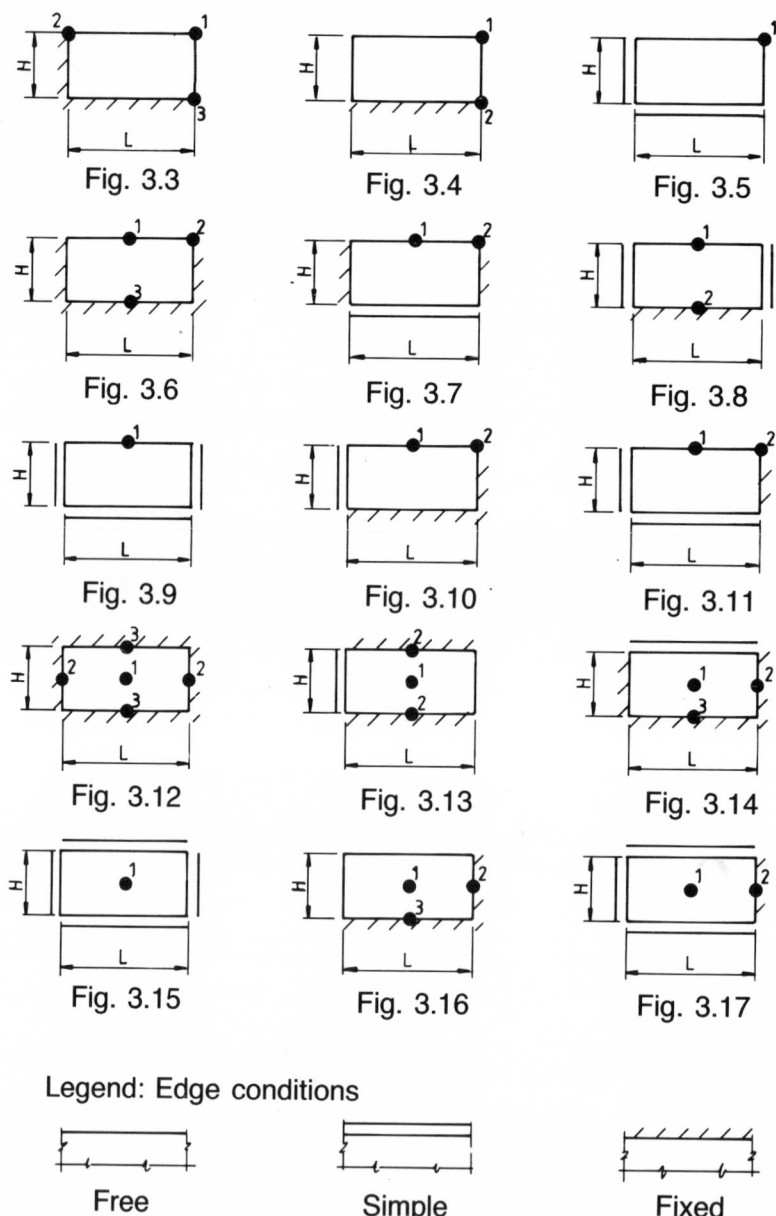

Fig. 3.3

Fig. 3.4

Fig. 3.5

Fig. 3.6

Fig. 3.7

Fig. 3.8

Fig. 3.9

Fig. 3.10

Fig. 3.11

Fig. 3.12

Fig. 3.13

Fig. 3.14

Fig. 3.15

Fig. 3.16

Fig. 3.17

Legend: Edge conditions

Free

Simple

Fixed

Table 3.2 Ultimate unit resistance for two-way elements (symmetrical yield-lines) (to be used in conjunction with Figs 3.18 to 3.23).

Edge conditions	Yield line locations	Limits	Ultimate unit resistance
Two adjacent edges supported and two edges free		$x \leq L$ $y \leq H$	$\dfrac{5(M_{HN} + M_{HP})}{x^2}$ or $\dfrac{6L\,M_{VN} + (5M_{VP} - M_{VN})x}{H^2(3L - 2x)}$ $\dfrac{5(M_{VN} + M_{VP})}{y^2}$ or $\dfrac{6H\,M_{HN} + (5M_{HP} - M_{HN})y}{L^2(3H - 2y)}$
Three edges supported and one edge free		$x \leq \dfrac{L}{2}$ $y \leq H$	$\dfrac{5(M_{HN} + M_{HP})}{x^2}$ or $\dfrac{2M_{VN}(3L - x) + 10\,x\,M_{VP}}{H^2(3L - 4x)}$ $\dfrac{5(M_{VN} + M_{VP})}{y^2}$ or $\dfrac{4(M_{HN} + M_{HP})(6H - y)}{L^2(3H - 2y)}$
Four edges supported		$x \leq \dfrac{L}{2}$ $y \leq \dfrac{H}{2}$	$\dfrac{5(M_{HN} + M_{HP})}{x^2}$ or $\dfrac{8(M_{VN} + M_{VP})(3L - x)}{H^2(3L - 4x)}$ $\dfrac{5(M_{VN} + M_{VP})}{y^2}$ or $\dfrac{8(M_{HN} + M_{HP})(3H - y)}{L^2(3H - 4y)}$

Table 3.3 Ultimate unit resistance for two-way elements (unsymmetrical yield-lines) (to be used in conjunction with Figs 3.18 to 3.33).

Edge conditions	Yield line locations	Limits	Ultimate unit resistance
Two adjacent edges supported and two edges free		$x \leq L$ $y \leq H$	Same as in Table 3.2
Three edges supported and one edge free		$x \leq \dfrac{L}{2}$	$\dfrac{5(M_{HN1}+M_{HP1})}{X_1^2}$ or $\dfrac{5(M_{HN3}+M_{HP})}{X_2^2}$ or $\dfrac{(5M_{VP}-M_{VN2})(X_1+X_2)+6M_{VN2}L}{H^2(3L-2X_1-2X_2)}$
		$y \leq H$	$\dfrac{(M_{HN1}+M_{HP})(6H-Y)}{X^2(3H-2Y)}$ or $\dfrac{(M_{HN2}+M_{HP})(6H-Y)}{(L-X)^2(3H-2Y)}$ or $\dfrac{5(M_{VN3}+M_{VP})}{Y^2}$
Four edges supported		$x \leq \dfrac{L}{2}$	$\dfrac{(M_{VN1}+M_{VP})(6L-X_1-X_2)}{Y^2(3L-2X_1-2X_2)}$ or $\dfrac{(M_{VN2}+M_{VP})(6L-X_1-X_2)}{(H-Y)^2(3L-2X_1-2X_2)}$ or $\dfrac{5(M_{HN1}+M_{HP})}{X_1^2}$ or $\dfrac{5(M_{HN2}+M_{HP})}{X_2^2}$
		$y \leq \dfrac{H}{2}$	$\dfrac{5(M_{VN1}+M_{VP})}{Y_1^2}$ or $\dfrac{5(M_{VN2}+M_{VP})}{Y_2^2}$ or $\dfrac{(M_{HN2}+M_{HP})(6H-Y_1-Y_2)}{(L-X)^2(3H-2Y_1-2Y_2)}$ or $\dfrac{(M_{HN1}+M_{HP})(6H-Y_1-Y_2)}{X^2(3H-2Y_1-2Y_2)}$

Table 3.4 Ultimate support shears for two-way elements (symmetrical yield-lines) (to be used in conjunction with Table 3.2).

Edge conditions	Yield line locations	Limits	Horizontal shear, V_{sH}	Vertical shear, V_{sV}
Two adjacent edges supported and two edges free		$x \leq L$	$\dfrac{3r_u x}{5}$	$\dfrac{3r_u H\left(2 - \dfrac{x}{L}\right)}{\left(6 - \dfrac{x}{L}\right)}$
		$y \leq H$	$\dfrac{3r_u L\left(2 - \dfrac{y}{H}\right)}{\left(6 - \dfrac{y}{H}\right)}$	$\dfrac{3r_u y}{5}$
Three edges supported and one edge free		$x \leq \dfrac{L}{2}$	$\dfrac{3r_u x}{5}$	$\dfrac{3r_u H\left(1 - \dfrac{x}{L}\right)}{\left(3 - \dfrac{x}{L}\right)}$
		$y \leq H$	$\dfrac{3r_u L\left(2 - \dfrac{y}{H}\right)}{2\left(6 - \dfrac{y}{H}\right)}$	$\dfrac{3r_u y}{5}$
Four edges supported		$x \leq \dfrac{L}{2}$	$\dfrac{3r_u x}{5}$	$\dfrac{3r_u H\left(1 - \dfrac{x}{L}\right)}{2\left(3 - \dfrac{x}{L}\right)}$
		$y \leq \dfrac{H}{2}$	$\dfrac{3r_u L\left(1 - \dfrac{y}{H}\right)}{2\left(3 - \dfrac{y}{H}\right)}$	$\dfrac{3r_u y}{5}$

Table 3.5 Ultimate support shears for two-way elements (unsymmetrical yield-lines) (to be used in conjunction with Table 3.3).

Edge conditions	Yield line locations	limits	Horizontal shear, V_{sH}	Vertical shear, V_{sV}
Two adjacent edges supported and two edges free		$x \leq L$ $y \leq H$	Same as in Table 3.4	Same as in Table 3.4
Three edges supported and one edge free		$x_1 \leq \dfrac{L}{2}$ $x_2 \leq \dfrac{L}{2}$ $y \leq H$	$\dfrac{3x_1 r_u}{5}$ $\dfrac{3x_2 r_u}{5}$ $\dfrac{3r_u x(2H - y)}{6H - y}$ $\dfrac{3r_u x\,(L - x)(2H - y)}{6H - y}$	$\dfrac{3r_u H(2L - x_1 - x_2)}{6L - x_1 - x_2}$ $\dfrac{3r_u y}{5}$
Four edges supported		$x_1 \leq \dfrac{L}{2}$ $x_2 \leq \dfrac{L}{2}$ $y_1 \leq \dfrac{H}{2}$ $y_2 \leq \dfrac{H}{2}$	$\dfrac{3r_u x_1}{5}$ $\dfrac{3r_u x_2}{5}$ $\dfrac{3r_u x\,(2H - y_1 - y_2)}{6H - y_1 - y_2}$ $\dfrac{3r_u(L - x)(2H - y_1 - y_2)}{6H - y_1 - y_2}$	$\dfrac{3r_u y(2L - x_1 - x_2)}{6L - x_1 - x_2}$ $\dfrac{3r_u(H - y)(2L - x_1 - x_2)}{6L - x_1 - x_2}$ $\dfrac{3r_u y_1}{5}$ $\dfrac{3r_u y_2}{5}$

Chapter 4
Design of Reinforced Concrete Columns

4.0 NOTATION

a_x	Deflection in column due to slenderness producing additional moment about x-axis
a_y	Deflection in column due to slenderness producing additional moment about y-axis
A_c	Net area of concrete in a column cross-section
A_{sc}	Total area of steel in a column cross-section
A_{sx}	Area of steel in tension to resist bending about x-axis
A_{sy}	Area of steel in tension to resist bending about y-axis
b	Width of rectangular column section − dimension perpendicular to y-axis
b'	Effective depth of tensile steel reinforcement resisting moment about y-axis
c	Coefficient of torsional stiffness
C	Torsional stiffness
d	Effective depth of tensile reinforcement
E_c	Modulus of elasticity of concrete
E_s	Modulus of elasticity of steel
f_y	Characteristic yield strength of steel
f_{cu}	Characteristic cube strength of concrete at 28 days
F	Coefficient for calculation of cracked section moment of inertia
G	Shear modulus
h	Overall depth of rectangular column section − dimension perpendicular to x-axis
h'	Effective depth to tensile steel reinforcement resisting moment about x-axis
h_s	Diameter to centreline of reinforcement in a circular column
h_{max}	Maximum overall dimension of a rectangular concrete section
h_{min}	Minimum overall dimension of a rectangular concrete section
I	Moment of inertia
K	Factor governing deflection of column due to slenderness
l_e	Effective height of column
l_o	Clear height of column
l_{ex}	Effective height for consideration of slenderness about x-axis
l_{ey}	Effective height for consideration of slenderness about y-axis
m	Modular ratio $= E_s/E_c$
M	Applied bending moment on a section

M_x	Applied bending moment about x-axis
M_y	Applied bending moment about y-axis
M_x'	Modified bending moment about x-axis to account for biaxial bending
M_y'	Modified bending moment about y-axis to account for biaxial bending
$M_{add\,x}$	Additional moment about x-axis due to slenderness
$M_{add\,y}$	Additional moment about the y-axis due to slenderness
N	Axial load on column
N_{uz}	Design ultimate capacity of a section subjected to axial load only
N_{bal}	Design axial load capacity of a balanced section ($=0.25f_{cu}bd$)
p	Percentage of tensile reinforcement
p'	Percentage of compressive reinforcement
p_x	Percentage of tensile reinforcement to resist moment about x-axis
p_y	Percentage of tensile reinforcement to resist moment about y-axis
T	Applied torsion
v_x	Shear stress in concrete due to bending about x-axis
v_y	Shear stress in concrete due to bending about y-axis
v_{cx}	Design concrete shear stress in concrete due to bending about x-axis (N/mm^2)
v_{cy}	Design concrete shear stress in concrete due to bending about y-axis (N/mm^2)
V_x	Shear force in concrete column due to bending about x-axis
V_y	Shear force in concrete column due to bending about y-axis
β	Coefficient to determine effective height of a column
β	Coefficient to determine modified bending moments in biaxial bending
ϕ	Diameter of reinforcing bar or equivalent diameter of a group of bars

4.1   ANALYSIS OF COLUMNS

4.1.1   Moment of inertia   See Section 2.1.3.

4.1.2   Modulus of elasticity   See Section 2.1.4.

4.1.3   Shear modulus   See Section 2.1.6.

Note: In normal framed construction Torsional Rigidity of RC columns may be ignored in the analysis and the torsional stiffness may be given a very small value in the computer analysis. Torsional rigidity becomes important only where torsion is relied on to carry the load as in curved beams.

4.1.4   Poisson's ratio   See Section 2.1.7.

4.1.5   Shear area   See Section 2.1.8.

4.1.6   Thermal strain   See Section 2.1.9.

4.1.7 Effective heights

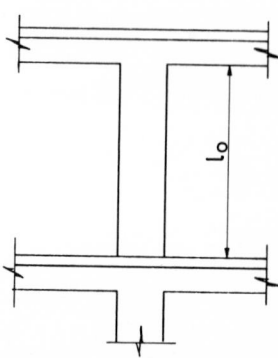

SK 4/1 Effective height of column.

Braced: All horizontal loads carried by shear walls or bracing system.
Unbraced: Horizontal loads carried by columns as parts of frame structure.

$$l_e = \beta l_o$$

where l_e = effective height
$\quad\quad l_o$ = clear height
$\quad\quad \beta$ = values given in Tables 4.1 and 4.2.

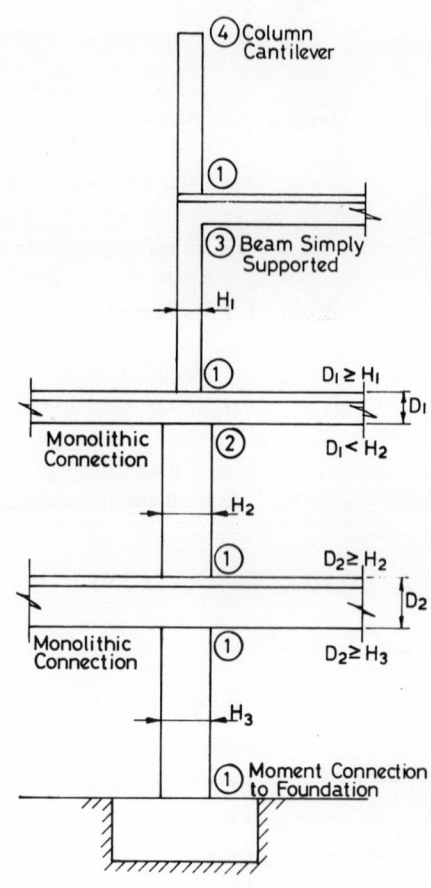

SK 4/2 Column end conditions.

Table 4.1 Values of β for braced columns.

End condition at top	End condition at bottom		
	1	2	3
1	0.75	0.80	0.90
2	0.80	0.85	0.95
3	0.90	0.95	1.00

Table 4.2 Values of β for unbraced columns.

End condition at top	End condition at bottom		
	1	2	3
1	1.2	1.3	1.6
2	1.3	1.5	1.8
3	1.6	1.8	—
4	2.2	—	—

Note: Foundations of columns designed to carry moments may be considered as end condition 1 for the column.

4.1.8   Analysis of columns

Find the following internal forces by analysis:

(1) Bending moments about principal axes: M_x and M_y
(2) Shear forces about principal axes: V_x and V_y
(3) Deflections at critical points: δ
(4) Rotations at joints (if required): θ
(5) Torsions (if relevant): T
(6) Direct axial loads: N

Use a general-purpose 2-D or 3-D skeletal member suite of a computer software for the analysis, if required.

4.2   LOAD COMBINATIONS

4.2.1   General rules

The following load combinations and partial load factors should be used in carrying out the analysis of columns:

LC_1:	$1.4DL + 1.6LL + 1.4EP + 1.4WP$
LC_2:	$1.0DL + 1.4EP + 1.4WP$
LC_3:	$1.4DL + 1.4WL + 1.4EP + 1.4WP$
LC_4:	$1.0DL + 1.4WL + 1.4EP + 1.4WP$
LC_5:	$1.2DL + 1.2LL + 1.2WL + 1.2EP + 1.2WP$

Note: Load combinations LC_2 and LC_4 should be considered only when the effect of dead and live load are considered to be beneficial.

where DL = dead load
$\qquad LL$ = live load or imposed load
$\qquad WL$ = wind load
$\qquad WP$ = water pressure
$\qquad EP$ = earth pressure.

The general principle of load combination is to leave out the loads which have beneficial effect. If the load is of a permanent nature, like dead load, earth load or water load, use the partial load factor of 1 for that load which produces a beneficial rather than adverse effect. This rule of combination will be used for design as well as for the check of stability of structure.

Note: No reduction or redistribution of loads is allowed from the columns.

4.2.2   Exceptional loads   See Section 2.2.4.

4.3   STEP-BY-STEP DESIGN PROCEDURE FOR COLUMNS

4.3.1   Rectangular columns

Step 1   Analysis

Moments, shear forces and axial forces should be determined manually or using computer software. Additional moments induced by deflection of slender columns are found in Step 5. For braced columns which are assumed to carry vertical loads only, a nominal eccentricity of vertical loads equal to 0.05 times the overall dimension in the plane of bending not exceeding 20 mm should be considered. For biaxial bending, minimum eccentricity should be considered about one axis at a time.

Step 2   Check slenderness of column

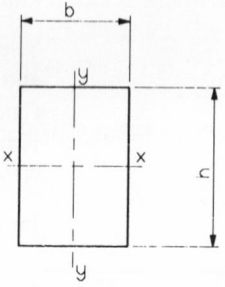

SK 4/3 Section through a column.

Find l_{ex}/h and l_{ey}/b.

See Section 4.1.7 for the determination of effective heights l_{ex} and l_{ey}.

Note: For short columns both ratios should be less than 15 for braced and 10 for unbraced.

For columns generally, $l_o \leq 60b$

For cantilever columns, $l_o \leq 100b^2/h \leq 60b$

Step 3 *Determination of cover*
Determine cover required to reinforcement, as per Tables 11.6 and 11.7.

Step 4 *Design of short columns*

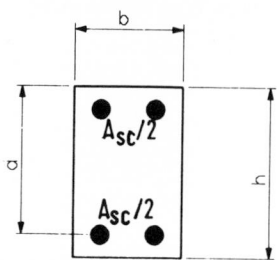

SK 4/4 Symmetrically reinforced column.

(1) No moment from analysis
Select reinforcement size and number.

Find $N = 0.4f_{cu}A_c + 0.75A_{sc}f_y$

where A_c = net area of concrete = $bh - A_{sc}$

Check $N >$ applied direct load

(2) Column supporting continuous beams where analysis does not allow for framing into columns (no moment in column)

Find $N = 0.35f_{cu}A_c + 0.67A_{sc}f_y$

Check $N >$ applied direct load

(3) Column subjected to uniaxial moment and direct load
Determine d/h corresponding to cover found in step 3.

Find $e = M/N$ and then e/h.

Select appropriate Table from Tables 11.8 to 11.17 corresponding to f_{cu}, and d/h.
Calculate N/bh.
Find from appropriate Table the value of p which satisfies the calculated N/bh against the e/h due to applied moment. From p calculate A_{sc}.

Find A_{sc}.

Note: For symmetrically reinforced columns as designed above, the total area of steel should be divided by 2 and placed at the two opposite faces of the column in relation to the axis about which the moment is applied. More reinforcement may be necessary at the other two faces from other considerations. The total percentage of reinforcement should be below 6%.

Step 5   Design of slender columns

Table 4.3 Summary of column additional moments.

Column type	Bending about major axis only	Bending about minor axis only	Bending about both axes
Braced $15 < \dfrac{l_{ex}}{h} \le 20$ $15 < \dfrac{l_{ey}}{b} \le 20$ $\dfrac{h}{b} < 3$	$a_{ux} = \beta_a K h$ $\beta_{ax} = \dfrac{1}{2000}\left(\dfrac{l_{ex}}{b}\right)^2$ $M_{addx} = N a_u$ $M_x = M_{xi} + {}^*M_{addx}$ $K = \dfrac{N_{uz} - N}{N_{uz} - N_{bal}} \le 1$	$a_{uy} = \beta_a K b$ $\beta_{ay} = \dfrac{1}{2000}\left(\dfrac{l_{ey}}{b}\right)^2$ $M_{addy} = N a_{uy}$ $M_y = M_{yi} + {}^*M_{addy}$	$a_{ux} = \beta_{ax} K h$ $\beta_{ax} = \dfrac{1}{2000}\left(\dfrac{l_{ex}}{h}\right)^2$ $M_{addx} = N a_{ux}$ $M_x = M_{xi} + {}^*M_{addx}$
Unbraced $10 < \dfrac{l_{ex}}{h} \le 20$ $10 < \dfrac{l_{ey}}{b} \le 20$ $\dfrac{h}{b} < 3$	$N_{uz} = 0.45 f_{cu} A_c + 0.87 f_y A_{sc}$ $N_{bal} = 0.25 f_{cu} b d$		$a_y = \beta_{ay} K b$ $\beta_{ay} = \dfrac{1}{2000}\left(\dfrac{l_{ey}}{b}\right)^2$ $M_{addy} = N a_{uy}$ $M_y = M_{yi} + {}^*M_{addy}$
Braced and unbraced $20 < \dfrac{l_{ex}}{h}$ and/or $20 < \dfrac{l_{ey}}{b}$ and/or $\dfrac{h}{b} \ge 3$	$a_{ux} = \beta_{ax} K h$ $\beta_{ax} = \dfrac{1}{2000}\left(\dfrac{l_{ex}}{b}\right)^2$ $M_{addx} = N a_{ux}$ $M_x = M_{xi} + {}^*M_{addx}$ $a_{uy} = \beta_{ay} K b$ $\beta_{ay} = \dfrac{1}{2000}\left(\dfrac{l_{ey}}{b}\right)^2$ $M_{addy} = N a_{uy}$ $M_y = M_{addy}$	$a_{ux} = \beta_{ax} K h$ $\beta_{ax} = \dfrac{1}{2000}\left(\dfrac{l_{ex}}{h}\right)^2$ $M_{addx} = N a_{ux}$ $M_x = M_{addx}$ $a_{uy} = \beta_{ay} K b$ $\beta_{ay} = \dfrac{1}{2000}\left(\dfrac{l_{ey}}{b}\right)^2$ $M_{addy} = N a_{uy}$ $M_y = M_{yi} + {}^*M_{addy}$	$a_{ux} = \beta_{ax} K h$ $\beta_{ax} = \dfrac{1}{2000}\left(\dfrac{l_{ex}}{h}\right)^2$ $M_{addx} = N a_{ux}$ $M_x = M_{xi} + {}^*M_{addx}$ $a_y = \beta_{ay} K b$ $\beta_{ay} = \dfrac{1}{2000}\left(\dfrac{l_{ey}}{b}\right)^2$ $M_{addy} = N a_{uy}$ $M_y = M_{yi} + {}^*M_{addy}$

* The addition of M_{add} will be done following sketches SK4/5 and SK4/6 as appropriate. M_{xi} is the initial moment and M_x is the final moment about x-axis. M_{addx} is the additional moment due to slenderness.

For unbraced columns at any storey find a_u for all columns in any orthogonal direction and then find a_{uav} given by

$$a_{uav} = \frac{\Sigma a_u}{n} \text{ where } n = \text{number of columns.}$$

Find additional moment for all columns using a_{uav} as deflection.
If any value of a_u for any individual column at a level is twice a_{uav}, then discard that column from the calculation of a_{uav}.

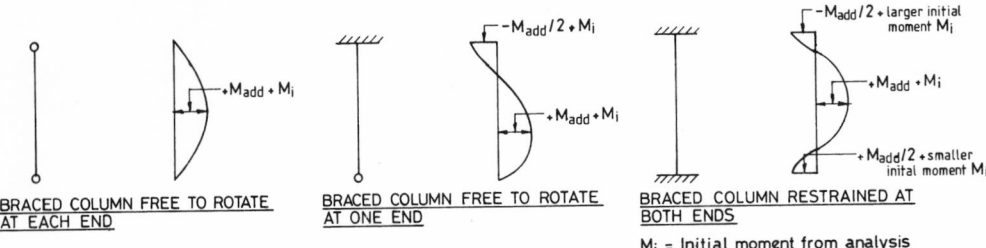

SK 4/5 Braced column − additional moments.

SK 4/6 Unbraced column − additional moments.

Braced column restrained at both ends:

the initial moment at mid height $M_i = 0.4\,M_1 + 0.6\,M_2 \geqslant 0.4\,M_2$

where M_1 = smaller end moment
$\quad\quad\; M_2$ = larger end moment

Unbraced column restrained at both ends:

the full additional moment may be combined with the initial end moment of stiffer joint. M_{add} for the other end may be reduced proportional to the joint stiffness.

Determine d/h corresponding to cover found in Step 3.
Find $e = M/N$ and then e/h.

Select appropriate Table from Tables 11.8 to 11.17 corresponding to f_{cu} and d/h.
Calculate N/bh.
Find from the appropriate Table the value of p which satisfies the calculated N/bh against the e/h due to applied moment. From p calculate A_{sc}.

See note in Step 4.

Step 6 *Design of column to biaxial bending and direct load*
Select diameter of reinforcement:

Find h' and b'.
Find M_x/h' and M_y/b'.

If $M_x/h' > M_y/b'$,

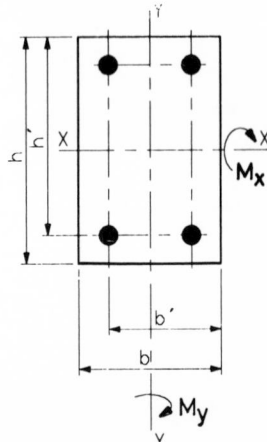

SK 4/7 Column subject to biaxial bending.

$$M'_x = M_x + \left(\frac{\beta h'}{b'}\right) M_y$$

If $M_y/b' > M_x/h'$,

$$M'_y = M_y + \left(\frac{\beta b'}{h'}\right) M_x$$

Find $N/f_{cu}bh$.

Values of β are given in the table below.

$N/f_{cu}bh$	0	0.1	0.2	0.3	0.4	0.5	>0.6
β	1.00	0.88	0.77	0.65	0.53	0.42	0.30

Note: Biaxial bending is reduced to uniaxial bending by the multiplier β.

Design as uniaxial bending, depending on which directional bending is predominant.

Find A_{sc} following the method in Step 5.

See note in Step 4.

Step 7 *Check shear stress*

Find design shear forces V_x and V_y from analysis.

Find M_x/N and M_y/N.

(1) If $M_x/N \leq 0.60h$ and $M_y/N \leq 0.60b$

$$V_x/bh' \leq 0.8\sqrt{f_{cu}} \leq 5\,\text{N/mm}^2$$
$$\text{and}\quad V_y/b'h \leq 0.8\sqrt{f_{cu}} \leq 5\,\text{N/mm}^2$$

No shear check is necessary.

(2) If M$_x$/N > 0.60h   and/or   M$_y$/N > 0.60b

Find $v_x = V_x/bh'$
$\quad\quad v_y = V_y/b'h$

$$p_x = \frac{100A_{sx}}{bh'}$$

$$p_y = \frac{100A_{sy}}{hb'}$$

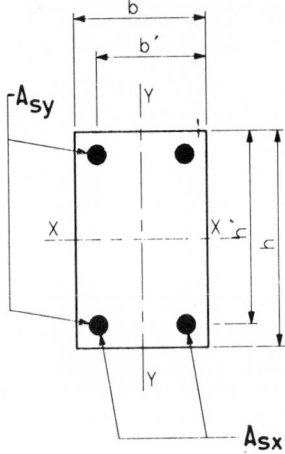

SK 4/8 Areas of steel for shear check of column.

From Figs 11.2 to 11.6, find v_{cx} and v_{cy} i.e. the design concrete shear stresses corresponding to p_x and p_y.
Modify v_{cx} and v_{cy} to take into account axial loading.

$$v'_{cx} = v_{cx} + \frac{0.6NV_xh}{A_cM_x}$$

$$v'_{cy} = v_{cy} + \frac{0.6NV_yb}{A_cM_y}$$

Note: N is +ve for compression and −ve for tension. V_xh/M_x and V_yb/M_y should not be greater than 1. Check: $(v_x/v'_{cx}) + (v_y/v'_{cy}) \leq 1$.
If this condition is not satisfied, then shear reinforcement in the form of links is required.

Design of shear reinforcement for columns

$$v''_{cx} = \frac{v'_{cx}v_x}{v_x + v_y}$$

$$v''_{cy} = \frac{v'_{cy}v_y}{v_x + v_y}$$

where v''_{cx} = available concrete shear strength for calculation of shear reinforcement for bending about x-axis
$\quad\quad\quad v''_{cy}$ = available concrete shear strength for calculation of shear reinforcement for bending about y-axis.

Note: To avoid shear cracking prior to ultimate limit state, modification of the design concrete shear stress to account for direct load should be according to the following formula: $v_c' = v_c[1 + N/(A_c v_c)]^{\frac{1}{2}}$.

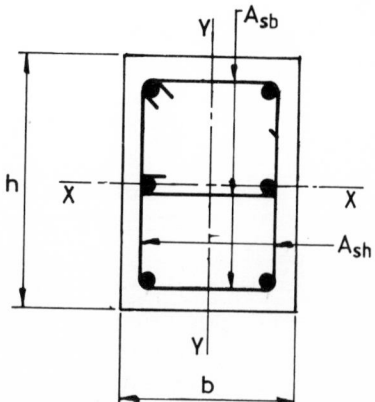

SK 4/9 Shear reinforcement in column section.

$$V_{cx}' = v_{cx}'' b h'$$
$$V_{cy}' = v_{cy}'' b' h$$

$$V_{sx} = \frac{0.87 f_{yv} A_{sh} h'}{S}$$

$$V_{sy} = \frac{0.87 f_{yv} A_{sb} b'}{S}$$

where f_{yv} = characteristic yield strength of link reinforcement
A_{sh} = area of all legs of link reinforcement in one set resisting shear due to bending about x-axis
A_{sb} = area of all legs of link reinforcement in one set resisting shear due to bending about y-axis
S = spacing of a set of link in the column.

Check $V_{sx} \geq V_x - V_{cx}'$ and $V_{sy} \geq V_y - V_{cy}'$

4.3.2 Circular columns

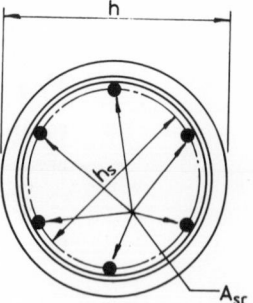

SK 4/10 Circular column – typical section with minimum of six bars.

Step 1   Analysis
Carry out analysis.

Step 2   Check slenderness of column
Find l_e/h, when h = diameter.
See Section 4.1.7 for the determination of effective height l_e.

Note: For short columns, the ratio l_e/h should be less than 15 for braced and 10 for unbraced.

Step 3   Determination of cover
Determine cover required to reinforcement, as per Tables 11.6 and 11.7.

Step 4   Design of short columns

(1) No significant moment from analysis
Select reinforcement size and at least six bars.

Find $A_c = 0.25 \pi h^2 - A_{sc}$
Find $N = 0.4 f_{cu} A_c + 0.75 A_{sc} f_y$

Check $N \geq$ applied direct load

(2) Column supporting continuous beams or flat slab where analysis does not allow for distribution of moment to the column

Find $N = 0.35 f_{cu} A_c + 0.67 A_{sc} f_y$

Check $N \geq$ applied direct load

(3) Column subjected to moment and direct load
Determine h_s/h corresponding to cover found in Step 3.
Find $e = M/N$ and then e/R, where R = radius of column.
Select appropriate table from Tables 11.18 to 11.27 corresponding to f_{cu} and h_s/h.
Calculate N/R^2.
Find, from the appropriate table, the value of p which satisfies the calculated N/R^2 against the e/R due to applied moment M.
Find A_{sc} from p and use at least six bars.

Step 5   Design of slender columns
$$a = \frac{l_e^2}{2000h}$$

$$M_{add} = NaK$$

Combine this additional moment, M_{add}, with the moments obtained from analysis following the figures of Step 5 (Section 4.3.1), assuming $K = 1$ for conservatism.

Otherwise $N_{bal} = 0.15 f_{cu} h^2$

$$N_{uz} = 0.45 f_{cu} A_c + 0.87 f_y A_{sc}$$

and $K = \dfrac{N_{uz} - N}{N_{uz} - N_{bal}}$

which may be found by iteration using successive assumptions of A_{sc}. Design the column for the combined moment M and direct load N following Step 4.

Step 6 Biaxial moment and direct load

If biaxial moments are present by analysis on the column, combine these two orthogonal moments by taking the square root of the sum of the squares and then adding M_{add} to the combined moment.

Design the column for the combined moment M and the direct load N following Step 4. $M = \sqrt{(M_x^2 + M_y^2)}$

Step 7 Check shear stress

Find design shear forces V_x and V_y from analysis.
Find M/N, where $M = \sqrt{(M_x^2 + M_y^2)}$.
Find $V = \sqrt{(V_x^2 + V_y^2)}$

(1) If M/N ≤ 0.60h

$V/0.75A_c \le 0.8 \sqrt{f_{cu}} \le 5 \, N/mm^2$

No shear check is necessary.

(2) If M/N > 0.60h, check shear stress

$v = \dfrac{V}{0.75A_c} \le 0.8 \sqrt{f_{cu}} \le 5 \, N/mm^2$

$p = \dfrac{50A_{sc}}{0.75A_c} = \dfrac{66.7A_{sc}}{A_c}$

Assuming only 50% of the total reinforcement is effective in tension.

Find v_c corresponding to p and f_{cu} from Figs 11.2 to 11.5.

$v_c' = v_c + \dfrac{0.6NVh}{A_cM}$

If $v \le v_c'$, no shear reinforcement is necessary.

When $v > v_c'$, find $V_c = 0.75v_c'A_c$.
$V_s = 0.87f_{yv}A_v(z/S)$ from truss analogy (see Section 1.6.1).
Find z/R from appropriate table from Tables 11.18 to 11.27 corresponding to f_{cu}, h_s/h, p, N/R^2 and e/R.

$2A_s = A_v$

where f_{yv} = characteristic yield strength of link reinforcement
A_s = area of the link reinforcement in the form of hoop
S = spacing of link.

Check $V_s \ge V - V_c$

See note in Step 7 of Section 4.3.1

4.3.3 Rectangular and circular columns

Step 8 *Minimum reinforcement*
For rectangular and circular columns,

$$\frac{100A_{sc}}{A_c} \geq 0.4$$

Step 9 *Maximum reinforcement*
For rectangular and circular columns,

vertically cast columns $\qquad \dfrac{100A_{sc}}{A_c} \leq 6$

horizontally cast columns $\qquad \dfrac{100A_{sc}}{A_c} \leq 8$

at laps of columns $\qquad \dfrac{100A_{sc}}{A_c} \leq 10$

Step 10 *Containment of reinforcement*

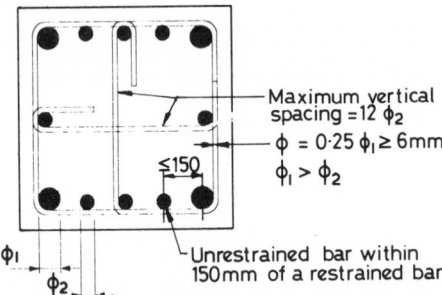

Maximum vertical spacing $= 12\,\phi_2$
$\phi = 0.25\,\phi_1 \geq 6\,\text{mm}$
$\phi_1 > \phi_2$

ϕ_1
ϕ_2

Unrestrained bar within 150mm of a restrained bar

SK 4/11 Typical arrangement of bars in a column section.

Minimum diameter of links = 0.25 times largest bar diameter $\geq 6\,\text{mm}$

Maximum spacing of links = 12 times smallest diameter of bar

Typical arrangement of bars is shown in SK4/11.

Step 11 *Check crack width (optional)*
No checks are necessary if applied ultimate load $\geq 0.2 f_{cu} A_c$

4.4 WORKED EXAMPLES

Example 4.1 *Design of a biaxially loaded slender column*
The column is braced in the $X-X$ direction, i.e. for bending about $Y-Y$ axis, and unbraced in the $Y-Y$ direction, i.e. for bending about $X-X$ axis.
Size of column: 400×600
Clear height of column $= 8\,\text{m}$.

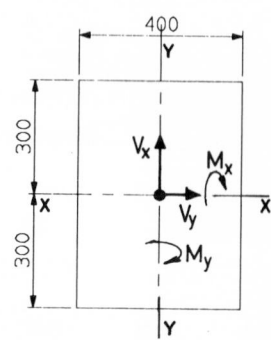

SK 4/12 Biaxially loaded column
section.

Beam size in the major direction = 400×500 at each floor.
Beam size in the minor direction = 300×350 at each floor.
Direct load on column = 2500 kN = N
Bending moment, $M_x = 150$ kNm $V_x = 150$ kN
Bending moment, $M_y = 80$ kNm $V_y = 80$ kN
All columns are of same size at each floor level.

Step 1   Analysis
Not required.

Step 2   Check slenderness of column (see Tables 4.1 and 4.2)
Effective height, $l_{ex} = 1.80 \times l_o$
$= 1.80 \times 8$
$= 14.4$ m for unbraced column

Assume end condition 2 at bottom and 3 at top for bending about x axis.

Effective height, $l_{ey} = 1.0 \times 8$
$= 8$ m for braced column

Assume end condition 3 at both top and bottom for bending about y axis.

$$\frac{l_{ex}}{h} = \frac{14.4}{0.6} = 24 > 10 \text{ for unbraced}$$

$$\frac{l_{ey}}{b} = \frac{8.0}{0.4} = 20 > 15 \text{ for braced}$$

Hence the column should be designed as slender about both axes.

Step 3   Determination of cover
Grade of concrete = 40 N/mm²
Exposure = moderate
Fire resistance = 2 hours
MSA = 20 mm
Minimum nominal cover = 30 mm, from Tables 11.6 and 11.7
Diameter of link = 10 mm assumed
Diameter of main bars = 40 mm assumed

$h' = h -$ cover $-$ dia. of link $- \frac{1}{2}$dia. of bar
$= 600 - 30 - 10 - 20$
$= 540\,\text{mm}$

$b' = 400 - 30 - 10 - 20$
$= 340\,\text{mm}$

Step 4   Design of short columns
Not required.

Step 5   Design of slender columns
Assume $100A_{sc}/bh = 5$

$A_c =$ net concrete area $= (1 - 0.05)bh = 0.95bh$

$$
\begin{aligned}
N_{uz} &= 0.45 f_{cu}A_c + 0.87 f_y A_{sc} \\
&= (0.95 \times 0.45 \times 40 + 0.87 \times 460 \times 0.05) \times 400 \times 600 \times 10^{-3} \\
&= 8906\,\text{kN}
\end{aligned}
$$

$$
\begin{aligned}
N_{bal} &= 0.25 f_{cu}bh \\
&= 0.25 \times 40 \times 400 \times 600 \times 10^{-3} \\
&= 2400\,\text{kN}
\end{aligned}
$$

$$
K = \frac{N_{uz} - N}{N_{uz} - N_{bal}} = \frac{8906 - 2500}{8906 - 2400}
$$

$= 0.98$ for assumed 5% reinforcement

$$
\begin{aligned}
a_x &= \frac{1}{2000}\left(\frac{l_{ex}}{b}\right)^2 hK \\
&= \frac{1}{2000} \times \left(\frac{14\,400}{600}\right)^2 \times 600 \times 0.98 \\
&= 169.3\,\text{mm}
\end{aligned}
$$

$$
\begin{aligned}
a_y &= \frac{1}{2000}\left(\frac{l_{ey}}{h}\right)^2 bK \\
&= \frac{1}{2000} \times \left(\frac{8000}{400}\right)^2 \times 400 \times 0.98 \\
&= 78.4\,\text{mm}
\end{aligned}
$$

$$
\begin{aligned}
M_{add\,x} &= Na_xK \\
&= 2500 \times 0.1693 \\
&= 423\,\text{kNm}
\end{aligned}
$$

$$
\begin{aligned}
M_{add\,y} &= Na_yK \\
&= 2500 \times 0.0784 \\
&= 196\,\text{kNm}
\end{aligned}
$$

Step 6   Biaxial moment and direct load
$M_x = 150 + 423 = 573\,\text{kNm}$ (see SK 4/6 — column free to rotate one end.)

$M_y = 80 + 196 = 276\,\text{kNm}$ (see SK 4/5 — column free to rotate both ends.)

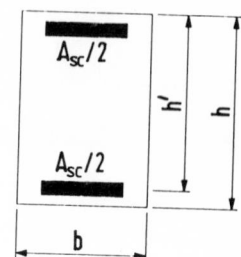

SK 4/13 Equivalent uniaxial
bending of columns.

$$\frac{M_x}{h'} = \frac{573}{0.54} = 1061 \text{ kN}$$

$$\frac{M_y}{b'} = \frac{276}{0.34} = 812 \text{ kN}$$

$$\frac{N}{bhf_{cu}} = \frac{2500 \times 10^3}{400 \times 600 \times 40} = 0.26$$

$\beta = 0.70$ from table in Step 6 of Section 4.3.1.

Biaxial bending: $M_x/h' > M_y/b'$

$$M'_x = M_x + \beta\left(\frac{h'}{b'}\right)M_y$$

$$= 573 + 0.70 \times \left(\frac{540}{340}\right) \times 276$$

$$= 880 \text{ kNm}$$

$$k = \frac{h'}{h} = \frac{540}{600} = 0.90 \qquad e = \frac{M}{N} = \frac{880}{2500} = 0.352 \text{ m}$$

$$\frac{e}{h} = \frac{0.352}{0.600} = 0.59$$

$$\frac{N}{bh} = \frac{2500 \times 10^3}{400 \times 600} = 10.4 \text{ N/mm}^2$$

Select Table 11.12 for $f_{cu} = 40 \text{ N/mm}^2$ and $k = 0.90$.
From Table 11.12: for $e/h = 0.6$ and $p = 2.0$, $N/bh = 9.05$, and for $p = 3.0$,
$N/bh = 10.95$.
By linear interpolation, $p = 2.69$ for $e/h = 0.59$, and $N/bh = 10.4$.

$$A_{sc} = \frac{2.69 \times 400 \times 600}{100}$$

$$= 6456 \text{ mm}^2$$

Use 4 no. 32 dia. bars on each face 400 wide (6434 mm²).

See Step 5: revised $N_{uz} = 6705 \text{ kN}$ and corresponding $K = 0.98$; no change.

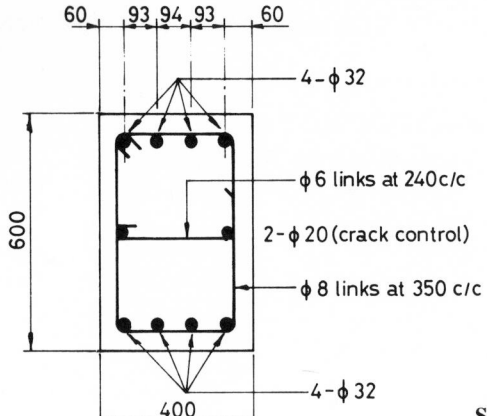

SK 4/14 Designed column section.

Step 7   Shear check

$$\frac{M_x}{N} = \frac{150}{2500}$$

$$= 0.06 \, \text{m} < 0.60h$$

$$\frac{M_y}{N} = \frac{80}{2500}$$

$$= 0.032 \, \text{m} < 0.60b$$

$$\frac{V_x}{bh'} = \frac{150 \times 10^3}{400 \times 540} = 0.69 \, \text{N/mm}^2 < 0.8\sqrt{f_{cu}} < 5 \, \text{N/mm}^2$$

$$\frac{V_y}{b'h} = \frac{80 \times 10^3}{600 \times 340} = 0.39 \, \text{N/mm}^2$$

No shear check is necessary.

Step 8   Minimum reinforcement
Minimum reinforcement = 0.4%   satisfied

Step 9   Maximum reinforcement
Maximum reinforcement = 6%   satisfied

Step 10   Containment of reinforcement
Minimum diameter of link = 0.25 × 32
= 8 mm

Maximum spacing of links = 12 × smallest bar diameter
= 12 × 32 = 384 mm

Step 11   Check crack width
$N = 2500 \, \text{kN} > 0.2 f_{cu} A_c = 1920 \, \text{kN}$
So no check necessary.

Example 4.2 *Design of a column with predominant moment about the major axis*

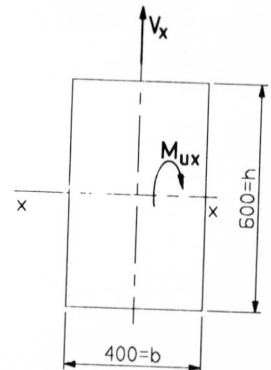

SK 4/15 Column with moment
about X–X axis.

Rectangular section.

$h = 600\,\text{mm}$ $b = 400\,\text{mm}$

Ultimate bending moment, $M_{ux} = 640\,\text{kNm}$

Ultimate direct load, $N_u = 1280\,\text{kN}$

Ultimate shear force, $V_x = 320\,\text{kN}$

Service bending moment, $M_{sx} = 400\,\text{kNm}$

Service direct load, $N_s = 800\,\text{kN}$

Clear height of column $= 4\,\text{m}$ between floors

End condition (1) at both ends of column in both directions of bending. Unbraced column in both directions of bending.

Step 1 *Analysis*
Not required.

Note: Minimum eccentricity $= 20\,\text{mm}$

$M_{uy} = 20 \times 1280\,\text{kNmm}$
$\qquad = 25.6\,\text{kNm}$

By inspection this moment in isolation will not cause a more onerous design than the predominant moment M_{ux}.

Step 2 *Check slenderness of column (see Table 4.2)*
Effective height, $l_{ex} = 1.2 \times 4 = 4.8\,\text{m}$
$\qquad\qquad\qquad\quad l_{ey} = 1.2 \times 4 = 4.8\,\text{m}$

$$\frac{l_{ex}}{h} = \frac{4.8}{0.6} = 8 < 10$$

$$\frac{l_{ey}}{b} = \frac{4.8}{0.4} = 12 > 10$$

The column is slender about minor axis.

Step 3   Determination of cover

Grade of concrete $= 40 \, \text{N/mm}^2$

Exposure = severe

Fire resistance = 2 hours

Maximum size of aggregates = 20 mm

Minimum nominal cover = 30 mm

Diameter of link = 10 mm   assumed

Diameter of main bars = 25 mm   assumed

$$d = h' = h - \text{cover} - \text{dia. of link} - \tfrac{1}{2}\text{dia. of bar}$$
$$= 600 - 40 - 10 - 12.5$$
$$= 537.5 \, \text{mm}$$

$$b' = 400 - 40 - 10 - 12.5$$
$$= 337.5 \, \text{mm}$$

Step 4   Design of short columns

Not required.

Step 5   Design of slender columns

$$\frac{h}{b} = 1.5 < 3$$

$$\frac{l_{ex}}{h} = 8 < 20$$

Additional moment about minor axis can be ignored (see Table 4.3).

$$a_x = \frac{1}{2000}\left(\frac{l_{ex}}{b}\right)^2 hK$$

$$= \frac{1}{2000} \times \left(\frac{4800}{400}\right)^2 \times 600 \times 1 \qquad \text{(assume } K = 1 \text{ for conservatism)}$$

$$= 43.2 \, \text{mm}$$

$$M_{add\,x} = Na_x$$
$$= 1280 \times 0.0432$$
$$= 55.3 \, \text{kNm}$$

$$M_x = 640 + 55.3 \quad \text{(see SK 4/6 - column restrained at both ends)}$$
$$= 695.3 \, \text{kNm}$$

Design as a beam following Step 10 of Section 2.3.

$$M_d = M + N\left(\frac{h}{2} - d_1\right)$$

$$= 695.3 + 1280\left(\frac{0.6}{2} - 0.0625\right)$$

$$= 999.3 \, \text{kNm}$$

$$K = \frac{M_d}{f_{cu}bd^2} = \frac{999.3 \times 10^6}{40 \times 400 \times 537.5^2}$$

$$= 0.216 > 0.156$$

Compression reinforcement is required.

$$z = 0.775d$$
$$= 444 \, \text{mm}$$

$$A_s' = \frac{(K - 0.156)f_{cu}bd^2}{0.87f_y(d - d')}$$

$$= \frac{(0.216 - 0.156) \times 40 \times 400 \times 537.5^2}{0.87 \times 460 \times (537.5 - 62.5)}$$

$$= 1459 \, \text{mm}^2$$

$$A_s = \left(\frac{0.156f_{cu}bd^2}{0.87f_yz}\right) + A_s' - \frac{N}{0.87f_y}$$

$$= \left(\frac{0.156 \times 40 \times 400 \times 537.5^2}{0.87 \times 460 \times 444}\right) + 1459 - \left(\frac{1280 \times 10^3}{0.87 \times 460}\right)$$

$$= 2319 \, \text{mm}^2$$

Use 3 no. 32 mm dia. bars each face (2412 mm²)

Design by using Table 11.12.

$$e = M/N = 0.543 \qquad e/h = 0.905$$

$$k = \frac{h'}{h} = \frac{537.5}{600} = 0.90$$

$$\frac{N}{bh} = \frac{1280 \times 10^3}{400 \times 600} = 5.33 \, \text{N/mm}^2$$

From Table 11.12 by linear interpolation, $p = 2\%$.

$$A_{sc} = \frac{2 \times 400 \times 600}{100} = 4800 \, \text{mm}^2$$

Use 3 no. 32 dia. bars on each face (2412 mm²).

Note: The two different design methods produce exactly the same result.

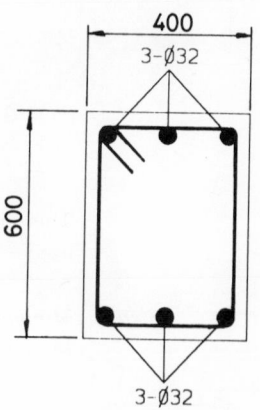

SK 4/16 Designed column section.

Step 6   *Biaxial moment and direct load*
Not required.

Step 7   *Check shear stress*

$$\frac{M_x}{N} = \frac{640}{1280} = 0.5 > 0.60h = 0.36\,\mathrm{m}$$

Shear check is required.

$$v = \frac{V_x}{h'b} = \frac{320 \times 10^3}{400 \times 536}$$

$$= 1.49\,\mathrm{N/mm^2} < 5\,\mathrm{N/mm^2}$$

$$h' = 600 - 40 - 8 - 16 = 536\,\mathrm{mm}$$

$$p = \frac{100A_s}{bh'}$$

$$= \frac{100 \times 2412}{400 \times 536}$$

$$= 1.125$$

From Fig. 11.5,

$$v_c = 0.77\,\mathrm{N/mm^2}$$

$$v_c' = v_c + \frac{0.6NVh}{A_cM}$$

$$\frac{Vh}{M} = \frac{320 \times 10^3 \times 600}{640 \times 10^6} = 0.30 < 1$$

$$v_c' = 0.77 + \frac{0.60 \times 1280 \times 10^3 \times 0.3}{400 \times 600}$$

$$= 1.73\,\mathrm{N/mm^2} > 1.49\,\mathrm{N/mm^2}$$

No shear reinforcement is necessary.

To avoid shear cracks at ultimate load, use the following modification formula:

$$v_c' = v_c\left(1 + \frac{N}{A_cv_c}\right)^{\frac{1}{2}}$$

$$= 0.77\left(1 + \frac{1280 \times 10^3}{400 \times 600 \times 0.77}\right)^{\frac{1}{2}}$$

$$= 2.167\,\mathrm{N/mm^2} > 1.73\,\mathrm{N/mm^2}$$

This modified higher value of design concrete shear strength may not be used.

Step 8   *Minimum reinforcement*
Minimum reinforcement $= 0.4\%$ satisfied

Step 9   ***Maximum reinforcement***
Maximum reinforcement $= 6\%$ satisfied

Step 10   ***Containment of reinforcement***
Minimum diameter of link $= 0.25 \times 32$
$$= 8\,\text{mm}$$

Maximum spacing of links $= 12 \times$ dia. of bar
$$= 12 \times 32$$
$$= 384\,\text{mm} > 350\,\text{mm}\quad\text{OK}$$

Centre-to-centre spacing of bars $= 136\,\text{mm} < 150\,\text{mm}$

Central 32 mm diameter bar need not be restrained.
Use 2-legged links 8 mm diameter at 350 mm centres.

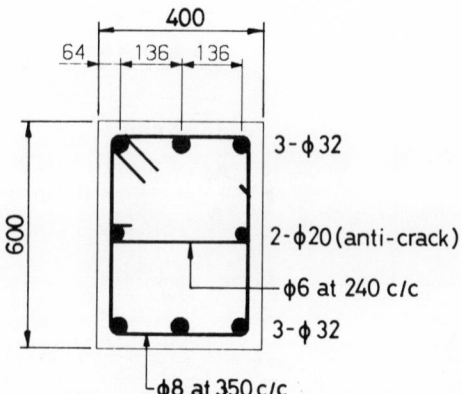

SK 4/17 Final column section.

Step 11   ***Check crack width (optional)***
$A_\text{s} = A_\text{s}' = 2412\,\text{mm}^2$

$d = 536\,\text{mm}$

$$m = \frac{E_\text{s}}{E_\text{c}} = 10$$

$d' = 64\,\text{mm}$

Service bending moment, $M_\text{sx} = 400\,\text{kNm}$

Service direct load, $N_\text{s} = 800\,\text{kN}$

The formulae used below are for a triangular concrete stress block (see Section 1.13.2).
Assume value of $x = d/2 = 260\,\text{mm}$, say.

First trial

$q_1 = bx = 400 \times 260 = 104\,000\,\text{mm}^2$

(See Section 1.13.2 for explanation of symbols.)

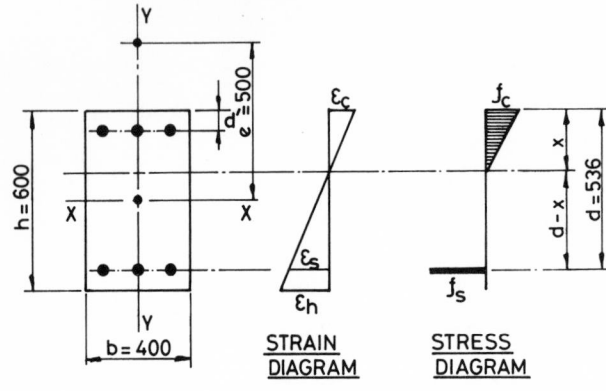

STRAIN
DIAGRAM

STRESS
DIAGRAM

SK 4/18 Calculation of crack width.

$$g = \frac{0.5q_1x + mA_sd + (m-1)A_s'd'}{q_1 + mA_s + (m-1)A_s'}$$

$$= \frac{0.5 \times 104\,000 \times 260 + (10 \times 2412 \times 536) + (9 \times 2412 \times 64)}{104\,000 + (10 \times 2412) + (9 \times 2412)}$$

$$= 185.8\,\text{mm}$$

$$e = \frac{M}{N} = \frac{400}{800} = 0.5\,\text{m} = 500\,\text{mm}$$

$$k_1 = \left(\frac{e-g}{d}\right) + 1$$

$$= \left(\frac{500 - 185.8}{536}\right) + 1$$

$$= 1.586$$

$$k_2 = \frac{x}{2d}\left(1 - \frac{x}{3d}\right)$$

$$= \left(\frac{260}{2 \times 536}\right)\left(1 - \frac{260}{3 \times 536}\right)$$

$$= 0.203$$

$$k_3 = (m-1)\left(1 - \frac{d'}{x}\right)$$

$$= 9\left(1 - \frac{64}{260}\right)$$

$$= 6.785$$

$$f_c = \frac{Nk_1}{k_2bd + k_3A_s'\left(1 - \frac{d'}{d}\right)}$$

$$= \frac{800 \times 10^3 \times 1.586}{0.203 \times 400 \times 536 + 6.785 \times 2412 \times \left(1 - \frac{64}{536}\right)}$$

$$= 21.90\,\text{N/mm}^2$$

$$f_s = \frac{f_c(0.5q_1 + k_3 A_s') - N}{A_s}$$

$$= \frac{21.90 \times (0.5 \times 104\,000 + 6.785 \times 2412) - 800 \times 10^3}{2412}$$

$$= 289.1\,\text{N/mm}^2$$

Check: $x = \dfrac{d}{1 + \left(\dfrac{f_s}{mf_c}\right)}$

$$= \frac{536}{1 + \left(\dfrac{289.1}{10 \times 21.9}\right)}$$

$$= 231\,\text{mm} < 260\,\text{mm}\quad\text{assumed}$$

Second trial

Assume $x = (260 + 231)/2 = 240\,\text{mm}$ say

$q_1 = 96\,000\,\text{mm}^2$

$g = 182.2\,\text{mm}$

$k_1 = 1.593$
$k_2 = 0.190$
$k_3 = 6.60$

$f_c = 23.27\,\text{N/mm}^2$
$f_s = 285.0\,\text{N/mm}^2$

$x = 240.8\,\text{mm}$ assumed $x = 240\,\text{mm}$, hence OK

$$\varepsilon_s = \frac{f_s}{E_s} = \frac{285}{200 \times 10^3} = 1.425 \times 10^{-3}$$

$$\varepsilon_h = \left(\frac{h - x}{d - x}\right)\varepsilon_s$$

$$= \left(\frac{600 - 240}{536 - 240}\right) \times 1.425 \times 10^{-3}$$

$$= 1.733 \times 10^{-3}$$

$$\varepsilon_{mh} = \varepsilon_h - \frac{b(h - x)^2}{3E_s A_s(d - x)}$$

$$= 1.733 \times 10^{-3} - \frac{400 \times (600 - 240)^2}{3 \times 200 \times 10^3 \times 2412 \times (536 - 240)}$$

$$= 1.612 \times 10^{-3}$$

$$a_{c1} = \sqrt{(64^2 + 64^2)} - 16$$
$$= 74.5\,\text{mm}$$

$$a_{c2} = \sqrt{(64^2 + 68^2)} - 16$$
$$= 77.4\,\text{mm}$$

$$a_{cr} = 77.4 \, \text{mm}$$

$$
\begin{aligned}
W_{cr} &= \frac{3 a_{cr} \, \varepsilon_m}{1 + 2\left(\dfrac{a_{cr} - C_{min}}{h - x}\right)} \\[2mm]
&= \frac{3 \times 77.4 \times 1.612 \times 10^{-3}}{1 + 2\left(\dfrac{77.4 - 48}{600 - 240}\right)} \\[2mm]
&= 0.32 \, \text{mm} > 0.3 \, \text{mm}
\end{aligned}
$$

Crack width slightly exceeded and may be allowed.

Example 4.3   ***Design of a member with uniaxial moment and tension***
Rectangular section.
Size: 600 mm × 400 mm

Ultimate direct load in tension = 250 kN

Ultimate bending moment, $M_x = 250 \, \text{kNm}$

Ultimate shear force, $V_x = 250 \, \text{kN}$

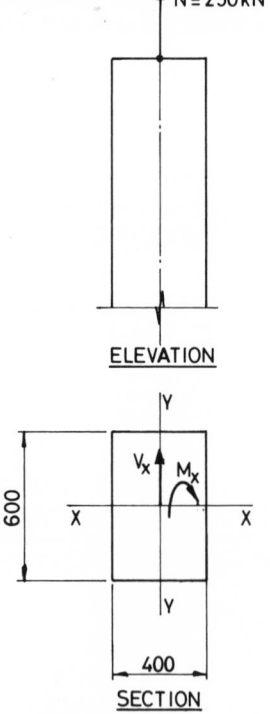

SK 4/19 Member subject to uniaxial bending and tension.

Step 1   *Analysis*
Not required.

Step 2   Check slenderness of member
Not required.

Step 3   Determination of cover
Grade of concrete $= 40\,\text{N/mm}^2$
Exposure $=$ moderate
Fire resistance required $= 1$ hour
Maximum size of aggregates $= 20\,\text{mm}$
Minimum nominal cover $= 30\,\text{mm}$ from Tables 11.6 and 11.7
Diameter of link $= 10\,\text{mm}$ assumed
Diameter of main bar $= 40\,\text{mm}$ assumed

$h' = h -$ cover $-$ dia. of link $- \frac{1}{2}$ dia. of bar
$\quad = 600 - 30 - 10 - 20$
$\quad = 540\,\text{mm}$

$b' = 400 - 30 - 10 - 20$
$\quad = 340\,\text{mm}$

Step 4   Design of short columns

Method 1   Design as RC beam (see Step 10 of Section 2.3)

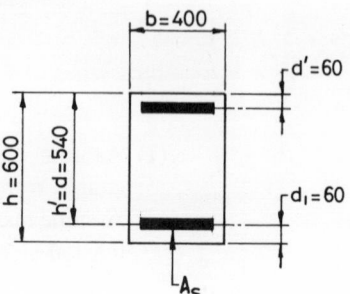

SK 4/20 Design of column section.

$M_x = 250\,\text{kNm}$

$N = -250\,\text{kN}$

$M_d = M_x - N\left(\dfrac{h}{2} - d_1\right)$

$\quad = 250 - 250 \times (0.3 - 0.06)$

$\quad = 190\,\text{kNm}$

$K = \dfrac{M_d}{f_{cu}bd^2} = \dfrac{190 \times 10^6}{40 \times 400 \times 540^2}$

$\quad = 0.04 < 0.156$ no compressive reinforcement

$z = h'\left[0.5 + \sqrt{\left(0.25 - \dfrac{K}{0.9}\right)}\right]$

$$= 540 \left[0.5 + \sqrt{\left(0.25 - \frac{0.04}{0.9} \right)} \right] \leq 0.95d$$

$$= 0.95 \times 540 = 513 \, \text{mm}$$

$$A_s = \frac{M}{0.87 f_y z} + \frac{N}{0.87 f_y}$$

$$= \frac{190 \times 10^6}{0.87 \times 513 \times 460} + \frac{250 \times 10^3}{0.87 \times 460}$$

$$= 1550 \, \text{mm}^2$$

Use 2 no. 32 dia. (1608 mm²) bars on each short face.

Method 2   Simple steel beam theory

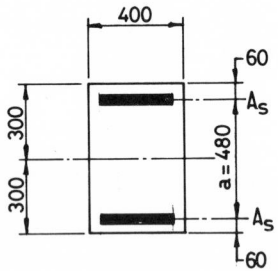

SK 4/21 Design by steel beam theory.

(1) Assume compression and tension steel in equal amount to form a couple to resist the moment.
(2) Assume axial tension carried equally by steel on compression and tension side.

Lever arm of steel (centre-to-centre distance) $= a = h' - 60 = 480 \, \text{mm}$

$$\text{Steel required for bending moment} = \frac{M}{0.87 f_y a}$$

$$= \frac{250 \times 10^6}{0.87 \times 460 \times 480} = 1301 \, \text{mm}^2$$

$$\text{Steel required for axial tension on each face} = \frac{0.5N}{0.87 f_y}$$

$$= \frac{0.5 \times 250 \times 10^3}{0.87 \times 460}$$

$$= 312 \, \text{mm}^2$$

Total steel required on each face $= 1301 + 312 = 1613 \, \text{mm}^2$

Again, 2 no. 32 dia. (1608 mm²) on each face will be adequate.

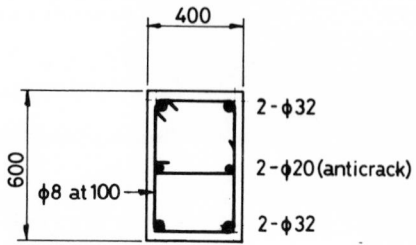

SK 4/22 Designed concrete
section.

Note: Both methods produce the same result but Method 2 is very conservative
usually.

Step 5 *Design of slender columns*
Not required.

Step 6 *Biaxial bending and direct load*
Not required.

Step 7 *Check shear stress*

$$v = \frac{V_x}{h'b} = \frac{250 \times 10^3}{540 \times 400}$$

$$= 1.16 \, \text{N/mm}^2 < 5 \, \text{N/mm}^2$$

$$p = \frac{100 A_s}{bh'} = \frac{100 \times 1608}{540 \times 400}$$

$$= 0.74\%$$

From Fig. 11.5,

$$v_c = 0.67 \, \text{N/mm}^2$$

$$v_c' = v_c + \frac{0.6 N V h}{A_c M}$$

$$\frac{Vh}{M} = \frac{250 \times 10^3 \times 600}{250 \times 10^6} = 0.60 < 1$$

$$v_c' = 0.67 - \frac{0.6 \times 0.6 \times 250 \times 10^3}{400 \times 600}$$

$$= 0.295 \, \text{N/mm}^2 < 1.16 \, \text{N/mm}^2$$

Note: *N* is −ve in tension.

Shear reinforcement is required.

$$V_c' = 0.295 \times 540 \times 400 \times 10^{-3} = 63.7 \, \text{kN}$$

Assume 8 mm diameter links ($f_y = 460\,\text{N/mm}^2$) at 100 mm centres.

$$V_s = \frac{0.87 f_{yv} A_{sh} h'}{S}$$

$$= \frac{10^{-3} \times (0.87 \times 460 \times 100 \times 540)}{100} = 216\,\text{kN}$$

$V_s > V - V_c' = 250 - 63.7 = 186.3\,\text{kN}$ okay

Step 8   **Minimum reinforcement**
Minimum reinforcement = 0.4%

Reinforcement provide = 3216 mm²

$$= \frac{3216 \times 100}{400 \times 600}$$

$$= 1.34\% \quad \text{okay}$$

Step 9   **Maximum reinforcement**
Maximum reinforcement = 6% satisfied

Step 10   **Containment of reinforcement**
Minimum diameter of link = $0.25 \times 32 = 8\,\text{mm}$ satisfied

Maximum spacing of links = 12 × dia. of bar
$$= 12 \times 32 = 384\,\text{mm} \quad \text{satisfied}$$

Step 11   **Check crack width**

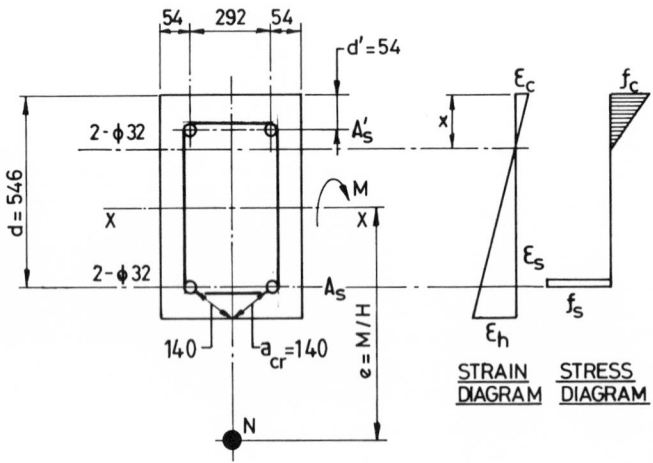

SK 4/23 Crack width calculations.

Service bending moment = 160 kNm

Service tension = 160 kN

Assume depth of neutral axis $x = h/2 = 300\,\text{mm}$

The formulae used below assume a triangular concrete stress block (see Section 1.13.2).

Assume eccentricity e from centre of stressed area, i.e. at g from extreme compressive fibre.

First trial

$$e = \frac{M}{N} = \frac{160 \times 10^3}{160} = 1000 \, mm$$

$$d' = 30 + 8 + 16 = 54 \, mm$$

$$d = 600 - 30 - 8 - 16 = 546 \, mm$$

$$\frac{x}{d} = \frac{300}{546} = 0.55$$

$$A_s = A_s' = 1608 \, mm^2$$

$$q_1 = bx = 400 \times 300 = 12 \times 10^4 \, mm^2$$

$$m = \frac{E_s}{E_c} = 10$$

$$g = \frac{0.5 q_1 x + m A_s d + (m - 1) A_s' d'}{q_1 + m A_s + (m - 1) A_s'}$$

$$= 183 \, mm$$

$$k_1 = \left(\frac{e + g}{d} \right) - 1$$

$$= 1.167$$

$$k_2 = \frac{x}{2d} \left(1 - \frac{x}{3d} \right)$$

$$= 0.224$$

$$k_3 = (m - 1) \left(1 - \frac{d'}{x} \right)$$

$$= 7.38$$

$$f_c = \frac{N k_1}{k_2 bd + k_3 A_s' \left(1 - \frac{d'}{d} \right)}$$

$$= 3.13 \, N/mm^2$$

$$f_s = \frac{f_c (0.5 q_1 + k_3 A_s') + N}{A_s}$$

$$= 239.4 \, N/mm^2$$

Check $\quad x = \dfrac{d}{1 + \left(\dfrac{f_s}{m f_c} \right)}$

$$= 62.8 \, mm < 300 \, mm \quad assumed$$

Second trial

Assume $x = 130\,\text{mm}$

$$q_1 = 52\,000\,\text{mm}^2$$

$$g = 157\,\text{mm}$$

$$k_1 = 1.119$$
$$k_2 = 0.11$$
$$k_3 = 5.26$$

$$f_c = 5.66\,\text{N/mm}^2$$
$$f_s = 221\,\text{N/mm}^2$$

$$x = 111\,\text{mm} \qquad \text{near enough to } 130\,\text{mm}$$

No more trials are required.

Tension in steel $= 221\,\text{N/mm}^2$

$x = 115\,\text{mm}$ say

$$\varepsilon_s = \frac{f_s}{E_s} = \frac{221}{200 \times 10^3} = 1.105 \times 10^{-3}$$

$$\varepsilon_h = \left(\frac{h - x}{d - x}\right)\varepsilon_s$$

$$= \left(\frac{600 - 115}{546 - 115}\right) \times 1.105 \times 10^{-3}$$

$$= 1.243 \times 10^{-3}$$

$$\varepsilon_{mh} = \varepsilon_h - \frac{b(h - x)^2}{3E_s A_s(d - x)}$$

$$= 1.016 \times 10^{-3}$$

$$a_{cr} = \sqrt{(54^2 + 146^2)} - 16$$
$$= 140\,\text{mm}$$

$$W_{cr} = \frac{3a_{cr}\,\varepsilon_m}{1 + 2\left(\dfrac{a_{cr} - c_{min}}{h - x}\right)}$$

$$= 0.29\,\text{mm} < 0.3\,\text{mm} \quad \text{OK}$$

Step 12   *Spacing of bars (required for members in tension)*
See Step 24 of Section 2.3.

$\text{MSA} + 5 = 25\,\text{mm}$

Dia. of bar $= 32\,\text{mm}$

Clear distance between bars $= 260\,\text{mm} > 32\,\text{mm}$ OK

Maximum clear spacing of bars in tension $\leq 47\,000/f_s \leq 300\,\text{mm}$

$f_s = 221\,\text{N/mm}^2$ from Step 10.

Maximum spacing $\leq 47\,000/221 \leq 213\,\text{mm}$

Note: Actual clear spacing is 260 mm which does not satisfy this condition. Since crack width calculations show that the crack of 0.3 mm may not be exceeded, this spacing of bars need not be changed.

Example 4.4   *Design of a member with biaxial moment and tension*

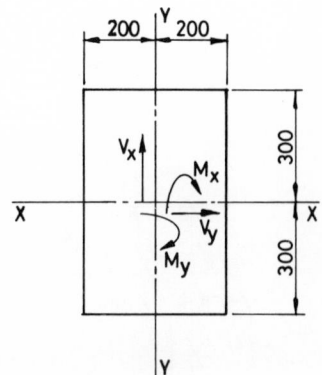

SK 4/24 Section subject to biaxial bending and tension.

Rectangular section.
Size: 600 mm × 400 mm

Ultimate direct load in tension = 250 kN

Ultimate bending moment, $M_x = 250\,\text{kNm}$

Ultimate bending moment, $M_y = 150\,\text{kNm}$

Ultimate shear force, $V_x = 250\,\text{kN}$

Ultimate shear force, $V_y = 150\,\text{kN}$

Step 1   **Analysis**
Not required.

Step 2   **Check slenderness of member**
Not required because the member is in tension.

Step 3   **Determination of cover**
Grade of concrete = 40 N/mm^2
Exposure = moderate
Fire resistance required = 1 hour
Maximum size of aggregates = 20 mm
Minimum nominal cover = 30 mm   from Tables 11.6 and 11.7
Diameter of link = 10 mm   assumed
Diameter of main bar = 40 mm   assumed

$$h' = h - \text{cover} - \text{dia. of link} - \tfrac{1}{2}\text{dia. of bar}$$
$$= 600 - 30 - 10 - 20$$
$$= 540\,\text{mm}$$

$$b' = 400 - 30 - 10 - 20$$
$$= 340\,\text{mm}$$

Step 4 Design of short columns
Not required.

Step 5 Design of slender columns
Not required.

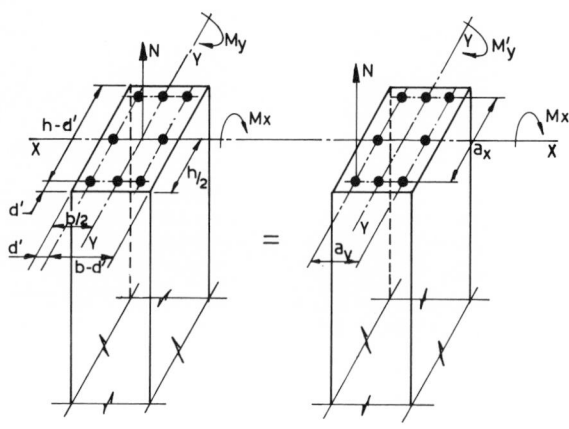

SK 4/25 Design as steel beam with
transferred tension.

Step 6 Biaxial bending and direct load

Method 1 Design as steel beam with transferred tension

$$M_x = 250\,\text{kNm}$$

$$N = -250\,\text{kN} \quad \text{(tension)}$$

$$M'_x = M_x - N\left(\frac{h}{2} - d'\right)$$
$$= 250 - 250(0.3 - 0.06)$$
$$= 190\,\text{kNm}$$

$$M'_y = M_y - N\left(\frac{b}{2} - d'\right)$$
$$= 150 - 250(0.2 - 0.6)$$
$$= 115\,\text{kNm}$$

Note: This operation means that the tension (250 kN) has been transferred to
one corner of the rectangular section.

Taking the steel beam approach, assume that the lever arm to resist bending moment about each axis is the distance between the centre of steel reinforcement on each face.

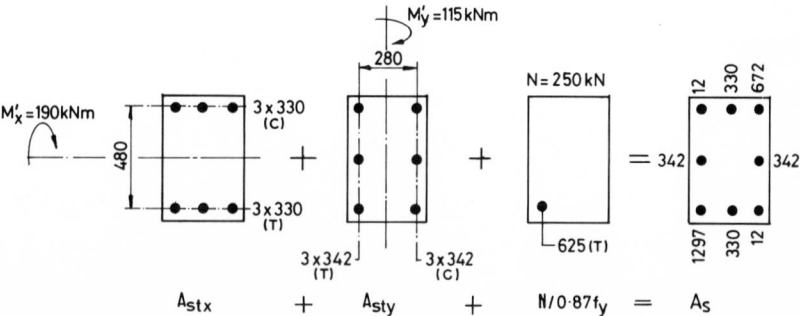

SK 4/26 Design as steel beam with transferred tension.

$$a_x = 600 - 2 \times 60 = 480 \, \text{mm}$$

$$a_y = 400 - 2 \times 60 = 280 \, \text{mm}$$

$$A_{stx} = \frac{M'_x}{0.87 f_y a_x}$$

$$= \frac{190 \times 10^6}{0.87 \times 460 \times 480}$$

$$= 989 \, \text{mm}^2$$

Assume 3 no. bars of $330 \, \text{mm}^2$ each on each short face.

$$A_{sty} = \frac{M'_y}{0.87 f_y a_y}$$

$$= \frac{115 \times 10^6}{0.87 \times 460 \times 280}$$

$$= 1026 \, \text{mm}^2$$

Assume 3 no. bars of $342 \, \text{mm}^2$ each on each long face.

Area of bar required at a corner of the member due to the transferred tension

$$= \frac{N}{0.87 f_y}$$

$$= \frac{250 \times 10^3}{0.87 \times 460} = 625 \, \text{mm}^2$$

Total area of bar required in one corner $= 330 + 342 + 625$
$$= 1297 \, \text{mm}^2$$

One no. 40 diameter bar at each corner ($1257\,\text{mm}^2$) with 1 no. 25 diameter bar at the centre of each face ($491\,\text{mm}^2$ each bar) will be adequate because $491\,\text{mm}^2$ is greater than $330\,\text{mm}^2$ or $342\,\text{mm}^2$ found before.

Method 2 Design as steel beam without transferred tension

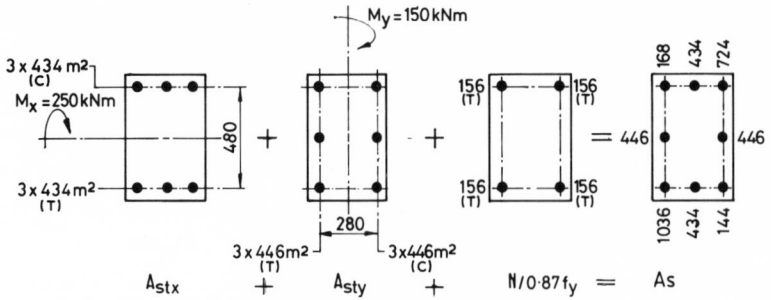

SK 4/27 Design as steel beam without transferred tension.

$a_x = 480\,\text{mm}$ as before

$a_y = 280\,\text{mm}$ as before

$$A_{\text{st}x} = \frac{M_x}{0.87 f_y a_x}$$

$$= \frac{250 \times 10^6}{0.87 \times 460 \times 480}$$

$$= 1301\,\text{mm}^2$$

Assume 3 no. bars of $434\,\text{mm}^2$ each on each short face.

$$A_{\text{st}y} = \frac{M_y}{0.87 f_y a_y}$$

$$= \frac{150 \times 10^6}{0.87 \times 460 \times 280}$$

$$= 1338.5\,\text{mm}^2$$

Assume 3 no. bars of $446\,\text{mm}^2$ each on each long face.

Area of steel required for tension $= 625\,\text{mm}^2$ as before

This area can be divided over the total number of 4 no. corner bars in the member. Hence, use 4 no. bars of $156\,\text{mm}^2$ each.

Area of corner bars $= 434 + 446 + 156$
$$\qquad\qquad\qquad = 1036\,\text{mm}^2 \quad\text{(use 40 mm dia. bars} = 1257\,\text{mm}^2\text{)}$$

The arrangement of reinforcement is exactly the same as before. Use 4 no. 40 mm dia. bars in the corners and 1 no. 25 mm dia. bar at the centre of each face because 1 no. 25 mm bar equal to $491\,\text{mm}^2$ is bigger than $434\,\text{mm}^2$ or $446\,\text{mm}^2$ found before.

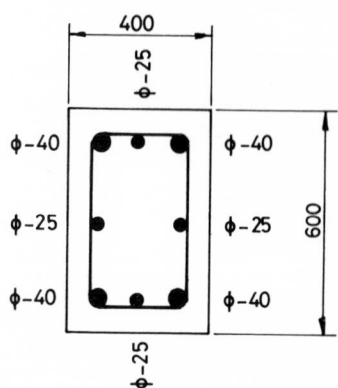

SK 4/28 Section designed by
Methods 1 and 2.

Method 3   Interaction curve method
(See Reference 13.)

Reinforcement required for M_x only.

$M_x = 250\,\text{kNm}$

$d = 540\,\text{mm}$

$f_{cu} = 40\,\text{N/mm}^2$

$$K = \frac{M}{f_{cu}bd^2} = \frac{250 \times 10^6}{40 \times 400 \times 540^2} = 0.05$$

$$z = d\left[0.5 + \sqrt{\left(0.25 - \frac{K}{0.9}\right)}\right]$$

$$= d\left[0.5 + \sqrt{\left(0.25 - \frac{0.05}{0.9}\right)}\right]$$

$$= 0.94d = 508\,\text{mm}$$

$$A_{st} = \frac{M}{0.87 f_{yz}} = \frac{250 \times 10^6}{0.87 \times 460 \times 508}$$

$$= 1230\,\text{mm}^2$$

Reinforcement required for M_y only

$M_y = 150\,\text{kNm}$

$d = 340\,\text{mm}$

$$K = \frac{150 \times 10^6}{40 \times 600 \times 340^2} = 0.05$$

$$z = 0.94d = 320\,\text{mm}$$

$$A_{st} = \frac{150 \times 10^6}{0.87 \times 460 \times 320} = 1171\,\text{mm}^2$$

$$\text{Area of steel required for tension only} = \frac{N}{0.87 f_y} = 625\,\text{mm}^2$$

Total reinforcement requirement $= 2 \times (1230 + 1171) + 625 = 5427 \, \text{mm}^2$

Try 1 no. 32 mm dia. bar at each corner and 1 no. 25 mm dia. bar at centre of each side.
Total $A_s = 5180 \, \text{mm}^2$.

P_N = applied ultimate tension = 250 kN

P_o = capacity of section in tension alone
$= A_s(0.87 f_y)$
$= 5180 \times 0.87 \times 460 \times 10^{-3}$
$= 2073 \, \text{kN}$

M_{ux} = ultimate moment in x direction = 250 kNm

M_{px} = ultimate moment capacity in x direction when tension and M_y are zero

A_{sx} = 2 no. ϕ 32 + 1 no. ϕ 25
$= 2099 \, \text{mm}^2$

Compression in concrete = tension in steel   (see Section 1.5.1)

or $0.402 f_{cu} b x = 0.87 f_y A_s$

or $x = \dfrac{0.87 f_y A_s}{0.402 f_{cu} b}$

$= \dfrac{0.87 \times 460 \times 2099}{0.402 \times 40 \times 400}$

$= 130 \, \text{mm} < 0.5 d = 270 \, \text{mm}$ OK

$z = d - 0.45x = 540 - 0.45 \times 130 = 481.5 \, \text{mm}$

$\therefore$ $M_{px} = 0.87 f_y A_s z = 0.87 \times 460 \times 2099 \times 481.5 \times 10^{-6}$
$= 404 \, \text{kNm}$

M_{uy} = ultimate moment in y direction = 150 kNm

M_{py} = ultimate moment capacity in y direction when tension and M_x are zero

$A_{sy} = 2099 \, \text{mm}^2$

$x = \dfrac{0.87 f_y A_s}{0.402 f_{cu} h} = \dfrac{0.87 \times 460 \times 2099}{0.402 \times 40 \times 600} = 87 \, \text{mm} < d/2 = 170 \, \text{mm}$ OK

$z = d - 0.45x = 340 - 0.45 \times 87 = 301 \, \text{mm}$

$M_{py} = 0.87 f_y A_s z = 0.87 \times 460 \times 2099 \times 301 \times 10^{-6} = 252.6 \, \text{kNm}$

Unity equation

$$\frac{P_N}{P_o} + \left(\frac{M_{ux}}{M_{px}}\right)^{1.5} + \left(\frac{M_{uy}}{M_{py}}\right)^{1.5} \leq 1$$

or $\dfrac{250}{2073} + \left(\dfrac{250}{404}\right)^{1.5} + \left(\dfrac{150}{252.6}\right)^{1.5} = 1.065 > 1$ unacceptable

Increase reinforcement to 8 no. 32 dia. bars instead of 4 no. 32 dia. and 4 no. 25 dia. No more checking is necessary.

Area provided by this method is 6432 mm² compared with 6992 mm² by the other two methods. This gives an 8% saving in reinforcement when the interaction formula is used. The interaction formula is not yet codified. The exponential changes from 1.5 for rectangular sections to 1.75 for square sections.

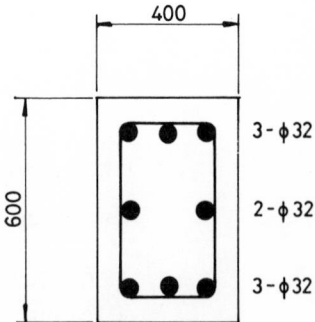

SK 4/29 Section designed by interaction curve.

Step 7   **Check shear stress**

Following ACI 318 – M83, Clause 11.3.2.3,[4] members subject to significant axial tension have a concrete shear resistance given by:

$$V_{cx} = 0.17\left(1 - 0.3\,\frac{N_u}{A_g}\right)\sqrt{f'_c}\,bd$$

$$= 0.17 \times \left(1 - \frac{0.3 \times 250 \times 10^3}{400 \times 600}\right) \times \sqrt{(0.8 \times 40)} \times 400 \times 540 \times 10^{-3}$$

$$= 142.8 \,\text{kN}$$

Similarly,

$$V_{cy} = 0.17 \times \left(1 - \frac{0.3 \times 250 \times 10^3}{400 \times 600}\right) \times \sqrt{(32)} \times 600 \times 340 \times 10^{-3}$$

$$= 134.9 \,\text{kN}$$

$\phi\,V_{cx} = 0.85 \times 142.8 = 121.4\,\text{kN} < V_{ux} = 250\,\text{kN}$
$\phi\,V_{cy} = 0.85 \times 134.9 = 114.7\,\text{kN} < V_{uy} = 150\,\text{kN}$

Shear reinforcement required for both orthogonal directions of shear. It is assumed that concrete shear resistance will be effective in the x direction only. In the y direction the total shear force will be carried by shear reinforcement.

Assume $S_{vx} = S_{vy} = 100\,\text{mm}$

$V_{sx} \geq V_{ux} - \phi\,V_{cx} = 250 - 121.4 = 128.6\,\text{kN}$

$$A_{svx} = \frac{V_{sx}S_{vx}}{0.85f_y d}$$

$$= \frac{128.6 \times 10^3 \times 100}{0.85 \times 400 \times 540}$$

$$= 70\,\text{mm}^2$$

$$V_{sy} \geq V_{uy} = 150\,\text{kN}$$

$$A_{svy} = \frac{V_{sy}S_{vy}}{0.85f_yd}$$

$$= \frac{150 \times 10^3 \times 100}{0.85 \times 400 \times 340}$$

$$= 130\,\text{mm}^2$$

A_{sv} is the larger of A_{svx} and A_{svy}, i.e. $130\,\text{mm}^2$ at $100\,\text{mm}$ spacing or $(A_{sv}/S_v) = 1.3$

Use 10 mm dia. links at 120 mm centres $(A_{sv}/S_v = 1.30)$.

Note: ACI 318[4] restricts stress in shear reinforcement to a maximum of $400\,\text{N/mm}^2$.

Design of Shear reinforcement using BS 8110: Part 1: 1985.[1]

$$v_x = \frac{V_x}{bh'} = \frac{250 \times 10^3}{400 \times 540} = 1.14\,\text{N/mm}^2$$

$$A_{sx} = 3 \text{ no. } 32 \text{ dia. bar} = 2412\,\text{mm}^2$$

$$p_x = \frac{100A_{sx}}{bh'} = \frac{100 \times 2412}{400 \times 540} = 1.12$$

$$v_{cx} = 0.76\,\text{N/mm}^2 \qquad \text{from Fig. 11.5}$$

$$A_{sy} = 3 \text{ no. } 32 \text{ dia. bar} = 2412\,\text{mm}^2$$

$$p_y = \frac{100A_{sy}}{b'h} = \frac{100 \times 2412}{340 \times 600} = 1.18$$

$$v_{cy} = 0.82\,\text{N/mm}^2$$

$$v_y = \frac{V_y}{bd} = \frac{150 \times 10^3}{600 \times 340} = 0.74\,\text{N/mm}^2$$

Modify v_{cx} and v_{cy} to take into account axial tension.

$$v'_{cx} = v_{cx} + \left(\frac{0.6NV_xh}{A_cM_x}\right) \qquad \frac{V_xh}{M_x} = 0.6 < 1$$

$$= 0.76 - \left(\frac{0.6 \times 250 \times 10^3 \times 0.6}{400 \times 600}\right)$$

$$= 0.385\,\text{N/mm}^2$$

$$v'_{cy} = v_{cy} + \left(\frac{0.6NV_yb}{A_cM_y}\right) \qquad \frac{V_yb}{M_y} = 0.4 < 1$$

$$= 0.82 - \left(\frac{0.6 \times 250 \times 10^3 \times 0.4}{400 \times 600} \right)$$

$$= 0.57\,\text{N/mm}^2$$

$$v''_{cx} = \frac{v'_{cx}v_x}{v_x + v_y} = \frac{0.385 \times 1.14}{1.14 + 0.74}$$

$$= 0.23\,\text{N/mm}^2$$

$$v''_{cy} = \frac{v'_{cy}v_y}{v_x + v_y} = \frac{0.57 \times 0.74}{1.14 + 0.74}$$

$$= 0.22\,\text{N/mm}^2$$

$$V'_{cx} = v''_{cx}bh' = 0.23 \times 400 \times 540 \times 10^{-3} = 49.7\,\text{kN}$$
$$V'_{cy} = v''_{cy}b'h = 0.22 \times 340 \times 600 \times 10^{-3} = 44.9\,\text{kN}$$

Assume 10 mm dia. bar ($f_y = 460\,\text{N/mm}^2$) used as links at a spacing of 150 mm. Area of two legs is 157 mm².

$$V_{sx} = \frac{0.87 f_{yv} A_{sh} h'}{S}$$

$$= \frac{0.87 \times 460 \times 157 \times 540 \times 10^{-3}}{150} = 226.2\,\text{kN}$$

$$V_{sy} = \frac{0.87 f_{yv} A_{sb} b'}{S}$$

$$= \frac{0.87 \times 460 \times 157 \times 340 \times 10^{-3}}{150} = 142.4\,\text{kN}$$

Check: $V_{sx} \ge V_x - V'_{cx} = 250 - 49.7 = 200.3\,\text{kN} < 226.2\,\text{kN}$ OK
$V_{sy} \ge V_y - V'_{cy} = 150 - 44.9 = 105.1\,\text{kN} < 142.4\,\text{kN}$ OK

Note: Slightly less shear reinforcement required when designed to BS 8110: Part 1: 1985.[1]

Step 8   ***Minimum reinforcement***
Reinforcement provided $= 6432\,\text{mm}^2 = 2.68\% > 0.4\%$

Step 9   ***Maximum reinforcement***
Maximum reinforcement $= 6\%$ not exceeded.

Step 10   ***Containment of reinforcement***
All reinforcement in tension. Containment rules do not apply.
Rules for minimum shear reinforcement in beams, as in Section 2.3 Step 13, should apply.

$$\text{Minimum } \frac{A_{sv}}{S_v} = \frac{0.4b}{0.87 f_{yv}}$$

$$= \frac{0.4 \times 600}{0.87 \times 460} = 0.6 < \frac{157}{150} = 1.04 \quad \text{OK}$$

Step 11 **Check spacing of bars for crack width**
See Section 2.3, Step 24.

MSA = 20 mm

Dia. of bar = 32 mm

Minimum clear distance between bars = 112 mm > 32 mm

Maximum clear distance between bars = 212 mm   with 3 no. 32 dia. on the long side

Service stress, $f_s = \dfrac{5}{8} f_y$ assumed

$$= \dfrac{5}{8} \times 460 = 287.5 \, \text{N/mm}^2$$

Maximum allowable clear spacing $= \dfrac{47000}{287.5} = 163 \, \text{mm} < 212 \, \text{mm}$ provided

This means that to reduce the probability of the crack width exceeding 0.3 mm, 4 bars should be used on the long face, i.e. 2 no. 32 dia. and 2 no. 25 dia. (total 6 no. 32 dia. and 4 no. 25 dia. in the member).

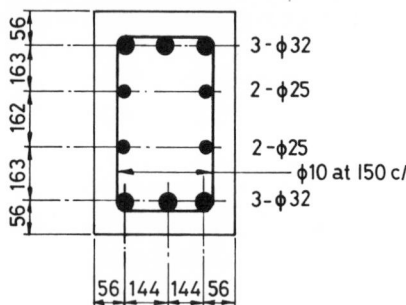

SK 4/30 Final designed section.

Chapter 5
Design of Corbels and Nibs

5.0 NOTATION

a_v	Distance from centre of load to nearest face of column for a corbel
a_v	Distance from free edge of nib to nearest link in beam
A_s	Area of steel reinforcement in tension to resist bending
A_{sh}	Area of horizontal steel reinforcement to resist shear in corbel
b	Width of corbel
d	Effective depth from bottom of corbel to centre of tensile reinforcement
d_b	Depth of corbel at edge of loaded area
f_s	Tensile stress in steel
f_y	Characteristic yield strength of steel
f_{cu}	Characteristic cube strength of concrete at 28 days
F_c	Concrete strut force in compression
F_t	Steel tensile force
F_{bt}	Tensile force in reinforcement at start of a bend
h	Overall depth of corbel
M	Applied moment on a section
p	Percentage of tensile reinforcement
r	Internal radius of a bend in a bar
S_h	Spacing of horizontal links in a corbel
T	Tension force applied to corbel along with vertical load
v	Shear stress in concrete (N/mm^2)
v_c	Design shear stress in concrete (N/mm^2)
v_c'	Modified design shear stress to account for a_v
V	Vertical load on corbel
x	Distance of neutral axis from bottom of corbel
z	Depth of lever arm
β	Angle of inclination to horizontal of concrete strut in a corbel
ε_s	Strain in steel reinforcement
ϕ	Diameter of reinforcing bar or equivalent diameter of a group of bars

5.1 LOAD COMBINATIONS

5.1.1 General rules   See Section 2.2.1.

5.1.2   Exceptional loads   See Section 2.2.4.

5.2   STEP-BY-STEP DESIGN PROCEDURE FOR CORBELS

Step 1   **Determine ultimate loads on the corbel**
           Follow load combination rules of Section 2.2.

Step 2   **Determination of corbel geometry**

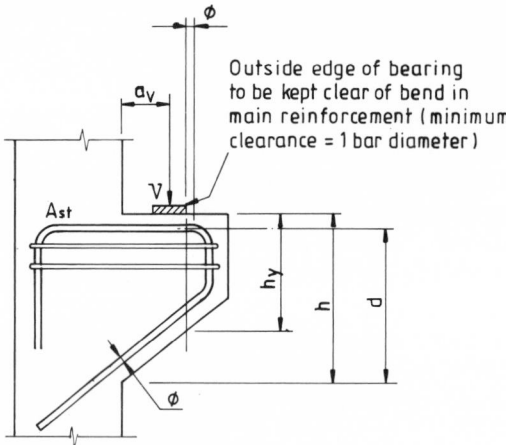

SK 5/1 Corbel geometry.

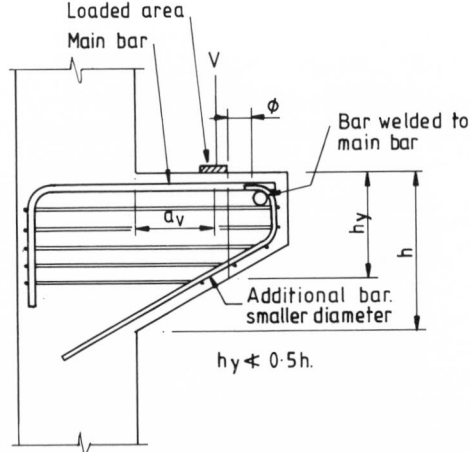

SK 5/2 Alternative corbel geometry.

Check the following:

(1) Bearing stress on concrete under bearing plate $\leq 0.8 f_{cu}$.
(2) Distance from end of loaded area to face of corbel should be as shown.
(3) Depth at root of corbel should be such that shear stress V/bd is less than $0.8\sqrt{f_{cu}}$ or $5 \, \text{N/mm}^2$, whichever is the lesser.

(4) Depth at outer edge of loaded area should be at least half the depth at the root.

(5) If a_v is greater than d, the corbel should be designed as a cantilever beam.

Step 3   Evaluation of internal forces

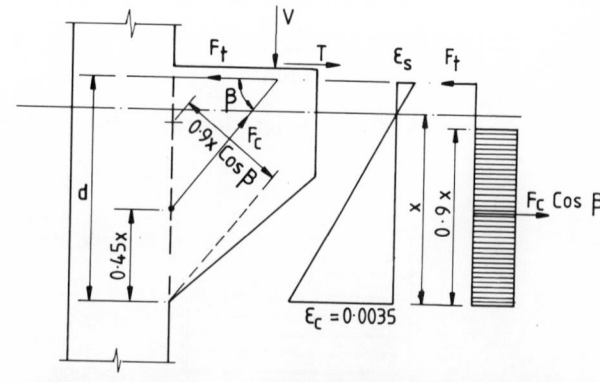

SK 5/3 Strut and tie diagram of a reinforced concrete corbel.

FORCE DIAGRAM   STRAIN DIAGRAM   STRESS DIAGRAM

b = WIDTH OF CORBEL

Draw strut and tie diagram as shown and find the following parameters.

$$v = \frac{V}{bd}$$

Find v/f_{cu} and a_v/d.
Find z/d from Fig. 5.1.
Find z and $x = (d - z)/0.45$
Find $F_t = T + Va_v/z$
F_t = tension in steel reinforcement
T = applied horizontal load along with V
z = depth of lever arm; x = depth of neutral axis

$$A_s = \frac{F_t}{0.87f_y} \geq \frac{0.5V}{0.87f_y} + \frac{T}{0.87f_y}$$

Alternatively,

$$F_t = F_c \cos \beta + T = \frac{Va_v}{z} + T$$

$$F_c = \left(\frac{0.67f_{cu}}{1.5}\right) b\, 0.9x \cos \beta = 0.402f_{cu}bx \cos \beta$$

$$V = F_c \sin \beta$$

$$z = d - 0.45x$$

By iteration, find x after assuming x in first trial. With final value of x, find z and F_t. From F_t, find A_s.

Step 4   **Check shear**

$$p = \frac{100A_s}{bd}$$

Find v_c from Figs 11.2 to 11.5 and multiply by $2d/a_v$ to get v'_c for corbel.

If $v < v'_c$, provide nominal shear reinforcement

Nominal reinforcement area $= 0.5A_s$

A_s is obtained in Step 3. Provide nominal links in upper two-thirds of effective depth d.

If $v > v'_c$, design shear reinforcement

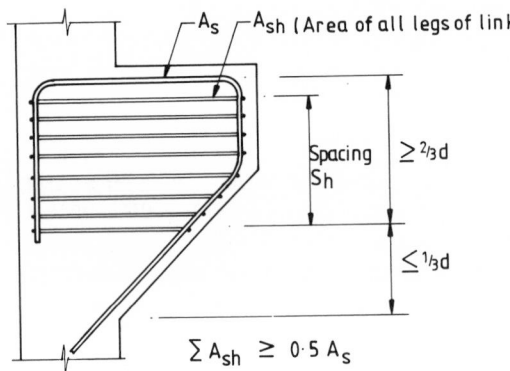

SK 5/4 Typical reinforcement arrangement in a corbel.

$$A_{sh} \geq \frac{bS_h(v - v_c)}{0.87f_y}$$

Provide A_{sh} in upper two-thirds of d at a spacing of S_h.

Note: Total area of all legs of links in a vertical plane should be more than or equal to $0.5A_s$.

Step 5   **Minimum tension reinforcement**
$A_s \geq 0.004bh$

Step 6   **Maximum tension reinforcement**
$A_s \leq 0.040bh$

Step 7   **Check bearing stress inside bend**
The following must be satisfied:

$$\text{bearing stress} = F_{bt}r\phi \leq \frac{2f_{cu}}{1 + 2\left(\dfrac{\phi}{a_b}\right)}$$

See step 22 of Section 2.3 for notation

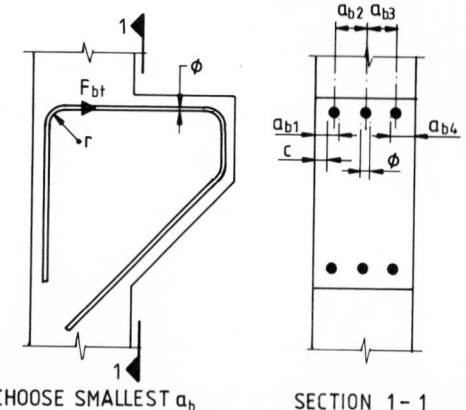

Residual tension in steel at bend.

SK 5/5 Bearing stress inside bend.   CHOOSE SMALLEST a_b SECTION 1-1

Step 8   Spacing of bars
Minimum clear spacing horizontally $= MSA + 5 \geq$ dia. of bar

where MSA = maximum size of aggregate.

Minimum clear spacing vertically $= \dfrac{2\,MSA}{3}$

Maximum clear spacing of bars in tension $\leq \dfrac{47\,000}{f_s} \leq 300$

f_s = service stress in bar

5.3   STEP-BY-STEP DESIGN PROCEDURE FOR NIBS

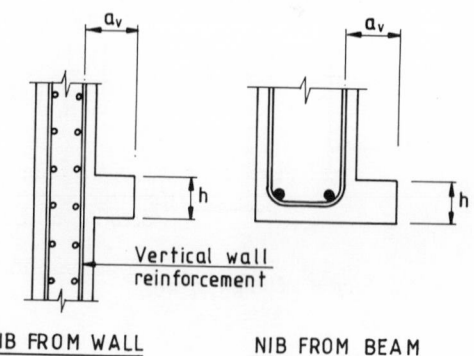

Vertical wall reinforcement

SK 5/6 Typical arrangement of nibs.

NIB FROM WALL   NIB FROM BEAM

Step 1   Determine cover to reinforcement
Determine cover required to reinforcement as per Tables 11.6 and 11.7.

Step 2   Determine ultimate loads on nib
Follow load combination rules of Section 2.2.

Step 3   Determine nib geometry

(1) Bearing stress under load $\leq 0.4f_{cu}$ for dry bearing
$\qquad\qquad\qquad\qquad\qquad\quad \leq 0.6f_{cu}$ for bedded bearing.

(2) Find effective bearing length which is the least of:
   (a) bearing length
   (b) one-half of bearing length plus 100 mm
   (c) 600 mm.

(3) Find net bearing width $= \dfrac{\text{design ultimate support reaction}}{(\text{effective bearing length}) \times 0.4f_{cu}} \geq 40\,\text{mm}$

(4) Find allowance for spalling, as per Tables 5.1 and 5.2.

(5) Find allowance for inaccuracies, as per Table 5.3.

(6) Nominal   bearing   width = (net   bearing   width) + (allowances   for spalling) + (allowances for inaccuracies)

(7) Nib projection = (nominal bearing width) + 25 mm
   Allow chamfer minimum 15 mm.

(8) Overall depth of nib should be less than 300 mm.

(9) Select diameter of reinforcement and find a_v and d.

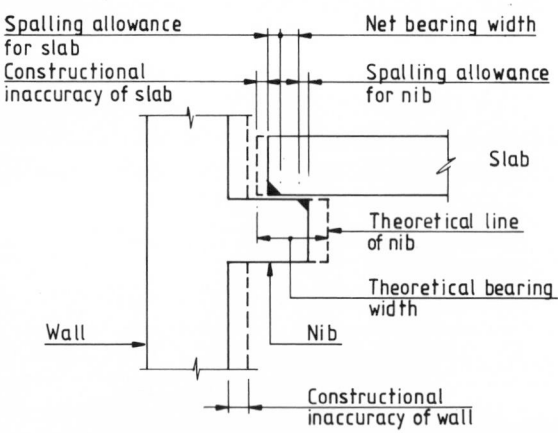

SK 5/7 Typical calculation for net bearing width of nib.

Step 4   Design of nib
$$M = Va_v$$

V = ultimate load per metre

$$K = \frac{M}{bd^2f_{cu}}$$

b = 1 metre

$$z = d\left[0.5 + \sqrt{\left(0.25 - \frac{K}{0.9}\right)}\right] \leq 0.95d$$

$$A_s = \frac{M}{0.87 f_y z} \text{ per metre}$$

Step 5 *Determine minimum reinforcement*
Minimum reinforcement $= 0.0013bh$

Step 6 *Maximum spacing of bars*
Maximum allowable spacing $= 3 \times$ (effective depth) + (diameter of bar) $\leq$ 750 mm

$$\text{Clear spacing} \leq \frac{47\,000}{f_s} \leq 300 \text{ mm}$$

where $f_s =$ service stress.

Step 7 *Check shear*
$V =$ ultimate load per metre

$$v = \frac{V}{bd}$$

$b = 1$ metre

$$p = \frac{100 A_s}{bd}$$

Find v_c from Figs 11.2 to 11.5.

Find $v_c' = \left(\frac{2d}{a_v}\right) v_c$

Check that $v \leq v_c'$

If not, increase depth of nib.

Note: If tensile reinforcement found in Step 3 is kept straight and exposed at end, shear stress v should be less than $v_c'/2$.

Step 8 *Extra vertical reinforcement in beam*

$$A_{sv} = \frac{V}{0.87 f_y} \text{ per metre length of beam}$$

Step 9 *Isolated loads on continuous nib*
To find effective width of load dispersal, assume a 45° angle of line of failure crack as shown.

$l_e =$ effective width for isolated load on continuous nibs

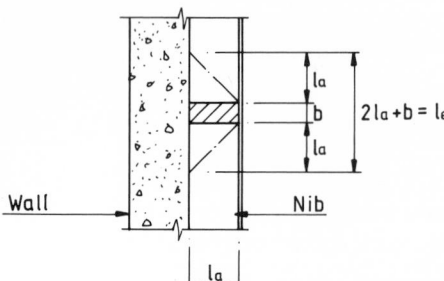

SK 5/8 Plan of wall and nib showing effective length of nib for a line load of width b.

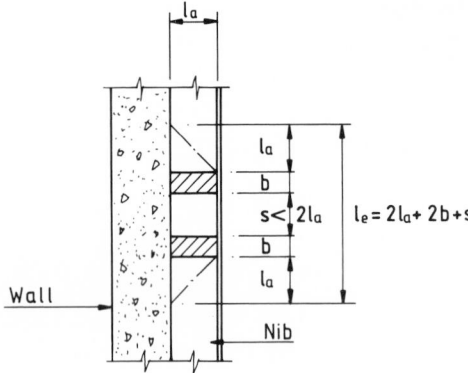

SK 5/9 Plan of wall and nib showing effective length with multiple line loads.

5.4   WORKED EXAMPLES

Example 5.1   Design of a corbel

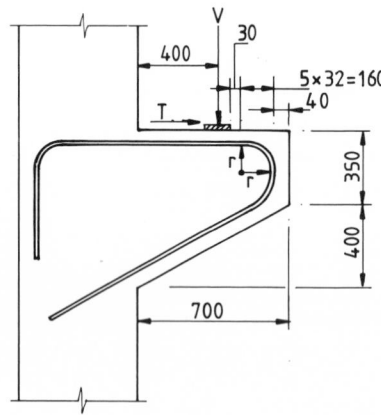

SK 5/10 Elevation of corbel.

Step 1   *Determine ultimate loads on the corbel*
Ultimate vertical load $= V = 800\,\text{kN}$
Ultimate horizontal load $= T = 80\,\text{kN}$

(Ignore small eccentricity of horizontal load from tension steel.)

Line of action of load at 400 mm from face of column.

Size of column = 600 mm × 400 mm

Corbel about the major axis of column.

Width of corbel = 400 mm

Step 2   Determination of corbel geometry
$$f_{cu} = 40\,\text{N/mm}^2$$
$$f_y = 460\,\text{N/mm}^2$$

Minimum cover to reinforcement = 30 mm

Assumed diameter of main reinforcement = 32 mm

Assumed diameter of horizontal links = 10 mm

Bearing plate used.

Maximum bearing stress = $0.8 f_{cu} = 32\,\text{N/mm}^2$

Length of bearing plate = 300 mm

$$\text{Minimum bearing width} = \frac{V}{32 \times 300} = \frac{800 \times 10^3}{32 \times 300} = 83\,\text{mm}$$

Actual width of bearing plate = 100 mm = l_w > 83 mm   OK

$$l = \text{length of corbel} = a_v + \frac{1}{2} l_w + \text{length of bend of bar} + \text{min. cover} +$$
$$\text{dia. of link} + \text{min. cover}$$
$$= 400 + 50 + 5 \times 32 + 30 + 10 + 30$$
$$= 680\,\text{mm}\quad \text{say } 700\,\text{mm}$$

Use $h = 750$ mm at column face.

$d = 750 - 30 - 16 = 704$ mm

Maximum allowable shear stress at column face = $5\,\text{N/mm}^2$

$$d > \frac{V}{5b} = \frac{800 \times 10^3}{5 \times 400} = 400\,\text{mm}$$

$$v = \frac{V}{bd}$$

$$= \frac{800 \times 10^3}{400 \times 704} = 2.84\,\text{N/mm}^2 < 0.8\sqrt{f_{cu}} = 5.05\,\text{N/mm}^2$$

Step 3   Evaluation of forces

First trial
From strut and tie diagram (Step 3 in Section 5.2),

$$F_t = F_c \cos\beta + T = \frac{Va_v}{z} + T$$

$$F_c = \left(\frac{0.67f_{cu}}{1.5}\right)b\ 0.9x\cos\beta = 0.402f_{cu}bx\cos\beta$$

$$V = F_c \sin\beta$$

$$z = d - 0.45x$$

Assume $x = 0.4d = 282$ mm, say.

$z = d - 0.45x$
$= 704 - (0.45 \times 282)$
$= 577$ mm

$$\cot\beta = \frac{a_v}{z} = \frac{400}{577} = 0.6932$$

$\sin\beta = 0.8218 \qquad \cos\beta = 0.5697$

$$F_c = \frac{V}{\sin\beta} = 973.5\,\text{kN}$$

$$x = \frac{F_c}{0.402f_{cu}b\cos\beta}$$

$$= \frac{973.5 \times 10^3}{0.402 \times 40 \times 400 \times 0.5697}$$

$= 265.7$ mm

Second trial

$x = 265$ mm

$z = 584.7$ mm

$\cot\beta = 0.6841$

$\sin\beta = 0.8254$

$\cos\beta = 0.5646$

$F_c = 969.2\,\text{kN}$

$x = 266.9$ mm   OK

Final $z = 585$ mm

$$F_t = \frac{Va_v}{z} + T$$

$$= \left(\frac{800 \times 10^3 \times 400}{585}\right) + 80 \times 10^3 = 627 \times 10^3\,\text{N}$$

$$\varepsilon_s = 0.0035 \times \left(\frac{704 - 265}{265}\right)$$

$= 5.798 \times 10^{-3} > 0.002$

So the steel will be at the yield stress level

$$f_y = 460 \, \text{N/mm}^2$$

$$A_s = \frac{F_t}{0.87 f_y} \geq \left(\frac{0.5V}{0.87 f_y}\right) + \left(\frac{T}{0.87 f_y}\right) = 1200 \, \text{mm}^2$$

$$= \frac{627 \times 10^3}{0.87 \times 460}$$

$$= 1567 \, \text{mm}^2 > 1200 \, \text{mm}^2 \quad \text{OK}$$

Use 2 no. 32 dia. bars as main tension reinforcement (1608 mm²). Alternatively by use of the chart in Fig. 5.1,

$$\frac{v}{f_{cu}} = \frac{2.84}{40} = 0.071$$

$$\frac{a_v}{d} = \frac{400}{704} = 0.568$$

From Fig. 5.1,

$$\frac{z}{d} = 0.83$$

$$z = 704 \times 0.83 = 584 \, \text{mm}$$

Note: The chart gives the same z as is obtained by iteration. Having found z from the chart, find F_t and A_s

Step 4   Check shear

$$p = \frac{100 A_s}{bd}$$

$$= \frac{100 \times 1608}{400 \times 704}$$

$$= 0.57$$

From Fig. 11.5,

$$v_c = 0.608 \, \text{N/mm}^2$$

$$v_c' = \left(\frac{2d}{a_v}\right) v_c = \frac{2 \times 704 \times 0.608}{400} = 2.14 \, \text{N/mm}^2 < 2.84 \, \text{N/mm}^2$$

Shear reinforcement is required.
Horizontal links are provided. Assume $S_h = 200 \, \text{mm}$.

$$A_{sh} \geq \frac{b S_h (v - v_c')}{0.87 f_y} = \frac{400 \times 200 \times (2.84 - 2.14)}{0.87 \times 460} = 140 \, \text{mm}^2$$

Required: 2-legged 10 mm diameter links at 200 centres for the upper two-thirds of d.

$$\frac{2}{3}d = \frac{2}{3} \times 704 = 470 \, \text{mm}$$

Required: 3 sets of links of 10 mm diameter at 200 mm centres.

Total area of legs $= 471\,\text{mm}^2 < 0.5 \times 1567 = 783.5\,\text{mm}^2$

Main tension steel required $= 1567\,\text{mm}^2$

Use 5 sets of links 10 mm diameter at 100 mm centres ($785\,\text{mm}^2$).

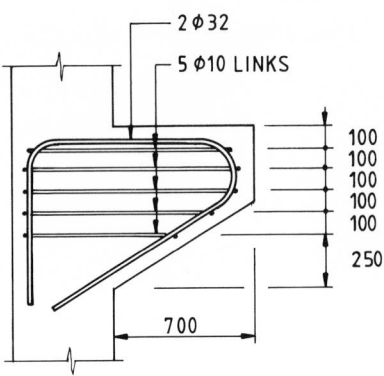

SK 5/11 Elevation of designed corbel.

Step 5   *Minimum tension reinforcement*
$A_s > 0.004bh = 1200\,\text{mm}^2$ satisfied

Step 6   *Maximum tension reinforcement*
$A_s < 0.040bh = 12\,000\,\text{mm}^2$

Not exceeded.

Step 7   *Check bearing stress inside bend*

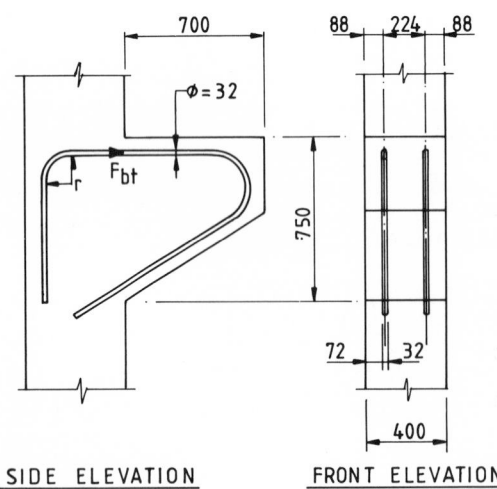

SIDE ELEVATION   FRONT ELEVATION

SK 5/12 Bearing stress inside bend.

$$F_{bt} = \left(\frac{F_t}{\text{no. of bars}}\right)\left(\frac{A_{s\ req}}{A_{s\ prov}}\right)$$

$$= \frac{627}{2} \times \frac{1567}{1608}$$

$$= 305.5\,\text{kN}$$

Ultimate anchorage bond stress

$$f_{bu} = 0.5\,\sqrt{f_{cu}} \qquad \text{(for Type 2 deformed bar as obtained from Table 3.28}$$
$$\text{of BS\,8110: Part 1)}$$
$$= 0.5\sqrt{40} = 3.16\,\text{N/mm}^2$$

$$\text{Anchorage bond length required} = \frac{F_{bt}}{\pi\phi f_{bu}}$$

$$= \frac{305.5 \times 10^3}{\pi \times 32 \times 3.16}$$

$$= 962\,\text{mm}$$

In the column, the straight length of bar before start of bend is taken as approximately equal to 350 mm which is say one-third of the required anchorage length. Hence

$$\text{Tension in bar at start of bend} = \frac{2}{3}\,F_{bt}$$

$$= \frac{2}{3} \times 305.5 = 203.7\,\text{kN}$$

r = internal radius of bend
$= 4 \times 32$ (minimum)
$= 128\,\text{mm}$ standard

$\phi = 32\,\text{mm}$

a_b = cover + bar diameter for corner bar
$= 72 + 32 = 104\,\text{mm}$

Centre-to-centre distance of bars $= 224\,\text{mm} > 104\,\text{mm}$

$\therefore\quad a_b = 104\,\text{mm}$

$$\frac{F_{bt}}{r\phi} = \frac{203.7 \times 10^3}{128 \times 32} = 49.73\,\text{N/mm}^2$$

$$\frac{2f_{cu}}{1 + \left(\dfrac{2\phi}{a_b}\right)} = \frac{2 \times 40}{1 + 2\left(\dfrac{32}{104}\right)} = 49.52\,\text{N/mm}^2 < 49.73\,\text{N/mm}^2$$

Standard radius bend will be adequate.
Calculation of anchorage bond length:

Anchorage value standard bend $= 12 \times 32 = 384\,\text{mm}$ (includes 4 diameter straight)

Straight before bend $= 350\,\text{mm}$

Bar should project vertically into column after standard bend by minimum of

$$962 - 384 - 350 + (4 \times 32) = 356 \, \text{mm}$$

Step 8 *Spacing of bars*

Minimum horizontal spacing $= 20 + 5 = 25 \, \text{mm}$

$$\text{Maximum clear spacing of bars in tension} < \frac{47\,000}{f_s} < 300$$

f_s = service stress   (from crack width calculations in Step 9)
   $= 226.4 \, \text{N/mm}^2$

$$\frac{47\,000}{f_s} = \frac{47\,000}{226.4} = 208 \, \text{mm}$$

Actual clear spacing $= 224 - 32 = 192 \, \text{mm} < 208 \, \text{mm}$ OK

Clear distance between the corner of corbel and the nearest tension bar should not be greater than 80 mm as per clause 3.12.11.2.5 of BS 8110: Part 1: 1985. Actual clear distance is 72 mm.

Note: No crack width calculation is required if maximum spacing of bars in tension does not exceed the recommendations of clause 3.12.11.2 of BS 8110: Part 1.

Step 9 *Crack width calculations*

Note: This step is optional and is included to show the method of calculation of crack width for a corbel.

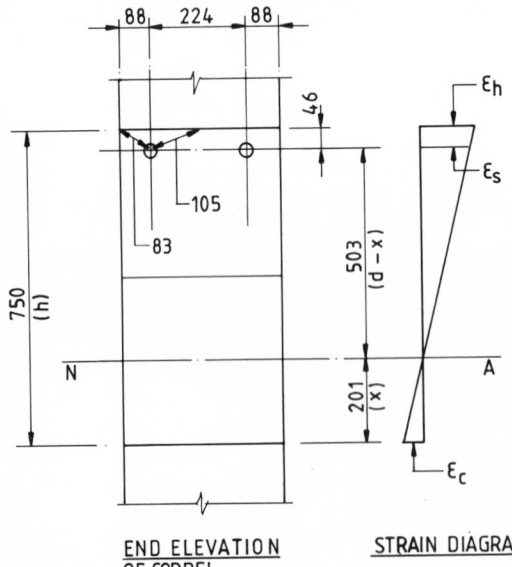

END ELEVATION
OF CORBEL

STRAIN DIAGRAM

SK 5/13 Crack width calculations.

Service horizontal load $= 50\,\text{kN}$

Service vertical load $= 500\,\text{kN}$

Moment at face of column $= 500 \times a_v = 200\,\text{kNm}$

See Section 1.13 and assume $A_s' = 0$

$$x = \frac{mA_s}{b}\left[\left(1 + \frac{2bd}{A_s m}\right)^{\frac{1}{2}} - 1\right]$$

$$\frac{10 \times 1608}{400}\left[\left(1 + \frac{2 \times 400 \times 704}{1608 \times 10}\right)^{\frac{1}{2}} - 1\right]$$

$$= 201\,\text{mm}$$

$$z = d - \frac{x}{3}$$

$$= 704 - \frac{201}{3}$$

$$= 637\,\text{mm}$$

$$f_{sb} = \frac{M}{A_s z}$$

$$= \frac{200 \times 10^6}{1608 \times 637}$$

$$= 195.3\,\text{N/mm}^2 \quad \text{due to flexure}$$

$$f_{sh} = \frac{50 \times 10^3}{1608} = 31.1\,\text{N/mm}^2 \quad \text{due to horizontal load}$$

$$f_s = f_{sb} + f_{sh}$$
$$= 195.3 + 31.1$$
$$= 226.4\,\text{N/mm}^2$$

$$\varepsilon_s = \frac{f_s}{E_s} = \frac{226.4}{200 \times 10^3} = 1.132 \times 10^{-3}$$

$$\varepsilon_h = \left(\frac{h - x}{d - x}\right)\varepsilon_s$$

$$= \left(\frac{750 - 201}{704 - 201}\right) \times 1.132 \times 10^{-3}$$

$$= 1.235 \times 10^{-3}$$

$$\varepsilon_{mh} = \varepsilon_h - \frac{b(h - x)^2}{3E_s A_s\,(d - x)}$$

$$= 1.235 \times 10^{-3} - \frac{400(750 - 201)^2}{3 \times 200 \times 10^3 \times 1608 \times (704 - 201)}$$

$$= 0.9866 \times 10^{-3}$$

$$a_{cl} = \sqrt{(88^2 + 46^2)} - 16 = 83.3\,\text{mm}$$

$$a_{c2} = \sqrt{(112^2 + 46^2)} - 16 = 105\,\text{mm}$$

$$a_{cr} = 105\,\text{mm}$$

$$W_{cr} = \frac{3a_{cr}\,\varepsilon_m}{1 + 2\left(\dfrac{a_{cr} - c_{min}}{h - x}\right)}$$

$$= \frac{3 \times 105 \times 0.9866 \times 10^{-3}}{1 + 2\left(\dfrac{105 - 30}{750 - 201}\right)}$$

$$= 0.244\,\text{mm} < 0.3\,\text{mm}$$

Crack width criterion is satisfied.

Example 5.2 *Design of concrete nib*

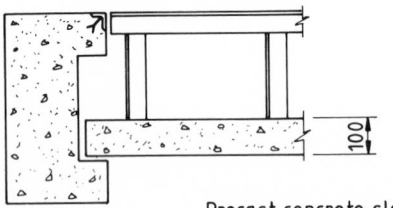

Precast concrete slab

SK 5/14 General arrangement of nib.

Reinforced concrete in-situ floor beams with nibs to carry precast floor units.

Clear gap between beams $= 4.5\,\text{m}$

Width of floor units $= 400\,\text{mm}$

Depth of floor units $= 100\,\text{mm}$

False floor + finish on units $= 2.5\,\text{kN/m}^2$

Imposed load on floor $= 5.0\,\text{kN/m}^2$

Grade of concrete for beam $= \text{C40}$

Assume dry bearing.

Step 1 *Determine cover to reinforcement*
Exposure $=$ mild

Fire resistance $= 1$ hour

Grade of concrete $= \text{C40}$

Maximum size of aggregate $= 20\,\text{mm}$

Minimum thickness of floor $= 95\,\text{mm}$

Nominal cover $= 20\,\text{mm}$

Step 2   *Determine loading*

Self-weight of precast unit $= 2.5\,\text{kN/m}^2$

False floor + finish $= 2.5\,\text{kN/m}^2$

Total dead load $= 5\,\text{kN/m}^2$

Imposed load $= 5\,\text{kN/m}^2$

Ultimate load $= 1.4 \times 5 + 1.6 \times 5 = 15\,\text{kN/m}^2$

Reaction at either end of precast floor unit (400 mm) $= 4.5 \times 15 \times 0.5 \times 0.4$
$$= 13.5\,\text{kN}$$

Step 3   *Determine nib geometry*

Allowable bearing stress $= 0.4 f_{cu} = 0.4 \times 40 = 16\,\text{N/mm}^2$

Effective bearing length is the least of:
(a) bearing length $= 400\,\text{mm}$
(b) one-half bearing length + 100 $= 300\,\text{mm}$
(c) 600 mm.
Effective bearing length $= 300\,\text{mm}$

$$\text{Net bearing width} = \frac{\text{ultimate support reaction}}{(\text{effective bearing length}) \times 0.4 f_{cu}} \geq 40$$

$$= \frac{13.5 \times 10^3}{300 \times 16} = 2.8\,\text{mm}$$

Net bearing width $= 40\,\text{mm}$

Allowance for spalling (from Tables 5.1 and 5.2) $= 20 + 0 = 20\,\text{mm}$

Allowance for inaccuracies (from Table 5.3) $= 25\,\text{mm}$

Nominal bearing width $= 40 + 20 + 25 = 85\,\text{mm}$

Nib projection $= 85\,\text{mm} + 15\,\text{mm}$ (chamfer) + 10 mm(clearance) $= 110\,\text{mm}$

Nominal length of precast units $= 4.5\,\text{m} - 2 \times 10\,\text{mm}$ (clearance)
$$= 4.48\,\text{m}$$

Minimum depth of nib $= 2 \times$ (minimum cover) + 8 × (diameter of bar)
$$= 2 \times 20 + 8 \times 8 = 104\,\text{mm} < 300\,\text{mm}$$

Minimum depth of nib 105 mm, say.

Note: The depth of the nib can be reduced if 6 mm diameter mild steel bars are used or welded anchor bars are used at straight ends of flexural bars.

Step 4   *Design of nib*

$a_v = 110 - 15$ (chamfer) + 20 (cover) + 5 (half dia. of link)
$= 120\,\text{mm}$

$d = 105 - 20 - 4 = 81\,\text{mm}$

M = bending moment per metre

= (load per metre run) $\times a_v$

= $43.5 \times 0.5 \times 15 \times 0.12$

= $4.05\,\text{kNm/m}$

$$K = \frac{M}{f_{cu}bd^2}$$

$$= \frac{4.05 \times 10^6}{40 \times 1000 \times 81^2}$$

$$= 0.0154$$

$$z = d\left[0.5 + \sqrt{\left(0.25 - \frac{K}{0.9}\right)}\right] \le 0.95d$$

$$= 0.95d = 77\,\text{mm}$$

$$A_s = \frac{M}{0.87f_yz} = \frac{4.05 \times 10^6}{0.87 \times 460 \times 77} = 131\,\text{mm}^2/\text{m}$$

Step 5 Determine minimum reinforcement
Minimum reinforcement = $0.0013bh$

$$= 0.0013 \times 1000 \times 105$$

$$= 137\,\text{mm}^2/\text{m}$$

Step 6 Maximum spacing of bars
Maximum spacing = $3 \times$ effective depth + bar dia.

$$= 3 \times 81 + 8$$

$$= 251\,\text{mm}\quad\text{centres}$$

Use 8 mm dia. bars at 250 centres ($201\,\text{mm}^2/\text{m}$).
(See Example 2.3, Step 25 for refinement.)

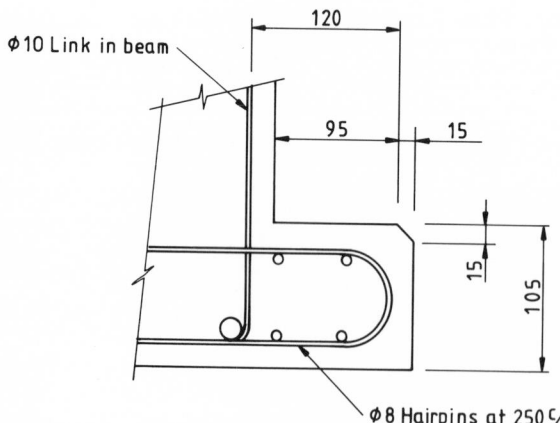

ϕ 10 Link in beam

120

95   15

15

105

ϕ 8 Hairpins at 250 $^c/_c$

SK 5/15 Typical reinforcement in nib.

Step 7   Check shear

V = ultimate load per metre length

$\quad = 4.5 \times 0.5 \times 15$

$\quad = 33.75 \, \text{kN/m}$

$$v = \frac{V}{bd} = \frac{33.75 \times 10^3}{1000 \times 81}$$

$\quad = 0.42 \, \text{N/mm}^2$

$$p = \frac{100 A_s}{bd}$$

$$= \frac{100 \times 201}{1000 \times 81}$$

$\quad = 0.25$

From Fig. 11.5,

$v_c = 0.62 \, \text{N/mm}^2$

$$v_c' = \frac{v_c 2d}{a_v}$$

$$= \frac{0.62 \times 2 \times 81}{120}$$

$\quad = 0.84 \, \text{N/mm}^2 > 0.42 \, \text{N/mm}^2$

Step 8   *Extra vertical reinforcement in beam*

In addition to links, an area of reinforcement is required in the beam to carry the load from the nib.

$$A_{sv} = \frac{V}{0.87 f_y}$$

$$= \frac{33.75 \times 10^3}{0.87 \times 460}$$

$\quad = 84 \, \text{mm}^2/\text{m}$

5.5 FIGURES AND TABLES FOR CHAPTER 5

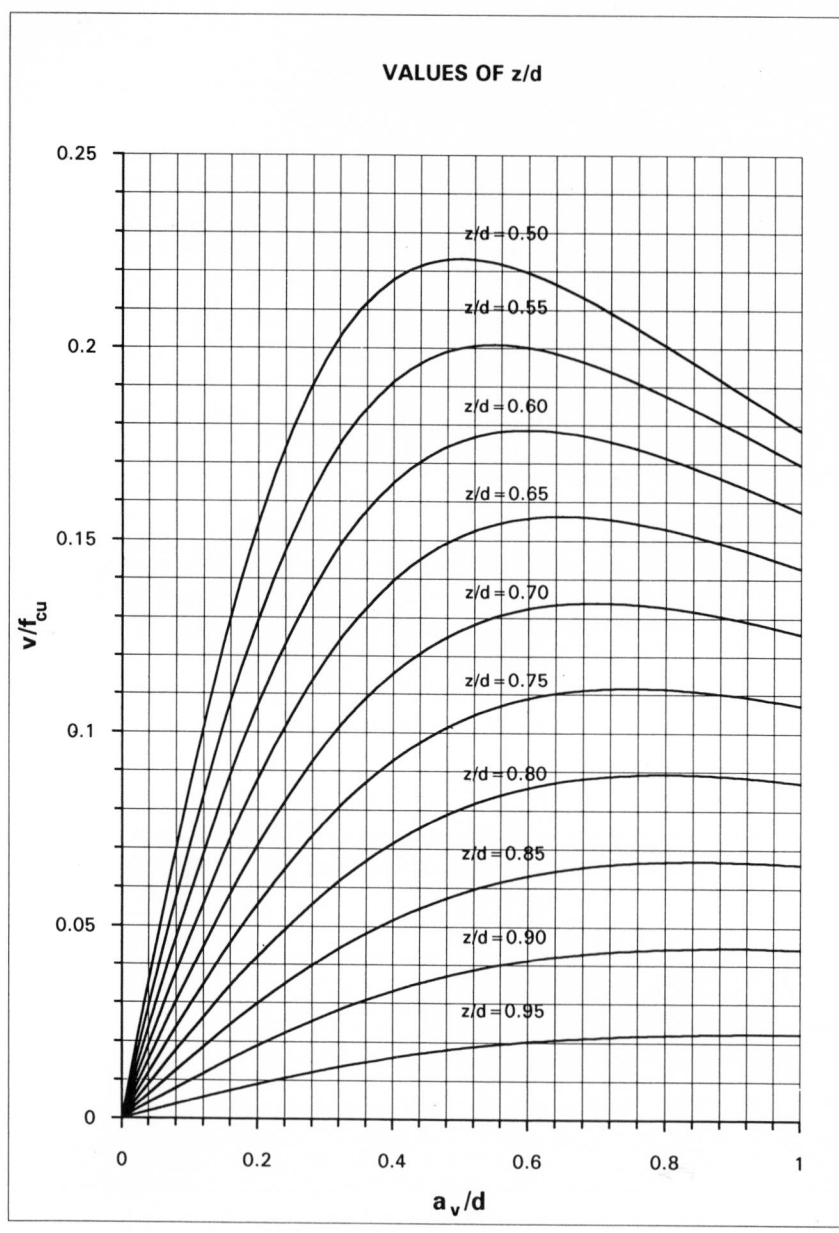

Fig. 5.1 Chart for determining z/d.

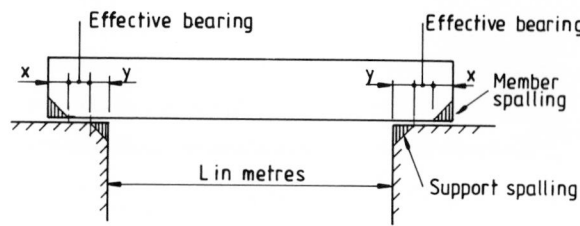

Table 5.1 Allowance for effects of spalling at supports.

Material of support	Distance y (mm)
Steel	0
Concrete Grade 30 or over	15
Brickwork or masonry	25
Concrete below Grade 30	25
Reinforced concrete nib less than 300 mm deep	Nominal cover to reinforcement
Reinforced concrete nib less than 300 mm deep with vertical loop reinforcement exceeding 12 mm in diameter	Nominal cover plus inner radius of bend

Table 5.2 Allowance for effects of spalling at supported members.

Reinforcement at bearing of supported member	Distance x (mm)
Straight bars, horizontal loop or vertical loop reinforcement not exceeding 12 mm diameter	10 or end cover, whichever is greater
Tendons or straight bars exposed at end of member	0
Vertical loop reinforcement of bar diameter exceeding 12 mm	End cover plus inner radius of bend of bar

Table 5.3 Allowance for construction inaccuracies.

Material of support	Construction inaccuracy (mm)
Steel or precast concrete support	15 or $3L$, whichever is greater
Masonry supports	20 or $4L$, whichever is greater
In-situ concrete supports	25 or $5L$, whichever is greater

Chapter 6

Design of Pad Foundations

6.0 NOTATION

a_{cr}	Point on surface of concrete to nearest face of a bar
A	Length of a side of a rectangular pad foundation
A_s	Area of tensile reinforcement
A_{sx}	Area of tensile reinforcement to resist bending about x-axis
A_{sy}	Area of tensile reinforcement to resist bending about y-axis
b	Width of reinforced concrete section
B	Width of a side of a rectangular pad foundation (least dimension)
c	Soil cohesion (kN/m^2)
c_{min}	Minimum cover to tensile reinforcement
C_x	Size of column in x-direction
C_y	Size of column in y-direction
C_{kd}	Cone resistance by static cone penetration tests (kg/cm^2)
d	Effective depth to tensile reinforcement of a concrete section
d_x	Effective depth to tensile reinforcement resisting moment about x-axis
d_y	Effective depth to tensile reinforcement resisting moment about y-axis
D	Depth of foundation below ground level
e_x	Resultant eccentricity of all column vertical loads in x-direction
e_y	Resultant eccentricity of all column vertical loads in y-direction
e_{xg}	Eccentricity of vertical end reaction from ground beams in the x-direction
e_{yg}	Eccentricity of vertical end reaction from ground beams in the y-direction
E_c	Modulus of elasticity of concrete
E_s	Modulus of elasticity of steel
f_s	Tensile stress in steel reinforcement
f_y	Characteristic yield strength of steel
f_{cu}	Characteristic cube strength of concrete at 28 days
F	Frictional resistance to horizontal movement under pad foundation
h	Overall depth of concrete section/thickness of pad
H	Effective depth of soil under foundation for computation of settlement
H_a	Active pressure on side of a foundation (kN)
H_p	Passive resistance on side of a foundation (kN)
H_u	Ultimate factored horizontal load at underside of a foundation
H_x	Unfactored horizontal shear from column on foundation in the x-direction

H_y	Unfactored horizontal shear from column on foundation in the y-direction
H_{xu}	Factored horizontal shear from column on foundation in x-direction
H_{yu}	Factored horizontal shear from column on foundation in y-direction
K_a	Active pressure coefficient of soil
K_h	Modulus of subgrade reaction for horizontal movement in soil
K_p	Rankine passive pressure coefficient of soil
l_x	Dimension of a rectangular pad footing in x-direction
l_y	Dimension of a rectangular pad footing in y-direction
m	Modular ratio E_s/E_c
m_v	Coefficient of volume compressibility of soil (m^2/MN)
M_x	Unfactored moment from column on foundation about x-axis
M_y	Unfactored moment from column on foundation about y-axis
M_x^*	Unfactored moment on foundation about x-axis due to eccentric surcharge
M_y^*	Unfactored moment on foundation about y-axis due to eccentric surcharge
M_{xg}	Unfactored fixed end moment from ground beams on foundation about x-axis
M_{xu}	Factored moment from column on foundation about x-axis
M_{xx}	Combined unfactored total moment on foundation about x-axis
M_{yg}	Unfactored fixed end moment from ground beams on foundation about y-axis
M_{yu}	Factored moment from column on foundation about y-axis
M_{yy}	Combined unfactored total moment on foundation about y-axis
M_{xu}^*	Factored moment on foundation about x-axis due to eccentric surcharge
M_{yu}^*	Factored moment on foundation about y-axis due to eccentric surcharge
M_{xgu}	Factored fixed-end moment from ground beams on foundation about x-axis
M_{xxu}	Combined factored total moment on foundation about x-axis
M_{ygu}	Factored fixed-end moment from ground beams on foundation about y-axis
M_{yyu}	Combined factored total moment on foundation about y-axis
n_h	Coefficient to determine horizontal modulus of subgrade reaction
N	Unfactored vertical load from column on foundation
N_c	Soil bearing capacity coefficient as per Terzaghi
N_q	Soil bearing capacity coefficient as per Terzaghi
N_u	Factored vertical load from column on foundation
N_γ	Soil bearing capacity coefficient as per Terzaghi
p	Total overburden pressure at foundation level
p_o	Effective overburden pressure at foundation level/centre of layer
p_x	Percentage of tensile reinforcement to resist moment about x-axis
p_y	Percentage of tensile reinforcement to resist moment about y-axis
P	Unfactored combined total vertical load on soil under a pad foundation
P_s	Sliding resistance of concrete pad foundation on soil

P_u	Factored combined total vertical load on soil under a pad foundation
P_v	Allowable vertical load on soil under a pad foundation
P_{Hx}	Sliding resistance of base in x-direction
P_{Hy}	Sliding resistance of base in y-direction
q_n	Net pressure on soil for settlement computation (MN/m^2)
q_u	Unconfined compressive strength (kN/m^2)
q_{ult}	Ultimate bearing capacity of soil under a pad foundation
r or R	Radius of circular footing
R	Restraint factor for computation of early thermal cracking
s	Shear strength of soil
s_u	Shear strength from unconfined tests ($= q_u/2$)
T_1	Differential temperature in a concrete pour for calculation of early thermal cracking
U_n	Perimeter of column at prescribed multiples of effective depth of pad
U_o	Perimeter of column footprint on pad foundation
v_c	Design concrete shear stress
v_n	Shear stress in concrete at perimeter defined by U_n
V	Shear force across critical section in a pad foundation
V_n	Shear force in a critical perimeter defined by U_n
V_u	Factored end shear of ground beam
w_{max}	Maximum crack width (mm)
x	Depth of neutral axis in a concrete section from compression face
z	Depth of lever arm
Z	Depth of top of pad foundation below ground level
α	Coefficient of thermal expansion of concrete/°C
γ	Unit weight of soil (kN/m^3)
γ_w	Unit weight of water (kN/m^3)
δ	Angle of friction between soil and concrete
Δ	Horizontal movement of foundation
Δ_{max}	Maximum allowable horizontal movement of foundation
ε_h	Calculated strain in concrete at a depth h from compression face
ε_m	Strain corrected for stiffening effect
ε_r	Tensile strain in concrete due to temperature gradient causing early thermal cracking
ε_s	Strain at centre of steel reinforcement
ε_{mh}	Strain at depth h corrected for stiffening effect
ρ_{crit}	Critical percentage of steel required to distribute early thermal cracking
σ_z	Vertical stress at centre of a layer of soil due to net foundation pressure
ϕ	Angle of internal friction

6.1 ANALYSIS FOR BEARING PRESSURE ON SOIL

6.1.1 Isolated single column pad (bearing pressure calculations)

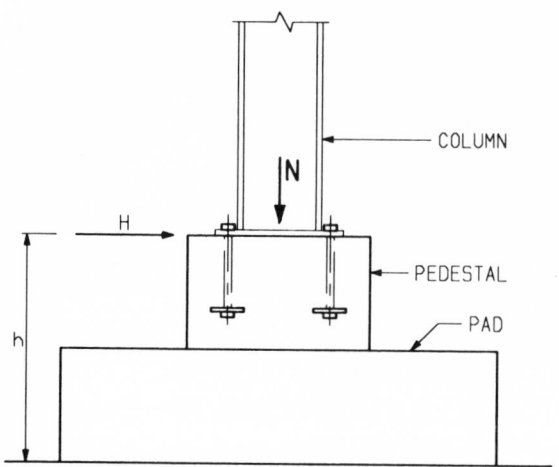

SK 6/1 Typical column foundation in reinforced concrete.

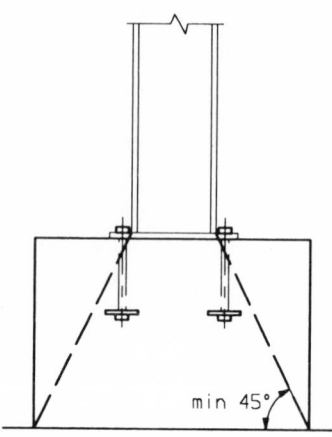

SK 6/2 Typical mass concrete foundation.

Loads from column

N = combined vertical load unfactored

M_x = combined moment about $x-x$ unfactored
M_y = combined moment about $y-y$ unfactored

H_x = combined horizontal shear in x direction unfactored
H_y = combined horizontal shear in y direction unfactored

e_x = eccentricity in x direction of vertical load N from CG of base
e_y = eccentricity in y direction of vertical load N from CG of base

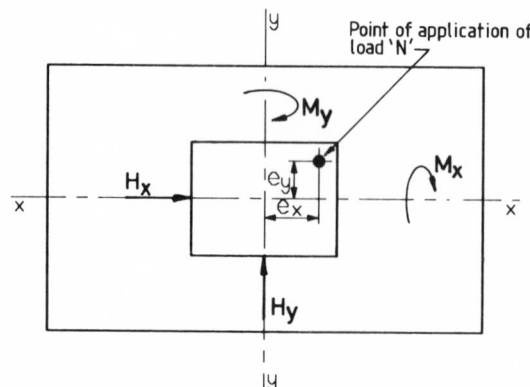

SK 6/3 Typical loads from column on foundation shown on plan.

Loads at underside of pad on soil

P = vertical load = N + weight of foundation + weight of backfill + surcharge on backfill

M_{xx} = moment about $x-x$ = $M_x + Ne_y + H_yh + M_x^*$
M_{yy} = moment about $y-y$ = $M_y + Ne_x + H_xh + M_y^*$

where M_x^* and M_y^* are moments with respect to CG of base due to eccentric surcharge on backfill.

Note: In finding the load on the soil at the underside of the pad footing the directions of the loads, eccentricities and moments must be taken into account. With reversible horizontal loads and moments, all possible combinations should be examined. Eccentric heavy surcharge on part of the backfill on foundation may in certain cases produce higher bearing pressure and should be investigated.

6.1.2 Single column pads connected by ground beams (bearing pressure calculations)

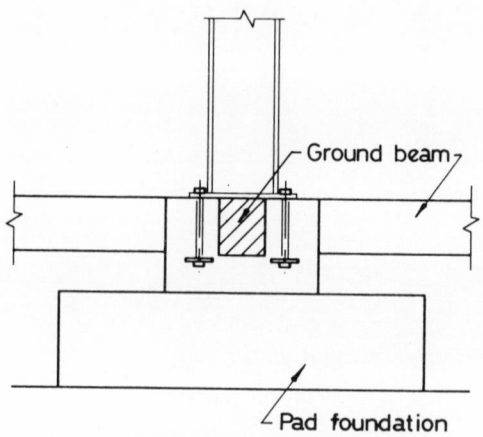

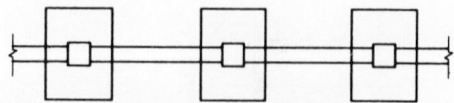

SK 6/5 Plan of foundations connected by ground beams.

SK 6/4 Typical arrangement of ground beams to column foundation.

Assumptions

(1) The pad foundation is assumed rigid and its rotation is very small.
(2) The ground beam may be designed as fixed to the foundation with zero rotation at the ends.
(3) The horizontal loads in any orthogonal direction from all columns with connected foundations will be algebraically added and then divided by the number of columns. The total horizontal load in any direction will be shared equally between connected foundations.
(4) Because of the very high rotational stiffness of the pad foundations relative to the ground beam, it is assumed that the horizontal loads, moments and load eccentricities at the top of the foundation will cause

cantilever moment on the soil—pad foundation interface and the ground beam will be unaffected.

(5) The pad foundation will be designed to resist the fixed-end moments from the connected ground beams. The ground beams may also be designed and detailed as pin-jointed to the foundation when there will be no fixed-end moments on the foundation.

(6) The pad foundation should be designed to resist the fixed-end moments from ground beams due to differential settlements, if any, of connected foundations. The ground beams may also be designed and detailed as pin-jointed to the foundation when the fixed-end moments due to differential settlements will be negligible.

Note: To avoid excessive stresses and serious damage, ground beams should preferably be cast on a compressible or rapidly degradable layer of material such that some free vertical movement is allowed to cater for vertical ground movements and differential settlements.

Loads from columns

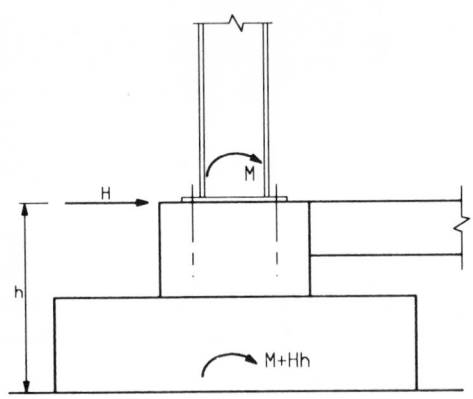

SK 6/6 Horizontal shear causing additional moment.

N = combined vertical load – unfactored

M_x = combined moment about x–x – unfactored

M_y = combined moment about y–y – unfactored

H_x = combined horizontal shear in x direction – unfactored

e_x = eccentricity in x direction of vertical load N from CG of base
e_y = eccentricity in y direction of vertical load N from CG of base

Loads from ground beams

ΣV = combined end shear (vertical) unfactored of all beams

ΣM_{xg} = combined fixed-end moment about x–x unfactored (beams running y–y direction)

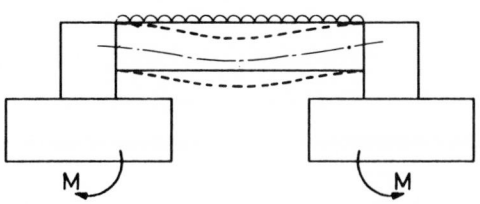

SK 6/7 Fixed-end moments from ground beams.

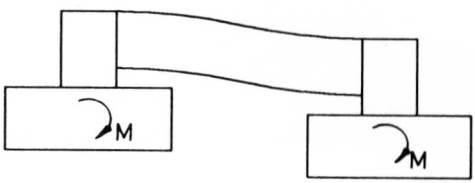

SK 6/8 Fixed-end moments due to differential settlement.

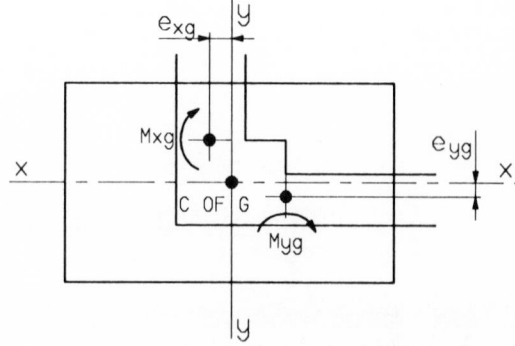

SK 6/9 Bending moment and eccentricity of load from ground beams.

ΣM_{yg} = combined fixed-end moment about $y-y$ unfactored (beams running $x-x$ direction)

e_{xg} = eccentricity of vertical shear V from CG of foundation

e_{yg} = eccentricity of vertical shear V from CG of foundation

Note: M_{xg} and M_{yg} should include the effects of dead load, live load and differential settlements on the ground beam — unfactored.

Loads at underside of pad on soil

P = vertical load = $N + \Sigma V$ + weight of foundation + weight of backfill + surcharge on backfill

M_{xx} = moment on $x-x$ = $M_x + \Sigma M_{xg} + Ne_y + \Sigma(Ve_{yg}) + H_y h + M_x^*$

M_{yy} = moment on $y-y$ = $M_y + \Sigma M_{yg} + Ne_x + \Sigma(Ve_{xg}) + H_x h + M_y^*$

H = horizontal shears = H_x and H_y

where M_x^* and M_y^* are moments with respect to CG of base due to eccentric surcharge on backfill.

Note: In finding the load on the soil at the underside of the pad footing the directions of the loads, eccentricities and moments must be taken into account. With reversible horizontal loads and moments, all possible combinations should be examined. Eccentric heavy surcharge on part of the

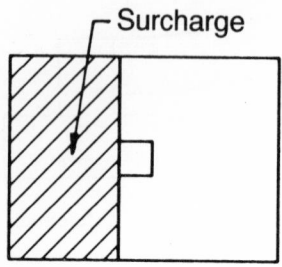

SK 6/10 Eccentricity of surcharge on plan of pad foundation.

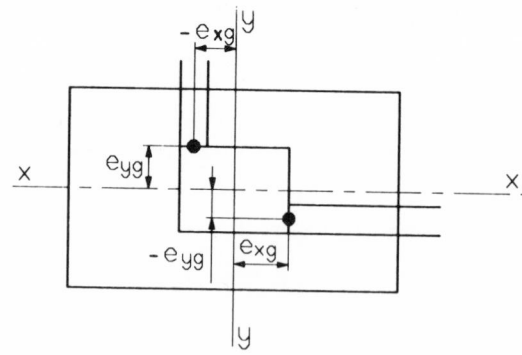

SK 6/11 Eccentricity of load from simply supported end of ground beam.

backfill on the foundation may in certain cases produce higher bearing pressure on the soil and should be investigated.

6.1.3 Isolated multiple column pad (bearing pressure calculation)

Loads from columns

ΣN = summation of all column vertical loads — unfactored

ΣM_x = algebraic summation of all column moments about x–x — unfactored

ΣM_y = algebraic summation of all column moments about y–y — unfactored

ΣH_x = algebraic summation of all column horizontal shears in x-direction — unfactored

ΣH_y = algebraic summation of all column horizontal shears in y-direction — unfactored

e_x = distance in the x-direction of CG of all column vertical loads from CG of base

e_y = distance in the y-direction of CG of all column vertical loads from CG of base

Loads at underside of pad on soil

P = vertical load = ΣN + weight of foundation + weight of backfill + surcharge on backfill

M_{xx} = moment about x–x = $\Sigma M_x + \Sigma N e_y + \Sigma H_y h + M_x^*$

M_{yy} = moment about y–y = $\Sigma M_y + \Sigma N e_x + \Sigma H_x h + M_y^*$

H = horizontal shears = ΣH_x and ΣH_y

where M_x^* and M_y^* are due to eccentric surcharge.

6.1.4   Multiple column pads connected by ground beams (bearing pressure calculations)

Assumptions   See Section 6.1.2.

Loads from columns   See Section 6.1.3.

Loads from ground beams   See Section 6.1.2.

Loads at underside of pad foundation

P = vertical load = $\Sigma N + \Sigma V$ + weight of foundation + weight of backfill + surcharge on backfill

M_{xx} = moment on $x-x$ = $\Sigma M_x + \Sigma Ne_y + \Sigma M_{xg} + \Sigma(Ve_{yg}) + \Sigma H_y h + M_x^*$

M_{yy} = moment on $y-y$ = $\Sigma M_y + \Sigma Ne_x + \Sigma M_{yg} + \Sigma(Ve_{xg}) + \Sigma H_x h + M_y^*$

H = horizontal shear = ΣH_x and ΣH_y

where M_x^* and M_y^* are due to eccentric surcharge on backfill.

6.2   ANALYSIS FOR ULTIMATE LOAD

6.2.1   Isolated single column pad

Note: Use load factors and combinations as stated in Section 6.3.

Load from column

N_u = combined vertical load − factored

M_{xu} = combined moment about $x-x$ − factored

M_{yu} = combined moment about $x-x$ − factored

H_{xu} = combined horizontal shear x-direction − factored

H_{yu} = combined horizontal shear y-direction − factored

e_x = eccentricity of N_u in x-direction

e_y = eccentricity of N_u in y-direction

Loads at underside of pad foundation

$P_u = N_u + 1.4$ (weight of foundation + weight of backfill) + 1.6 (surcharge on backfill)

$M_{xxu} = M_{xu} + N_u e_y + H_{yu} h + M_{xu}^*$

$$M_{yyu} = M_{yu} + N_u e_x + H_{xu}h + M^*_{yu}$$

$$H_u = H_{xu} \quad \text{and} \quad H_{yu}$$

where M^*_{xu} and M^*_{yu} are ultimate moments on base due to eccentric surcharge on backfill.

6.2.2   Single column pads connected by ground beams

Note: Use load factors and combinations as stated in Section 6.3. For assumptions, see Section 6.1.2.

Load from columns   See Section 6.2.1.

Load from ground beams

ΣV_u = combined factored end shear of all beams

ΣM_{xgu} = combined factored end moment about $x-x$ (beams running $y-y$)

ΣM_{ygu} = combined factored end moment about $y-y$ (beams running $x-x$)

e_{xg} = eccentricity of V_u from CG of base in x-direction (beams $x-x$)

e_{yg} = eccentricity of V_u from CG of base in y-direction (beams $y-y$)

Note: M_{xgu} and M_{ygu} should include the effects of dead load, live load and differential settlements on the ground beam.

Loads at underside of pads foundation on soil

$$P_u = N_u + \Sigma V_u + 1.4 \text{ (weight of foundation + weight of backfill)} + 1.6 \text{ (surcharge on backfill)}$$

$$M_{xxu} = M_{xu} + \Sigma M_{xgu} + N_u e_y + \Sigma(V_u e_{yg}) + H_{yu}h + M^*_{xu}$$

$$M_{yyu} = M_{yu} + \Sigma M_{xgu} + N_u e_x + \Sigma(V_u e_{xg}) + H_{xu}h + M^*_{yu}$$

$$H_u = H_{xu} \quad \text{and} \quad H_{yu}$$

where M^*_{xu} and M^*_{yu} are ultimate moments on base due to eccentric surcharge on backfill.

6.2.3   Multiple column pads

Note: For multiple column pad foundations with or without ground beams, use the same philosophy as in Sections 6.2.1 and 6.2.2 but with the following loads from all columns on the base summed up, algebraically.

ΣN_u = summation of all ultimate vertical loads from columns

ΣM_{xu} = summation of all ultimate moments about $x-x$ axis

ΣM_{yu} = summation of all ultimate moments about $y-y$ axis

ΣH_{xu} = summation of all horizontal shears in x-direction

ΣH_{yu} = summation of all horizontal shears in y-direction

e_x = resultant eccentricity of all column vertical loads in x-direction

e_y = resultant eccentricity of all column vertical loads in y-direction

6.3 LOAD COMBINATIONS

Loads from the columns will be combined using the following principles.

6.3.1 Bearing pressure calculations

LC_1: $1.0DL + 1.0IL + 1.0EP + 1.0CLV + 1.0CLH$
No increase in allowable bearing capacity.
LC_2: $1.0DL + 1.0EP + 1.0CLV + 1.0CLH + 1.0WL$ (or $1.0EL$)
25% increase in allowable bearing capacity.
LC_3: $1.0DL + 1.0IL + 1.0EP + 1.0WL$ (or $1.0EL$)
25% increase in allowable bearing capacity.
LC_4: $1.0DL + 1.0WL$ (or $1.0EL$)
25% increase in allowable bearing capacity.

where DL = dead load
IL = imposed load
EP = earth pressure and water pressure
CLV = crane vertical loads
CLH = crane horizontal loads
WL = wind load
EL = earthquake load

6.3.2 Bending moment and shear calculations

LC_5: $1.4DL + 1.6IL + 1.4EP$
LC_6: $1.2DL + 1.2IL + 1.2EP + 1.2WL$ (or $1.2EL$)
LC_7: $1.4DL + 1.4WL$ (or $1.4EL$) $+ 1.4EP$
LC_8: $1.0DL + 1.4WL$ (or $1.4EL$) $+ 1.4EP$ (if adverse)
LC_9: $1.4DL + 1.4CLV + 1.4CLH + 1.4EP$
LC_{10}: $1.4DL + 1.6CLV + 1.4EP$
LC_{11}: $1.4DL + 1.6CLH + 1.4EP$
LC_{12}: $1.2DL + 1.2CLV + 1.2CLH + 1.2EP + 1.2\ WL$ (or $1.2EL$)

6.3.3 Settlement computation

$$LC_{13}: \qquad 1.0DL + 0.5IL$$
$$\text{(vertical direct loads only)}$$

6.4 SIGN CONVENTION

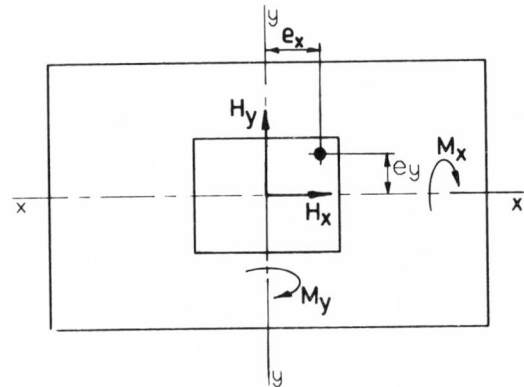

SK 6/12 Sign convention showing positive loads and eccentricity.

To avoid mistakes the following sign convention should be followed:

Vertical loads: downwards positive
Moments on base: clockwise positive
Horizontal shears: left to right positive: +ve x-direction
                              bottom to top positive: +ve y-direction
Eccentricities: +ve for +ve x and +ve for +ve y

6.5 ESSENTIALS OF SOIL MECHANICS

6.5.1 Ultimate bearing capacity

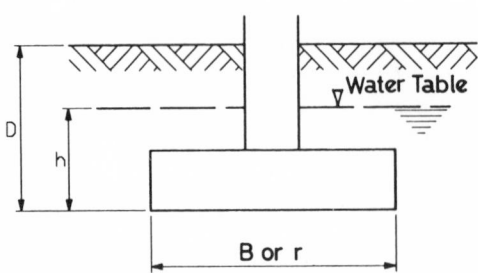

B = Least plan dimension of pad foundation
r = Radius of circular foundation

SK 6/13 Typical parameters for the calculation of bearing capacity.

From soil investigation and laboratory tests the following parameters should be available:

c = soil cohesion (kPa)

ϕ = angle of internal friction (degrees or radians)

γ = unit weight of soil (kN/m^3)

p = total overburden pressure at foundation level

$p_o = p - \gamma_w h$

h = height of water above foundation level

γ_w = unit weight of water (kN/m^3)

q_{ult} = ultimate bearing capacity as per Terzaghi (conservative approach)

Bearing capacity calculations for cohesionless and (c-ϕ) soils.

For continuous foundation

$q_{ult} = cN_c + p_o(N_q - 1) + 0.5\gamma BN_\gamma + p$

For square foundation

$q_{ult} = 1.3cN_c + p_o(N_q - 1) + 0.4\gamma BN_\gamma + p$

For circular foundation

$q_{ult} = 1.3cN_c + p_o(N_q - 1) + 0.3\gamma BN_\gamma + p$

$a = e^{(0.75\pi - \phi/2)\tan\phi}$

$$N_q = \frac{a^2}{2\cos^2(45° + \phi/2)}$$

$N_c = (N_q - 1)\cot\phi$

$$N_\gamma = 0.5\tan\phi \left(\frac{K_{py}}{\cos^2\phi} - 1\right)$$

Table 6.1 Values of K_{py} (as per Reference 6).

ϕ	0	5	10	15	20	25	30	35	40	45	50
K_{py}	10.8	12.2	14.7	18.6	25.0	35.0	52.0	82.0	141.0	298.0	800.0

N_q, N_c and N_γ may be obtained from Fig. 6.1.
Allowable bearing capacity = $q_{ult}/3$

Bearing capacity calculations for cohesive soils

$q_{ult} = cN_c + p$

N_c may be obtained from Table 6.2.
Allowable bearing capacity = $q_{ult}/3$

Table 6.2 Values of N_c for cohesive soils (as per Reference 6).

Types of footing	D/B or $D/2r$	Values of N_c
Circular or	0	6.2
square footing	0.5	7.3
	1.0	8.2
	1.5	9.1
	2.0	9.3
	2.5	9.3
	3.0	9.3
	3.5	9.3
	4.0	9.3
Strip footing	0	5.2
	0.5	6.2
	1.0	7.1
	1.5	7.7
	2.0	8.1
	2.5	8.2
	3.0	8.2
	3.5	8.2
	4.0	8.2

D = depth below ground to underside of pad foundation
B = least plan dimension of pad foundation
r = radius of circular pad foundation

Note: There are many different ways of calculating ultimate bearing capacity which take into account depth, water tables, load inclinations, various shapes of foundations, soil layers, etc. It is normal practice to have the allowable bearing capacities for various sizes of foundations determined by specialists carrying out the site investigation.

In the absence of laboratory tests for finding c, γ and ϕ, values from Tables 6.3 and 6.4 may be used to determine allowable bearing capacity if the description of the soil is known.

Table 6.3 Typical values of angle of internal friction, ϕ.

Soil type	Angle of internal friction, ϕ (degrees)
Medium gravel	40–55
Sandy gravel	45–50
Loose dry sand	28–34
Loose saturated sand	28–34
Dense dry sand	35–46
Dense saturated sand	33–44
Loose silty sand	20–22
Dense silty sand	25–30
Saturated clay	0

Table 6.4 Typical values of cohesive strength, c.

Soil type	Cohesive strength, c (kN/m^2)
Hard boulder clays	>300
Hard fissured clays	>300
Deep London and gault clays	>300
Hard weathered shales	>300
Hard weathered mudstones	>300
Very stiff boulder clay	150–300
Very stiff blue London clay	150–300
Very stiff weathered Keuper Marl	150–300
Stiff boulder clay	75–150
Stiff blue London clay	75–150
Stiff weathered Keuper Marl	75–150
Firm normally consolidated clay	40–75
Upper weathered 'brown' London clay	40–75
Soft normally consolidated clay	20–40

Note: Presumed allowable bearing capacities for various types of soil and grades of chalk and Keuper Marl may be obtained from Tables 2 and 3 of BS 8004: 1986.[2]

6.5.2   Settlement of foundation

Method 1   Quick approximate method

Soil parameter: Coefficient of volume compressibility (in m^2/MN) = m_v

Values of m_v should be available after soil investigation tests. Approximate values of m_v for clays may be obtained from Table 6.5.

Table 6.5 Typical values of coefficient of volume compressibility, m_v.

Soil type	Coefficient of volume compressibility, m_v (m^2/MN)
Heavily overconsolidated boulder clays	<0.05
Stiff weathered rocks	<0.05
Hard London clays	<0.05
Boulder clays	0.05–0.1
Very stiff London clays	0.05–0.1
Upper blue London clays	0.10–0.3
Weathered boulder clay	0.10–0.3
Weathered Keuper Marl	0.10–0.3
Normally consolidated alluvial clays	0.30–1.5
Estuarine clays	0.30–1.5
Organic alluvial clays and peats	>1.5

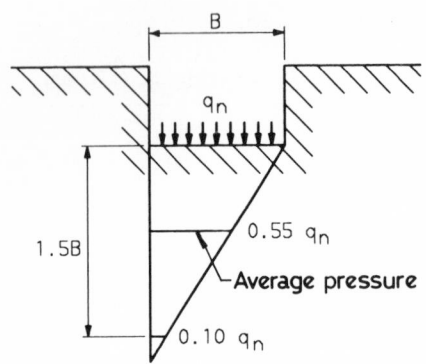

SK 6/14 Pressure distribution for settlement computation.

Consolidation settlement $= m_v \, \sigma_z \, H$

where m_v = average m_v of all layers up to a depth of $1.5B$
B = width of foundation (least dimension)
$\sigma_z = 0.55q_n$ (average pressure in centre of layers)
q_n = net pressure on the soil (MN/m^2)
$H = 1.5B$ (metres)

Note: Immediate settlement is ignored in these calculations.

Method 2   Settlement from static cone penetration tests

Soil parameter: Cone resistance (in kg/cm^2) $= C_{kd}$

Constant of compressibility $= C = \dfrac{147C_{kd}}{p_0}$

where p_0 = effective pressure at the centre of layer $= p - \gamma_w h$
(in kN/m^2)
p = total overburden pressure at the centre of layer (kN/m^2)
h = height of water to the centre of layer (metres)
γ_w = unit weight of water (kN/m^3)

Settlement of layer $= S = \dfrac{H}{C}\log_e\left(\dfrac{p + \sigma_z}{p_0}\right)$ metres

where σ_z = vertical stress at centre of layer (kN/m^2) as a result of net foundation pressure (q_n)
H = thickness of the layer of soil (metres)

σ_z may be obtained from Fig. 6.2 (see Reference 5).

Total estimated settlement = summation of settlement of each layer

Note: The cone penetrometer curve should be broken down into separate layers, each having approximately the same value of cone resistance within the layer. Even if the cone penetrometer curve does not indicate any layering of soil, the settlement should be computed in layers because the value of σ_z falls off rapidly with depth.

6.5.3   Sliding resistance

6.5.3.1   *Sliding of concrete on soil*

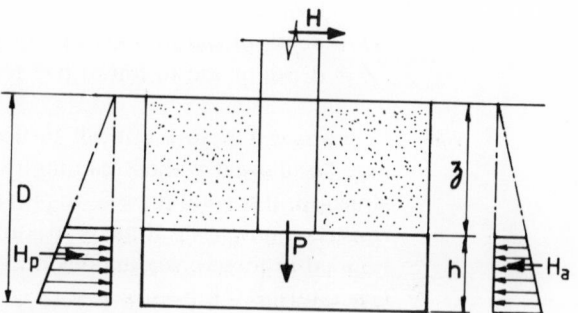

SK 6/15 Sliding resistance of pad foundation.

Sliding resistance of concrete foundation on soil, $P_s = F + H_p - H_a$

where F = frictional resistance under base
H_p = passive resistance due to horizontal movement of foundation
H_a = active pressure due to horizontal movement of foundation

$F = P \tan \delta$

Δ = horizontal movement of foundation into soil (metres)

Δ_{max} = maximum allowable horizontal movement on the basis of soil shear strength (metres)

For cohesionless soil $\Delta_{max} = \left(\dfrac{K_p}{K_h}\right) \gamma$

For cohesive soil $\Delta_{max} = \dfrac{\gamma D + 2c}{K_h}$

$K_h = n_h/B$ for cohesionless soil
$\quad = k_{si}/1.5B$ for cohesive soil

(See Table 6.7 for typical values of n_h and k_{si}.)

h = thickness of concrete pad foundation
δ = friction angle between concrete and soil (see Table 6.6).
ϕ = angle of internal friction of backfill material (see Table 6.3)
$K_p = \tan^2 (45° + \phi/2)$

γ = unit weight of backfill material (kN/m^3)
B = width of foundation over which horizontal soil pressure is active (metres)

H_p = $0.5\Delta BhK_h\ (Z + D)$ for cohesionless soil
$= K_h\Delta Bh$ for cohesive soil

K_a = active pressure coefficient of the backfill material
$= \tan^2\ (45° - \phi/2)$ for cohesionless soil
$=$ negligible   for cohesive soils

H_a = $0.5K_a\ (Z + D)\ Bh\gamma$ for cohesionless soil
P = total vertical load on the soil including the weight of foundation and backfill
D = depth of soil to underside of pad foundation
Z = depth of soil to top of pad foundation

Note: In practice it is very difficult to decide how much horizontal movement may be allowed without causing excessive stresses in other parts of the structure. It is good practice to provide total sliding resistance by frictional resistance only. The Rankine passive pressure coefficient, K_p, should not be used in these computations because a large movement is necessary to generate the full passive resistance. A factor of safety of 1.5 should be allowed against sliding.

Check: $P_s \geq 1.5H$

6.5.3.2   *Horizontal bearing capacity of soil*

Allowable horizontal bearing capacity of soil, $P_H = qA \tan\left(\dfrac{\phi}{1.5}\right) + \dfrac{2}{3}cA$

where c = cohesion of foundation soil (kN/m^2)
A = area of foundation
q = average unit pressure under foundation (kN/m^2)
ϕ = angle of internal friction of foundation soil

Check: $P_H \geq F$

6.6   BEARING PRESSURE CALCULATIONS

6.6.1   Rectangular pad – uniaxial bending – no loss of contact

$e_x = \dfrac{M_{yy}}{P}$ $e_y = 0$

$p_1 = \left(\dfrac{P}{AB}\right) + \left(\dfrac{6M_{yy}}{A^2B}\right)$ $p_2 = \left(\dfrac{P}{AB}\right) - \left(\dfrac{6M_{yy}}{A^2B}\right)$

Table 6.6 Typical values of friction angle between concrete and soil, δ.

Concrete on the following soil types	Friction angle, δ (degrees)
Clean sound rock	35
Clean gravel, gravel−sand mixtures	29−31
Coarse sand	29−31
Clean fine-to-medium sand	24−29
Silty medium-to-coarse sand	24−29
Clayey gravel	24−29
Clean fine sand	19−24
Silty-to-clayey fine-to-medium sand	19−24
Fine sandy silt	17−19
Very stiff and hard residual clay	22−26
Medium stiff and stiff clay	17−19
Bituminous or water-proofing membrane	0−5

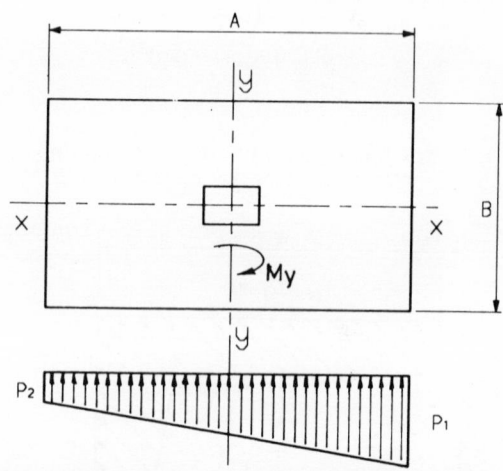

PRESSURE DIAGRAM FOR NO LOSS OF CONTACT

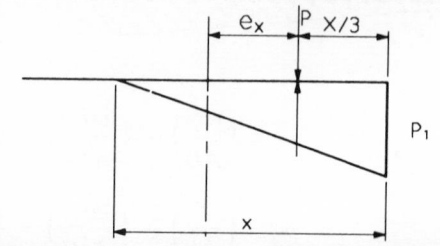

SK 6/16 Pressure diagrams for uniaxial bending and direct load on base.

PRESSURE DIAGRAM FOR LOSS OF CONTACT

Table 6.7 Typical coefficients of horizontal modulus of subgrade reaction.

	Values of n_h (cohesionless) (MN/m³)		
	Loose	Medium	Dense
Dry or moist sand	2.2	6.6	17.6
Submerged sand	1.26	4.4	10.7
	Values of k_{si} (cohesive) (MN/m³)		
Types of clay	Stiff	Very stiff	Hard
	7.2	14.4	28.8

6.6.2   Rectangular pad − uniaxial bending − loss of contact

$$p_1 = \frac{2P}{(1.5A - 3e_x)B}$$

$$x = 1.5A - 3e_x$$

6.6.3   Rectangular pad − biaxial bending − no loss of contact

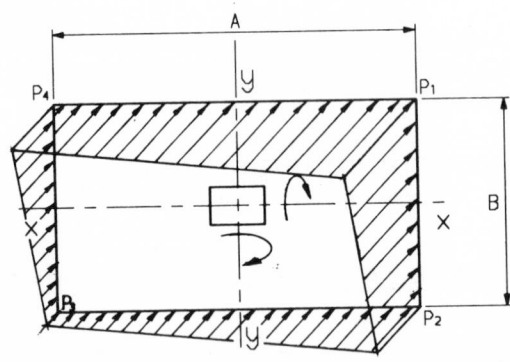

SK 6/17 Typical pressure diagram for biaxial bending and no loss of contact.

$$p_1 = \left(\frac{P}{AB}\right) + \left(\frac{6M_{yy}}{A^2B}\right) + \left(\frac{6M_{xx}}{AB^2}\right)$$

$$p_2 = \left(\frac{P}{AB}\right) + \left(\frac{6M_{yy}}{A^2B}\right) - \left(\frac{6M_{xx}}{AB^2}\right)$$

$$p_3 = \left(\frac{P}{AB}\right) - \left(\frac{6M_{yy}}{A^2B}\right) - \left(\frac{6M_{xx}}{AB^2}\right)$$

$$p_4 = \left(\frac{P}{AB}\right) - \left(\frac{6M_{yy}}{A^2B}\right) + \left(\frac{6M_{xx}}{AB^2}\right)$$

6.6.4 Rectangular pad – biaxial bending – loss of contact

Partial contact of soil/foundation (see Fig. 6.3).
The resultant of soil pressure diagram under base has co-ordinates e_x and e_y.

6.6.4.1 *Resultant in Zone 1 of base* (see Fig. 6.3)
Factor of safety for overturning is less than 1.5. Redesign size of base.

6.6.4.2 *Resultant in Zone 2 of base* (see Fig. 6.3)
No loss of contact of base. Calculate pressures as in Section 6.6.3.

6.6.4.3 *Resultant in Zone 3 of base* (see Fig. 6.3)

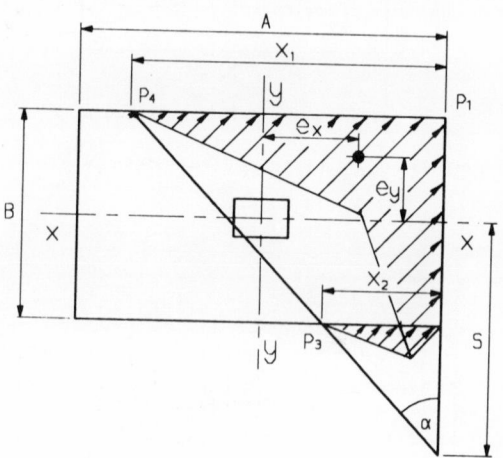

SK 6/18 Biaxial bending – loss of
contact. Zero pressure on line
p_3-p_4.

$$e_x = \frac{M_{yy}}{P} \qquad e_y = \frac{M_{xx}}{P}$$

$$S = \frac{B}{12}\left[\frac{B}{e_y} + \left(\frac{B^2}{e_y^2} - 12\right)^{\frac{1}{2}}\right]$$

$$\tan \alpha = \frac{3(A - 2e_x)}{2(S + e_y)}$$

$$p_1 = \left(\frac{12P}{B\tan\alpha}\right)\left(\frac{B + 2S}{B^2 + 12S^2}\right)$$

$$p_2 = \left[\frac{S - \dfrac{B}{2}}{S + \dfrac{B}{2}}\right] p_1$$

$$p_3 = p_4 = 0$$

$$x_1 = \left(S + \frac{B}{2}\right)\tan\alpha \qquad x_2 = \left(S - \frac{B}{2}\right)\tan\alpha$$

6.6.4.4 Resultant in Zone 4 of base (see Fig. 6.3)

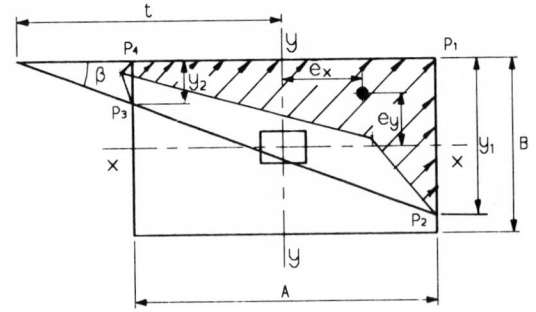

SK 6/19 Biaxial bending — loss of contact. Zero pressure on line p_2–p_3.

$$t = \frac{A}{12}\left[\frac{A}{e_x} + \left(\frac{A^2}{e_x^2} - 12\right)^{\frac{1}{2}}\right]$$

$$\tan\beta = \frac{3(B - 2e_y)}{2(t + e_x)}$$

$$p_1 = \left(\frac{12P}{A\tan\beta}\right)\left(\frac{A + 2t}{A^2 + 12t^2}\right)$$

$$p_4 = \left[\frac{t - \dfrac{A}{2}}{t + \dfrac{A}{2}}\right] p_1$$

$$p_2 = p_3 = 0$$

$$y_1 = \left(t + \frac{A}{2}\right)\tan\beta \qquad y_2 = \left(t - \frac{A}{2}\right)\tan\beta$$

6.6.4.5   *Resultant in Zone 5 of base* (see Fig. 6.3)

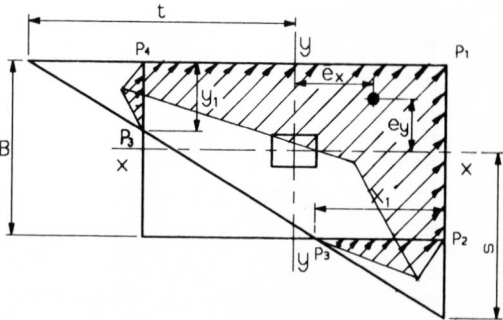

SK 6/20 Biaxial bending – loss of contact. Zero pressure on line p_3–p_3.

$$k = \frac{e_x}{A} + \frac{e_y}{B}$$

$$p_1 = \left(\frac{P}{AB}\right)k[12 - 3.9(6k - 1)(1 - 2k)(2.3 - 2k)]$$

S and t are as in Sections 6.6.4.3 and 6.6.4.4.

$$p_2 = \left[\frac{S - \dfrac{B}{2}}{S + \dfrac{B}{2}}\right] p_1$$

$$p_3 = 0$$

$$p_4 = \left[\frac{t - \dfrac{A}{2}}{t + \dfrac{A}{2}}\right] p_1$$

$$x_1 = \left(S - \frac{B}{2}\right)\left[\frac{t + \dfrac{A}{2}}{S + \dfrac{B}{2}}\right] \qquad y_1 = \left(t - \frac{A}{2}\right)\left[\frac{S + \dfrac{B}{2}}{t + \dfrac{A}{2}}\right]$$

Note: To find maximum pressure at a corner the design chart in Fig. 6.4 may be used. At the initial design stage when the size of the foundation is being determined, this design chart becomes very useful.

Find e_x/A and e_y/B.
Read from Fig. 6.4 the value of K.

$$\text{Maximum pressure} = \frac{PK}{AB}$$

6.6.5 Multiple column − biaxial bending − no loss of contact

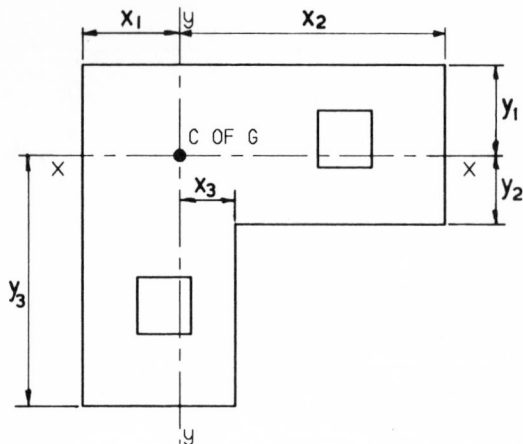

SK 6/21 Typical unsymmetrical pad foundation and co-ordinates of corners to find bearing pressure.

(1) Find area of foundation, A.
(2) Find centre of gravity of plan area of foundation.
(3) Find second moments of area about two orthogonal axes going through CG of area of foundation (I_{xx} and I_{yy}).
(4) Find maximum ordinates X and Y of corner points of foundation with respect to CG of area of foundation.
(5) Find eccentricities of all vertical loads from CG of area of foundation.
(6) Find total vertical load through CG of foundation and total moments about orthogonal axes passing through CG of area of foundation (P, M_{xx} and M_{yy}).
(7) Find maximum and minimum pressures at various points on foundation using the equation below.

$$P_n = \frac{P}{A} + \frac{M_{xx}y_n}{I_{xx}} + \frac{M_{yy}x_n}{I_{yy}}$$

Note: This method is valid only when there is no loss of contact between the foundation and the soil. Use a consistent sign convention as in Section 6.4.

6.6.6 Circular pad − biaxial bending − no loss of contact

$$A = \frac{\pi D^2}{4} = 0.7854D^2$$

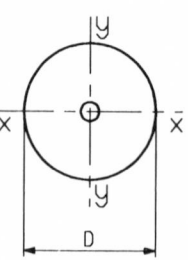

SK 6/22 Circular pad foundation.

$$Z = \frac{\pi D^3}{32} = 0.0982D^3$$

$$M = (M_{xx}^2 + M_{yy}^2)^{\frac{1}{2}}$$

$$e = \frac{M}{P} \leq \frac{D}{8}$$

$$p_{max} = \frac{P}{A} + \frac{M}{Z}$$

Design forces in pad foundation

Maximum shear force across diameter $D = V = 1.285pR^2 + 1.571qR^2$

Maximum bending moment across diameter $D = M_1 = 0.595pR^3 + 0.667qR^3$

where R = radius of circular pad = $D/2$
q = minimum pressure = $(P/A) - (M/Z)$
$p + q$ = maximum pressure = $(P/A) + (M/Z)$
$p = 2M/Z$.

6.7   Step-by-step design procedure for pads

Step 1   *Select type and depth of foundation*
The types of foundations are as follows:

(A) Reinforced concrete pad with single column.
(B) Reinforced concrete pad with multiple column.
(C) Reinforced concrete pad with single column and ground beams.
(D) Reinforced concrete pad with multiple column and ground beams.
(E) Mass concrete pad with single column.

Note: Type E may be used for light single-storey structures only.

The depth of the foundation is governed by the following:

- Shrinking and swelling of clay.
- Frost attacks.
- Holding-down bolt arrangement of columns.
- Suitable bearing stratum.
- Water table and soluble sulphates.
- Width of foundation which is normally kept more than depths for shallow foundations.

Step 2   *Select approximate size*
From the ground investigation report and from Tables 1–3 of BS 8004: 1986,[2] find the presumed allowable bearing capacity.
Find total maximum unfactored vertical load from column.
Find maximum unfactored moments M_x and M_y from column.

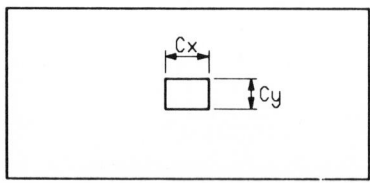

SK 6/23 Equivalent eccentricity of direct load on pad foundation.

Find eccentricities $e_x = \dfrac{M_{yy}}{P}$ and $e_y = \dfrac{M_{xx}}{P}$

Assume $A \geq 6e_x$ and $B \geq 6e_y$

For biaxial bending,

find e_x/A and e_y/B, and from Fig. 6.4 find the value of K.

Maximum approximate pressure $= \dfrac{PK}{AB}$

Check whether the maximum pressure is lower than presumed allowable bearing capacity.

Note: At this stage some of the loads from the self-weight of the foundation, ground beams, backfill, eccentricities of surcharges, etc. have not been included and hence a margin has to be left in the bearing pressure to account for these. Moreover, the actual bearing pressure computations and settlement computations may further enhance the size of the foundation.

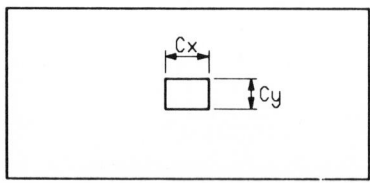

SK 6/24 Dimensions of column or pedestal on pad foundation.

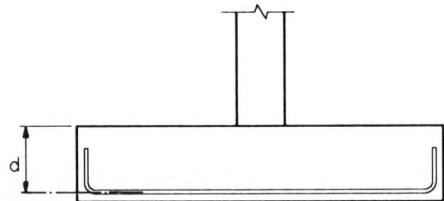

SK 6/25 Effective depth of pad.

Determine minimum thickness of pad:
Find $v_{max} = 0.8\sqrt{f_{cu}}$ or $5\,\text{N/mm}^2$ whichever is lesser
Find column perimeter $= U_o = 2(C_x + C_y)$.
Find total ultimate vertical load from column $= N_u$
Find $d \geq N_u/(v_{max}\,U_o)$
also $d \geq 0.5[(C_1^2 + 4C_2)^{0.5} - C_1]$ whichever gives larger d.

$$C_1 = U_o/6$$
$$C_2 = N_u/12v_c \quad \text{(assume } v_c \text{, which is dependent on percentage of tensile reinforcement)}$$

Choose overall depth of pad allowing for cover.

Note: Consideration need not be given to anchorage length of column bars in pad foundation if all column bars are in compression.

Step 3 **Calculate bearing capacity of soil**
Follow Section 6.5.

Step 4 **Calculate column load combinations**
Follow Section 6.3.

Step 5 **Calculate approximate settlement**
Follow Method 1 of Section 6.5.2.

Note: The approximate settlement computation will help to determine the level of differential settlement for which the building should be designed.

Step 6 **Carry out analysis for bearing pressure**
Follow Section 6.1.

Step 7 **Calculate bearing pressures under foundation**
Follow Section 6.6.

Step 8 **Calculate sliding resistance of foundation**
Follow Section 6.5.3.

Step 9 **Check combined sliding and bearing**
Check: $\left(\dfrac{P}{P_v}\right) + \left(\dfrac{H_x}{P_{Hx}}\right) + \left(\dfrac{H_y}{P_{Hy}}\right) \leq 1$

where P = total vertical load unfactored
P_v = area of base $\times$ allowable bearing capacity
H_x, H_y = total horizontal load unfactored in x- and y-directions
P_{Hx}, P_{Hy} = sliding resistance of base (Step 8) in x- and y-directions (see Section 6.5.3.)

Step 10 **Carry out analysis of bearing pressure for bending moment and shear**
Follow Section 6.2.

Step 11 **Calculate bearing pressure for bending moment and shear**
Follow Section 6.6.

Step 12 **Calculate bearing moments and shears in pad**
Critical sections for bending moments and shears:

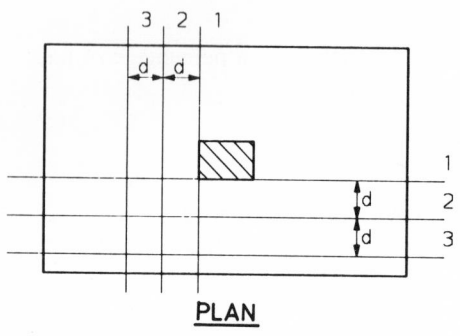

SK 6/26 Critical section for checking bending moments and shears in pad foundation.

SK 6/27 Critical section for bending moment.

- Sections 1, 4 − design bending moment.
- Section 1 − shear.
- Section 2 − shear.
- Section 3 − shear.

Note: When calculating bending moments and shears the downward loads on the pad will be considered with the upwards loads.

In a complicated unsymmetrical combined column foundation the bending moment and shear force diagram based on the pressure distribution should be drawn and critical sections determined from these diagrams.

Step 13 **Determine cover to reinforcement**

From the soil investigations report, find the concentration of sulphates expressed as SO_3.

From Table 17 of BS 8004: 1986,[2] find the appropriate type of concrete.

Class of exposure	Total SO_3 (%)	Minimum cover on blinding concrete (mm)	Minimum cover elsewhere (mm)
1	<0.2	35	75
2	0.2−0.5	40	80
3	0.5−1.0	50	90
4	1.0−2.0	60	100
5	>2.0	60	100

Note: Concrete in 'class of exposure 5' needs protective coating.

Step 14 **Calculate area of tension reinforcement and distribution**

M = bending moment due to ultimate loads in pad

Find effective depth, d.

b = total width of section

$$K = \frac{M}{f_{cu}bd^2} \le 0.156$$

$$z = d\left[0.5 + \sqrt{\left(0.25 - \frac{K}{0.9}\right)}\right] \le 0.95d$$

$$A_s = \frac{M}{0.87f_y z}$$

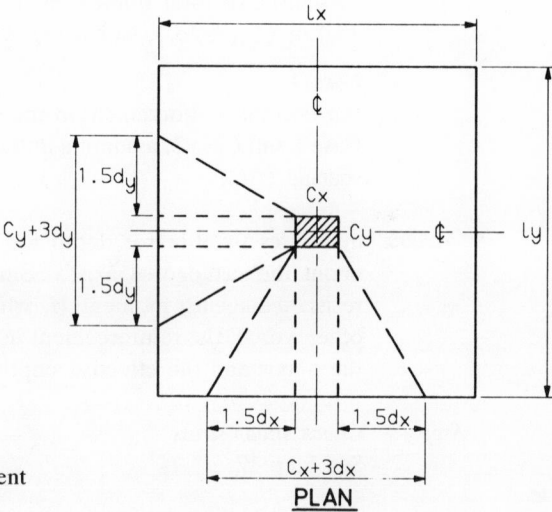

PLAN

SK 6/28 Rules for distribution of reinforcement in pad foundation.

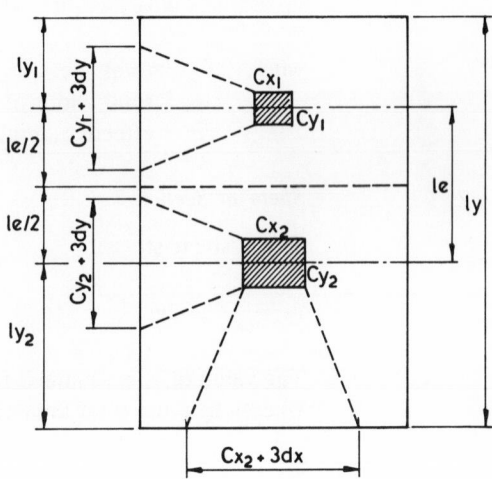

SK 6/29 Detailing rules for pad foundation with two columns.

Note: Increase depth of pad foundation if K is greater than 0.156.

Distribution of tension reinforcement

Case 1

If $l_x > 1.5 \ (C_x + 3d_x)$, distribute two-thirds of total reinforcement in y-direction in the zone $(C_x + 3d_x)$.

Case 2

If $l_y > 1.5 \ (C_y + 3d_y)$, distribute two-thirds of total reinforcement in x-direction in the zone $(C_y + 3d_y)$.

Case 3

If $l_x > 1.5 \ (C_{x1} + 3d_x)$ or $1.5(C_{x2} + 3d_x)$, whichever is the lesser, distribute two-thirds of total reinforcement (top) in y-direction in the zone $(C_{x1} + 3d_x)$ or $(C_{x2} + 3d_x)$, whichever is the lesser.

Case 4

For bottom reinforcement in the combined foundation, follow the rules in Case 1 and Case 2, assuming individual foundations to the centre of column spacing $(l_c/2)$.

Note: d_x relates to effective depth for resistance against moment M_x which is about the orthogonal axis x. Similarly, d_y relates to effective depth for resistance against moment M_y which is about the orthogonal axis y, or, in other words, the reinforcement in the x-direction is to resist moment about the y axis and the effective depth is d_y.

Step 15   ***Check shear stress***
See Step 12.

Shear at Section 1

Check shear stress:

$$v_1 = \frac{V_1}{bd} \le 0.8\sqrt{f_{cu}} \quad \text{or} \quad 5 \,\text{N/mm}^2$$

where $V_1 = $ total shear at Section 1
$b = $ total width of Section 1
$d = $ effective depth of Section 1.

Shear at Section 2

Check shear stress:

$$v_2 = \frac{V_2}{bd} \le 2v_c$$

The value of v_c is obtained from Figs 11.2 to 11.5, depending on $100A_s/bd$ where A_s is the total area of tensile reinforcement in the section.

Shear at Section 3

Check shear stress:

$$v_3 = \frac{V_3}{bd} \le v_c$$

Note: Change the thickness of the pad if the shear stress at any section exceeds the allowable limit. It is not cost-effective to provide shear reinforcement in the pad foundation.

Step 16 Check punching shear

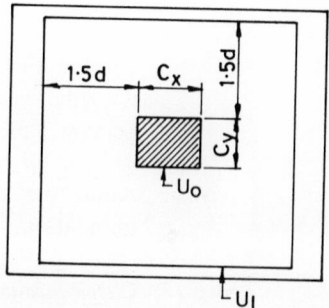

SK 6/30 Punching shear perimeters in pad foundation.

N_u = maximum ultimate vertical load from column

Find perimeters U_o, U_1 and U_2.

$$d = 0.5(d_x + d_y)$$

$$U_o = 2(C_x + C_y)$$
$$U_1 = (U_o + 12d)$$

Check: $\quad v_o = \dfrac{N_u}{U_o d} \le 0.8\sqrt{f_{cu}} \quad \text{or} \quad 5\,\text{N/mm}^2$

$$v_1 = \frac{N_u - p_1 A_1}{U_1 d} \le v_c$$

$$A_1 = (C_x + 3.0d)(C_y + 3.0d)$$

where p_1 = average upwards pressure over area A_1 enclosed by perimeter U_1.

The value of v_c is obtained from Figs 11.2 to 11.5 corresponding to p_x or p_y, whichever is the lesser.

$$p_x = \frac{100A_{sx}}{l_x d_x}$$

$$p_y = \frac{100A_{sy}}{l_y d_y}$$

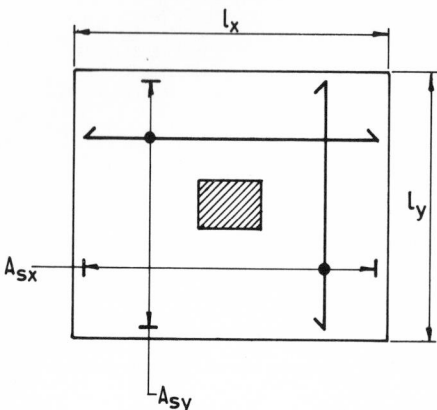

SK 6/31 Typical tensile reinforcement plan for pad foundation.

Change the thickness of the pad if the punching shear stress exceeds v_c, otherwise shear reinforcement will be required as per Step 7 of Section 3.3.

Note: Apply the same principle individually to each column for a combined foundation.

Step 17   **Check minimum reinforcement for flexure**
Minimum tensile reinforcement $= 0.0013bh$ in both directions ($f_y = 460 \, \text{N/mm}^2$)

Note: Provide this minimum reinforcement also at the top of the foundation where top reinforcement is required for flexure.

Step 18   **Check spacing of reinforcement**
Clear spacing of bars should not exceed $3d$ or 750 mm.

Percentage of reinforcement $100A_s/bd$ in pad (%)	Maximum clear spacing of bars for $f_y = 460 \, \text{N/mm}^2$ (mm)
1 or over	160
0.75	210
0.5	320
0.3	530
less than 0.3	$3d$ or 750

Note: The above rules for spacing of bars in tension will in most cases ensure adequate control of crack widths to 0.3 mm where the cover does not exceed 50 mm.

Step 19   **Check early thermal cracking**
Determine pour configuration:

(1) Massive pour cast on blinding concrete: $R = 0.1$ to 0.2
(2) Massive pour cast on existing mass concrete: $R = 0.3$ to 0.4 at base
$R = 0.1$ to 0.2 at top

where $R =$ restraint factor.

Determine the value of temperature, T_1, for OPC concrete cast on ground from the table.

Section thickness (mm)	T_1(°C)
300	17
500	28
700	28
1000	28

Calculate $\varepsilon_r = 0.8 T_1 \alpha R$

where $\alpha = 12 \times 10^{-6}/°C$, or values from Table 2.3 may be used

$$W_{max} = \frac{3 a_{cr}\, \varepsilon_r}{1 + 2 \left(\dfrac{a_{cr} - C_{min}}{h - x} \right)}$$

Note: The design crack width is 0.3 mm. If this is exceeded, closer spacings of bars may be used.

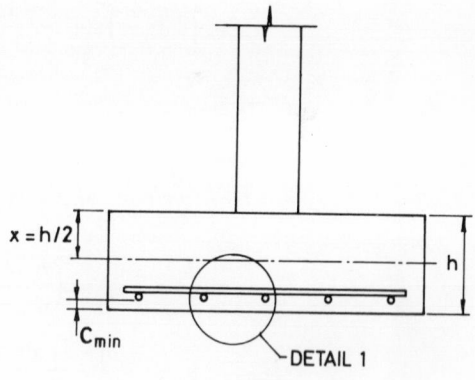

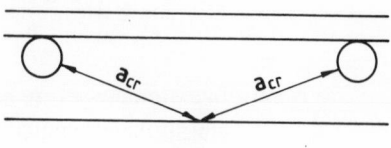

SK 6/32 Early thermal crack width calculation.

Step 20 **Check minimum reinforcement to distribute thermal cracking**

$\rho_{crit} = 0.0035$ for $f_y = 460\,\text{N/mm}^2$

Up to 300 mm thickness of pad foundation

$A_{st} = 0.00175bh$ near top surface in each direction

From 300 mm to 500 mm thickness of pad foundation

$A_{st} = 0.001\,75bh\,\text{mm}^2$ near top surface in each direction

$A_{sb} = 0.35b\,\text{mm}^2$ near bottom surface in each direction

Over 500 mm thickness of pad foundation

$A_{st} = 0.875b\,\text{mm}^2$ near top surface in each direction

$A_{sb} = 0.35b\,\text{mm}^2$ near bottom surface in each direction

where b = width of pad perpendicular to the direction of reinforcement (mm)

h = overall thickness of pad (mm).

Step 21 **Check crack width due to flexure**

Serviceability limit state
Loading conditions LC_1 to LC_4 in Section 6.3.
Find bending moment M across a critical section, as in Step 12.

$$m = \frac{E_s}{E_c} = 15 \qquad \text{for long-term loading}$$

$$x = \left[\left(\frac{mA_s}{b}\right)^2 + \frac{2mA_sd}{b}\right]^{\frac{1}{2}} - \frac{mA_s}{b}$$

$$z = d - \frac{x}{3}$$

$$f_s = \frac{M}{A_sz}$$

$$\varepsilon_s = \frac{f_s}{E_s} \qquad \varepsilon_h = \left(\frac{h-x}{d-x}\right)\varepsilon_s$$

$$\varepsilon_{mh} = \varepsilon_h - \frac{b(h-x)^2}{3E_sA_s(d-x)}$$

$$W_{max} = \frac{3a_{cr}\,C_{min}}{1 + \dfrac{2(a_{cr} - C_{min})}{h-x}}$$

Note: In extremely severe exposure conditions it is prudent to check crack widths and provide adequate reinforcement to limit it to an allowable value.

Step 22 Design mass concrete foundation

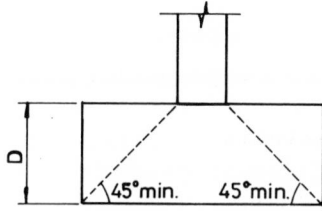

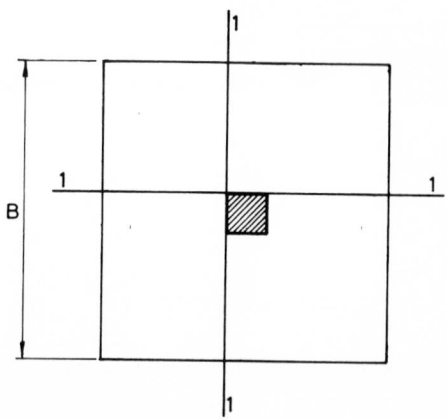

SK 6/34 Critical sections in mass concrete foundation.

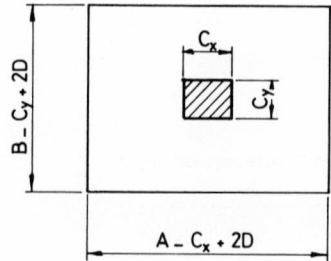

SK 6/33 Typical plan and elevation of mass concrete pad foundation.

The size and depth will be determined based on a 45° dispersion of load from column through mass concrete on to soil.

D = depth of foundation.

At critical section 1, find the bending moment and shear from the soil pressure diagram for unfactored loads: serviceability limit state.

M = bending moment across critical section

V = shear across critical section

Check: f_{tb} = stress in concrete in bending tension

$$= \frac{6M}{BD^2} \quad \text{or} \quad \frac{6M}{AD^2} \leq 0.37\sqrt{f_{cu}}$$

v_d = shear stress in concrete

$$= \frac{V}{BD} \quad \text{or} \quad \frac{V}{AD} \leq 0.037 f_{cu}$$

Punching shear stress $= \dfrac{V}{2(C_x + C_y)D} \leq 0.037 f_{cu}$

Bearing stress under column base plate $= \dfrac{V}{C_x C_y} \leq 0.275 f_{cu}$

Step 23 ***Calculate settlement***
Follow Method 2 of Section 6.5.2.
Use load combination LC_{13} of Section 6.3.

Note: The settlement calculations should be carried out to give a better under-
standing of the global effects on the structure. It may be necessary to alter
the sizes of some of the pad foundations in a structure in order to even out
the differential settlements. It is also important in certain cases to feed
back these settlements in the analysis of the structure.

6.8 WORKED EXAMPLES

Example 6.1 ***RC pad with single column***
Internal column.

Column size $= 400\,\text{mm} \times 400\,\text{mm}$
Column spacing $= 6\,\text{m} \times 6\,\text{m}$ on plan

The unfactored column loads are given in the following table:

	Dead	Imposed	Wind
Vertical load N (kN)	610	480	—
Horizontal shear H_x (kN)	—	—	42
Horizontal shear H_y (kN)	—	—	38
Moment M_x (kNm)	—	—	95
Moment M_y (kNm)	—	—	105

Suitable bearing stratum at 1000 mm below ground level. Medium dense
silty sand.

Step 1 ***Select type and depth of foundation***

Type: Reinforced concrete pad with single RC column

Depth: 1000 mm below finished ground level
      1150 mm below finished floor level.

Depth selected from considerations of:

• Frost attack.
• Swelling of soil.
• Suitable bearing stratum.

Step 2 ***Select approximate size***

Presumed allowable bearing capacity from BS 8004: 1986[2] $= 150\,\text{kN/m}^2$

Maximum Vertical load $V = 610 + 480 = 1090\,\text{kN}$

Maximum eccentricity $= e_x = M_y/V = 105/1090 = 0.1\,\text{m}$

$6e_x = 0.6\,\text{m} \le A$

$$\frac{V}{150} = 7.3\,\text{m}^2$$

Assume a $3\,\text{m} \times 3\,\text{m}$ foundation pad with area of $9\,\text{m}^2$: $A = 3\,\text{m}$ and $B = 3\,\text{m}$.

Determine minimum thickness of pad
Assume grade of concrete $= \text{C30}$

$v_{max} = 0.8\sqrt{f_{cu}}$ or $5\,\text{N/mm}^2$ (whichever is lesser) $= 4.38\,\text{N/mm}^2$

$U_o = 2(C_x + C_y) = 2(400 + 400) = 1600\,\text{mm}$

Total factored load from column

$N_u = 1.4 \times 610 + 1.6 \times 480 = 1622\,\text{kN}$

$$d \ge \frac{N_u}{v_{max}U_o} = \frac{1622 \times 10^3}{4.38 \times 1600} = 231\,\text{mm}$$

or $\quad d \ge \frac{1}{2}[(C_1^2 + 4C_2)^{\frac{1}{2}} - C_1] = 430\,\text{mm}$

where $\quad C_1 = \dfrac{U_o}{6} = \dfrac{1600}{6} = 267\,\text{mm}$

$\qquad\quad C_2 = \dfrac{N_u}{12v_c} = \dfrac{1622 \times 10^3}{12 \times 0.45} = 300\,370\,\text{mm}^2$

Assumed $v_c = 0.45\,\text{N/mm}^2$ which corresponds to about 0.3% tension reinforcement for $f_{cu} = 30\,\text{N/mm}^2$. Choose overall depth of pad equal to $500\,\text{mm}$ allowing for adequate cover.

Step 3 ***Calculate bearing capacity of soil***
(See Section 6.5.1.)

Note: This step may not be necessary if the allowable bearing capacity for the selected size of foundation is already available from the soils investigation report. However, for completeness of a foundation design problem, this step is included.

From field and laboratory tests the following soil parameters of the bearing stratum are known:

ground water table $= 2.0\,\text{m}$ below ground level

$h = 0$, i.e. height of water table above foundation level is zero.

γ = unit weight of soil $= 18\,\text{kN/m}^3$

$\quad p$ = total overburden pressure at the foundation level.
$\qquad = 18 \times 1$ (below ground)
$\qquad = 18\,\text{kN/m}^2$

$$p_o = p - \gamma_w h = 18 \, \text{kN/m}^2$$

γ_w = unit weight of water

c = soil cohesion = $10 \, \text{kN/m}^2$

ϕ = angle of internal friction = $22° = 0.384$ radian

$$a = e^{(0.75\pi - \phi/2) \tan \phi}$$
$$= e^{(0.75\pi - 0.192) \tan 22°}$$
$$= 2.3974$$

$$N_q = \frac{a^2}{2 \cos^2 \left(45° + \dfrac{\phi}{2}\right)} = 9.19$$

$$N_c = (N_q - 1) \cot \phi = 20.3$$

$K_{py} = 29.0$ from Table 6.1 in Section 6.5.1

$$N_\gamma = \frac{\tan \phi}{2} \left(\frac{K_{py}}{\cos^2 \phi} - 1\right) = 6.6$$

$$\begin{aligned}
q_{ult} &= 1.3cN_c + p_o(N_q - 1) + 0.4\gamma B N_\gamma + p \\
&= 1.3 \times 10 \times 20.3 + 18 \times (9.19 - 1) + 0.4 \times 18 \times 3 \times 6.6 + 18 \\
&= 572 \, \text{kN/m}^2
\end{aligned}$$

$$\text{Allowable bearing capacity} = \frac{q_{ult}}{3} = \frac{572}{3} = 190 \, \text{kN/m}^2$$

Step 4   Calculate column load combinations
(See Section 6.3.)

Bearing pressure calculations

$LC_1 = 1.0DL + 1.0IL$
$LC_3 = 1.0DL + 1.0IL + 1.0WL$

LC_1: Combined vertical column load, $N = 610 + 480 = 1090 \, \text{kN}$
$\qquad\quad H_x = 0 \qquad H_y = 0 \qquad M_x = 0 \qquad M_y = 0$

LC_3: N = vertical load = $1090 \, \text{kN}$
$\qquad\quad H_x = 42 \, \text{kN} \qquad M_y = 105 \, \text{kNm} \qquad H_y = 0 \qquad M_x = 0$

Note: By inspection, wind in one direction only may be checked for a square
foundation.

Bending moment and shear calculations

$LC_5 = 1.4DL + 1.6IL$
$LC_6 = 1.2DL + 1.2IL + 1.2WL$
$LC_7 = 1.4DL + 1.4WL$

LC_5: $N_u = 1.4 \times 610 + 1.6 \times 480 = 1622 \, \text{kN}$
$\qquad\quad H_{xu} = 0 \qquad H_{yu} = 0 \qquad M_{xu} = 0 \qquad M_{yu} = 0$

LC_6: $N_u = 1.2 \times 610 + 1.2 \times 480 = 1308\,\text{kN}$
$H_{xu} = 1.2 \times 42 = 50.4\,\text{kN}$ $H_{yu} = 0$
$M_{xu} = 0$ $M_{yu} = 1.2 \times 105 = 126\,\text{kNm}$

LC_7: $N_u = 1.4 \times 610 = 854\,\text{kN}$
$H_{xu} = 1.4 \times 42 = 58.8\,\text{kN}$ $H_{yu} = 0$
$M_{yu} = 1.4 \times 105 = 147\,\text{kNm}$

Step 5 *Calculate approximate settlement*
This step may be ignored since the foundations are not connected by ground beams and the differential settlements will have little effect on the design of this foundation.

Step 6 *Carry out analysis for bearing pressure*

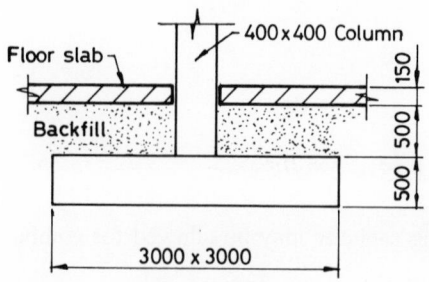

SK 6/35 Section through pad foundation.

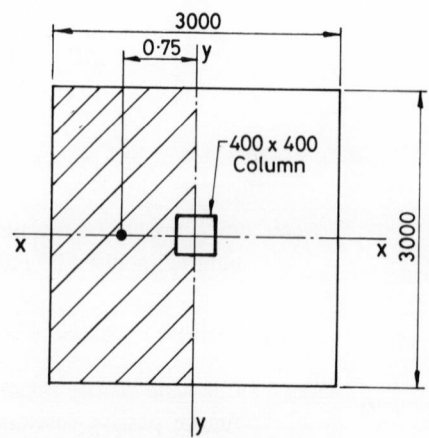

SK 6/36 Eccentric surcharge on pad foundation on plan.

Weight of foundation $= 9\,\text{m}^2 \times 0.50\,\text{m} \times 24\,\text{kN/m}^3$
$= 108\,\text{kN}$

Weight of backfill + ground slab $= 9\,\text{m}^2 \times 0.50\,\text{m} \times 18\,\text{kN/m}^3 + 9\,\text{m}^2$
$\times 0.15\,\text{m} \times 24\,\text{kN/m}^3$
$= 113.4\,\text{kN}$

Surcharge on ground slab $= 5\,\text{kN/m}^2$

Weight of surcharge on half foundation $= 4.5\,\text{m}^2 \times 5\,\text{kN/m}^2 = 22.5\,\text{kN}$

Eccentricity of surcharge $= 0.75\,\text{m}$

Weight of surcharge on full foundation $= 45\,\text{kN}$

LC_1: $p =$ total vertical load
$= 1090 + 108 + 113.4 + 45$
$= 1356.4\,\text{kN}$

$$H_x = 0 \qquad H_y = 0 \qquad M_x = 0 \qquad M_y = 0$$

LC_3: $P = 1090 + 108 + 113.4 + 22.5$
$= 1333.9 \, \text{kN}$
$H_x = 42 \, \text{kN} \qquad M_y = 105 \, \text{kNm} \qquad H_y = 0$
$M_{xx} = 0 \qquad M_{yy} = 105 + 42 \times 0.45 + 22.5 \times 0.75 = 140.8 \, \text{kNm}$

Step 7 *Calculate bearing pressure under foundation*
(See Section 6.6.)

LC_1: $\quad p = \dfrac{1356.4}{9} = 150.7 \, \text{kN/m}^2 < 190 \, \text{kN/m}^2$

LC_2: $\quad e_x = \dfrac{M_{yy}}{P}$

$\qquad = 0.104 \, \text{m} < A/6 = 0.50 \, \text{m}$

$\qquad p_1 = \dfrac{P}{AB} + \dfrac{6M_{yy}}{A^2 B}$

$\qquad = \dfrac{1333.9}{9} + \dfrac{6 \times 140.8}{9 \times 3}$

$\qquad = 179.5 \, \text{kN/m}^2 < 190 \times 1.25 = 237.5 \, \text{kN/m}^2$

Note: 25% overstress on allowable bearing capacity may be allowed for combinations including wind.

Bearing pressures are within allowable limits.

Step 8 *Calculate sliding resistance of foundation*
Ignore passive resistance because horizontal movement of the foundation should be avoided.
(See Section 6.5.3.)
Assume $\delta = 17°$ from Table 6.6.

$P = 610 + 108 + 113.4 \text{ (dead load only)} = 831.4 \, \text{kN}$

$P_s = P \tan \delta = 831.4 \times \tan 17° = 254 \, \text{kN} > 1.5H = 1.5 \times 42 = 63 \, \text{kN}$

$P_H = qA \tan \phi + cA$
$\quad = 831.4 \times \tan 22° + 10 \times 9$
$\quad = 426 \, \text{kN} > P_s = 254 \, \text{kN}$

Step 9 *Check combined sliding and bearing*
$P = 1356.4 \, \text{kN}$

$P_v = 190 \, \text{kN/m}^2 \times 9 \, \text{m}^2 = 1710 \, \text{kN}$

$H_x = 42 \, \text{kN}$

$P_{Hx} = 426 \, \text{kN}$

$$\frac{P}{P_v} + \frac{H_x}{P_{Hx}} = \frac{1356.4}{1710} + \frac{42}{426}$$

$$= 0.89 < 1 \quad \text{OK}$$

Step 10 *Carry out analysis of bearing pressure for bending moment and shear*

LC_5: $N_u = 1622\,\text{kN}$

$\qquad P_u = N_u + 1.4\,(\text{foundation} + \text{backfill}) + 1.6\,(\text{surcharge on backfill})$

$\qquad\quad = 1622 + 1.4 \times (108 + 113.4) + 1.6 \times 45$

$\qquad\quad = 2004\,\text{kN}$

$\qquad H_{xu} = 0 \qquad H_{yu} = 0 \qquad M_{xxu} = 0 \qquad M_{yyu} = 0$

LC_6: $P_u = N_u + 1.2\,(\text{foundation} + \text{backfill} + \text{surcharge})$

$\qquad\quad = 1308 + 1.2\,(108 + 113.4 + 22.5)$

$\qquad\quad = 1601\,\text{kN}$

$\qquad M_{xxu} = 0$

$\qquad M_{yyu} = M_{yu} + H_{xu}h + M_{yu}^*$

$\qquad\quad = 126 + 50.4 \times 0.45 + 1.2 \times 22.5 \times 0.75$

$\qquad\quad = 168.9\,\text{kNm}$

LC_7: $P_u = N_u + 1.4\,(\text{foundation} + \text{backfill})$

$\qquad\quad = 854 + 1.4\,(108 + 113.4)$

$\qquad\quad = 1164\,\text{kN}$

$\qquad M_{xxu} = 0$

$\qquad M_{yyu} = M_{yu} + H_{xu}h$

$\qquad\quad = 147 + 58.8 \times 0.45$

$\qquad\quad = 173.5\,\text{kNm}$

Step 11 *Calculate bearing pressure for bending moment and shear*

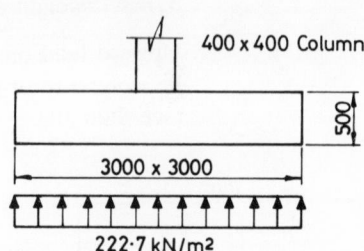

SK 6/37 Uniform bearing pressure for load combination LC_5.

LC_5: $p = \dfrac{P_u}{AB} = \dfrac{2004}{9} = 222.7\,\text{kN/m}^2$

LC_6: $p_1 = \dfrac{P_u}{AB} + \dfrac{6M_{yyu}}{A^2B}$

$\qquad\quad = \dfrac{1601}{9} + \dfrac{6 \times 168.9}{27}$

$\qquad\quad = 177.9 + 37.5 = 215.4\,\text{kN/m}^2$

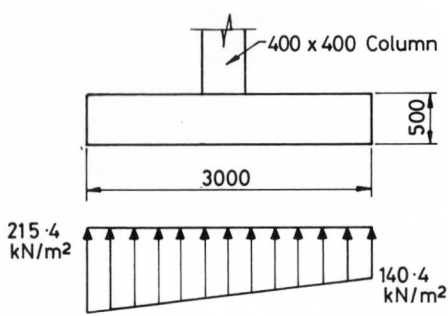

-400 x 400 Column

3000

500

215·4 kN/m²

140·4 kN/m²

SK 6/38 Bearing pressure for load combination LC_6.

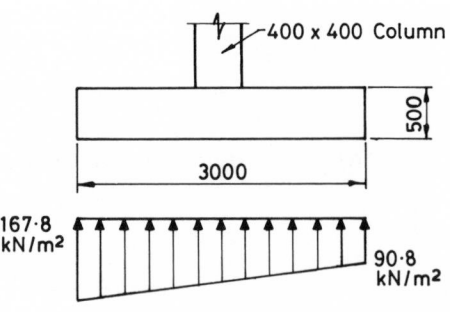

-400 x 400 Column

3000

500

167·8 kN/m²

90·8 kN/m²

SK 6/39 Bearing pressure for load combination LC_7.

$$p_2 = \frac{P_u}{AB} - \frac{6M_{yyu}}{A^2B}$$

$$= 140.4 \, \text{kN/m}^2$$

$$LC_7: \quad p_1 = \frac{P_u}{AB} + \frac{6M_{yyu}}{A^2B}$$

$$= \frac{1164}{9} + \frac{6 \times 173.5}{27}$$

$$= 129.3 + 38.5 = 167.8 \, \text{kN/m}^2$$

$$p_2 = 129.3 - 38.5 = 90.8 \, \text{kN/m}^2$$

Step 12   Calculate bending moments and shears in pad
LC_5:   Downward load on pad $= p_d$
$p_d =$ self-weight of pad + backfill + surcharge

Upward load on pad $= p_u$
$p_u =$ pressure of ground on pad
(see Step 10).

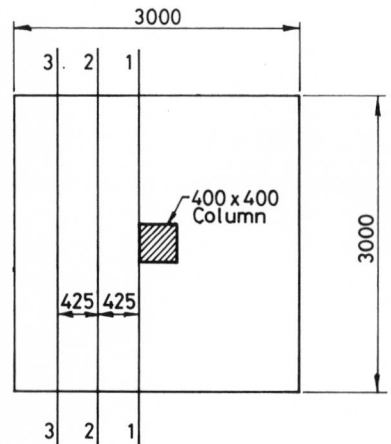

3000

3 . 2 1

-400 x 400 Column

3000

425 425

3 2 1

SK 6/40 Critical sections for bending moment and shears on plan of pad foundation.

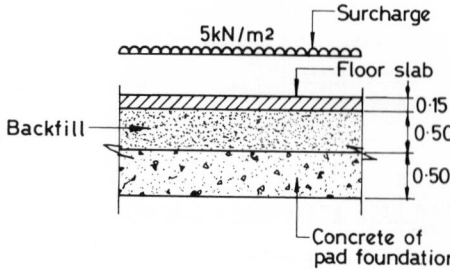

SK 6/41 Calculation of p_d for downward load on overhang.

$$p_d = (0.50 \times 24 + 0.50 \times 18 + 0.15 \times 24) \times 1.4 + (5\,\text{kN/m}^2) \times 1.6$$
$$= 42.4\,\text{kN/m}^2$$

$$p_u = 222.7\,\text{kN/m}^2 \quad \text{constant}$$

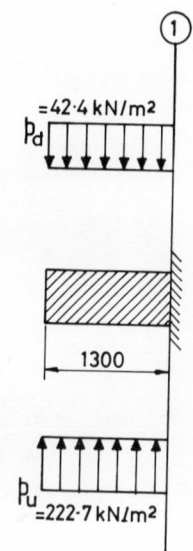

SK 6/42 Typical loading on pad foundation overhang at section 1−1.

Cantilever overhang at section $1-1 = 1500 - 200 = 1300\,\text{mm} = l$

Bending moment at section $1 = M_1 = \dfrac{(p_u - p_d)Bl^2}{2}$

$$= \dfrac{180.3 \times 3 \times 1.3^2}{2}$$

$$= 457.1\,\text{kNm}$$

Shear at section $1 = V_1 = (p_u - p_d)Bl$
$$= 180.3 \times 3 \times 1.3$$
$$= 703.2\,\text{kN}$$

Assume $d = 425\,\text{mm}$

Shear at section $2 = V_2 = (p_u - p_d)B(l - 0.425) = 473.3\,\text{kN}$

Shear at section $3 = V_3 = (p_u - p_d)B(l - 2 \times 0.425) = 243.4\,\text{kN}$

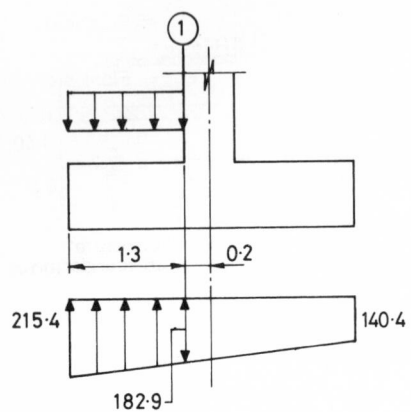

SK 6/43 Loading on overhang of pad foundation due to LC_6.

LC_6: Pressure at section $1-1 = 177.9 + \dfrac{37.5 \times 0.2}{1.5} = 182.9 \, \text{kN/m}^2$

$p_d = 1.2(0.50 \times 24 + 0.50 \times 18 + 0.15 \times 24 + 5) = 35.5 \, \text{kN/m}^2$

Bending moment $= M_1 = (182.9 - 35.5) \times 3 \times \dfrac{1.3^2}{2}$

$$+ \, (215.4 - 182.9) \times 0.5 \times 1.3 \times 3 \times \frac{2}{3} \times 1.3$$

$$= 428.6 \, \text{kNm}$$

The shears at sections 1, 2 and 3 need not be checked. By inspection they will be less critical than LC_5.

LC_7 need not be checked. By inspection it will not be critical.

Step 13   **Determine cover to reinforcement**
From SI report, total $SO_3 = 0.5\%$
Class of exposure $= 3$
(See write-up of Step 13 in Section 6.7.)
75 mm blinding concrete will be used.
Minimum cover on blinding concrete $= 50 \, \text{mm}$
Assume 16 mm diameter HT Type 2 deformed bars.
Effective depth of top layer (symmetrical reinforcement in both directions), $d = 500 - 50 - 16 - 8 = 426 \, \text{mm}$

Step 14   **Calculate area of tensile reinforcement**
Maximum bending moment on section $1-1 = 457.1 \, \text{kNm}$ (LC_5)

$$K = \frac{M}{f_{cu}bd^2} = \frac{457.1 \times 10^6}{30 \times 3000 \times 426^2} = 0.028$$

$$z = d\left[0.5 + \sqrt{\left(0.25 - \frac{K}{0.9}\right)}\right] \le 0.95d$$

$$= d\left[0.5 + \sqrt{\left(0.25 - \frac{0.028}{0.9}\right)}\right] = 0.95 \times 426 = 405\,\text{mm}$$

$$A_{st} = \frac{M}{0.87 f_y z} = \frac{457.1 \times 10^6}{0.87 \times 460 \times 405}$$

$$= 2820\,\text{mm}^2$$

Use 15 no. 16 dia. Type 2 HT bars in each direction ($3015\,\text{mm}^2$).

Distribution of tension reinforcement
(See write-up of Step 14 in Section 6.7.)

$$C_x = C_y = 400\,\text{mm}$$

$$d_x = 500 - 50 - 8 = 442\,\text{mm}$$
$$d_y = 500 - 50 - 16 - 8 = 426\,\text{mm}$$

$$1.5(C_y + 3d_y) = 2517\,\text{mm} < l_y = 3000\,\text{mm}$$
$$1.5(C_x + 3d_x) = 2589\,\text{mm} < l_x = 3000\,\text{mm}$$

$$\frac{2}{3} A_{st} = \frac{2}{3} \times 2820 = 1880\,\text{mm}^2$$

Reinforcement over central $C_y + 3d_y$(1678 mm) and $C_x + 3d_x$(1726 mm)

$$= \frac{1880}{1.678} = 1120\,\text{mm}^2/\text{m}$$

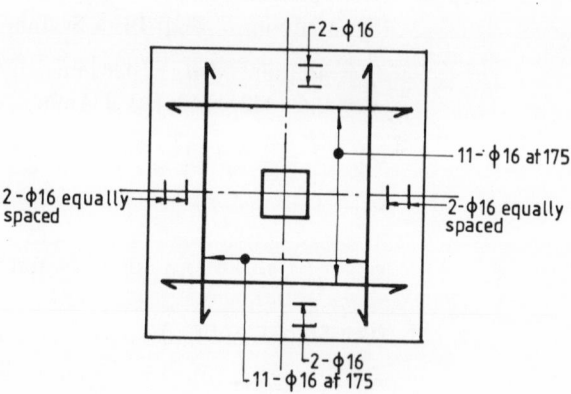

SK 6/44 Distribution of
reinforcement in pad foundation.

Use 11 no. 16 mm dia. bars at 175 mm centres ($1149\,\text{mm}^2/\text{m}$) over the central zone in each direction.
Use 2 no. 16 mm dia. bars on each side outside the central zone.
Total number of 16 mm bars used $= 15 (3015\,\text{mm}^2)$
All bars are HT Type 2.

Step 15 *Check shear stress*

(See Step 12 − LC_5.)

Check $v_1 = \dfrac{V_1}{bd} \le 0.8\sqrt{f_{cu}}$ or $5\,\text{N/mm}^2$

$$= \frac{703.2 \times 10^3}{3000 \times 426} = 0.55\,\text{N/mm}^2 < 0.8\sqrt{f_{cu}}$$

Check $v_2 = \dfrac{V_2}{bd} \le 2v_c$

$$= \frac{473.3 \times 10^3}{3000 \times 426} = 0.37\,\text{N/mm}^2$$

$A_{st} = $ 15 no. 16 mm dia. bars $= 3015\,\text{mm}^2$

Use larger d (442 mm) for calculation of p.

$$p = \frac{100 A_{st}}{bd}$$

$$= \frac{100 \times 3015}{3000 \times 442} = 0.23\%$$

From Fig. 11.3 for $f_{cu} = 30\,\text{N/mm}^2$,

$v_c = 0.42\,\text{N/mm}^2 > v_2$

No more shear checks are necessary.

Step 16 *Check punching shear*

(See write-up of Step 14 in Section 6.7.)

$d_x = 442\,\text{mm}$ $d_y = 426\,\text{mm}$
$d = 0.5\,(442 + 426) = 434\,\text{mm}$

$U_o = 2\,(C_x + C_y) = 2(400 + 400) = 1600\,\text{mm}$

$U_1 = (U_o + 12d) = 1600 + 12 \times 434 = 6808\,\text{mm}$

$$v_o = \frac{N_u}{U_{od}} \le 0.8\sqrt{f_{cu}}   \text{or}   5\,\text{N/mm}^2$$

(See Step 4 − LC_5.)

$N_u = 1622\,\text{kN}$

$$v_o = \frac{1622 \times 10^3}{1600 \times 434} = 2.34\,\text{N/mm}^2 < 0.8\sqrt{f_{cu}}$$

$p_1 = p_u - p_d = 180.3\,\text{kN/m}^2$ (see Step 12)

$A_1 = (C_x + 3.0d_x)(C_y + 3.0d_y)$
$\quad = (400 + 3.0 \times 426)(400 + 3.0 \times 442) \times 10^{-6}$
$\quad = 2.90\,\text{m}^2$

$$v_1 = \frac{N_u - p_1 A_1}{U_1 d}$$

$$= \frac{(1622 - 180.3 \times 2.9) \times 10^3}{6808 \times 434}$$

$$= 0.37 \, \text{N/mm}^2 \quad \text{OK}$$

$$v_c = 0.42 \, \text{N/mm}^2 \qquad \text{(from Step 15)}$$

Step 17 *Check minimum reinforcement for flexure*
Minimum tensile reinforcement $= 0.0013 bh = 0.0013 \times 3000 \times 500$
$$= 1950 \, \text{mm}^2 < 3015 \, \text{mm}^2$$
$$\text{provided}$$

No top tension in pad foundation.

Step 18 *Check spacing of reinforcement*
Percentage reinforcement $p = 0.23\%$ \qquad (see Step 15)
Maximum spacing $= 750 \, \text{mm}$ \quad not exceeded

Step 19 *Check early thermal cracking*
(See write-up of Step 19 in Section 6.7.)

$R = 0.15, \quad \text{say}$

$T_1 = 28°C$

$\alpha = 12 \times 10^{-6}/°C$

$\varepsilon_r = 0.8 T_1 \alpha R$
$$= 0.8 \times 28 \times 12 \times 10^{-6} \times 0.15$$
$$= 4.032 \times 10^{-5}$$

$x = h/2 = 250 \, \text{mm} \quad \text{(assumed)}$

$a_{cr} = \sqrt{(74^2 + 137.5^2)} - 8 = 148 \, \text{mm}$

$$W_{max} = \frac{3 a_{cr} \varepsilon_r}{1 + \left(\dfrac{2(a_{cr} - C_{min})}{h - x} \right)}$$

$$= \frac{3 \times 148 \times 4.032 \times 10^{-5}}{1 + \left(\dfrac{2(148 - 66)}{500 - 250} \right)}$$

$$= 0.01 \, \text{mm} < 0.3 \, \text{mm}$$

Step 20 *Check minimum reinforcement to distribute thermal cracking*
(See write-up of Step 20 in Section 6.7.)

Top reinforcement $= 0.00175 bh$
$$= 0.00175 \times 3000 \times 500$$
$$= 2625 \, \text{mm}^2 \quad \text{over } 3000 \, \text{mm}$$

Bottom reinforcement $= 0.35b$
$$= 0.35 \times 3000$$
$$= 1050\,\text{mm}^2$$

Note: If thermal cracking has to be avoided, then a top mesh of 16 dia. bars at 200 c/c should be provided. This may seem unnecessary in the present circumstances since the pad will be fully buried in the ground.

Step 21 Check crack width due to flexure

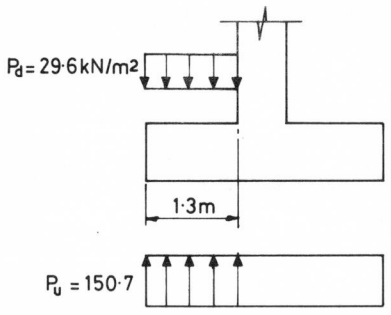

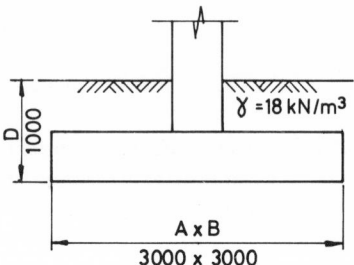

SK 6/46 Calculation of net foundation pressure at serviceability limit state.

SK 6/45 Pressure diagram for serviceability limit state.

Serviceability limit state
Loading condition LC_1. (See Step 7.)

$p_u = 150.7\,\text{kN/m}^2$
$p_d = 0.5 \times 24 + 0.5 \times 18 + 0.15 \times 24 + 5 = 29.6\,\text{kN/m}^2$

$$M = \frac{(p_u - p_d)Bl^2}{2}$$

$$= \frac{(150.7 - 29.6) \times 3 \times 1.3^2}{2} = 307\,\text{kNm}$$

$m = 15 \qquad A_s = 3015\,\text{mm}^2 \qquad b = 3000\,\text{mm} \qquad d = 426\,\text{mm}$

$$x = \left[\left(\frac{mA_s}{b}\right)^2 + \frac{2mA_sd}{b}\right]^{\frac{1}{2}} - \frac{mA_s}{b}$$

$$= \left[\left(\frac{15 \times 3015}{3000}\right)^2 + \frac{2 \times 15 \times 3015 \times 426}{3000}\right]^{\frac{1}{2}} - \frac{15 \times 3015}{3000}$$

$$= 99\,\text{mm}$$

$$z = d - \frac{x}{3} = 426 - \frac{99}{3} = 393\,\text{mm}$$

$$f_s = \frac{M}{zA_s} = \frac{307 \times 10^6}{393 \times 3015} = 259\,\text{N/mm}^2$$

$$\varepsilon_s = \frac{f_s}{E_s} = \frac{259}{200 \times 10^3} = 1.295 \times 10^{-3}$$

$$\varepsilon_h = \left(\frac{h-x}{d-x}\right)\varepsilon_s = \left(\frac{500-99}{426-99}\right) \times 1.295 \times 10^{-3} = 1.588 \times 10^{-3}$$

$$\varepsilon_{mh} = \varepsilon_h - \frac{b(h-x)^2}{3E_sA_s(d-x)}$$

$$= 1.588 \times 10^{-3} - \frac{3000 \times (500-99)^2}{3 \times 200 \times 10^3 \times 3015 \times (426-99)} = 0.773 \times 10^{-3}$$

$c_{min} = 50 + 16 = 66\,mm$ for second layer

$a_{cr} = 148\,mm$ (see Step 19)

$$W_{max} = \frac{3a_{cr}\,\varepsilon_{mh}}{1 + \dfrac{2(a_{cr} - c_{min})}{h-x}}$$

$$= \frac{3 \times 148 \times 0.773 \times 10^{-3}}{1 + \dfrac{2(148-66)}{500-99}}$$

$$= 0.24\,mm < 0.3\,mm \quad OK$$

Note: The crack width should be checked if the foundation level is below the water table and the total SO_3 is higher than 1%.

Step 22 **Design mass concrete foundation**
Not required.

Step 23 **Calculate settlement**
Load combination LC_{13}: (See Section 6.3.)

$LC_{13} = 1.0DL + 0.5IL$ (vertical loads only)

$P = 610 + 0.5 \times 480 + 108 + 113.4 + 22.5 = 1093.9\,kN$

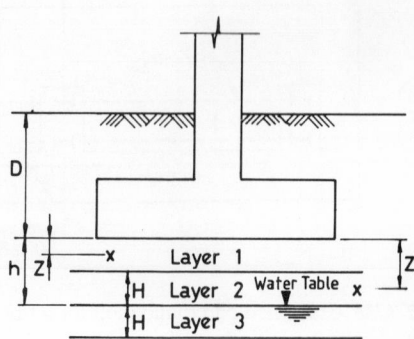

SK 6/47 Calculation of settlement.

Gross foundation pressure $= \dfrac{P}{AB} = 121.5\,\text{kN/m}^2$

Weight of soil removed $= ABD\gamma = 3 \times 3 \times 1 \times 18 = 162\,\text{kN}$

$q_n = $ net foundation pressure $= 121.5 - \dfrac{162}{9} = 103.5\,\text{kN/m}^2$

$\dfrac{A}{B} = 1 \qquad A = B = 3\,\text{m}$

Complete the settlement computation table up to $Z = 2.5B$.

Ground water table at 2.0 m below ground level

Total settlement $= 24.53\,\text{mm}$

$C_{kd} = $ cone resistance (kg/cm^2)
$C = 147\,C_{kd}/p_o$
$\sigma_z = $ vertical stress at centre of layer (kN/m^2)
$Z = $ depth to centre of layer
$h = $ height of ground water above centre of layer
$p_o = p - \gamma_w h$

σ_z is obtained from Fig. 6.2.

Note: Check whether this long-term predicted settlement is going to cause any problem elsewhere in the structural system.

Step 24   ***Design connection of pad to column (see Chapter 10)***

Example 6.2   ***RC pad with multiple columns***
Foundation for the braced columns.

Unfactored loads from the columns.

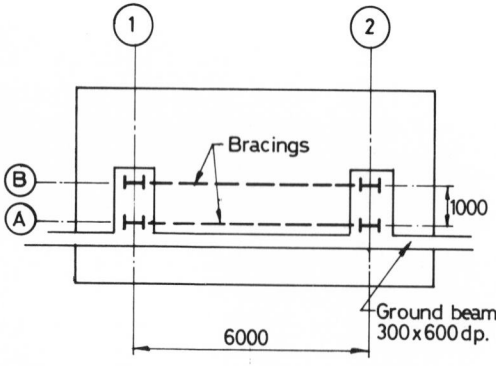

SK 6/48 Combined pad foundation for a braced bay.

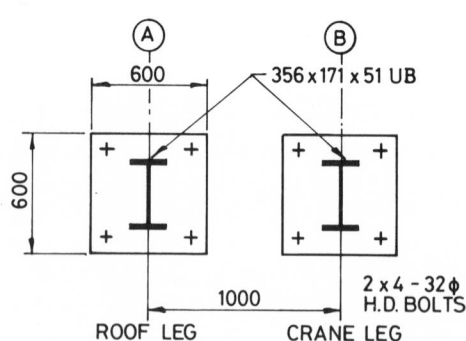

SK 6/49 Typical arrangement of steel columns at braced bay.

Settlement computation table for Example 6.1

H	Z	Z/B	σ_z/q_n	σ_z	C_{kd}	h	p_o	C	S (mm)
1.0	0.5	0.167	0.8	82.8	35	0	27	190	7.38
1.0	1.5	0.5	0.52	53.82	35	0.5	40	129	6.61
1.0	2.5	0.833	0.35	36.22	35	1.5	48	107	5.25
1.0	3.5	1.167	0.22	22.77	65	2.5	56	170	2.00
1.0	4.5	1.5	0.16	16.56	65	3.5	64	149	1.54
1.0	5.5	1.8333	0.10	10.35	90	4.5	72	184	0.73
1.0	6.5	2.167	0.08	8.28	90	5.5	80	165	0.60
1.0	7.5	2.5	0.075	7.76	120	6.5	88	200	0.42

Loading table for Example 6.2

Columns	DL		IL		CLV		CLH		WL_1		WL_2	
	V	H	V	H	V	H	V	H	V	H	V	H
A1	+80	—	+40	—	—	—	∓200	±12	∓105	±9	∓50	±25
A2	+80	—	+40	—	—	—	—	—	∓105	±9	±50	±25
B1	+50	—	+20	—	+900	—	±200	±12	±105	±9	∓50	±25
B2	+50	—	+20	—	—	—	—	—	±105	±9	±50	±25

WL_1 = transverse wind WL_2 = longitudinal wind

600 mm × 300 mm wide ground beam to carry 230 mm brickwork 3 m high.
Suitable bearing stratum at 1200 mm below ground level.
Finished floor level is 150 mm above finished ground level.
Stiff to very stiff clay layer.

Step 1 **Select type and depth of foundation**
Type: Reinforced concrete pad foundation – combined 2 sets of columns.
Length of 32 mm diameter HD bolts anchorage assembly = 400 mm.
The bottom of grout under base plate will be 500 mm below finished floor
level.
It is easier for construction if the HD bolt is in the pedestal.

Height of pedestal = 1000 mm   say
Thickness of pad = 650 mm   assumed

Underside of base at 1650 mm below floor level which is 1500 mm below
finished ground level.

Step 2 **Select approximate size**
Size of pedestal = 1400 mm × 2500 mm

Presumed allowable bearing capacity from BS 8004: 1986[2] = 200 kN/m^2

Assume a projection of 1750 mm around the pedestals. The trial size

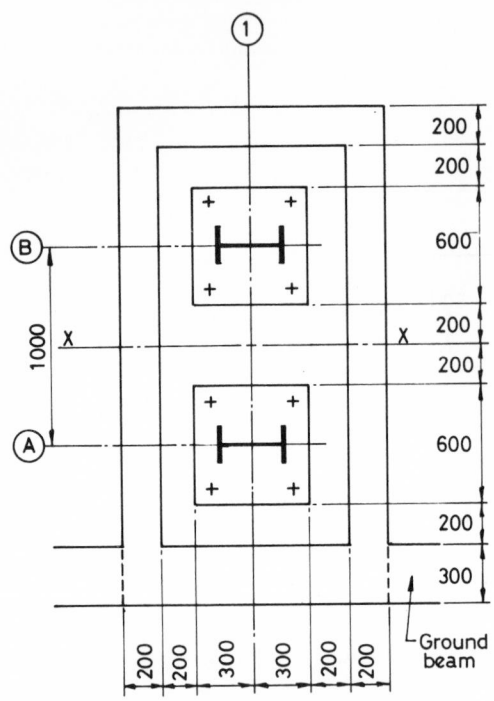

SK 6/50 Typical detail of column base and pedestal on plan.

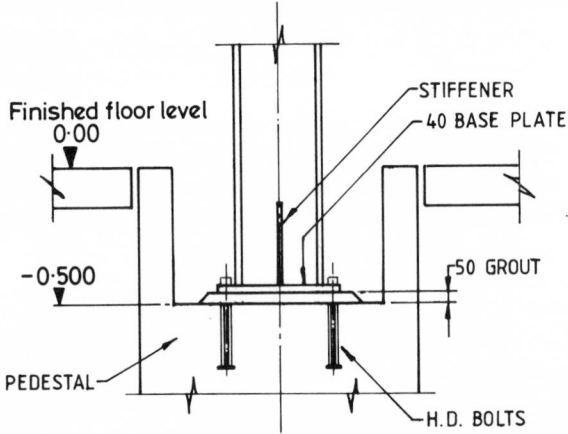

SK 6/51 Typical section through pedestal column connection.

becomes $11\,000 \times 6000$. This is based on experience and may need some revision after all the calculations are carried out.
Determine minimum thickness of pad.

Ultimate vertical load on pad through one pedestal =
$N_u = 1.4 \times (80 + 80 + 50 + 50) + 1.6 \times (900)$
$\quad = 1850\,kN \quad$ including weight of pedestal

$U_o = 2(C_x + C_y) = 2 \times (1400 + 2500) = 7800\,mm$

$C_1 = U_o/6 = 1300\,mm$
$C_2 = N_u/v_c = 1850/0.4 = 4265\,mm^2$

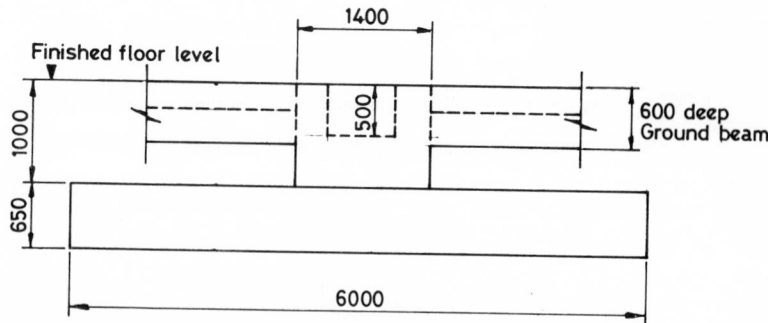

SK 6/52 Elevation of pad foundation.

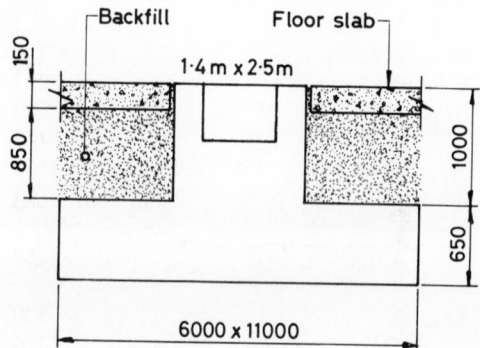

SK 6/53 Section through pad
foundation.

$v_c = 0.4 \, \text{N/mm}^2$ assumed for $f_{cu} = 30 \, \text{N/mm}^2$

$d = 0.5[(C_1^2 + 4C_2)^{0.5} - C_1] = 3.5 \, \text{mm}$

Punching shear will not be critical.

Step 3   Calculate bearing capacity of soil

Note: This step may not be necessary if the soils investigation report includes the allowable bearing capacity calculations for different sizes of foundation.

From field and laboratory tests the following soil parameters of the bearing stratum are known.

γ = unit weight of soil = $19 \, \text{kN/m}^3$

Ground water table = 2.0 m   below ground level

$h = 0$ (ground water table below the level of foundation)

Purely cohesive bearing stratum.

$q_{ult} = cN_c + p$

p = overburden pressure = $\gamma D = 19 \times 1.5 \, \text{m} = 28.5 \, \text{kN/m}^2$

c = cohesive strength = $75 \, \text{kN/m}^2$

$$D = 1.5\,\text{m} \qquad B = 6.0\,\text{m} \qquad \frac{D}{B} = 0.25$$

$$N_c = 6.7 \qquad \text{from Table 6.2 in Section 6.5.1}$$

$$q_{ult} = 75 \times 6.7 + 28.5 = 531\,\text{kN/m}^2$$

$$\text{Allowable bearing capacity} = \frac{q_{ult}}{3} = 177\,\text{kN/m}^2$$

Step 4 Calculate column load combinations

$$LC_1 = 1.0DL + 1.0IL + 1.0EP + 1.0CLV + 1.0CLH$$

Load case no.	Combination no.	Column no.	Vertical load, V (kN)	Ecc., e_x (m)	Ecc., e_y (m)	Ve_x (kNm)	Ve_y (kNm)	H_x (kN)	H_y (kN)
LC_1	1	A1	−80	−3.0	−0.5	+240	+40	−	+12
	$DL + IL + CLV$	A2	120	3.0	−0.5	+360	−60	−	−
	$+ CLH$	B1	1170	−3.0	+0.5	−3510	+585	−	+12
		B2	70	3.0	+0.5	+210	+35	−	−
	Totals		1280			−2700	+600	−	+24
LC_1	2	A1	320	−3.0	−0.5	−960	−160	−	−12
	$DL + IL + CLV$	A2	120	3.0	−0.5	+360	−60	−	−
	$- CLH$	B1	770	−3.0	0.5	−2310	+385	−	−12
		B2	70	3.0	0.5	+210	+35	−	−
	Totals		1280			−2700	+200	−	−24

$$LC_2 = 1.0DL + 1.0EP + 1.0CLV + 1.0CLH + 1.0WL_1 \ (\text{or} \pm 1.0WL_2)$$
(total 8 possible combinations)

Load case no.	Combination no.	Column no.	Vertical load, V (kN)	Ecc., e_x (m)	Ecc., e_y (m)	Ve_x (kNm)	Ve_y (kNm)	H_x (kN)	H_y (kN)
LC_2	1	A1	−225	−3.0	−0.5	+675	+113	−	+21
	$DL + CLV$	A2	−25	3.0	−0.5	−75	+13	−	+9
	$+ CLH + WL_1$	B1	1255	−3.0	0.5	−3765	+628	−	+21
		B2	155	3.0	0.5	+465	+78	−	+9
	Totals		1160			−2700	+832	−	60
LC_2	2	A1	−70	−3.0	−0.5	+210	+35	−25	+12
	$DL + CLV$	A2	+30	3.0	−0.5	+90	−15	−25	
	$+ CLH - WL_2$	B1	1200	−3.0	0.5	−3600	+600	−25	+12
		B2	0	3.0	0.5	−	−	−25	
	Totals		1160			−3300	+620	−100	+24

Note: Most of the other combinations can be ignored by inspection. Also by inspection it is clear that load cases LC_3 and LC_4 will not produce more onerous design.

Step 5 *Calculate approximate settlement*
(See Method 1 in Section 6.5.2.)

Soil parameter required from soil investigation report = m_v m^2/MN
Assume this is not available.
From Table 6.5 of Section 6.5.2, assume

$m_v = 0.15\,\text{m}^2/\text{MN}$

Consolidation settlement = $m_v\sigma_z H$

q_n = average pressure = $50\,\text{kN/m}^2$ assumed

$\sigma_z = 0.55q_n = 27.5\,\text{kN/m}^2$

B = width of foundation = $6\,\text{m}$

$H = 1.5B = 9.0\,\text{m}$

$$\text{Settlement} = m_v\sigma_z H = \frac{0.15}{1000} \times 27.5 \times 9.0 \times 1000 = 37\,\text{mm}$$

Assume quarter of this predicted settlement as differential settlement. This is because the predicted maximum settlement is 40 mm, say, the minimum settlement is half (i.e. 20 mm) and the average settlement is 30 mm, so the average differential settlement is 10 mm (40 − 30).

Differential settlement = 10 mm

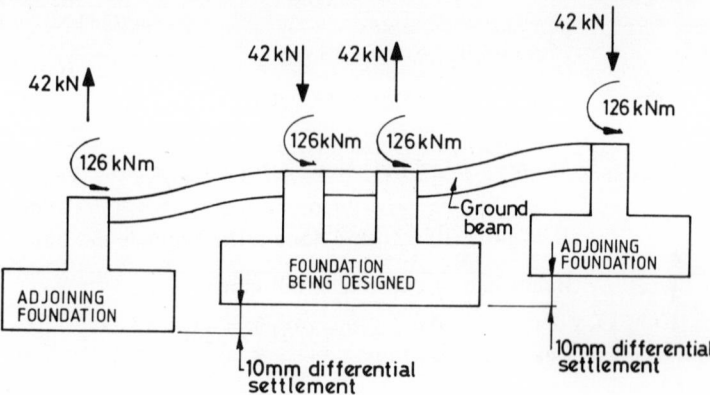

SK 6/54 Loads and moments due to differential settlement of pad foundation.

Step 6 *Carry out analysis for bearing pressure*

Self-weight of foundation = $2 \times 1.4 \times 2.5 \times 1.0 \times 25\,\text{kN/m}^3$ (pedestal) +
$6 \times 11 \times 0.65 \times 25\,\text{kN/m}^3$ (base)
= 1248 kN

Weight of backfill $= (6 \times 11 - 1.4 \times 2.5) \times 0.85\,\text{m} \times 18\,\text{kN/m}^3$
$$= 956\,\text{kN}$$

Weight of ground slab (150 mm) $= (4 \times 11) \times 0.15 \times 25\,\text{kN/m}^3$
$$= 165\,\text{kN}$$

Eccentricity, $e_y = +1.0\,\text{m}$

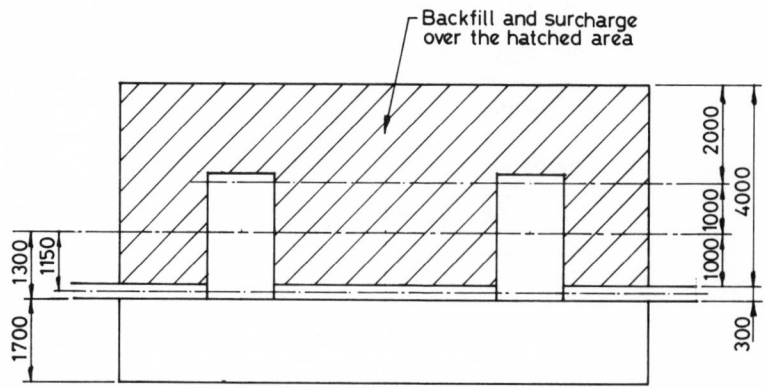

SK 6/55 Plan of foundation showing areas of superimposed loads.

Surcharge on ground slab @ $25\,\text{kN/m}^2 = 4 \times 11 \times 25 = 1100\,\text{kN}$
Eccentricity, $e_y = 1.0\,\text{m}$

Weight of ground beams + brickwork $= (12 - 2.8)\,\text{m} \times 0.3\,\text{m} \times 0.6\,\text{m} \times$
$$25\,\text{kN/m}^3\ (\text{ground beams})$$
$$+3\,\text{m} \times 0.23\,\text{m} \times 22\,\text{kN/m}^3$$
$$(\text{brickwork})$$
$$= 224\,\text{kN}$$

Eccentricity, $e_y = -1.15\,\text{m}$

Differential settlement $= 10\,\text{mm}$ assumed (see Step 5)

$$M_y = \frac{6\delta EI}{l^2} = 126\,\text{kNm} \qquad \text{for each beam}$$

$E = 14 \times 10^6\,\text{kN/m}^2$ long-term Young's modulus

Beam size $= 300\,\text{mm} \times 600\,\text{mm}$

Beam end reactions $= \dfrac{12\delta EI}{l^3} = \dfrac{2M_y}{l}$

$$= \frac{252}{6} = 42\,\text{kN}$$

Moment on the foundation $= 2 \times (-126) + (-42 \times 6) = -504\,\text{kNm}$

Note: There are many possible alternatives of the differential settlement. The worst, from the point of view of bearing pressure considering other loadings, is found by inspection.

Load case	Load type	Vertical, V(kN)	M_x (kNm)	M_y (kNm)	H_x (kN)	H_y (kN)
	Column vertical	1280	+600	−2700	−	−
	Column horizontal	−	+28	−	−	+24
	Foundation self-weight	1248	−	−	−	−
LC_1	Backfill	956	−	−	−	−
Combination 1	Ground slab	165	+165	−	−	−
	Surcharge on slab	1100	+1100	−	−	−
	Ground beam	224	−258	−	−	−
	Differential settlement	−	−	−504	−	−
LC_1	Totals	4973	+1635	−3204	−	+24

Load case	Load type	Vertical, V(kN)	M_x (kNm)	M_y (kNm)	H_x (kN)	H_y (kN)
	Column vertical	1160	+620	−3300	−	−
	Column horizontal	−	+28	−115	−100	+24
	Foundation self-weight	1248	−	−	−	−
LC_2	Backfill	956	−	−	−	−
Combination 2	Ground slab	165	+165	−	−	−
	Surcharge on slab	1100	+1100	−	−	−
	Ground beam	224	−258	−	−	−
	Differential settlement	−	−	−504	−	−
LC_2	Totals	4853	+1655	−3919	−100	+24

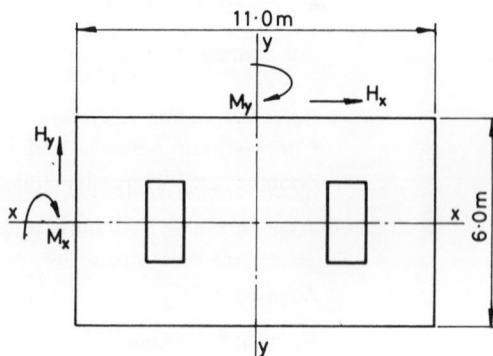

SK 6/56 Sign convention of positive forces on foundation.

Note: The foundations are connected by ground beams in the x-direction. The horizontal force H_x may be distributed equally among all connected foundations. For the sake of conservatism this has not been done.

Step 7   Calculate bearing pressures

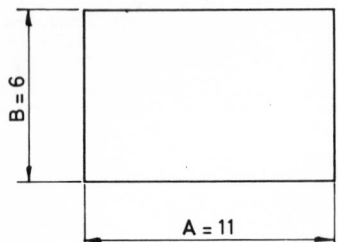

SK 6/57 Footprint of foundation.

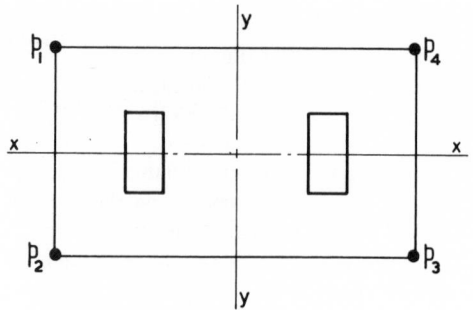

SK 6/58 Bearing pressure locations.

$$LC_1: \quad p = \left(\frac{P}{AB}\right) \pm \left(\frac{6M_x}{AB^2}\right) \pm \left(\frac{6M_y}{A^2B}\right)$$

$$p_1 = \left(\frac{4973}{6 \times 11}\right) + \left(\frac{6 \times 1635}{11 \times 6^2}\right) + \left(\frac{6 \times 3204}{11^2 \times 6}\right)$$

$$= 75 + 25 + 26$$

$$= 126\,\text{kN/m}^2 < 177\,\text{kN/m}^2 \qquad \text{(see Step 3)}$$

Similarly,

$$p_2 = 75 - 25 + 26 = 76\,\text{kN/m}^2$$
$$p_3 = 75 - 25 - 26 = 24\,\text{kN/m}^2$$
$$p_4 = 75 + 25 - 26 = 74\,\text{kN/m}^2$$

$$LC_2: \quad p_1 = 74 + 25 + 32 = 131\,\text{kN/m}^2$$
$$p_2 = 81\,\text{kN/m}^2$$
$$p_3 = 17\,\text{kN/m}^2$$
$$p_4 = 67\,\text{kN/m}^2$$

All bearing pressures are within allowable limit.

Step 8   Calculate sliding resistance of foundation
Check sliding between concrete and soil: ignore passive resistance.
Assume very aggressive soil and membrane tanking is used.

Assume $\delta = 5°$ $\tan \delta = 0.09$ $H_x = 100\,\text{kN}$ $H_y = 24\,\text{kN}$

Load case LC_2

Frictional resistance $= P\tan\delta = 4853 \times 0.09 = 437\,\text{kN} > H_x$ and H_y

$$\text{Factor of safety against sliding} = \frac{437}{\sqrt{(H_x^2 + H_y^2)}} = \frac{437}{\sqrt{(100^2 + 24^2)}}$$

$$= 4.2 \quad \text{OK}$$

Check horizontal bearing capacity of soil:
For cohesive soil,
$$P_{Hu} = cA = 75 \times 6 \times 11 = 4950\,\text{kN}$$

$$P_{Hx} = P_{Hy} = \frac{P_{Hu}}{1.5} = 3300 \, kN$$

Step 9 Check combined sliding and bearing

Load case LC$_2$

$$P = 4853 \, kN \qquad H_x = 100 \, kN \qquad H_y = 24 \, kN$$

$$P_v = \text{allowable bearing capacity} \times \text{area} = 177 \times 66 = 11\,682 \, kN$$

$$\frac{P}{P_v} + \frac{H_x}{P_{Hx}} + \frac{H_y}{P_{Hy}} = \frac{4853}{11\,682} + \frac{100}{3300} + \frac{24}{3300}$$

$$= 0.45 < 1 \quad \text{okay}$$

Step 10 Carry out analysis of bearing pressure for bending moment and shear
Ultimate column load combinations

Load case no.	Combination no.	Column no.	Vertical load, V(kN)	Ecc., e_x (m)	Ecc., e_y (m)	Ve_x (kNm)	Ve_y (kNm)	H_x (kN)	H_y (kN)
LC$_9$		A1	−168	−3.0	−0.5	+504	+84	—	+16.8
	1.4DL	A2	+112	3.0	−0.5	+336	−56	—	—
	+1.4CLV	B1	+1610	−3.0	+0.5	−4830	+805	—	+16.8
	+1.4CLH	B2	+70	3.0	+0.5	+210	+35	—	—
	Totals		+1624			−3780	+868	—	+33.6

Ultimate loads on foundation

Load case	Combination	Load type	Vertical, V (kN)	M_x (kNm)	M_y (kNm)	H_x (kN)	H_y (kN)
		Column vertical	1624	+868	−3780	—	—
		Column horizontal	—		+38.6	—	33.6
	1.4DL	Foundation self-weight	1747				
	+1.4CLV	Backfill	1338				
LC$_9$	+1.4CLH	Ground slab	231	+231			
		Ground beam	314	−361			
		Differential settlement			−706		
		Totals	5254	+776.6	−4486	—	33.6

Note: It is most important that a complicated loading system on a multiple column foundation should be investigated in a structured manner using tables as shown. Otherwise mistakes will creep in.

Only one load case has been analysed to show the method. All other load cases should be similarly investigated.

Step 11   Calculate bearing pressure for bending moment and shear

Load case LC₉

$$p_1 = \left(\frac{P}{AB}\right) + \left(\frac{6M_x}{AB^2}\right) + \left(\frac{6M_y}{A^2B}\right)$$

$$= \left(\frac{5254}{66}\right) + \left(\frac{6 \times 776.6}{11 \times 6^2}\right) + \left(\frac{6 \times 4486}{11^2 \times 6}\right)$$

$$= 80 + 12 + 37 = 129\,\text{kN/m}^2$$

$$p_2 = 80 - 12 + 37 = 105\,\text{kN/m}^2$$
$$p_3 = 80 - 12 - 37 = 31\,\text{kN/m}^2$$
$$p_4 = 80 + 12 - 37 = 55\,\text{kN/m}^2$$

Note: Pressures for load case LC_9 only have been calculated to show the method. In an actual design, other load cases should also be investigated.

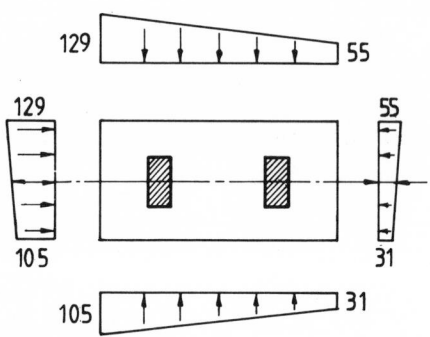

SK 6/59 Bearing pressure diagrams for load case LC_9.

Step 12   Calculate bending moments and shears in pad

Load case LC₉
Find pressures at critical sections 1, 2, 3, 4, 5, 6, 7 as shown. They are p_1, p_2, p_3, p_4, etc.

Average pressures at section 1 = p_1
Assume $d = 550\,\text{mm}$
Distance from $y-y$ axis = $4.8\,\text{m} = x_1$

$$p_{1,av} = \left(\frac{P}{AB}\right) + \left(\frac{12M_y x_1}{A^3B}\right)$$

$$= 80 + \frac{12 \times 4486 \times 4.8}{11^3 \times 6}$$

$$= 112\,\text{kN/m}^2$$

Similarly,

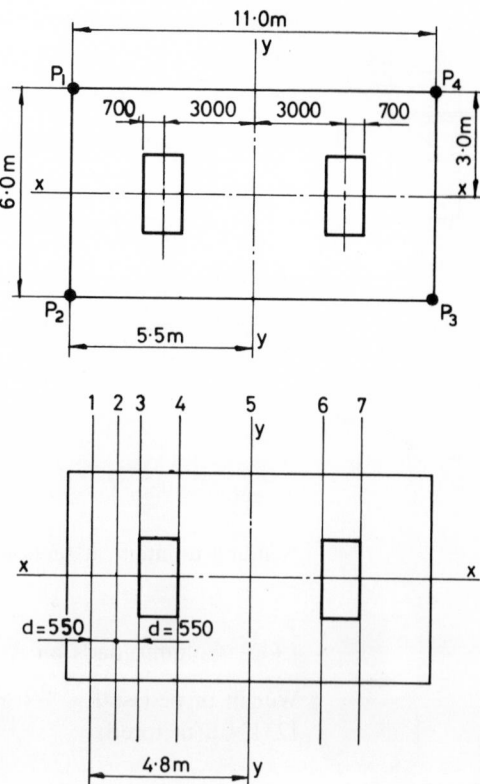

SK 6/60 Bearing pressure locations on plan and critical sections for bending moment and shear.

$$p_{2,av} = 109\,\text{kN/m}^2$$
$$p_{3,av} = 105\,\text{kN/m}^2$$
$$p_{4,av} = 96\,\text{kN/m}^2$$
$$p_{5,av} = 80\,\text{kN/m}^2$$
$$p_{6,av} = 64\,\text{kN/m}^2$$
$$p_{7,av} = 55\,\text{kN/m}^2$$

Average pressures at edges are $117\,\text{kN/m}^2$ and $43\,\text{kN/m}^2$.

p_d = downward load on pad (ultimate)
  = self-weight of foundation + backfill
  = $(0.65\,\text{m} \times 25\,\text{kN/m}^3 + 0.85\,\text{m} \times 18\,\text{kN/m}^3) \times 1.4$
  = $44\,\text{kN/m}^2$ uniform excluding ground slab

Equivalent weight of ground slab acting on half the foundation width
  = $0.5 \times 0.15\,\text{m} \times 25\,\text{kN/m}^3 \times 1.4$
  = $3\,\text{kN/m}^2$

Total downward load = $47\,\text{kN/m}^2$

Draw net pressure diagram.

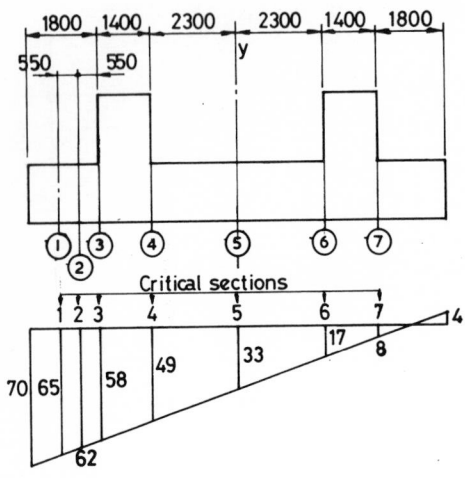

SK 6/61 Net bearing pressure diagram to find bending moments and shears − load case LC_9.

Bending moment at section 3 $= 6 \times \left[\dfrac{58 \times 1.8^2}{2} + \dfrac{(70 - 58)1.8^2}{3} \right]$

$= 642 \, \text{kNm}$

LC_9:   Column loads on $A_1 + B_1 = (1610 - 168) = 1442 \, \text{kN}$ (factored)

Weight of pedestal $= 1.4 \times 1.4 \, \text{m} \times 2.5 \, \text{m} \times 1.0 \, \text{m} \times 25 \, \text{kN/m}^3 =$
$123 \, \text{kN}$ (factored)

Bending moment at section 5 $= 6 \, \text{m} \times \left[\dfrac{33 \times 5.5^2}{2} + \dfrac{(70 - 33) \times 5.5^2}{3} \right]$

$- (1442 + 123) \times 3 \, \text{m}$

$= 538 \, \text{kNm}$ (no top tension)

Bending moment at section 7 $= 6 \, \text{m} \times \left(\dfrac{8 \times 1.8^2}{2} - \dfrac{12 \times 1.8^2}{3} \right)$

$= 0 \, \text{kNm}$

Shear at section 1 $= 6 \, \text{m} \times \left[(70 + 65) \times \dfrac{0.7}{2} \right]$

$= 284 \, \text{kN}$

Shear at section 2 $= 6 \, \text{m} \times \left[(62 + 70) \times \dfrac{1.25}{2} \right]$

$= 495 \, \text{kN}$

Shear at section 3 $= 6 \, \text{m} \times \left[(70 + 58) \times \dfrac{1.8}{2} \right]$

$= 691 \, \text{kN}$

Shear at section 4 $= 6 \, \text{m} \times \left[(70 + 49) \times \dfrac{3.2}{2} \right] - (1442 + 123)$

$= -423 \, \text{kN}$

Note: It is useful to draw the bending moment diagram for the load case. Similarly bending moments and shears should be calculated for all load cases and all critical sections parallel to the $x-x$ and $y-y$ axes following the recommendations in Step 12 of Section 6.7.

Step 13   ***Determine cover to reinforcement***
Follow Step 13 of Section 6.7.

Step 14   ***Calculate area of tension reinforcement and distribution***
Follow Step 14 of Section 6.7.

Step 15   ***to Step 23***
Similar to Example 6.1.

Note: The numerous other checks required in Step 15 to Step 23 in this example are not shown for brevity. They have already been shown in Example 6.1.

Example 6.3   ***Mass concrete pad − side bearing in cohesive soils***
Foundation for roadside signpost.

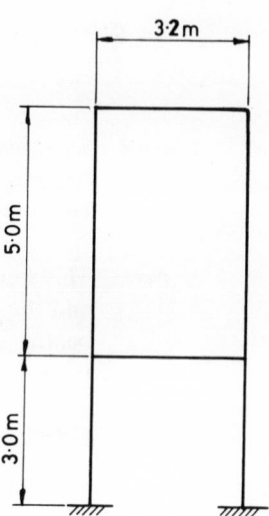

SK 6/62 Roadside signpost.

Vertical load = 18 kN
Horizontal wind shear = 4.5 kN
Bending moment due to wind = 25 kNm

Size of columns = 203 × 203 × 46 kg/m UC
Size of base plate = 400 × 400

Size of foundation bolts = 4 no. M24
Bolt spacing = 300 mm
Soil condition is stiff to very stiff clay.

Step 1   ***Select type and depth of foundation***

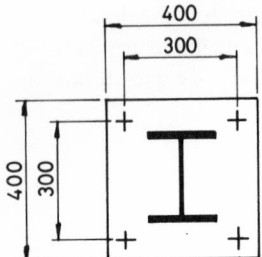

SK 6/63 Details of column base plate.

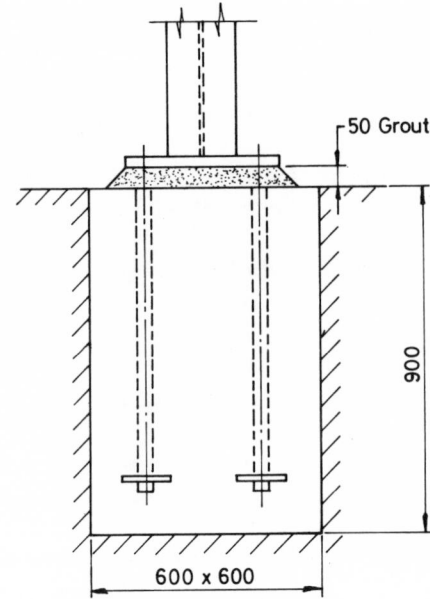

SK 6/64 First trial of a side bearing mass concrete foundation.

Step 2   **Select approximate size**

Assume 100 mm projection on all sides of base plate.
The foundation size selected is $600 \times 600 \times 900$ deep.

Note: This foundation will be designed as side bearing for the horizontal load and the applied moment. The vertical load will be carried by the direct bearing on the soil. It is very important that this foundation is cast against the soil.

Step 3   **Calculate bearing capacity of soil**

γ = unit weight of soil = $19 \, \text{kN/m}^3$

p = total overburden pressure at foundation level
   = $\gamma D = 19 \times 0.9 = 17.1 \, \text{kN/m}^2$

h = height of water above foundation level = $0 \, \text{m}$

$p_\text{o} = p = 17.1 \, \text{kN/m}^2$

c = minimum undrained soil cohesion = $150 \, \text{kN/m}^2$

$q_\text{ult} = c \, N_c + p$

$\dfrac{D}{B} = \dfrac{0.9}{0.6} = 1.5$

$N_c = 9.1$ from Table 6.2 in Section 6.5.1

$$q_{ult} = 9.1 \times 150 + 17.1 = 1382.1 \, kN/m^2$$

Allowable vertical bearing capacity $= \dfrac{q_{ult}}{3} = 460 \, kN/m^2$

Maximum horizontal bearing capacity is the ultimate passive resistance of soil given by the following equation:

$$
\begin{aligned}
q_h &= p_o \tan^2 (45° + \phi/2) + 2c \tan (45° + \phi/2) \\
&= 2c \quad \text{at ground level for } \phi = 0° \text{ and } p_o = 0 \\
&= 300 \, kN/m^2
\end{aligned}
$$

Step 4 Calculate column load combination
(See Section 6.3.)
For bearing pressure calculations:

$$LC_3 = 1.0DL + 1.0IL + 1.0WL$$

$N = 18 \, kN \qquad H_x = 4.5 \, kN \qquad M_x = 25 \, kNm \qquad M_y = 0$

For bending moment and shear calculations:

$$LC_7 = 1.4DL + 1.4WL$$

$N = 25.2 \, kN \qquad H_x = 6.3 \, kN \qquad M_x = 35 \, kNm$

Step 5 Calculate approximate settlement
This step can be ignored.

Step 6 Carry out analysis for bearing pressure

Self-weight of foundation $= 0.6 \, m \times 0.6 \, m \times 0.9 \, m \times 25 \, kN/m^3 = 8.1 \, kN$

$P = 18 + 8.1 = 26.1 \, kN \qquad H_x = 4.5 \, kN \qquad M_x = 25 \, kNm$

Step 7 Calculate bearing pressures
Soil is stiff to very stiff clay.
Determine horizontal modulus of subgrade reaction.

Assume $k_{si} = 14 \, MN/m^3$ (see Table 6.7 in Section 6.5.1)

$B = $ width of foundation $= 0.6 \, m$

$$K_h = \frac{k_{si}}{1.5B} = \frac{14}{1.5 \times 0.6} = 15.5 \, MN/m^3$$

Note: Horizontal load and vertical load are treated separately to find the bearing pressures. It is assumed that the vertical load will be carried uniformly on the base; size 600 mm × 600 mm. The horizontal load and moment will be carried by side bearing in the manner shown.

Assumptions:

(1) Foundation block is fully rigid.

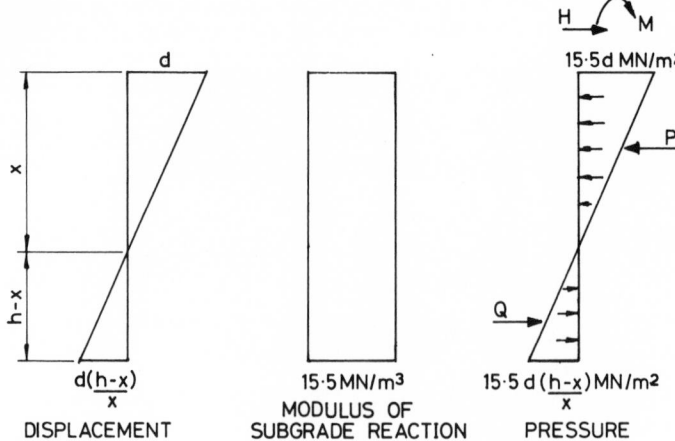

DISPLACEMENT MODULUS OF   PRESSURE
       SUBGRADE REACTION

SK 6/65 Displacement and pressure diagrams of side-bearing foundation in cohesive soil.

(2) Displacement d at top of foundation when subjected to bending moment M_x and horizontal shear H_x.

(3) Conservatively, that there is no net horizontal movement of whole foundation block. Moment and horizontal force are resisted by rotation only. The pressure diagram will be as shown if x is depth from ground level of point of rotation.

(4) Neglect friction on sides of foundation block.

(5) Neglect contribution from non-uniform pressure distribution on bottom surface of foundation.

$$P = 0.5 K_h Bxd = 4.65 dx \text{ MN}$$

$$Q = 0.5 K_h B(h - x)^2 \frac{d}{x} = 4.65(h - x)^2 \frac{d}{x} \text{ MN}$$

$$h = 0.9 \text{ m}$$

H = applied horizontal load = 4.5 kN = 0.0045 MN

Considering horizontal load equilibrium:

$$P = H + Q \quad \text{or} \quad P - Q = H$$

$$\text{or} \quad 4.65d \left[x - \frac{(0.9 - x)^2}{x} \right] = 4.5 \times 10^{-3}$$

Taking moment about the foundation level:

$$P\left(h - \frac{x}{3}\right) - \frac{Q(h - x)}{3} - M - Hh = 0$$

$$\text{or} \quad 4.65dx\left(h - \frac{x}{3}\right) - \frac{4.65d(h - x)^3}{3x} = 25 \times 10^{-3} + 4.5 \times 10^{-3} \times 0.9$$

or $4.65d\left[x\left(0.9 - \dfrac{x}{3}\right) - \dfrac{(0.9 - x)^3}{3x}\right] = 29.05 \times 10^{-3}$

Solving for the two unknowns d and x using a computer-assisted equation solver.

Displacement $= d = 22\,\text{mm}$
Point of rotation, $x = 460\,\text{mm}$

Maximum allowable shear stress in mass concrete $= 0.037 f_{cu} = 0.925\,\text{N/mm}^2$

$P = 4.65dx = 47.3\,\text{kN}$

$Q = 4.65(h - x)^2\left(\dfrac{d}{x}\right) = 42.8\,\text{kN}$

Maximum horizontal pressure on the soil $= 15.5d\,\text{MN/m}^2$

$$= 15.5 \times 1000 \times \dfrac{22}{1000}\,\text{kN/m}^2$$

$$= 341\,\text{kN/m}^2$$

This pressure is higher than the unconfined compressive strength of soil, which is $300\,\text{kN/m}^2$. To prevent local heave of soil, revise size of foundation to $900\,\text{mm} \times 900\,\text{mm} \times 1300\,\text{mm}$.

$K_h = \dfrac{k_{si}}{1.5B} = \dfrac{14}{1.5 \times 0.9} = 10.4\,\text{MN/m}^3$

$P = 0.5K_h Bxd$

$Q = 0.5K_h B(h - x)^2\,\dfrac{d}{x}$

$P - Q = H$

$h = 1.3\,\text{m} \qquad M = 25\,\text{kNm} \qquad H = 4.5\,\text{kN} \qquad B = 0.9\,\text{m}$

By solving the above equations,

$P = 34.5\,\text{kN}$
$d = 11.0\,\text{mm}$
$x = 673\,\text{mm}$
$Q = 30.0\,\text{kN}$
$p = 114\,\text{kN/m}^2$ maximum pressure $< 300\,\text{kN/m}^2$ ultimate pressure

Factor of safety on ultimate $= \dfrac{300}{114} = 2.63$

Note: The displacement of 11 mm at ground level will mean 130 mm displacement at top of an 8000 mm high structure. This should be checked for clearances or other obstructions.

Revised weight of foundation $= 26.3\,\text{kN}$

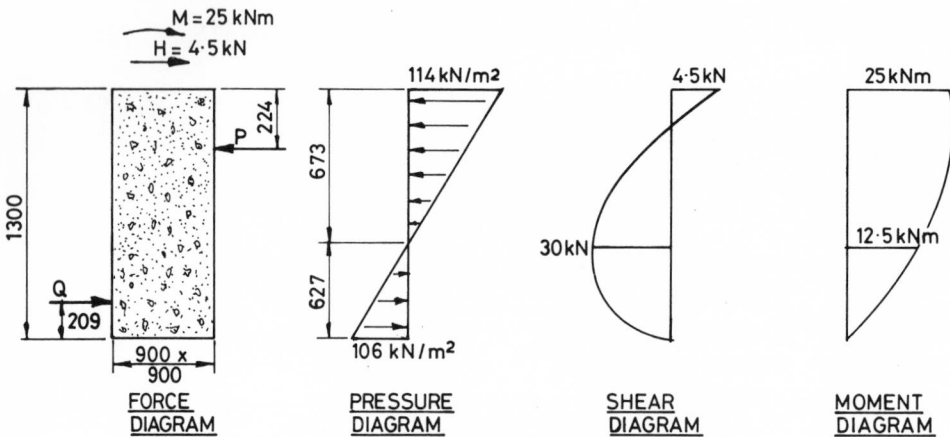

SK 6/66 Diagrams at serviceability limit state.

$$\text{Vertical Bearing Pressure} = \frac{P}{A} = \frac{44.3}{0.9 \times 0.9} = 54.7 \, \text{kN/m}^2 < 460 \, \text{kN/m}^2$$

Step 8 **Calculate sliding resistance**
Not required.

Step 9 **Check combined sliding and bearing**
Not required.

Step 10 **Carry out analysis of bearing pressure for bending moment and shear**
Not required.

Step 11 **to Step 21**
Not required.

Step 22 **Design mass concrete foundation**
Use C25 concrete $f_{cu} = 25 \, \text{N/mm}^2$

Maximum allowable tension in bending $= 0.37\sqrt{f_{cu}} = 0.37 \times (25)^{\frac{1}{2}} = 1.85 \, \text{N/mm}^2$

Maximum bending moment $= 25 \, \text{kNm}$ approximately

Maximum shear $= 30.0 \, \text{kN}$

Section modulus, $Z = \left(\frac{1}{6}\right) \times 900^3 = 1.215 \times 10^8 \, \text{mm}^3$

Bending tensile stress in mass concrete $= \dfrac{M}{Z} = \dfrac{25 \times 10^6}{1.215 \times 10^8}$
$$= 0.20 \, \text{N/mm}^2$$

Allowable bending tensile stress $= 1.85 \, \text{N/mm}^2$ OK

$$\text{Shear stress} = \frac{30 \times 10^3}{900 \times 900} = 0.037\,\text{N/mm}^2 < 0.925\,\text{N/mm}^2 \quad \text{OK}$$

Example 6.4 Mass concrete pad – side bearing in cohesionless soils
Same loading and example as in Example 6.3. Soil is dense sandy gravel with $\phi = 35°$.

Step 1 Select type and depth of foundation
Similar to Step 1 in Example 6.3.

Step 2 Select approximate size
Choose $1000 \times 1000 \times 1500$ deep foundation.

Note: In cohesionless soil the size of the foundation will be larger than in cohesive soil because the allowable horizontal bearing capacity at higher levels is lower. It will be very difficult to cast this foundation against the soil because the sides of the excavation may not stay vertical. It will be necessary to have well compacted granular backfill using mechanical compactors.

Step 3 Calculate bearing capacity of soil

$B = 1\,\text{m}$

$\gamma = $ unit weight of soil $= 18\,\text{kN/m}^3$

$p = $ total overburden pressure at foundation level $= 18 \times 1.5\,\text{m} = 27\,\text{kN/m}^2$

$p_o = p - \gamma_w h = 27\,\text{kN/m}^2 \quad$ as $h = 0$

$c = 0$

$\phi = $ angle of internal friction $= 35°$

$a = e^{(0.75\pi - \phi/2)\tan\phi} = 4.20$

$$N_q = \frac{a^2}{2\cos^2\left(45° + \dfrac{\phi}{2}\right)} = 41.4$$

$K_{py} = 82.0 \qquad$ from Table 6.1 in Section 6.5.1

$$N_\gamma = 0.5\tan\phi\left(\frac{K_{py}}{\cos^2\phi} - 1\right) = 42.4$$

$$
\begin{aligned}
q_{ult} &= 1.3cN_c + p_o(N_q - 1) + 0.4\gamma BN_\gamma + p \\
&= 27\,(41.4 - 1) + 0.4 \times 18 \times 1 \times 42.4 + 27 \\
&= 1423\,\text{kN/m}^2
\end{aligned}
$$

Allowable vertical bearing capacity $= \dfrac{q_{ult}}{3} = 474\,\text{kN/m}^2$

Maximum horizontal bearing capacity is ultimate passive resistance of soil given by the following equation:

$$q_h = p_o \tan^2\left(45° + \frac{\phi}{2}\right) + 2c \tan\left(45° + \frac{\phi}{2}\right) = \gamma h \tan^2\left(45° + \frac{\phi}{2}\right)$$

$$= 66h \text{ kN/m}^2$$

Step 4 *Calculate column load combination*
See Step 4 of Example 6.3.

Step 5 *Calculate approximate settlement*
Can be ignored.

Step 6 *Carry out analysis for bearing pressure*
Self-weight of foundation $= 1 \times 1 \times 1.5 \times 25 \text{ kN/m}^3 = 37.5 \text{ kN}$

$P = 37.5 + 18 = 55.5 \text{ kN}$

$H_x = 4.5 \text{ kN} \qquad M_x = 25 \text{ kNm}$

Step 7 *Calculate bearing pressures*
Assumptions:

(1) Foundation is rigid.
(2) Foundation carries part of moment and total horizontal shear by side bearing up to ultimate horizontal bearing pressure of $66Z$ kN/m^2 (see Step 3).
(3) Residual moment is carried by bottom surface of foundation as a variable pressure on surface. Factor of safety against overturning will be 1.5 or more.
(4) Rotational deformation of foundation is proportional to distance from point of rotation.
(5) There is no net horizontal movement.
(6) Point of rotation is assumed at depth x from ground level.

Modulus of subgrade reaction $= K_h = n_h \dfrac{Z}{B}$

$B = 1.0 \text{ m}$
$n_h = 6.6 \text{ MN/m}^3$ assumed

(See Table 6.7 in Section 6.5.3.)

$K_h = 6.6Z \text{ MN/m}^3$

$Z = $ depth from ground level (metres)

$P = 1.65dx^2 \dfrac{B}{2}$ for idealised triangular pressure distribution

$Q = 6.6hdB \dfrac{(h - x)^2}{2x}$ for idealised triangular pressure distribution

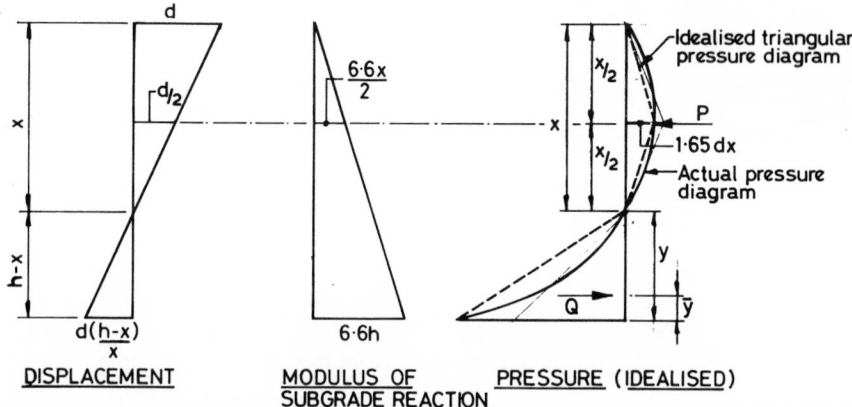

SK 6/67 Displacement and pressure diagrams of side-bearing foundation in cohesionless soil.

$P = Q + H$

$$M + Hh + Q\left(h - \frac{x}{3}\right) - P\left(h - \frac{x}{2}\right) = 0 \qquad \text{for idealised triangular}$$

distribution

$H = 4.5\,\text{kN}$

Maximum pressure $= p = 1.65dx$ from pressure diagram

$$\text{also } p = \frac{66x}{2000}\ \text{MN/m}^2 \qquad \text{(from Step 3)}$$

Equating the two gives

$$1.65dx = \frac{66x}{2000}$$

$d = 20\,\text{mm}$

Ultimate horizontal passive pressure on side of foundation is reached at a depth $x/2$ when horizontal deformation at top reaches 20 mm. It is assumed that moment-carrying capacity of foundation through side bearing will have a limiting value when deformation reaches 20 mm at top.

Find M when
$h = 1.5\,\text{m}$
$d = 20 \times 10^{-3}\,\text{m}$
$H = 4.5 \times 10^{-3}\,\text{MN}$
$B = 1.0\,\text{m}$

or $P = 0.0165x^2$ $Q = 0.099(1.5 - x)^2/x$ $P = Q + H$

and $M = P(1.5 - 0.5x) - \dfrac{Q(1.5 - x)}{3} - Hh$ (MNm)

Solving the equations:

$$P = 19.4 \times 10^{-3} \, \text{MN}$$
$$x = 1.092 \, \text{m}$$
$$Q = 14.9 \times 10^{-3} \, \text{MN}$$
$$M = 9.7 \, \text{kNm} \qquad \text{(maximum allowed by side bearing)}$$

Residual moment $= 25 - 9.7 = 15.3 \, \text{kNm}$

By rigorous analysis:
$d' = $ displacement at bottom
$d = $ displacement at top
$X = $ depth of neutral axis
$Y = $ bottom of foundation to neutral axis
$\bar{Y} = $ point of application of Q from bottom of foundation

$$P = \int_0^X \frac{(X - x)}{X} \, dB6.6x \, dx = 6.6 \, Bd\left(\frac{X^2}{6}\right) = 1.1 \, BdX^2$$

$$Q = 6.6d'B \int_0^Y \frac{y(X + y)}{Y} \, dy = 6.6d'B\left[\left(\frac{XY}{2}\right) + \frac{Y^2}{3}\right]$$
$$= 1.1d'B(3XY + 2Y^2)$$

$$X + Y = 1.5 \qquad \frac{X}{d} = \frac{Y}{d'}$$

$$M + 4.5 \times 1.5 + Q\bar{Y} - P\left(h - \frac{X}{2}\right) = 0$$

$$\bar{Y} = Y - \frac{\displaystyle\int_0^Y y^2(X + y) \, dy}{\displaystyle\int_0^Y y(X + y) \, dy}$$
$$= Y\left[1 - \frac{(X/3 + Y/4)}{(X/2 + Y/3)}\right]$$

Solving the above equations:

$X = 1.0317 \, \text{m} \qquad Y = 0.4683 \, \text{m}$

$d' = 8.9 \, \text{mm} \qquad$ pressure $= 88.1 \, \text{kN/m}^2$
$d = 19.7 \, \text{mm}$

$Q = 18.57 \, \text{kN} \qquad P = 23.07 \, \text{kN}$

$M = 13.25 \, \text{kNm}$
$\bar{Y} = 0.314Y$

This gives a higher value of M and hence is less conservative.

Vertical load on foundation $= P = 55.5 \, \text{kN} \qquad$ (see Step 6)

$$e = \frac{M}{P} = \frac{15.3}{55.5} = 0.276 \, \text{m} > \frac{A}{6} = 0.167 \, \text{m}$$

(See Section 6.6.2.)

$$P_1 = \frac{2P}{(1.5A - 3e)B}$$

$$= \frac{2 \times 55.5}{(1.5 \times 1 - 3 \times 0.276) \times 1}$$

$$= 165 \, \text{kN/m}^2 < 474 \, \text{kN/m}^2 \quad \text{(see Step 3)} \quad \text{OK}$$

$$x = 1.5A - 3e = 0.672 \, \text{m}$$

$$\text{Restraining moment} = 55.5 \times \frac{A}{2} = 55.5 \times 0.5$$

$$= 27.75 \, \text{kNm}$$

Overturning moment $= 15.3 \, \text{kNm}$

$$\text{Factor of safety against overturning} = \frac{27.75}{15.3} = 1.8 > 1.5 \quad \text{OK}$$

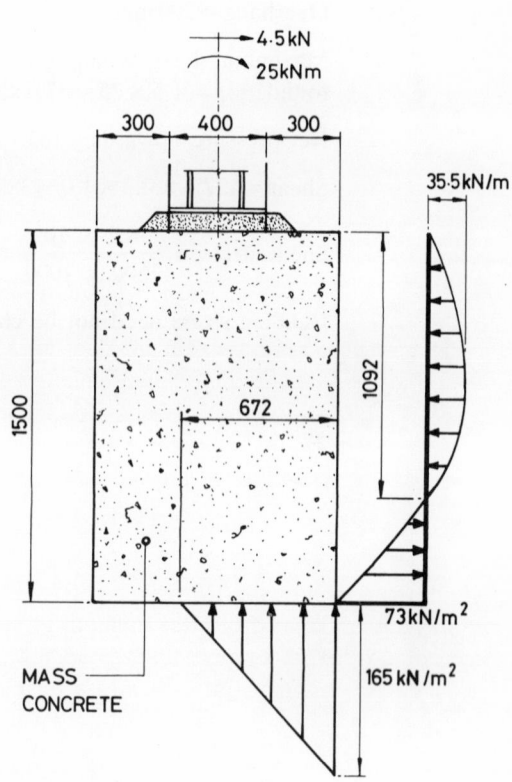

SK 6/68 Soil pressure diagram on mass concrete side-bearing foundation in cohesionless soil.

Note: This is a very conservative estimate of factor of safety against overturning because in practice the value of moment resisted by side bearing will not be restricted to 9.7 kNm but will increase till the pressure diagram becomes rectangular and not triangular as assumed in the analysis.

Maximum pressure at $x/2$ from ground level $= 1.65dx\,\text{MN/m}^2$
$$= 1.65 \times 20 \times 10^{-3} \times 1.092$$
$$\times\ 10^3\,\text{kN/m}^2$$
$$= 36\,\text{kN/m}^2$$

Maximum passive pressure that can be generated at $x/2 = 66 \times 1.092/2 = 36\,\text{kN/m}^2$

Step 8 to Step 21
Not required.

Step 22 Design mass concrete foundation
Use C25 mass concrete $f_{cu} = 25\,\text{N/mm}^2$

Bending about vertical plane:
follow same principle as in Example 6.3.

Bending about horizontal plane:

Overhang $= 200\,\text{mm}$

Maximum shear assuming uniform pressure of $165\,\text{kN/m}^2$ less weight of foundation $= 1.5 \times 25 = 37.5\,\text{kN/m}^2$

Net pressure upwards $= 165 - 37.5 = 127.5\,\text{kN/m}^2$

Shear $= 127.5 \times 0.3 \times 1.0 = 38.2\,\text{kN}$

Shear stress $= \dfrac{38.2 \times 10^3}{1500 \times 1000} = 0.025\,\text{N/mm}^2$ negligible   OK

Bending stress need not be checked.

6.9 FIGURES FOR CHAPTER 6

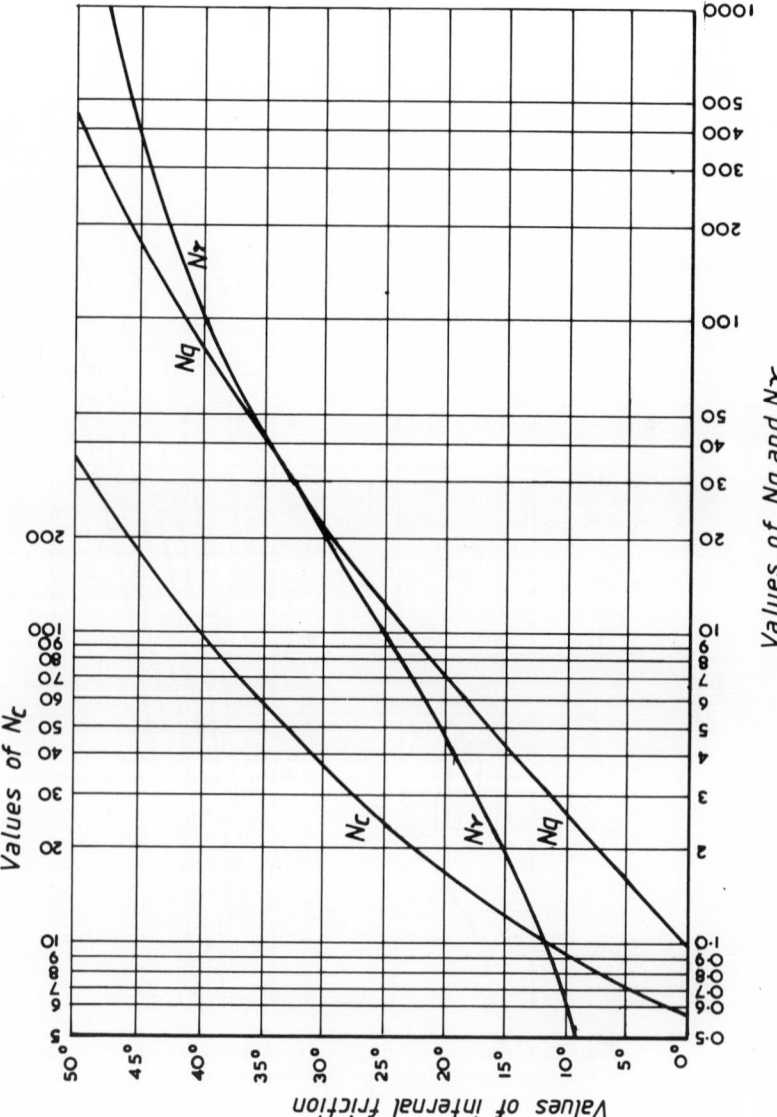

Fig. 6.1 Values of N_c, N_q and N_γ.

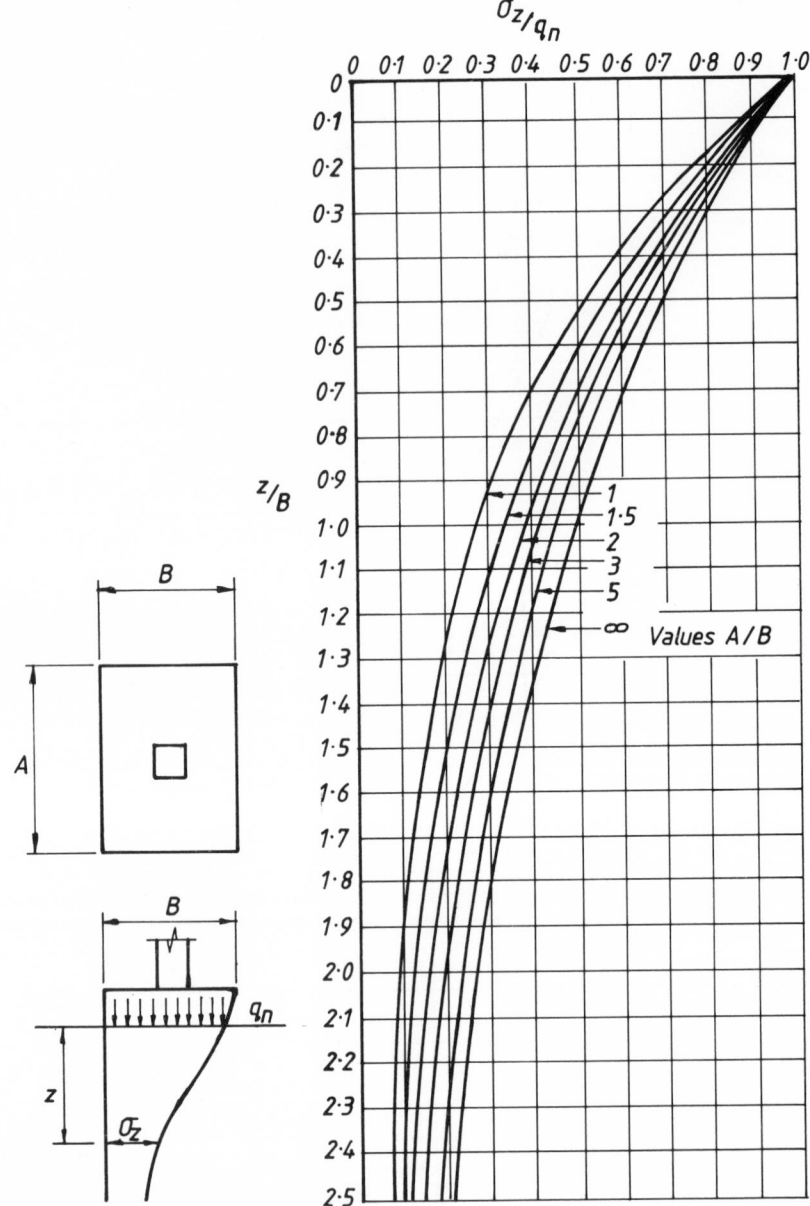

Fig. 6.2 Calculation of mean vertical stress (σ_z) at depth z beneath rectangular area $a \times b$ on surface, loaded at uniform pressure q_n.

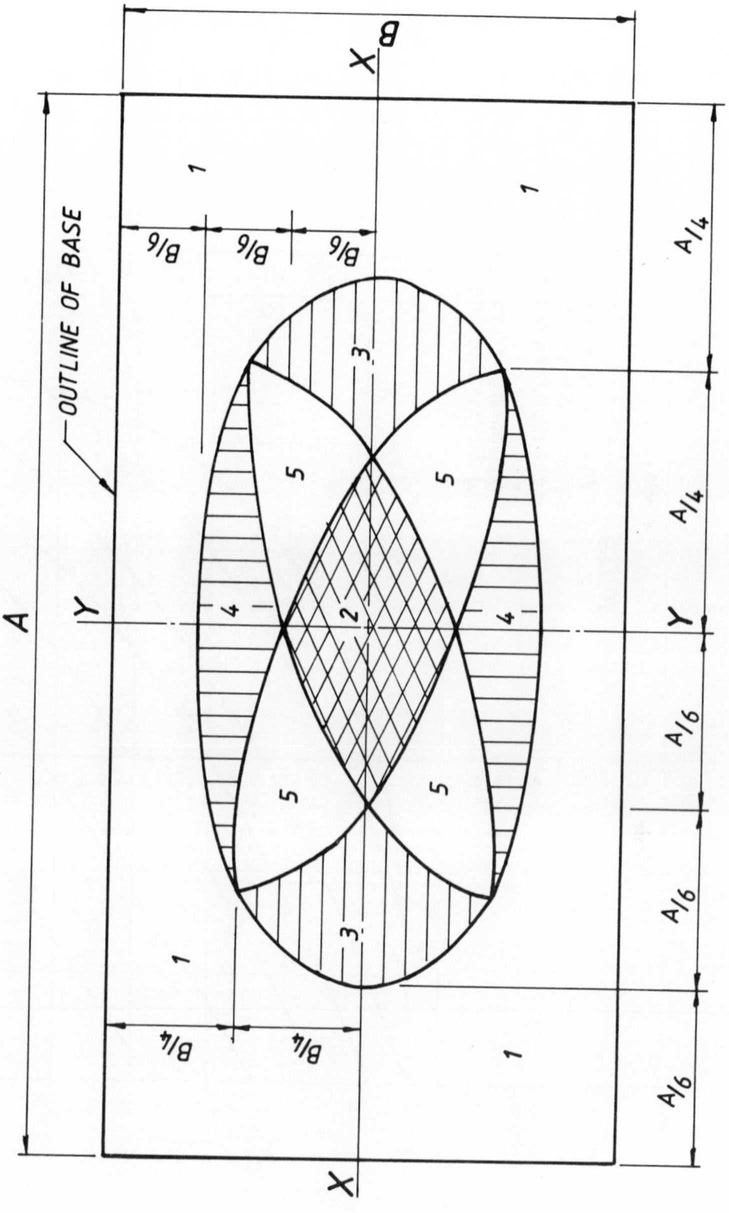

Fig. 6.3 Plan on base showing different zones.

Fig. 6.4 Chart for calculation of maximum pressure under a rectangular base subject to moments in two directions.

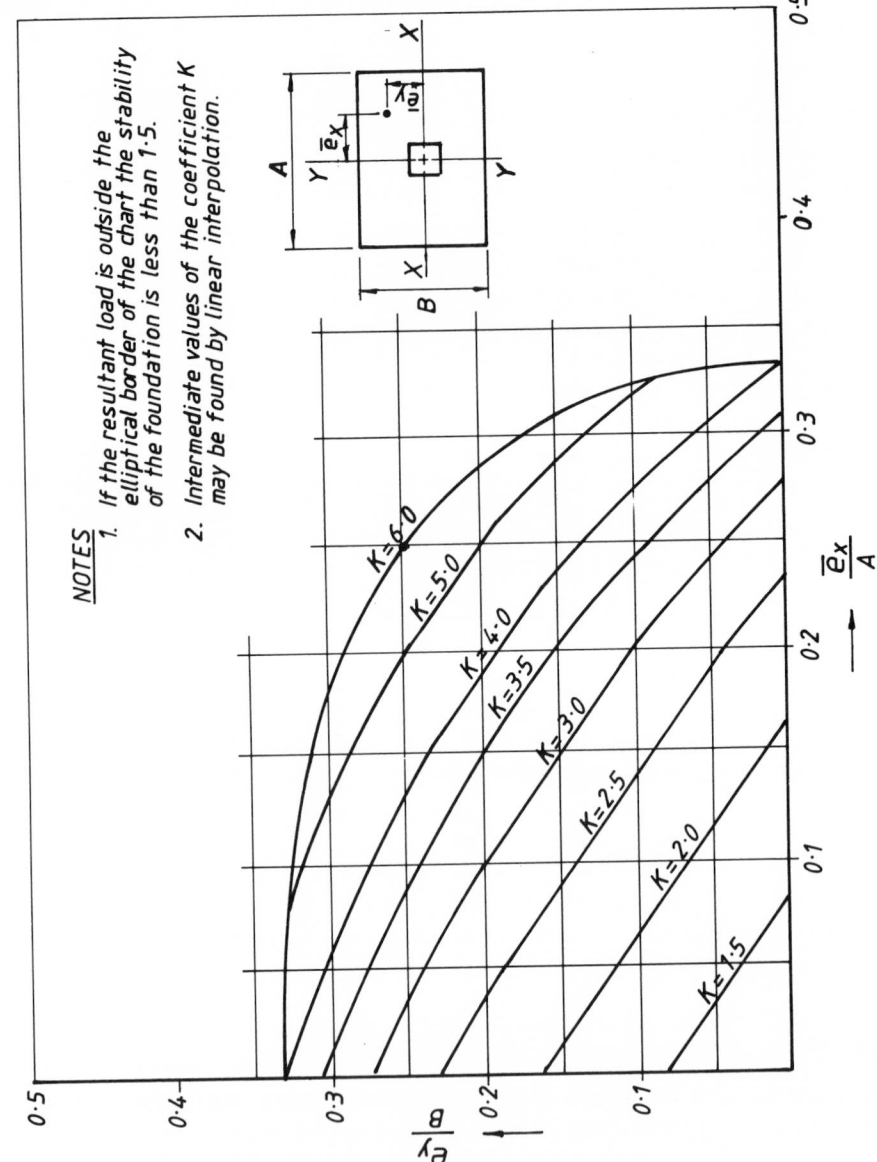

NOTES

1. If the resultant load is outside the elliptical border of the chart the stability of the foundation is less than 1·5.

2. Intermediate values of the coefficient K may be found by linear interpolation.

Chapter 7

Design of Piled Foundations

7.0 NOTATION

a	Deflection due to slenderness of a circular pile
a_v	Distance of shear plane from nearest support
a_x	Deflection due to slenderness producing additional moment about x-axis
a_y	Deflection due to slenderness producing additional moment about y-axis
A_c	Net area of concrete in a pile cross-section
A_p	Cross-sectional area of pile (m^2)
A_s	Surface area of pile in contact with soil
A_v	Total area of link bars perpendicular to longitudinal bars
A_{sc}	Total area of steel reinforcement in a pile
A_{st}	Area of tensile reinforcement in pile cap
A_{sv}	Area of steel effective in resisting shear in a pile
A_{sx}	Area of tensile steel in a pile section resisting moment about b-axis
A_{sy}	Area of tensile steel in a pile section resisting moment about h-axis
b	Width of reinforced concrete section
b	Overall dimension of rectangular pile section
b'	Effective depth of tensile reinforcement in b direction
B	Width or diameter of pile
B	Overall width of a group of piles
c	Soil cohesion for a stratum (kN/m^2)
C_H	Horizontal load-carrying capacity of a single pile
C_V	Vertical load-carrying capacity of a single pile
d	Effective depth to tensile reinforcement in a concrete section
D	Depth of a group of piles below ground
D_r	Relative density
e_x	Eccentricity of combined unfactored vertical load on pile cap in x-direction
e_y	Eccentricity of combined unfactored vertical load on pile cap in y-direction
e_{hx}	Eccentricity in x-direction of combined unfactored horizontal load H_y
e_{hy}	Eccentricity in y-direction of combined unfactored horizontal load H_x
E_f	Stress–strain modulus of pile material (kN/m^2)
E_s	Stress–strain modulus of soil (kN/m^2)
f_c	Stress in concrete due to prestress alone
f_s	Skin resistance at soil/pile interface
f_t	Maximum design principal tensile stress in concrete
f_y	Characteristic yield strength of steel reinforcement
f_{ci}	Cube strength of concrete at transfer of prestress

f_{cp}	Average concrete stress in a prestressed concrete section after losses
f_{cu}	Characteristic cube strength of concrete at 28 days
f_{pe}	Average tensile stress in steel tendons after all losses
f_{pu}	Characteristic ultimate strength of steel tendons
f_{yv}	Characteristic yield strength of shear reinforcement
h	Overall depth of pile cap
h	Overall dimension of a rectangular pile
h	Overall diameter of a circular pile
h'	Effective depth of tensile reinforcement in a rectangular pile in h-direction
H	Unfactored horizontal load on a single circular pile
H_x	Unfactored combined horizontal loads on pile cap in x-direction
H_y	Unfactored combined horizontal loads on pile cap in y-direction
H_{px}	Unfactored horizontal load on a single pile in x-direction
H_{py}	Unfactored horizontal load on a single pile in y-direction
H_{xu}	Ultimate horizontal load on pile cap in x-direction
H_{yu}	Ultimate horizontal load on pile cap in y-direction
H_{pxu}	Ultimate horizontal load on a single pile in x-direction
H_{pyu}	Ultimate horizontal load on a single pile in y-direction
I_f	Moment of inertia of pile (m^4)
I_z	Polar moment of inertia of a group of piles about z-axis through CG
I_{xx}	Moment of inertia of a group of piles about $x-x$ axis through CG of group
I_{yy}	Moment of inertia of a group of piles about $y-y$ axis through CG of group
k_s	Modulus of subgrade reaction of soil (kN/m^3)
K_s	Coefficient of friction
K_t	Factor used to determine transmission length of prestressing wires or strand
l_e	Effective length of pile for calculation of slenderness ratio
l_o	Unsupported length of pile
l_t	Transmission length of prestressing wires or strands
L	Depth of penetration of pile
L	Overall length of a group of piles
L_b	Average depth of pile in ground
m	Modular ratio E_s/E_c
m_v	Coefficient of volume compressibility (m^2/kN)
M	Factored bending moment in a circular pile section
M_o	Moment to produce zero stress at tension fibre of a prestressed section with $0.8f_{cp}$ (average uniform prestress)
M_p	Unfactored bending moment in a single circular pile
M_x	Unfactored combined moment on pile cap about x-axis
M_y	Unfactored combined moment on pile cap about y-axis
M'_x	Modified bending moment about x-axis to account for biaxial bending
M'_y	Modified bending moment about y-axis to account for biaxial bending
M^*_x	Unfactored moment about x-axis due to eccentric surcharge on pile cap
M^*_y	Unfactored moment about y-axis due to eccentric surcharge on pile cap
M_{px}	Unfactored bending moment in a single pile about x-axis due to H_{py}
M_{py}	Unfactored bending moment in a single pile about y-axis due to H_{px}
M_{xx}	Unfactored combined moment on pile group about x-axis
M_{yy}	Unfactored combined moment on pile group about y-axis

M_{pxu}	Ultimate bending moment in pile about x-axis
M_{pyu}	Ultimate bending moment in pile about y-axis
$M_{add\,x}$	Additional bending moment in pile about x-axis due to slenderness
$M_{add\,y}$	Additional bending moment in pile about y-axis due to slenderness
n	Slenderness ratio in a prestressed pile
N	Statistical average of SPT number for a soil stratum
N	Combined vertical load on pile cap – unfactored
N_q	Soil bearing capacity coefficient as per Terzaghi
N_u	Ultimate vertical load on a circular pile
N_γ	Soil bearing capacity coefficient as per Terzaghi
N_c'	Adjusted bearing capacity factor for cohesion
N_q'	Adjusted bearing capacity factor for $L/B > 1$
N_{uz}	Design ultimate capacity of a concrete section subjected to axial load only
N_{bal}	Design axial load capacity of a balanced section ($= 0.25f_{cu}bd$)
p	Percentage of tensile reinforcement in a circular pile
p_x	Percentage of tensile reinforcement in a pile section to resist bending about x-axis
p_y	Percentage of tensile reinforcement in a pile section to resist bending about y-axis
P	Total vertical load on a group of piles
P_a	Allowable unfactored vertical load on pile
P_u	Ultimate axial compressive load on pile
P_{pu}	End-bearing resistance of pile
P_{si}	Skin friction resistance of pile
$\bar{q}$	Effective vertical stress at pile point
q_c	Statistical average of cone resistance of soil in a stratum (kN/m^2)
q_u	Unconfined compressive strength (kN/m^2)
q_{cs}	Side friction resistance in a cone penetrometer
R	Number of piles in a group
R_{iH}	Initial estimate of number of piles based on total horizontal load
R_{iV}	Initial estimate of number of piles based on total vertical load
s	Spacing of nodes in pile for finite element analysis
S_v	Spacing of links used as shear reinforcement
T	Unfactored torsion on a group of piles
T_a	Allowable unfactored tension load on pile
T_u	Ultimate axial tensile load on pile
U	Perimeter at punching shear plane in a pile cap
v	Shear stress in concrete in pile cap
v_c	Design concrete shear stress in concrete
v_x	Shear stress in concrete for shear due to bending about x-axis
v_y	Shear stress in concrete for shear due to bending about y-axis
v_c'	Modified design shear stress to take into account axial compression
v_{cx}	Design shear stress in concrete for shear due to bending about x-axis
v_{cy}	Design shear stress in concrete for shear due to bending about y-axis
V	Ultimate shear force in a circular pile section
V_c	Shear resistance of a concrete section
V_{co}	Shear resistance of uncracked prestressed section
V_{cr}	Shear resistance of cracked prestressed section

W	Weight of pile (kN)
z	Depth of lever arm
α	Coefficient for calculation of skin resistance of a pile
β	Factor for computation of effective length of a pile
β	Factor for conversion of biaxial bending moment into uniaxial bending
γ	Unit weight of soil (kN/m³)
δ	Angle of friction between soil and concrete
μ	Poisson's ratio
ϕ	Angle of internal friction
ϕ	Nominal diameter of tendon in prestressed concrete section

7.1 VERTICAL LOAD – SINGLE PILE CAPACITY

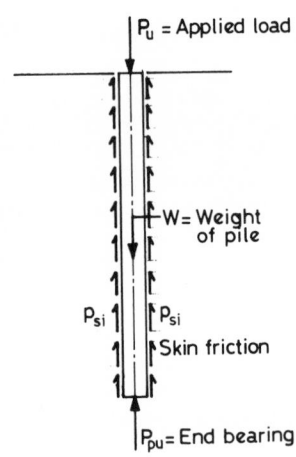

SK 7/1 Single pile capacity.

$$P_u = P_{pu} + \Sigma P_{si} - W$$
$$T_u = \Sigma P_{si} + W$$

where P_u = ultimate compressive load on pile
T_u = ultimate tensile load on pile
ΣP_{si} = skin friction resistance
P_{pu} = end-bearing resistance
W = weight of pile

First method for point resistance

$$P_{pu} = A_p(38N)\left(\frac{L_b}{B}\right) \leq 380N\,(A_p) \qquad \text{(see Reference 6, page 602)}$$

where A_p = cross-sectional area of pile (m²)
N = statistical average of the SPT number in a zone of about $8B$
above to $3B$ below the pile point

B = width or diameter of pile
L_b = average depth of pile in the ground

Second method for point resistance

$P_{pu} = A_p q_c$ (see Reference 6, page 602)

where A_p = cross-sectional area of pile (m^2)
         q_c = statistical average of cone point resistance in a zone of about $8B$ above to $3B$ below pile point (kN/m^2)

Third method for point resistance

$P_{pu} = A_p(N_c'c + \bar{q} N_q')$ (see Reference 6, page 598)

where A_p = cross-sectional area of pile (m^2)
        c = cohesion or undrained shear strength $S_u = q_u/2\,\text{kN/m}^2$
        q_u = unconfined compressive strength
        $\bar{q}$ = effective vertical stress at pile point
        N_c' = adjusted bearing capacity factor for cohesion (see Fig. 7.2)
        N_q' = bearing capacity factor adjusted for $L/b > 1$ dependent on initial angle of shearing resistance ϕ (see Fig. 7.2). (See Reference 8, page 600.)
        L = depth of penetration
        B = width or diameter of pile
L/B should be greater than L_c/B as obtained from Fig. 7.2 for the value of ϕ.

Note: Find point resistance by more than one method if soil test data allow and take the lowest for a conservative estimate.

Determination of skin resistance

$\Sigma P_{si} = \Sigma A_s f_s$

where A_s = pile perimeter × pile length over which f_s acts (m^2)
        f_s = skin resistance (kN/m^2)

First method of skin resistance

$f_s = 2N\,\text{kN/m}^2$ for large volume displacement piles
$f_s = N\,\text{kN/m}^2$ for small volume displacement piles

where N = statistical average blow count in stratum for SPT.

Second method of skin resistance

$f_s = 0.005q_c\,\text{kN/m}^2$

where q_c = cone penetration resistance (kN/m^2).

Third method of skin resistance

$f_s = q_{cs}\,\text{kN/m}^2$ for small volume displacement piles

$f_s = 1.5q_{cs}$ to $2.0q_{cs}$ for large volume displacement piles

where q_{cs} = side friction resistance in cone penetrometer.

Fourth method of skin resistance

$f_s = \alpha c + 0.5\, \bar{q}\, K_s \tan \delta$ (see Reference 8, page 603)

where c = average cohesion or S_u of stratum (kN/m^2)
$\quad\quad\quad \bar{q}$ = effective vertical stress (kN/m^2)
$\quad\quad\quad \delta$ = angle of friction between soil and pile
$\quad\quad\quad K_s$ = coefficient of friction
$\quad\quad\quad D_r$ = relative density of sand.

Table 7.1 Values of K_s (Reference 8, page 603).

Pile type	δ	K_s for low D_r	K_s for high D_r
Steel	20°	0.5	1.0
Concrete	0.75ϕ	1.0	2.0
Wood	0.67ϕ	1.5	4.0

(See Reference 7, page 136.)

Table 7.2 Values of α (Reference 7, page 126).

Soil condition		Values of α				
	D/B	$c=50$	$c=100$	$c=150$	$c=200$	$c=250$
Sands or sandy gravel	<10	1.0	1.0	1.0	1.0	1.0
overlying stiff to very	20	1.0	0.9	0.75	0.75	0.75
stiff cohesive soil	>40	0.9	0.65	0.4	0.4	0.4
Soft clays or silts	10	0.35	0.30	0.25	0.2	0.2
overlying stiff to very	>20	0.75	0.70	0.63	0.55	0.5
stiff cohesive soil						
Stiff to very stiff	10	0.9	0.7	0.3	0.2	0.2
cohesive soils without	>40	1.0	0.9	0.3	0.3	0.3
overlying strata						

The units of c are kN/m^2

Note: Find skin resistance by more than one method if soil test data allow and take an average.

$$P_a = \frac{P_u}{2.5} \quad T_a = \frac{T_u}{2.5}$$

where P_a = allowable pile load in compression
$\quad\quad\quad T_a$ = allowable pile load in tension

7.2 HORIZONTAL LOAD – SINGLE PILE CAPACITY

Method 1 Cohesive soils

$$k_s B = 1.3 \left(\frac{E_s B^4}{E_f I_f} \right)^{\frac{1}{12}} \left(\frac{E_s}{1 - \mu^2} \right)$$

as per Vesic, 1961 (see Reference 6).

where k_s = modulus of subgrade reaction (kN/m³)
  B = width or diameter of pile (m)
  E_s = stress–strain modulus of soil (kN/m²)
  E_f = stress–strain modulus of pile material (kN/m²)
  I_f = moment of inertia of pile (m⁴)
  μ = Poisson's ratio of soil

E_s may be obtained by the following methods:

(1) Triaxial tests.
(2) Borehole pressuremeter tests.
(3) $E_s = 650N \,(\text{kN/m}^2)$
  N = SPT number of blows.
(4) $E_s = 3 \,(1 - 2\mu)/m_v$ where m_v = coefficient of volume compressibility (m²/kN).

Method 2 Cohesive soils

$$k_s = 240 q_u \ \text{kN/m}^3$$

where q_u = unconfined compression strength (kN/m²).

Cohesionless soils

$$k_s = 80 \,[C_2 \bar{q} N_q + C_1 \,(0.5 \ \gamma \ B N_\gamma)] \ \text{kN/m}^3$$

as per Vesic (see Reference 8, page 631 and page 323, equation 9–8).

where $C_1 = C_2 = 1.0$ for square piles
  $C_1 = 1.3$ to 1.7 for circular piles
  $C_2 = 2.0$ to 4.4 for circular piles
  $\bar{q}$ = effective stress (kN/m²)
  γ = unit weight of soil
  B = width or diameter of pile

N_q and N_γ may be obtained from the following table (Hansen equations) – see Reference 8, page 137, Table 4–4:

Finite element model of vertical pile

Spring stiffness $= SBk_s$ kN/m

where S = node spacing not greater than B
  B = width or diameter of pile (m)
  k_s = modulus of subgrade reaction (kN/m³)

Table 7.3 Values of N_q and N_γ (Reference 8, page 137).

φ (degrees)	N_q	N_γ
0	1.0	0
5	1.6	0.1
10	2.5	0.4
15	3.9	1.2
20	6.4	2.9
25	10.7	6.8
30	18.4	15.1
35	33.3	33.9
40	64.2	79.5
45	134.9	200.8
50	319.0	568.5

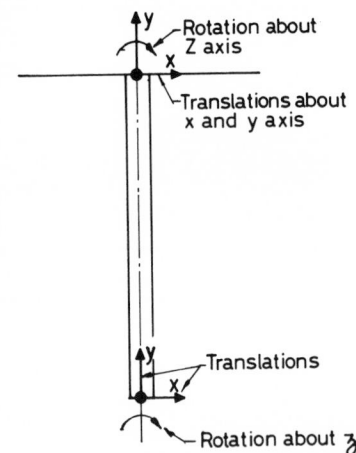

SK 7/2 Two-dimensional model of pile in soil (degrees of freedom — top and bottom of pile).

Note: For horizontal loads which are not constant and are reversible or repetitive, the top $1.5B$ of pile may be assumed unsupported by soil.

Boundary conditions
(1) Free head pile
Translations x, y Free at top
Rotation z Free at top
Translations y Restrained at bottom
Rotation z Free at bottom
(2) Fixed head pile
Translations x, y Free at top
Rotation z Rigid at top
Translations y Restrained at bottom
Rotation z Free at bottom

Material type
For sustained horizontal load due to dead load, water pressure, earth

pressure, etc., use short-term Young's modulus of concrete for bending moment computations but long-term Young's modulus of concrete for pile head deformation.

For short-term horizontal loads due to wind, earthquake, crane surge, etc., use short-term Young's modulus of concrete for bending moment and deflection computations.

Software
Use any fully validated software which has a suite for analysis of 2-D plane frame with sprung boundaries.

Member type
For rectangular pile use minimum width B in all computations involving B. A cracked section moment of inertia may be used for reinforced concrete piles based on Section 2.1.

7.3 PILE GROUP EFFECTS

7.3.1 Spacing of piles

$$S \geq 2B \quad \text{for end-bearing piles}$$
$$S \geq 3B \quad \text{for friction piles}$$

where S = spacing of piles
B = least width or diameter of pile.

Note: Piles carrying horizontal load should not be spaced at less than $3B$.

7.3.2 Pile group capacity

Ultimate group capacity = group friction capacity + group end-bearing capacity

Ultimate group friction capacity = $2D(B + L)c\alpha$

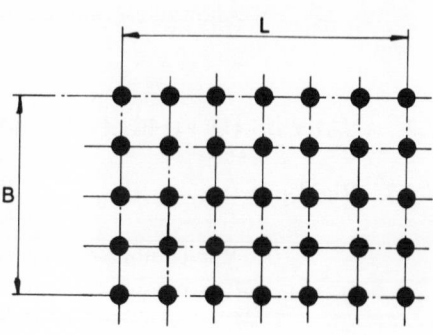

SK 7/3 Group of piles − plan of overall dimensions of group.

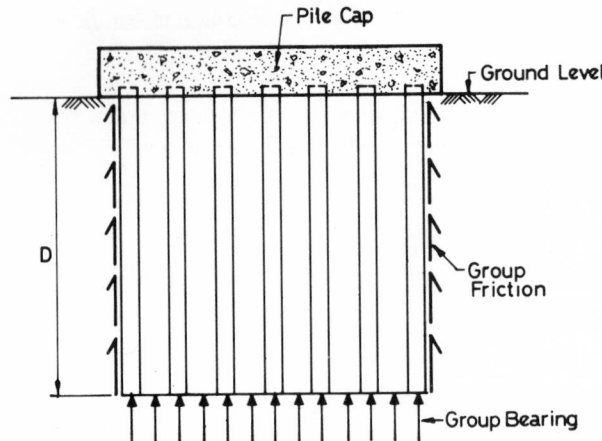

SK 7/4 Elevation of group of piles showing group capacity.

where c = average cohesion of clay
      = average S_u = average $q_u/2$
     α = coefficient (from Section 7.1, Table 7.2)
     D = depth of pile group below ground
     B = overall width of group
     L = overall length of group.

Ultimate group end-bearing capacity $= BL\ (N'_c c + \bar{q} N'_q)$

where c = cohesion or undrained shear strength $S_u = q_u/2$ at bottom of pile group
   q_u = unconfined compressive strength
   $\bar{q}$ = effective stress at bottom of pile group
   N'_q = bearing capacity factor (see Fig. 7.2)
   N'_c = bearing capacity factor (see Fig. 7.2)

Note: Total vertical load on a group of piles should not exceed the group capacity. Individual pile loads inside the group will be limited by the single pile capacity. Piles carrying horizontal load and spaced at $3B$ or more need not be checked for group effects due to horizontal load.

$$\text{Allowable group capacity} = \frac{\text{ultimate group capacity} + \text{ultimate group end-bearing capacity}}{2.5}$$

7.4 ANALYSIS OF PILE LOADS AND PILE CAPS

7.4.1 Rigid pile cap

N = combined vertical load on pile cap − unfactored
M_x = combined moment about $x–x$ − unfactored
M_y = combined moment about $y–y$ − unfactored

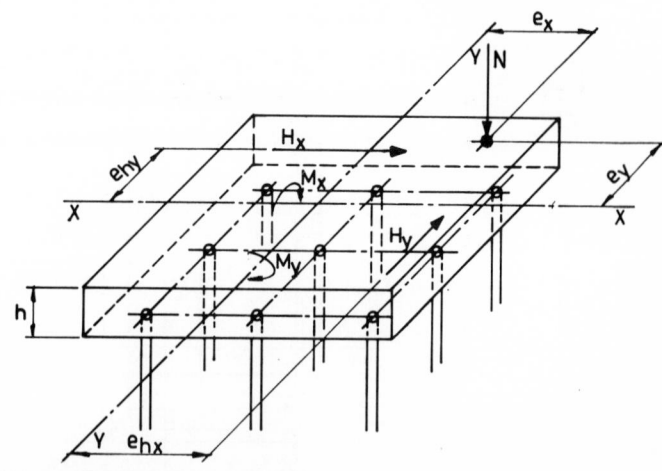

SK 7/5 Loads and eccentricity on pile cap.

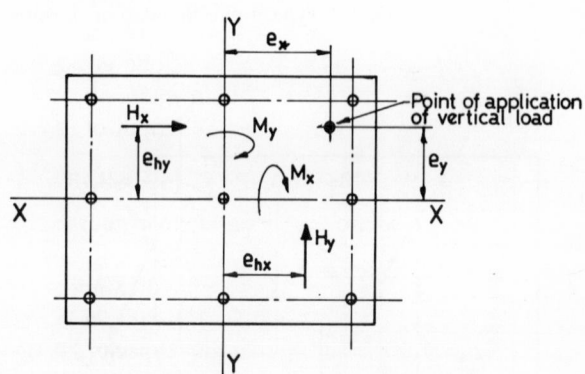

SK 7/6 Plan view of loads and eccentricity on pile cap.

H_x = combined horizontal load on pile cap − unfactored in $x-x$ direction
H_y = combined horizontal load on pile cap − unfactored in $y-y$ direction
e_x = eccentricity of N from CG of pile group in $x-x$ direction
e_y = eccentricity of N from CG of pile group in $y-y$ direction
e_{hx} = eccentricity of H_y from CG of pile group in $x-x$ direction
e_{hy} = eccentricity of H_x from CG of pile group in $y-y$ direction
h = depth of pile cap.

Loads on pile group

P = vertical load on pile group
  = N + weight of pile cap + weight of backfill on pile cap + surcharge on backfill
M_{xx} = moment about $x-x$ on pile group
  = $M_x + Ne_y + H_yh + M_x^*$
M_{yy} = moment about $y-y$ on pile group
  = $M_y + Ne_x + H_xh + M_y^*$

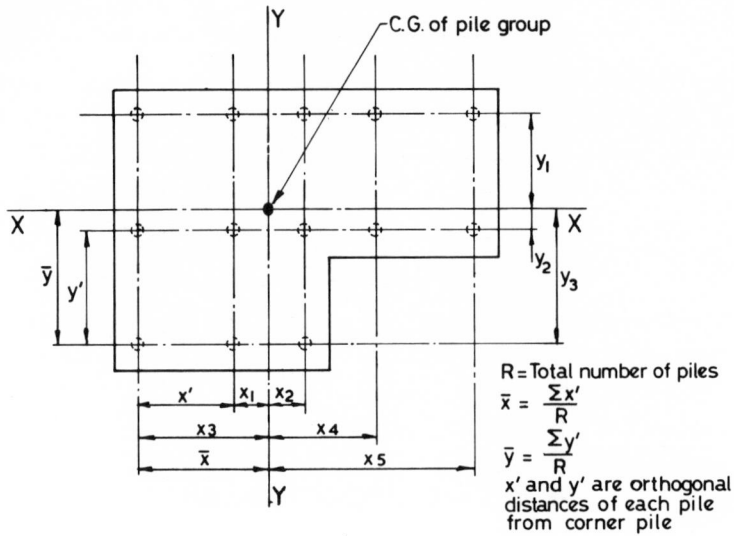

SK 7/7 Typical pile foundation showing CG of group and co-ordinates of piles.

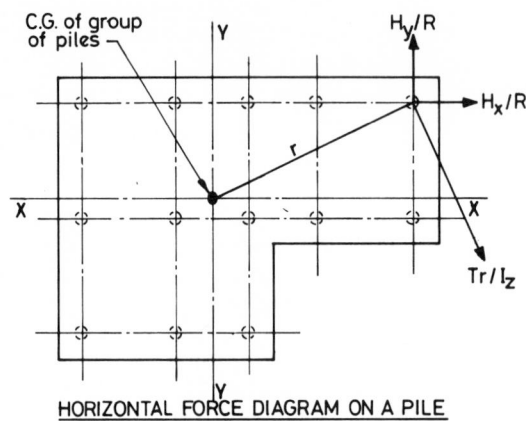

HORIZONTAL FORCE DIAGRAM ON A PILE

SK 7/8 Group of piles subject to horizontal loads and torsion.

where M_x^* and M_y^* are moments with respect to CG of pile group due to eccentric surcharge on backfill or pile cap.

T = torsion on pile group
 = $H_x e_{\mathrm{hy}} + H_y e_{\mathrm{hx}}$

$I_{xx} = \Sigma y^2$ about $x-x$ axis passing through CG of pile group
$I_{yy} = \Sigma x^2$ about $y-y$ axis passing through CG of pile group
$I_z = I_{xx} + I_{yy}$

R = number of piles in group.

$$\text{Vertical load on a pile} = \left(\frac{P}{R}\right) \pm \left(\frac{M_{xx}y}{I_{xx}}\right) \pm \left(\frac{M_{yy}x}{I_{yy}}\right)$$

$$\text{Horizontal load on any pile} = \text{resultant of } \frac{(H_x^2 + H_y^2)^{\frac{1}{2}}}{R} \text{ and } \frac{T(x^2 + y^2)^{\frac{1}{2}}}{I_z}$$

Sign convention

Vertical loads:	downwards positive
Torsion on pile group:	clockwise positive
Moments on pile group:	clockwise positive

+ve M_{xx} produces compression in piles which have +ve y ordinates.

+ve M_{yy} produces compression in piles which have +ve X ordinates.

H_x is positive in direction of increasing x in positive direction.

H_y is positive in direction of increasing y in positive direction.

Eccentricities are +ve for +ve x and +ve for +ve y.

Bending moments in pile cap

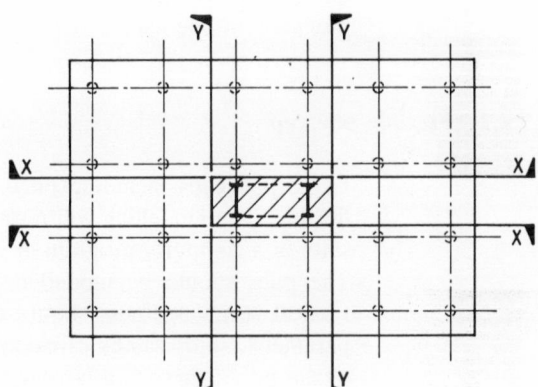

SK 7/9 Critical sections for bending moment in a pile cap.

Take sections $X-X$ or $Y-Y$ through pile cap at faces of columns or base plates. Find pile reactions due to combined and load factored basic load cases. Consider all upward and downward loadings across sections $X-X$ and $Y-Y$. Find bending moments across section. Find horizontal load on each pile by using the following expressions:

$$H_{pxu} = \frac{H_{xu}}{R}$$

$$H_{pyu} = \frac{H_{yu}}{R}$$

where R is number of piles in pile cap. Find bending moments in pile M_{pxu} corresponding to H_{pyu} and M_{pyu} corresponding to H_{pxu} assuming an end fixity to pile cap following the method in Section 7.2. H_{xu} and H_{yu} are combined factored ultimate horizontal loads.

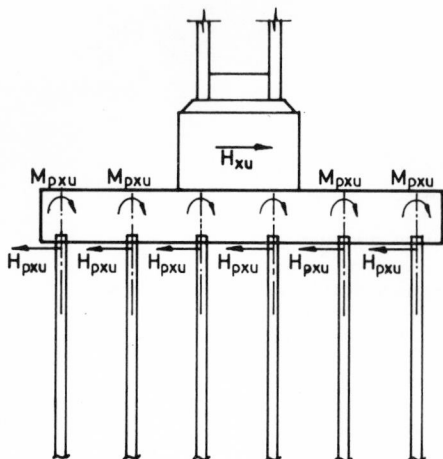

SK 7/10 Additional bending moment in pile cap due to pile fixity.

Algebraically add the bending moments in pile cap due to vertical load and pile fixity moments due to horizontal load to find design bending moments in pile cap.

7.4.2 Flexible pile cap

Large pile caps including piled raft foundations should be modelled as flexible. The modelling will normally be carried out using either a grillage suite of a computer program or a general-purpose finite element program. The piles should be modelled as springs in the vertical direction. The vertical spring stiffness should be obtained from test results on site. A parametric study can be carried out using minimum and maximum stiffness of the pile if there is a large variation.

Grillage model

(1) Divide pile cap into an orthogonal grillage network of beams. Ensure that piles are located at crossing of orthogonal beams. Each grillage beam represents a certain width of pile cap.
(2) Use short-term Young's modulus for concrete material properties.
(3) Full section concrete stiffness properties may be used for hypothetical grillage beams (hypothetical width × depth of pile cap).
(4) Piles will be modelled as sprung supports vertically.
(5) Vertical loads on pile cap may be dispersed at 45° up to central depth of pile cap.
(6) Apply at each node with a pile, the moments given by the following formulae:

$$M_x = \frac{H_y h}{R} \qquad \text{about } x\text{-axis}$$

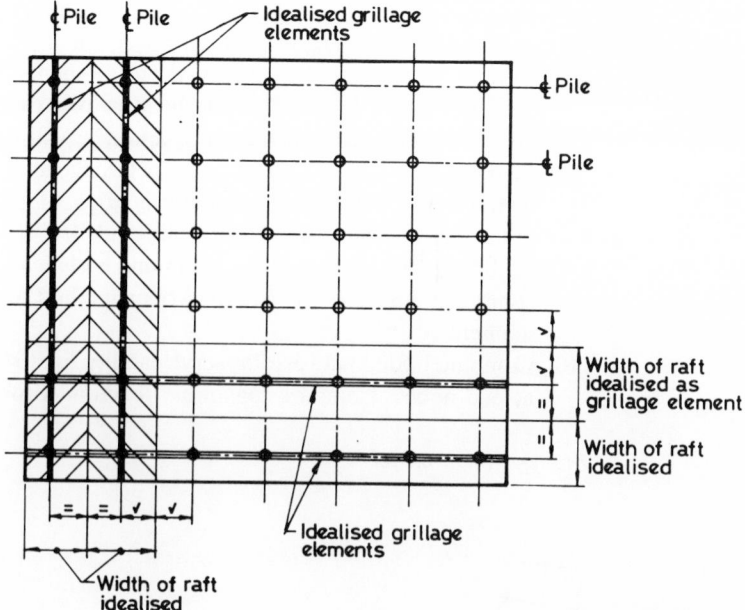

SK 7/11 Plan of raft on piles showing idealised grillage elements – flexible analysis.

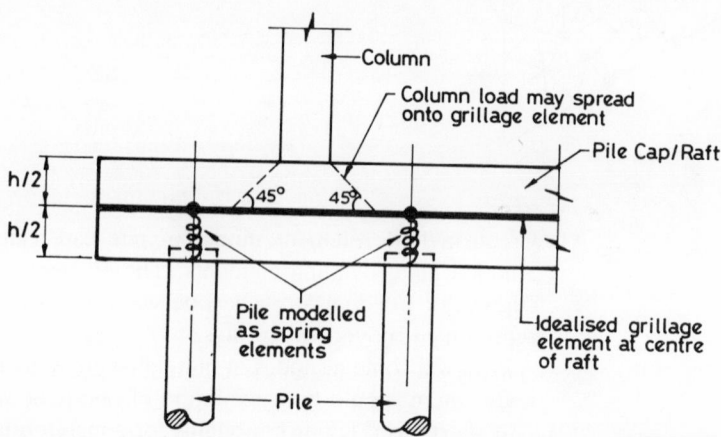

SK 7/12 Part section through raft showing details of grillage idealisation.

$$M_y = \frac{H_x h}{R} \quad \textit{about } y\text{-axis}$$

(7) Find horizontal load on each pile by using the following expressions:

$$H_{px} = \frac{H_x}{R} \quad \text{and} \quad H_{py} = \frac{H_y}{R}$$

where R is total number of piles in group.

(8) Find bending moments in pile, M_{px} corresponding to H_{py} and M_{py} corresponding to H_{px}, assuming an end fixity to pile cap following method in Section 7.2. Apply these moments to pile cap grillage model as nodal loads. The pile head to pile cap connection may be assumed as hinged and then M_{px} and M_{py} will be zero.

(9) Find bending moments in pile cap by grillage analysis. Divide bending moments by width of hypothetical strips of pile cap representing grillage beams and obtain M_x, M_y and M_{xy} in pile cap per metre width. Apply load factors and combine basic load cases. Modify these combined moments by Wood−Armer method to find design bending moments.[11,12]

(10) Combine basic load cases at serviceability limit state to find reactions at pile nodes. Compare maximum reaction with pile capacity.

Finite-element model

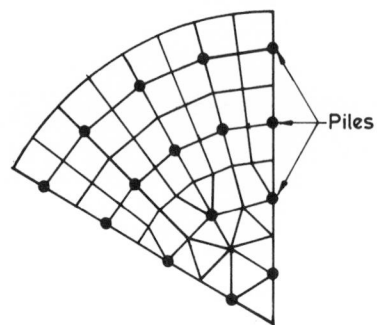

SK 7/13 Typical finite element modelling of a circular raft on piles.

(1) Create a finite element model of pile cap using either 4-noded or 8-noded plate bending elements. The elements may only have three degrees of freedom at each node viz z, θx and θy. The piles will be represented by vertical springs.

   Piles will come at nodes in finite element model. Between two piles' nodes there should be a minimum of one plate node without pile.

(2) Use short-term Young's modulus for concrete material properties.

(3) Full section concrete section properties may be used in the analysis.

(4) Vertical loads on pile cap may be dispersed at 45° up to central depth of pile cap. These loads may be applied as nodal loads or uniformly distributed loads on plate elements depending on software used.

(5) Apply at each node with a pile, the moments given by the following formulae.

$$M_x = \frac{H_y h}{R} \quad \text{about } x\text{-axis}$$

$$M_y = \frac{H_x h}{R} \quad \text{about } y\text{-axis}$$

(6) Find horizontal load on each pile by using the following expressions:

$$H_{px} = \frac{H_x}{R} \quad \text{and} \quad H_{py} = \frac{H_y}{R}$$

where R is total number of piles in group.

(7) Find bending moments in pile, M_{px} corresponding to H_{py} and M_{py} corresponding to H_{px}, assuming an end fixity to pile cap following method in Section 7.2. Apply these moments as nodal loads in finite element model at nodes with piles. These moments will be zero in the case of a hinged connection of pile to pile cap.

(8) Carry out analysis using a validated general-purpose finite element software. Apply load factors to combine basic load cases. Modify the combined M_x, M_y and M_{xy} using the Wood−Armer method to find design bending moments.[11,12]

(9) Combine basic load cases at serviceability limit state to find reactions at pile nodes. Compare maximum reaction with rated pile capacity.

7.5 LOAD COMBINATIONS

Applied loads on pile cap will be combined using the following principles.

7.5.1 Pile load calculations

LC_1: $1.0DL + 1.0IL + 1.0EP + 1.0CLV + 1.0CLH$
LC_2: $1.0DL + 1.0EP + 1.0CLV + 1.0CLH + 1.0WL$ (or $1.0EL$)
LC_3: $1.0DL + 1.0IL + 1.0EP + 1.0WL$ (or $1.0EL$)
LC_4: $1.0L + 1.0WL$ (or $1.0EL$)

where DL = dead load
IL = imposed load
EP = earth pressure and water pressure
CLV = crane vertical loads
CLH = crane horizontal loads
WL = wind load
EL = earthquake load.

7.5.2 Bending moment and shear calculations in pile cap or piles

LC_5: $1.4DL + 1.6IL + 1.4EP$
LC_6: $1.2DL + 1.2IL + 1.2EP + 1.2WL$ (or $1.2EL$)
LC_7: $1.4DL + 1.4WL$ (or $1.4EL$) $+ 1.4EP$
LC_8: $1.0DL + 1.4WL$ (or $1.4EL$) $+ 1.4EP$ (if adverse)
LC_9: $1.4DL + 1.4CLV + 1.4CLH + 1.4EP$
LC_{10}: $1.4DL + 1.6CLV + 1.4EP$
LC_{11}: $1.4DL + 1.6CLH + 1.4EP$
LC_{12}: $1.2DL + 1.2CLV + 1.2CLH + 1.2EP + 1.2WL$ (or $1.2EL$)

7.6 STEP-BY-STEP DESIGN PROCEDURE FOR PILED FOUNDATIONS

Step 1 *Select type of pile*
The type of pile will depend on the following principal factors:

- Environmental issues like noise, vibration.
- Location of structure.
- Type of structure.
- Ground conditions.
- Durability requirements.
- Programme duration.
- Cost.

The commonly available types of piles can be broadly classified as below.

Large-displacement piles (driven)

- Precast concrete.
- Prestressed concrete.
- Steel tube with closed end.
- Steel tube filled with concrete.

Small-displacement piles (driven)

- Precast concrete tube with open end.
- Prestressed concrete tube with open end.
- Steel H-section.
- Screw pile.

Non-displacement piles

- Bored and cast-in-situ concrete pile.
- Steel tube in bored hole filled with concrete.
- Steel or precast section in drilled hole.

Step 2 *Determine vertical capacity of single pile*
Follow Section 7.1.

Step 3 *Determine horizontal capacity of single pile*
Follow Section 7.2.

Note: Horizontal capacity of a single pile is limited by maximum deflection of pile cap that structure can accommodate and also by pile structural capacity.

Step 4 *Determine approximate number of piles and spacing*

$$R_{iV} = \frac{P}{C_V}$$

$$R_{iH} = \frac{H}{C_H}$$

$$R_i = R_{iV} \quad \text{or} \quad R_{iH}, \text{ whichever is greater}$$

where R_i = approximate number of piles
   P = total vertical load on pile cap – unfactored
   C_V = rated working load capacity of pile – vertical load
   C_H = rated working load capacity of pile – horizontal load
   H = total horizontal load on pile cap – unfactored
   $= (H_x^2 + H_y^2)^{\frac{1}{2}}$

Spacing of piles should be according to Section 7.3. To minimise the cost of pile cap, the spacing should be kept close to minimum allowed. Larger spacing increases the pile group capacity and pile group moment capacity.

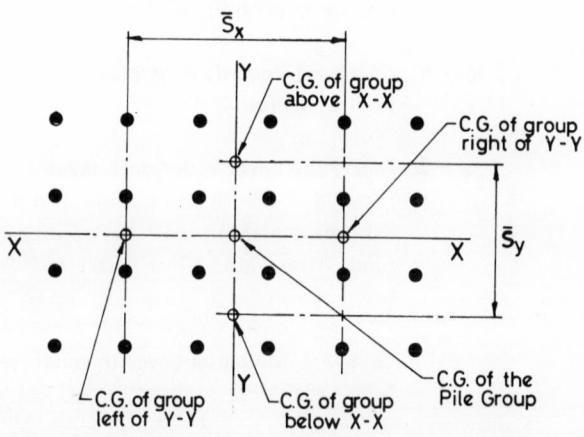

SK 7/14 Determination of approximate number of piles.

(1) Select a group of piles with approximate number of piles = R_i.
(2) Find CG of pile group and locate orthogonal axes $x-x$ and $y-y$ through the CG.
(3) Find CG of group of piles on left of axis $y-y$ and right of axis $y-y$.
(4) Find the x-axis distance between these two CGs and call it $\bar{S}_x$.
(5) Similarly, find $\bar{S}_y$ about y-axis.
(6) Find $M_x/P = e_y$ and $M_y/P = e_x$, where M_x and M_y are total combined applied moments on pile cap about $x-x$ and $y-y$ respectively.
(7) Find $e_x/\bar{S}_x$ and $e_y/\bar{S}_y$.
(8) Find E_x and E_y from Fig. 7.1.
(9) $R = \dfrac{1.1 \, R_{iV}}{E_x E_y} \geq R_{iH}$

where R = number of piles in group for checking pile load.

Note: The factor 1.1 is introduced to cater for additional vertical loads from self-weight of pile cap, surcharge on pile caps, backfilling, etc.

Revise the number of piles in group from R_i to R.

Step 5   *Determine size of pile cap*

Allow 1.5B from centre of pile to edge of pile cap.

Depth of pile cap is governed by the following:

- Shrinking and swelling of clay.
- Frost attacks.
- Holding down bolt assemblies for columns.
- Water table and soluble sulphates.
- Pile anchorage.
- Punching shear capacity of pile cap.

Step 6   *Carry out load combination*

Follow Section 7.5.

Step 7   *Check pile group effects*

Follow Section 7.3.

Step 8   *Carry out analysis of pile cap*

Follow Section 7.4.

Step 9   *Determine cover to reinforcement*

From the soils investigations report, find the concentration of sulphates expressed as SO_3.

Find, from Table 17 of BS 8004: 1986[2], the appropriate type of concrete.

Table 7.4 Minimum cover to reinforcement for class of exposure.

Class of exposure	Total SO_3 percentage	Minimum cover on blinding (mm)	Minimum cover elsewhere (mm)
1	<0.2	35	75
2	0.2 to 0.5	40	80
3	0.5 to 1.0	50	90
4	1.0 to 2.0	60	100
5	>2.0	60	100

Note: Concrete in 'class of exposure 5' needs protective membrane, or coating. The uneven heads of piles normally necessitate a minimum 75 mm cover over blinding for pile caps. The concrete piles will have minimum cover as specified elsewhere.

Step 10   *Calculate area of reinforcement in pile cap*

M = bending moment as found in Step 8 at ultimate limit state

$$K = \frac{M}{f_{cu}bd^2} \leq 0.156$$

where f_{cu} = concrete characteristic cube strength at 28 days

b = width of section over which moment acts

d = effective depth to tension reinforcement.

If K is greater than 0.156, increase depth of pile cap.

$$A_{st} = \frac{M}{0.87 f_y z}$$

$$z = d\left[0.5 + \sqrt{\left(0.25 - \frac{K}{0.9}\right)}\right] \leq 0.95d$$

Distribute this area of reinforcement uniformly across the section.

Note: The effective depth to tension reinforcement will be different in the two orthogonal directions.

Step 11   Check shear stress in pile cap

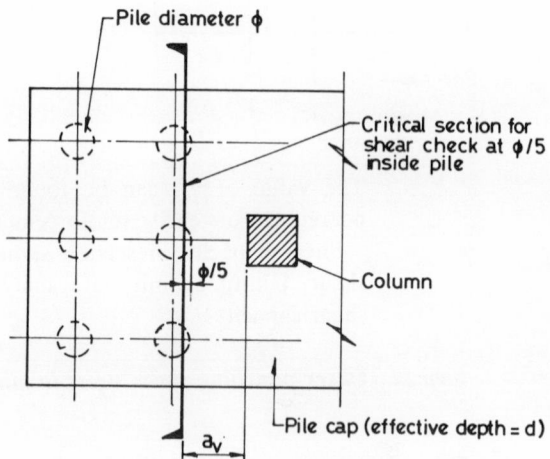

SK 7/15 Critical section for checking shear stress in pile cap.

Enhancement of shear stress is allowed if $a_v \leq 1.5d$

The critical section for checking shear stress in a pile cap is $\phi/5$ into the pile. All piles with centres outside this line should be considered for calculating shear across this section in pile cap. For shear enhancement, a_v is from face of column to this critical section. No enhancement of shear stress is allowed if a_v is greater than $1.5d$. Where pile spacing is more than 3ϕ then enhancement of shear should be applied only on strips of width 3ϕ. The rest of the section will be limited to unenhanced shear stress.

$$V = \frac{\Sigma P}{Bd} \leq v_c \qquad \text{or enhanced } v_{c1} \text{ if applicable}$$

where ΣP = sum of all pile reactions at ultimate loading on left of section

B = width of pile cap at critical section

$$d = \text{average effective depth at critical section}$$

$$v_{c1} = v_c \left(\frac{2d}{a_v}\right) \leq 0.8\sqrt{f_{cu}} \quad \text{or} \quad 5\,\text{N/mm}^2$$

For rectangular piles the critical section may be considered at face of pile.

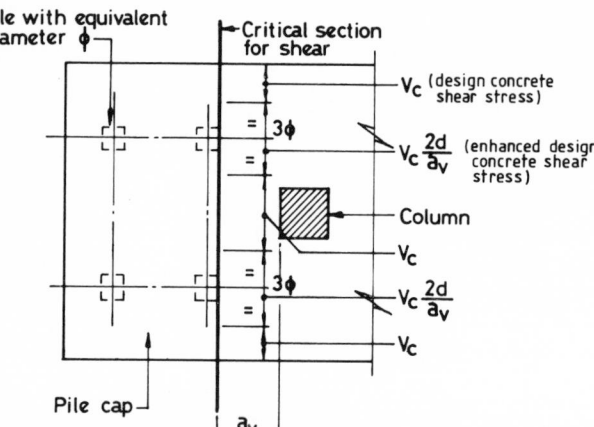

SK 7/16 Diagram showing zones of enhanced shear stress on critical section.

The value of v_{c1} can be found from Figs 11.2 to 11.5 depending on percentage of tensile reinforcement and f_{cu}.

Shear capacity of section should be greater than or equal to applied shear. Ultimate limit state analysis results should be used for checking shear capacity.

Step 12   Check punching shear stress in pile cap

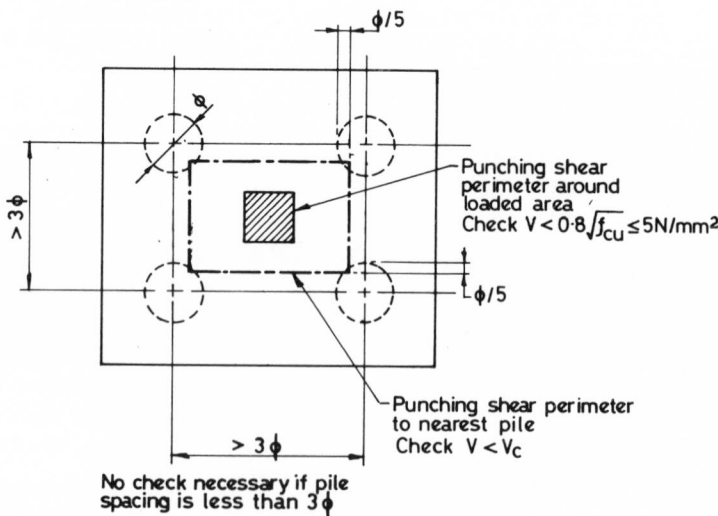

SK 7/17 Perimeters for punching shear checks.

When the spacing of piles is greater than 3 times the diameter of a pile then the punching shear plane for column should be considered. For rectangular piles the plane can be considered at face of pile. The stress on this punching shear plane should not exceed v_c depending on the percentage of tensile reinforcement in pile cap.

Check of punching shear stress is also required at perimeter at face of column or pile. This shear stress should not exceed $0.8\sqrt{f_{cu}}$ or $5\,\text{N/mm}^2$.

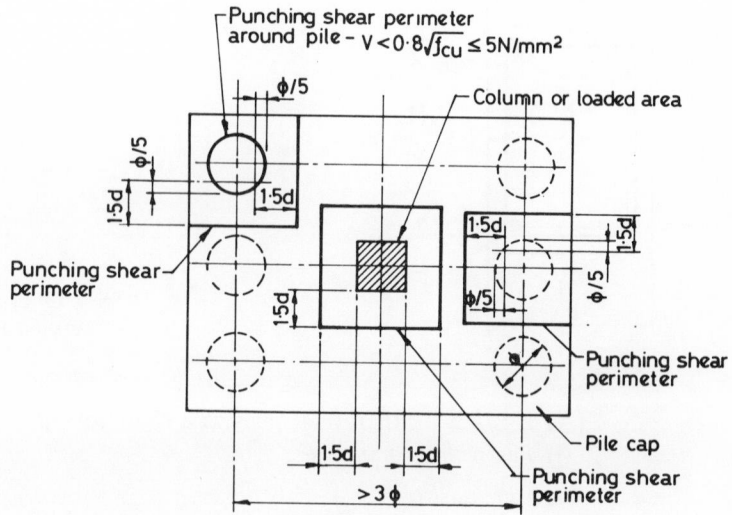

SK 7/18 Further perimeters for punching shear checks in a pile cap.

The punching shear planes for piles will depend on location of pile with respect to edge of pile cap.

Find the perimeter U at punching shear plane.

$$v = \frac{P}{Ud} \leq v_c$$

where P = ultimate vertical column load or ultimate vertical pile reaction
   v_c = design concrete shear stress obtained from Figs 11.2 to 11.5.

Percentage area of tensile reinforcement for computation of design concrete shear stress will be average percentage across punching shear planes.

Step 13   ***Check area of reinforcement in pile***

Effective length of pile, $l_e = \beta l_o$

where l_o = unsupported length of pile (piles which are not subjected to horizontal load may be assumed fully supported by ground from ground level; piles subjected to horizontal load may be assumed supported by ground at a depth of $1.5b$ below ground level where b is width of pile or diameter of pile)

$$\beta = 1.2 \qquad \text{for piles with head fixed to pile cap}$$
$$= 1.6 \qquad \text{for piles with head free to rotate.}$$

Rectangular piles

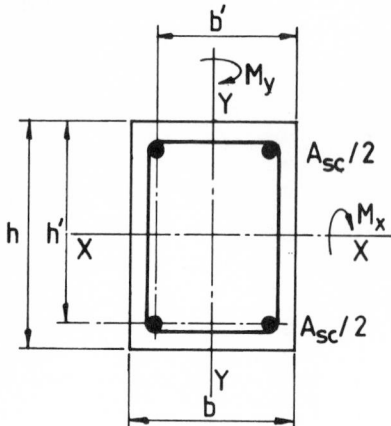

SK 7/19 Typical section through a rectangular pile.

(A) If $l_e/b \leq 10$, then treat piles as a short column.
 (i) *Pile with no moment*

$$N = 0.4 f_{cu}bh + 0.75 A_{sc}f_y$$

Check $N \geq$ applied direct load on pile.
 (ii) *Pile subjected to uniaxial moment*
Find $e = M/N$ and then e/h.
Find N/bh and select appropriate table from Tables 11.8 to 11.17 depending on f_{cu} and $k = d/h$.
From appropriate table find p which satisfies value of N/bh for given e/h.
Find $A_{sc} = pbh/100$.
Put $A_{sc}/2$ on each face of pile equidistant from axis of moment.

Note: The moment M in pile is due to horizontal load as obtained in Step 3 following Section 7.2.
(iii) *Pile subjected to biaxial moment*
Assuming diameter of reinforcement and finding cover from Step 9, find h' and b'.
Find M_x/h' and M_y/b'.
If $M_x/h' > M_y/b'$, then

$$M'_x = M_x + \beta M_y \left(\frac{h'}{b'}\right)$$

If $M_y/b' > M_x/h'$, then

$$M'_y = M_y + \beta M_x \left(\frac{b'}{h'}\right)$$

Find $N/f_{cu}bh$.
The values of β are given in the table below.

Table 7.5 Values of β for biaxial bending of pile.

$N/f_{cu}bh$	0	0.1	0.2	0.3	0.4	0.5	≥ 0.6
β	1.00	0.88	0.77	0.65	0.53	0.42	0.30

Design as uniaxial bending with N and M'_x or M'_y whichever is more prominent. Find A_{sc} in manner described in (ii) for pile subjected to uniaxial moment.

(B) If $l_e/b > 10$, then treat pile as a slender column.

$$a_x = \frac{1}{2000}\left(\frac{l_e}{h}\right)^2 hK$$

$$a_y = \frac{1}{2000}\left(\frac{l_e}{b}\right)^2 bK$$

Select A_{sc}.

$$K = \frac{N_{uz} - N}{N_{uz} - N_{bal}} \leq 1$$

$$N_{uz} = 0.45f_{cu}A_c + 0.87f_yA_{sc}$$

$$N_{bal} = 0.25f_{cu}bh$$

$$A_c = bh - A_{sc}$$

$$M_{add\,x} = Na_x$$
$$M_{add\,y} = Na_y$$

Combine these additional moments with moments obtained from analysis as in Step 3 following Section 7.2. Design pile subjected to biaxial bending as described previously.

Circular piles

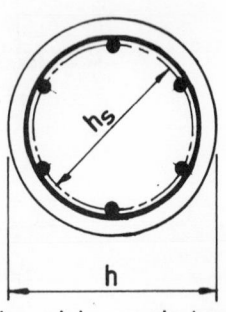

SK 7/20 Typical section through a circular pile.

Use minimum six bars

(A) If $l_e/h \leq 10$, then treat pile as a short column.

(i) *Pile with no moment*

Assume size of reinforcement and at least six bars.

$$A_c = 0.25\pi h^2 - A_{sc}$$

$$N = 0.4f_{cu}A_c + 0.75A_{sc}f_y$$

Check $N \geq$ applied vertical load on pile.

(ii) *Pile with moment*

Find $e = M/N$ and the e/R, where $2R = h$.

Find N/h^2 and select appropriate table from Tables 11.18 to 11.27 corresponding to f_{cu} and $k = h_s/h$.

Find p from appropriate table which satisfies N/h^2 for given value of e/R.

Find $A_{sc} = p\pi R^2/100$.

Use at least six bars.

(B) If $l_e/h > 10$, then treat pile as a slender column.

$$a = \frac{l_e^2}{2000h}K \quad \text{(assume } K = 1 \text{ conservatively)}$$

$$M_{add} = Na$$

Combine this additional moment with moment obtained by analysis in Step 3 following Section 7.2. Design pile with moment as described in (ii) above.

Step 14 Check stresses in prestressed concrete piles

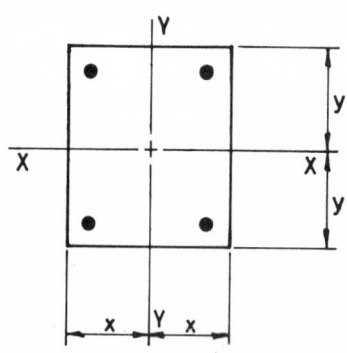

SK 7/21 Typical section of a pretensioned prestressed pile.

Stresses may be checked at the serviceability limit state only as per BS 8110: Part 1, Section 4.[1]

Permissible maximum compressive fibre stress in concrete $= 0.4f_{cu}$

Assume pile as Class 3 member with a limiting crack width of 0.1 mm.

Hypothetical flexural tensile stress in concrete $= 4.1\,\text{N/mm}^2$
for Grade 40
$= 4.8\,\text{N/mm}^2$
for Grade 50 and above

Depth factors to modify tensile stress are shown in the following table.

Depth (mm)	Factor
Up to 400	1.0
500	0.95
600	0.9

N = direct service load on pile

M_{xx} = bending moment as obtained from Step 3 about axis $x-x$
M_{yy} = bending moment as obtained from Step 3 about axis $y-y$.

Assume the pile section is uncracked.

Find A_c = area of concrete
I_{xx} = moment of inertia about $x-x$ axis
I_{yy} = moment of inertia about $y-y$ axis
P = residual prestress after all losses.

Maximum compressive stress in concrete $= \left(\dfrac{P+N}{A_c}\right) + \left(\dfrac{M_x y}{I_{xx}}\right)$
$+ \left(\dfrac{M_y x}{I_{yy}}\right)$

Maximum tensile stress in concrete $= \left(\dfrac{P+N}{A_c}\right) - \left(\dfrac{M_x y}{I_{xx}}\right)$
$- \left(\dfrac{M_y x}{I_{yy}}\right)$

m = modular ratio
f_s = strand stress prior to release
f_c = stress in concrete due to prestress alone.

(1) Loss due to elastic shortening $= \left(\dfrac{100 m f_c}{f_s}\right)\%$

(2) Loss due to relaxation of steel — refer to strand manufacturer's brochure.

(3) Loss due to creep of concrete — follow clause 4.8.5 of BS 8110: Part 1.[1]

(4) Loss due to shrinkage of concrete — follow clause 4.8.4 of BS 8110: Part 1.[1]

Note: Prestressed piles designed as fixed to pile cap must extend into pile cap by a minimum distance equal to transmission length given by the following equation:

$$l_t = \frac{K_t \phi}{\sqrt{f_{cu}}} \text{ (mm)}$$

where f_{cu} = concrete cube strength at 28 days

K_t = 600   for plain or indented wire

      = 400   for crimped wire

      = 240   for 7-wire standard or super strand

      = 360   for 7-wire drawn strand

ϕ = nominal diameter of tendon.

Step 15   Check shear capacity of RC pile

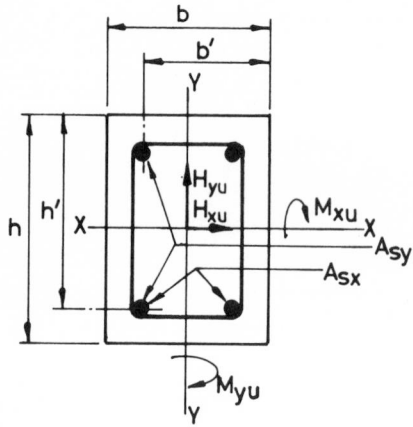

SK 7/22 Typical section through a rectangular pile subject to biaxial bending and shear.

Ultimate limit state shear forces in pile are H_{pxu} and H_{pyu}. Corresponding bending moments in pile are M_{pyu} and M_{pxu}. The ultimate coexistent direct load on pile is N_u.

Rectangular piles

No shear check is necessary if:
$M_{pxu}/N_u \leq 0.6h$
and $M_{pyu}/N_u \leq 0.6b$
and $H_{pyu}/bh' \leq 0.8\sqrt{f_{cu}} \leq 5\,\text{N/mm}^2$
and $H_{pxu}/hb' \leq 0.8\sqrt{f_{cu}} \leq 5\,\text{N/mm}^2$

Shear check is necessary if:
$M_{pxu}/N_u > 0.6h$ and/or $M_{pyu}/N_u > 0.6b$
Find $v_x = H_{pyu}/bh'$ and $v_y = H_{pxu}/hb'$
Find $p_x = 100A_{sx}/bh'$ and $p_y = 100A_{sy}/hb'$

Find v_{cx} and v_{cy} corresponding to p_x and p_y from Figs 11.2 to 11.5.

Check $\dfrac{v_x}{v_{cx}} + \dfrac{v_y}{v_{cy}} \leq 1$

If this check fails, provide shear reinforcement in the form of links.

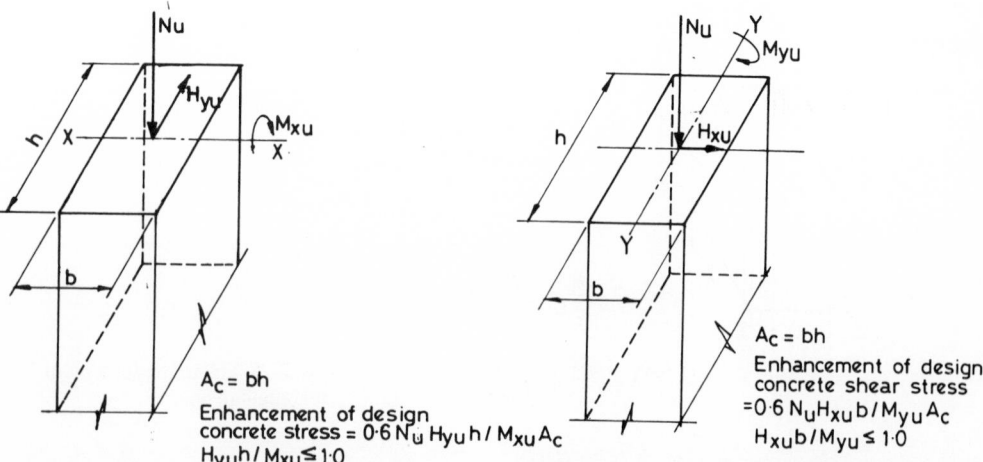

SK 7/23 Shear stress enhancement due to
presence of axial load.

SK 7/24 Shear stress enhancement due to
presence of axial load.

Note: v_{cx} and v_{cy} may be enhanced by using the following formulae due to presence of an axial load N_u:

$$v'_{cx} = v'_{cx} + \frac{0.6N_uH_{pyu}h}{A_cM_{pxu}} \leq 0.8\sqrt{f_{cu}} \leq 5\,\text{N/mm}^2$$

$$v'_{cy} = v'_{cy} + \frac{0.6N_uH_{pxu}b}{A_cM_{pyu}} \leq 0.8\sqrt{f_{cu}} \leq 5\,\text{N/mm}^2$$

$H_{pyu}h/M_{pxu}$ and $H_{pxu}b/M_{pyu}$ should be less than or equal to 1.0.

Shear reinforcement

$$A_{sv} = \frac{bS_v(v - v'_c)}{0.87f_{yv}}$$

where A_{sv} = total area of legs in direction of shear
        b = width of section perpendicular to direction of shear
        S_v = spacing of links
        $f_{yv} \leq 460\,\text{N/mm}^2$ for links.

Circular piles

    N_u = ultimate vertical load with H_{pu}
   H_{pu} = combined ultimate horizontal load
   M_{pu} = moment in pile due to H_{pu}

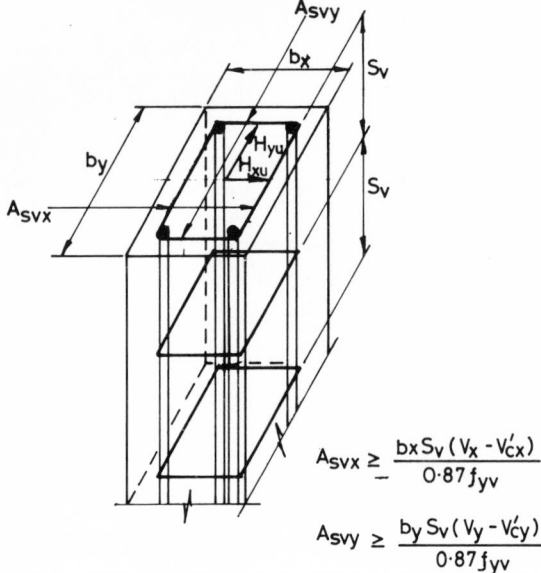

$$A_{svx} \geq \frac{b_x S_v (V_x - V'_{cx})}{0.87 f_{yv}}$$

$$A_{svy} \geq \frac{b_y S_v (V_y - V'_{cy})}{0.87 f_{yv}}$$

SK 7/25 Shear reinforcement in a rectangular pile.

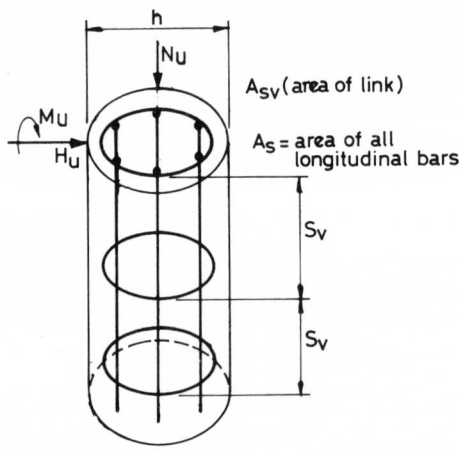

A_{sv} (area of link)

A_s = area of all longitudinal bars

SK 7/26 Shear reinforcement in a circular pile.

No shear check is necessary if:

$M_{pu}/N_u \leq 0.60h$ and $H_{pu}/0.75A_c \leq 0.8\sqrt{f_{cu}} \leq 5\,\text{N/mm}^2$

where $A_c = 0.25\pi h^2$.

Shear check is necessary if:

$M_{pu}/N_u > 0.60h$

Shear stress, $v = H_{pu}/0.75A_c$

$p = 100A_s/1.5A_c$ assuming 50% of bars effectively in tension

where A_s = total area of steel in pile.

Find v_c corresponding to p from Figs 11.2 to 11.5.

The shear stress v_c may be enhanced by using the following formula due to presence of an axial load N_u:

$$v_c' = v_c + \frac{0.6 N_u H_{pu} h}{A_c M_{pu}} \leq 0.8\sqrt{f_{cu}} \leq 5\,\text{N/mm}^2$$

$H_{pu}h/M_{pu}$ should be less than or equal to 1.0.

If $v > v_c'$, then use shear reinforcement.

$$V_s = 0.87 f_{yv} A_v \left(\frac{z}{S}\right) \qquad V_c = 0.75 v_c' A_c$$

where A_v = total area of link bars perpendicular to longitudinal bars, i.e. the two legs of hoop reinforcement
$\quad\quad\quad f_{yv}$ = characteristic yield strength of link reinforcement
$\quad\quad\quad S$ = spacing of links.

Find z/R from appropriate table from Tables 11.18 to 11.27 corresponding to f_{cu}, h_s/h, p, N/R^2 and e/R.

Check $H_{pu} \leq V_s + V_c$

The total shear resistance for inclined links =
$V_s = [0.87 f_y A_{sv} (\cos\alpha + \sin\alpha \cot\beta)(z/S)]$

where A_{sv} = total area of link bars i.e. the two legs of hoop reinforcement.
$\quad\quad\quad \beta$ may be taken as 45° when α is angle of inclination of link.

Step 16   Check shear capacity of prestressed pile

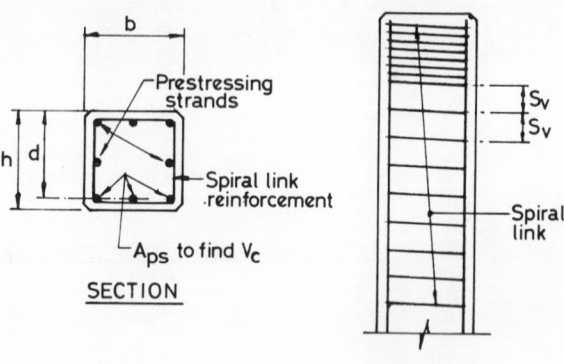

SK 7/27 Typical section and elevation of a prestressed concrete pile.

$$V_{co} = 0.67 bh (f_t^2 + 0.8 f_{cp} f_t)^{\frac{1}{2}}$$

$$V_{cr} = \left(1 - \frac{0.55 f_{pe}}{f_{pu}}\right) v_c bd + \frac{M_o V}{M} \geq 0.1 bd\sqrt{f_{cu}}$$

$V_c = V_{co}$ or V_{cr} as the case may be (kN) − design ultimate shear resistance

V_{co} = shear resistance of section uncracked (kN)

V_{cr} = shear resistance of section cracked (kN)

f_t = maximum design principal stress at the centroidal axis = $0.24\sqrt{f_{cu}}$

f_{cp} = design compressive stress at centroidal axis of concrete section due to prestress alone

f_{pe} = design effective prestress in tendons after all losses $\leq 0.6 f_{pu}$

f_{pu} = characteristic ultimate strength of tendons

v_c = design concrete shear strength from Figs 11.2 to 11.5 where percentage of steel reinforcement should include tendons plus any ordinary untensioned longitudinal steel reinforcement in tensile zone of section

d = effective depth to centroid of reinforcing steel in tension zone where reinforcing steel should include tendons and any untensioned reinforcement

f_{cu} = characteristic cube concrete strength at 28 days

M_o = moment to produce zero stress at tension fibre with $0.8 f_{cp}$ on section.

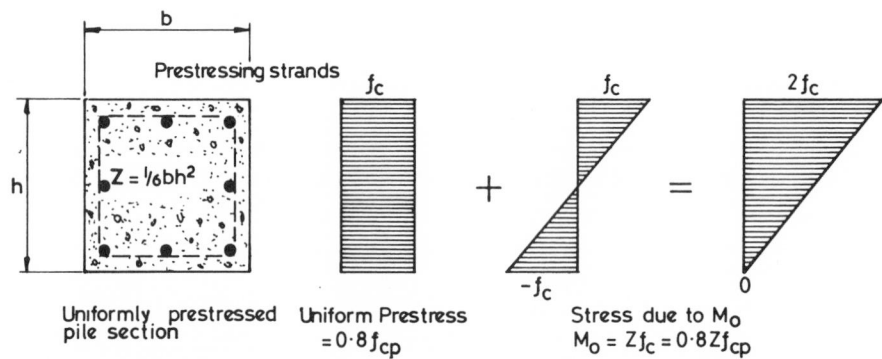

SK 7/28 Stress diagram for a symmetrical rectangular prestressed pile due to M_o.

If $H_{pu} < 0.5 V_c$, no shear reinforcement is required.
If $H_{pu} \geq 0.5 V_c$, then provide shear reinforcement as follows.
Shear reinforcement
If horizontal shear on pile, H_{pu}, is less than or equal to $(V_c + 0.4bd)$, then,

$$\frac{A_{sv}}{S_v} = \frac{0.4b}{0.87 f_{yv}}$$

If horizontal shear on pile, H_{pu}, is more than $(V_c + 0.4bd)$ then,

$$\frac{A_{sv}}{S_v} = \frac{H_{pu} - V_c}{0.87 f_{yv} d}$$

Note: For biaxial bending and shear, check requirement for shear reinforcement for each direction of bending separately, but allow for contribution of concrete shear resistance V_c in one direction of loading only for calculation of shear reinforcement. (See Step 7 of Section 4.3.1.)

Step 17 ***Check minimum reinforcement in RC pile***
For rectangular and circular piles, $100 A_{sc}/A_c \geq 0.4$.

Step 18 ***Check minimum prestress in prestressed pile***

Find slenderness ratio of pile $= n = \dfrac{l}{b}$

where b = minimum width of pile
l = total length of prestressed pile at commencement of driving.

Minimum prestress after losses = 60n psi
or $= 0.4n \, \text{N/mm}^2$

If diesel hammer is used,

minimum prestress in concrete $= 5 \, \text{N/mm}^2$

Step 19 ***Maximum reinforcement in pile***
$100 A_{sc}/A_c \leq 6$

Step 20 ***Containment of reinforcement in pile***
Minimum dia. of links $= 0.25 \times$ largest bar $\geq 6 \, \text{mm}$

Maximum spacing of links $= 12 \times$ smallest dia. of bar

Step 21 ***Links in prestressed piles***
At top and bottom $3B$ length of pile, provide 0.6% of volume of pile in volume of link.

Step 22 ***Minimum tension reinforcement in pile cap***
$A_s \geq 0.0013bh$ in both directions

Step 23 ***Curtailment of bars in pile cap***
A minimum anchorage of 12 times diameter of bar should be provided at ends by bending bar up vertically. Additionally check that full tension anchorage bond length is provided from critical section for bending in a pile cap where design for flexure and requirement for flexural steel in tension is determined. In finding anchorage bond length beyond that section, actual area of steel provided may be taken into account.

Step 24 ***Spacing of bars in pile cap***
Clear spacing of bars should not exceed $3d$ or 750 mm.

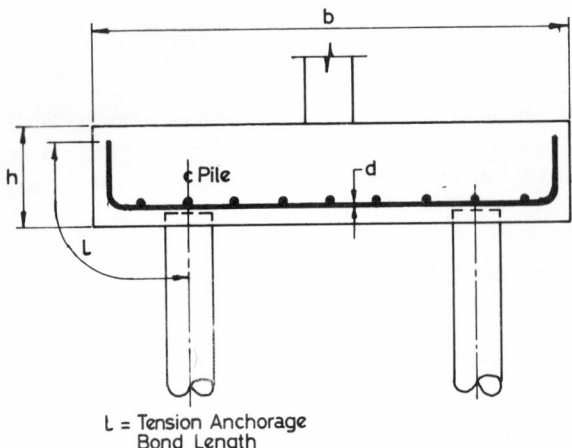

L = Tension Anchorage
Bond Length

SK 7/29 Typical section through a pile cap.

Percentage of reinforcement, $100A_s/bd$ (%)	Maximum clear spacing of bars in pile cap (mm)
1 or over	160
0.75	210
0.5	320
0.3	530
Less than 0.3	3d or 750

Note: This will deem to satisfy a crack width limitation of 0.3 mm.

Step 25   Early thermal cracking
See Chapter 3.

Step 26   Assessment of crack width in flexure
See Chapter 3.

Step 27   Connections
See Chapter 10 for connection of pile to pile cap and column to pile cap.

7.7   WORKED EXAMPLE

Example 7.1   Pile cap for an internal column of a building
Size of column = 800 mm × 800 mm
Spacing of column = 8 m × 8 m   on plan

Unfactored column loads

	Dead	Imposed	Wind
Vertical load, N (kN)	1610	1480	—
Horizontal shear, H_x (kN)	28	18	156
Horizontal shear, H_y (kN)	—	—	112
Moment, M_x (kNm)	—	—	448
Moment, M_y (kNm)	112	72	624

Geotechnical information (see SK 7/30)

Stratum 1

Average thickness of layer = 1.5 m

Classification: very loose yellow brown to brownish grey sandy silt.

Average N = 3 (SPT)

$c = 11.3\,\text{kN/m}^2$

$\phi = 4°$

$\gamma = 26\,\text{kN/m}^3$

Stratum 2

Average thickness of layer = 9 m

Classification: soft to medium bluish-grey clayey silt.

Average N = 5 (SPT)

$c = 20.2\,\text{kN/m}^2$

$\phi = 5°$

$\gamma = 24\,\text{kN/m}^3$

$\gamma_{sat} = 27\,\text{kN/m}^3$

Stratum 3

Average thickness of layer = 2 m

Classification: stiff to very stiff bluish-grey silty clay.

Average N = 14 (SPT)

$c = 60\,\text{kN/m}^2$

$\phi = 6°$

$\gamma_{sat} = 26\,\text{kN/m}^3$

Stratum 4

Average thickness of layer = 7 m

Classification: dense to very dense mottled brown sandy silt.

Average N = 24 (SPT)

$c = 13.8\,\text{kN/m}^2$

$\phi = 31°$

$\gamma_{sat} = 27\,\text{kN/m}^3$

STRATUM 1

STRATUM 2

STRATUM 3

STRATUM 4

STRATUM 5

AVERAGE DEPTH OF LAYERS

VERY LOOSE YELLOW BROWN
SANDY SILT N=3

1500

SOFT TO MEDIUM BLUISH-GREY
CLAYEY SILT N=5 AVERAGE

9000

STIFF TO VERY STIFF BLUISH-GREY
SILTY CLAY N=14 AVERAGE

2000

DENSE TO VERY DENSE MOTTLED BROWN
SANDY SILT N=24 AVERAGE

7000

VERY STIFF TO HARD
SILTY CLAY N=31 AVERAGE

15000

SK 7/30 Average ground condition
soil strata.

Stratum 5

Average thickness of layer $= 15$ m

Classification: very stiff to hard silty clay.

Average $N = 31$ (SPT)
$$c = 71.5 \, \text{KN/m}^2$$
$$\phi = 8°$$
$$\gamma_{sat} = 28 \, \text{kN/m}^3$$

Water table at 3.0 m below ground level.

Step 1 *Select type of pile*
Considering all the factors as described in Step 1 of Section 7.6 it is decided to use a non-displacement pile.
Choose 600 mm diameter bored and cast-in-situ concrete pile.

Step 2 *Determine vertical capacity of pile*
Follow Section 7.1.

$$P_u = P_{pu} + \Sigma P_{si} - W$$

First method of point resistance

$$P_{pu} = A_p (38N) \left(\frac{L_b}{B} \right)$$

Assume pile to go into Stratum 5 and stop at 8.0 m within Stratum 5.

L_b = average length of pile = $(1.5 + 9 + 2 + 7 + 8)$ m = 27.5 m

$$A_p = \text{cross-sectional area of pile} = \pi \times \frac{0.6^2}{4} = 0.283 \, \text{m}^2$$

$B = 0.60$ m

N = statistical average of SPT in a zone of about $8B$ above to $3B$ below pile point = 31

$$P_{pu} = 0.283 \times 38 \times 31 \times \frac{27.5}{0.6} = 15280 \, \text{kN}$$
$$\leq 380N(A_p) = 380 \times 31.0 \times 0.283 = 3334 \, \text{kN}$$

Second method of point resistance

$$P_{pu} = A_p (N'_c c + \bar{q} N'_q)$$
$$A_p = 0.283 \, \text{m}^2$$
$$c = 71.5 \, \text{kN/m}^2$$
$$\gamma_w = 10 \, \text{kN/m}^3$$

$\bar{q}$ = effective vertical stress at pile point
$\quad = 1.5 \times 26 + 1.5 \times 24 + 7.5 \times 27 + 2 \times 26 + 7 \times 27 + 8 \times 27$
$\quad - (27.5 - 3) \times 10$
$\quad = 489.5 \, \text{kN/m}^2$

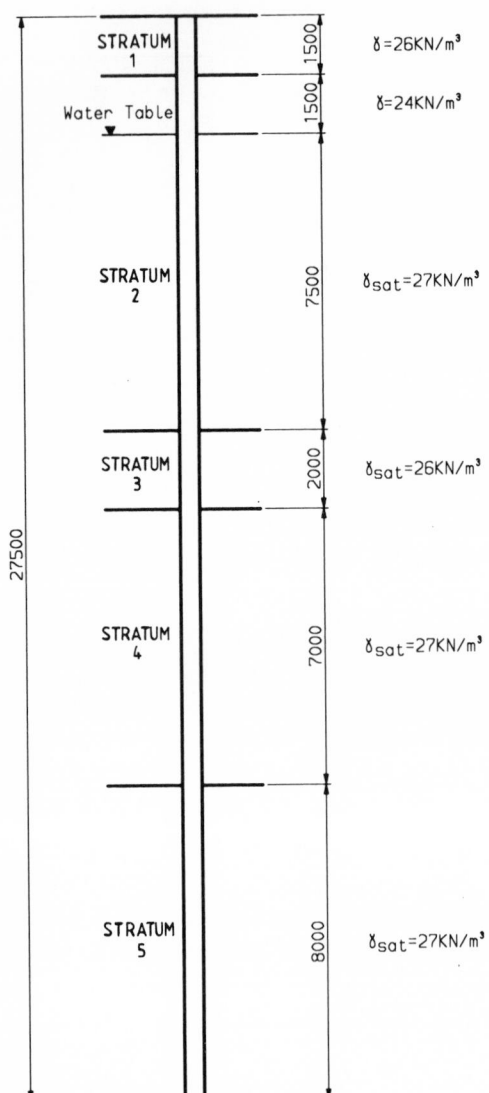

STRATUM 1 $\delta = 26 \text{KN/m}^3$

Water Table $\delta = 24 \text{KN/m}^3$

STRATUM 2 $\delta_{sat} = 27 \text{KN/m}^3$

STRATUM 3 $\delta_{sat} = 26 \text{KN/m}^3$

STRATUM 4 $\delta_{sat} = 27 \text{KN/m}^3$

STRATUM 5 $\delta_{sat} = 27 \text{KN/m}^3$

SK 7/31 The pile penetrating different strata.

$L = 27.5 \,\text{m} \qquad B = 0.60 \,\text{m}$

$L/B = 46 \qquad \phi = 8°$

From Fig. 7.2,

$N_q' = 3 \qquad N_c' = 15 \quad \text{and} \quad L_c/B = 3.5$

$$\frac{L}{B} >> \frac{L_c}{B}$$

$P_{\text{pu}} = 0.283 \,[(15 \times 71.5) + (3 \times 489.5)] = 719 \,\text{kN}$

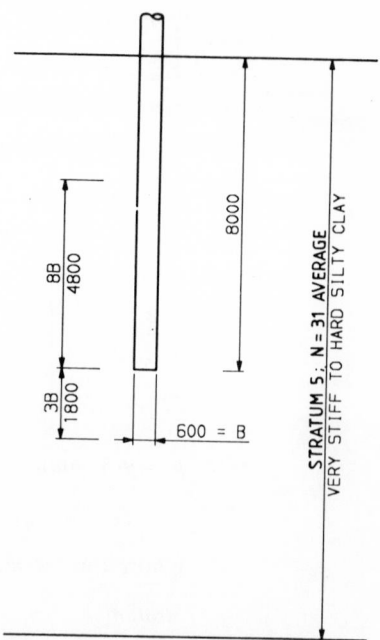

SK 7/32 Condition at bottom of
pile.

Determination of skin resistance

$$\Sigma P_{si} = \Sigma A_s f_s$$

Used non-displacement pile of 600 mm diameter.

First method of skin resistance

$$f_s = N\,kN/m^2$$

Stratum 1

$$A_{s1} = \text{perimeter} \times \text{depth of stratum}$$
$$= \pi \times 0.60 \times 1.5$$
$$= 2.83\,m^2$$

$$f_{s1} = 3\,kN/m^2$$

$$P_{si1} = 3 \times 2.83 = 8.5\,kN$$

Stratum 2

$$A_{s2} = \pi \times 0.60 \times 9 = 17\,m^2$$

$$f_{s2} = 5\,kN/m^2$$

$$P_{si2} = 5 \times 17 = 85\,kN$$

Stratum 3

$$A_{s3} = \pi \times 0.60 \times 2 = 3.8\,m^2$$

$$f_{s3} = 14\,kN/m^2$$

$$P_{si3} = 14 \times 3.8 = 53.2\,kN$$

Stratum 4

$A_{s4} = \pi \times 0.60 \times 7 = 13.2\,m^2$

$f_{s4} = 24\,kN/m^2$

$P_{si4} = 13.2 \times 24 = 316.8\,kN$

Stratum 5

$A_{s5} = \pi \times 0.60 \times 8 = 15.1\,m^2$

$f_{s5} = 31\,kN/m^2$

$P_{si5} = 15.1 \times 31 \times 468.1\,kN$

$\Sigma P_{si} = 931.6\,kN$

Fourth method of skin resistance

$f_s = \alpha c + 0.5\bar{q}K_s \tan\delta$

Ignore the second term because δ is very small.

Stratum 1

$\alpha = 0.75 \qquad c = 11.3\,kN/m^2$

$P_{si1} = A_{s1} \times f_{s1}$
$P_{si1} = 0.75 \times 11.3 \times 2.83 = 24\,kN$

$A_{s1} = 2.83\,m^2$

Stratum 2

$\alpha = 0.75 \qquad c = 20.2\,kN/m^2$

$P_{si2} = 0.75 \times 20.2 \times 17 = 257.2\,kN$

$A_{s2} = 17\,m^2$

Stratum 3

$\alpha = 0.75 \qquad c = 60\,kN/m^2$

$P_{si3} = 0.75 \times 60 \times 3.8 = 171\,kN$

$A_{s3} = 3.8\,m^2$

Stratum 4

$A_{s4} = 13.2\,m^2$

$\alpha = 2.0 \qquad$ say with high D_r

$c = 13.8\,kN/m^2 \qquad \phi = 31°$

$K_s = 2.0 \qquad$ from chart

$\delta = 0.75\phi = 23.25°$

$\tan\delta = 0.43$

$\bar{q}$ = effective vertical stress at middle of layer

= 1.5 × 26 + 1.5 × 24 + 7.5 × 27 + 2 × 26 + 3.5 × 27 − (16 − 3) × 10

= 294 kN/m²

$f_s = \alpha c + 0.5\bar{q}\, K_s \tan \delta$

P_{si4} = 13.2 [2 × 13.8) + (0.5 × 294 × 2 × 0.43)] = 2033 kN

The fourth method of skin resistance is giving much higher values than the first method and may be ignored from the point of view of conservatism.

$P_u = P_{pu} + P_{su}$

= 719 + 932

= 1651 kN

Allowable working load on pile = $\dfrac{1651}{2.5}$ = 660 kN

Designed pile is 600 mm diameter bored and cast in-situ concrete pile with an average length of 27.5 m to carry a working load of 660 kN. This is a conservative theoretical estimate of single pile vertical load capacity and must be verified by actual pile tests on site.

Step 3 *Determine horizontal capacity of single pile*
See Section 7.2.
Assume cohesive soil.

Method 1

$E_s = 650N$ where N = SPT No.

E_s of Stratum 1 = 650 × 3   =   1950 kN/m²
E_s of Stratum 2 = 650 × 5   =   3250 kN/m²
E_s of Stratum 3 = 650 × 14 =   9100 kN/m²
E_s of Stratum 4 = 650 × 24 = 15 600 kN/m²
E_s of Stratum 5 = 650 × 31 = 20 150 kN/m²

$$k_s B = 1.3 \left(\frac{E_s B^4}{E_f I_f} \right)^{\frac{1}{12}} \left(\frac{E_s}{1 - \mu^2} \right)$$

$E_f = 28 \times 10^6$ kN/m²   for pile concrete

$$I_f = \left(\frac{\pi}{64} \right) D^4 = \left(\frac{\pi}{64} \right) \times 0.60^4 = 6.36 \times 10^{-3} \, \text{m}^4$$

$k_{s1}B$ =   1672 kN/m² k_{s1} =   2787 kN/m³
$k_{s2}B$ =   2909 kN/m² k_{s2} =   4848 kN/m³
$k_{s3}B$ =   8875 kN/m² k_{s3} = 14 792 kN/m³
$k_{s4}B$ = 15 914 kN/m² k_{s4} = 26 523 kN/m³
$k_{s5}B$ = 20 999 kN/m² k_{s5} = 34 998 kN/m³

Method 2

$$k_s = 240q_u \text{ kN/m}^2$$
$$= 480c \text{ kN/m}^2$$

$$k_{s1} = 480 \times 11.3 = 5424 \text{ kN/m}^3$$
$$k_{s2} = 480 \times 20.2 = 9696 \text{ kN/m}^3$$
$$k_{s3} = 480 \times 60 = 28\,800 \text{ kN/m}^3$$
$$k_{s4} = 480 \times 13.8 = 6624 \text{ kN/m}^3$$
$$k_{s5} = 480 \times 71.5 = 34\,320 \text{ kN/m}^3$$

The values given by Method 1 are smaller or softer which will produce larger deflection and bending moments in pile.
For the sake of conservatism use values given by Method 1.

S = node spacing for finite element analysis = 0.60 m
B = 0.60 m
spring stiffness = SBk_s kN/m

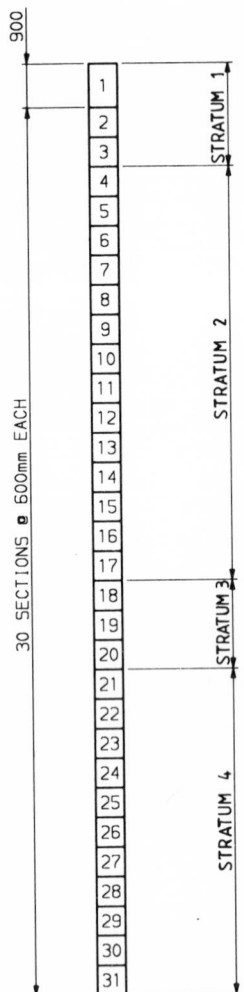

SK 7/33 Finite element model of pile.

Ignore top $1.5B$ of pile for lateral support from soil.
The whole length of pile need not be modelled.

Stratum 1
Spring stiffness = $0.60 \times 0.60 \times 2787$
$\qquad\qquad\qquad = 1003\,kN/m$

Stratum 2
Spring stiffness = $0.60 \times 0.60 \times 4848$
$\qquad\qquad\qquad = 1745\,kN/m$

Stratum 3
Spring stiffness = $0.60 \times 0.60 \times 14\,792$
$\qquad\qquad\qquad = 5325\,kN/m$

Stratum 4
Spring stiffness = $0.60 \times 0.6 \times 26\,523 = 9548\,kN/m$

Assume full fixity of pile with pile cap.
Apply unit load at top of pile and find pile stiffness and bending moment
and shear in pile using a two-dimensional computer program.

$$A = 0.283\,m^2 \qquad I = 6.36 \times 10^{-3}\,m^4$$

Results of computer run

Maximum moment = $2.48\,kNm/kN$

Pile top deflection = $0.12\,mm/kN$

Single pile horizontal stiffness = $\dfrac{1000}{0.12} = 8333\,kN/m$

Step 4 *Determine approximate number of piles and spacing*

Maximum vertical load on pile cap = $1610 + 1480 = 3090\,kN = P$

$$R_{iV} = \frac{P}{C_V} = \frac{3090}{660} = 4.7$$

Assume maximum allowable horizontal displacement of pile cap is 10 mm.

Maximum horizontal load = $28 + 18 + 156 = 202\,kN = H$

Maximum horizontal load on pile to limit deflection to 10 mm
$= 8333 \times 0.010$
$= 83\,kN$ per pile

$$R_{iH} = \frac{H}{C_H} = \frac{202}{83} = 2.4$$

R_i = greater of R_{iV} and R_{iH} = 4.7

$1.1R_i = 4.7 \times 1.1 = 5.17$

Use 6 no. piles.

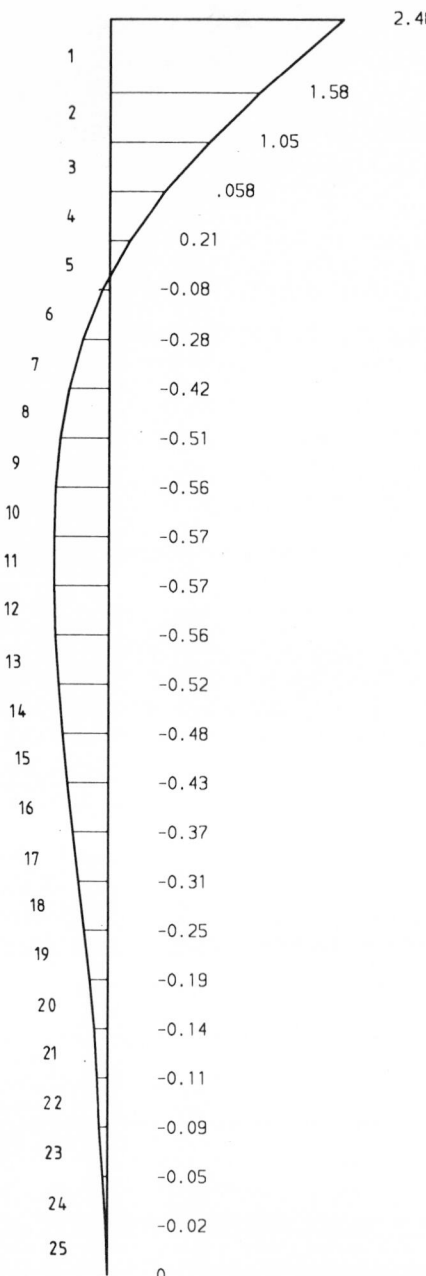

2.48

1.58

1.05

.058

0.21

-0.08

-0.28

-0.42

-0.51

-0.56

-0.57

-0.57

-0.56

-0.52

-0.48

-0.43

-0.37

-0.31

-0.25

-0.19

-0.14

-0.11

-0.09

-0.05

-0.02

0

SK 7/34 Bending moment (kNm) due to 1 kN horizontal load at top of pile.

Step 5   *Determine size of pile cap*

B = diameter of pile = 0.6 m

$1.5B = 1.5 \times 0.6 = 0.9$ m

Allow 0.9 m from centre of pile to edge of pile cap.
Assume 0.9 m depth of pile cap.

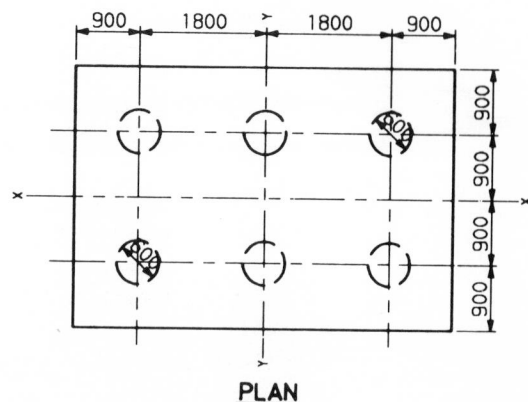

SK 7/35 Layout of piles under pile cap.

PLAN

Spacing of piles $\geq 3B \geq 3 \times 0.6 = 1.8\,\text{m}$

Size of pile cap assumed is $5.4\,\text{m} \times 3.6\,\text{m} \times 0.9\,\text{m}$.

Step 6   Carry out load combination

Estimation of load on pile

$LC_1 = 1.0DL + 1.0IL$

$N = 1610 + 1480 = 3090\,\text{kN}$

$H_x = 28 + 18 = 46\,\text{kN}$
$H_y = 0\,\text{kN}$

$M_x = 0\,\text{kNm}$
$M_y = 112 + 72 = 184\,\text{kNm}$

$LC_3 = 1.0DL + 1.0IL + 1.0WL$

$N = 3090\,\text{kN}$

Wind in $x-x$ direction

$H_x = 46 + 156 = 202\,\text{kN}$
$H_y = 0\,\text{kN}$

$M_x = 0\,\text{kNm}$
$M_y = 184 + 624 = 808\,\text{kNm}$

Wind in $y-y$ direction

$H_x = 46\,\text{kN}$
$H_y = 112\,\text{kN}$

$M_x = 448\,\text{kNm}$
$M_y = 184\,\text{kNm}$

$LC_4 = 1.0DL + 1.0WL$

$N = 1610\,\text{kN}$

Wind in x−x direction

$H_x = 28 + 156 = 184\,\text{kN}$
$H_y = 0\,\text{kN}$

$M_x = 0\,\text{kNm}$
$M_y = 112 + 624 = 736\,\text{kNm}$

Wind in y−y direction

$H_x = 28\,\text{kN}$
$H_y = 112\,\text{kN}$

$M_x = 448\,\text{kNm}$
$M_y = 112\,\text{kNm}$

Estimation of loads on piles for bending moment and shear calculations in pile cap

$LC_5 = 1.4DL + 1.6IL$

$N = 1.4 \times 1610 + 1480 \times 1.6 = 4622\,\text{kN}$

$H_x = 1.4 \times 28 + 1.6 \times 18 = 68\,\text{kN}$
$H_y = 0\,\text{kN}$

$M_x = 0\,\text{kNm}$
$M_y = 1.4 \times 112 + 1.6 \times 72 = 272\,\text{kNm}$

$LC_6 = 1.2DL + 1.2IL + 1.2WL$

$N = 1.2 \times 1610 + 1.2 \times 1480 = 3708\,\text{kN}$

Wind in x−x direction

$H_x = 1.2 \times (28 + 18 + 156) = 242.4\,\text{kN}$
$H_y = 0\,\text{kN}$

$M_x = 0\,\text{kNm}$
$M_y = 1.2 \times (112 + 72 + 624) = 969.6\,\text{kNm}$

Wind in y−y direction

$H_x = 1.2 \times (28 + 18) = 55.2\,\text{kN}$
$H_y = 1.2 \times 112 = 134.4\,\text{kN}$

$M_x = 1.2 \times 448 = 537.6\,\text{kNm}$
$M_y = 1.2 \times (112 + 72) = 220.8\,\text{kNm}$

$LC_7 = 1.4DL + 1.4WL$

$N = 1.4 \times 1610 = 2254\,\text{kN}$

Wind in x−x direction

$H_x = 1.4\,(28 + 156) = 257.6\,\text{kN}$
$H_y = 0\,\text{kN}$

$M_x = 0\,\text{kNm}$

$M_y = 1.4\,(112 + 624) = 1030.4\,\text{kNm}$

Wind in y−y direction

$H_x = 1.4 \times 28 = 39.2\,\text{kN}$

$H_y = 1.4 \times 112 = 156.8\,\text{kN}$

$M_x = 1.4 \times 448 = 627.2\,\text{kNm}$

$M_y = 1.4 \times 112 = 156.8\,\text{kNm}$

$LC_8 = 1.0DL + 1.4WL$

This condition may be ignored because it is highly unlikely that wind will cause any tension in piles.

Step 7 Check pile group effects

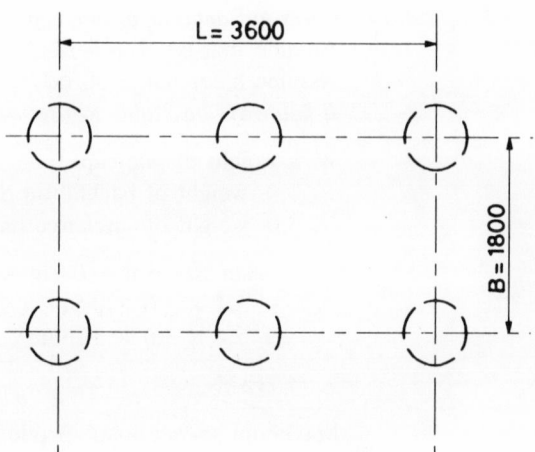

SK 7/36 Layout of piles − overall dimensions.

Group friction capacity $= 2D(B + L)c\alpha$

$B = 1.8\,\text{m} \qquad L = 3.6\,\text{m}$

c = average cohesion of clay

$$= \frac{(11.3 \times 1.5) + (20.2 \times 9.0) + (60 \times 2) + (13.8 \times 7) + (71.5 \times 8)}{27.5}$$

$$= 35.9\,\text{kN/m}^2$$

$\alpha = 0.75$ say (See Section 7.1, Table 7.2)

$D = 27.5\,\text{m}$

Group friction capacity $= 2 \times 27.5\,(1.8 + 3.6) \times 35.9 \times 0.75$

$\qquad\qquad\qquad\qquad\qquad = 7996\,\text{kN}$

Group end-bearing capacity $= BL(N'_c c + \bar{q}N'_q)$

$c = 71.5\,\text{kN/m}^2$ at bottom of group

$\bar{q}$ = effective stress at bottom of group = $489.5\,\text{kN/m}^2$ (see Step 2)

$\left.\begin{array}{l} N_q' = 3 \\ N_c' = 15 \end{array}\right\}$ for $\phi = 8°$

Group end-bearing capacity = $1.8 \times 3.6 \times (15 \times 71.5 + 489.5 \times 3)$
$$= 16\,465\,\text{kN}$$

Ultimate group capacity = $7996 + 16\,465 = 24\,461\,\text{kN}$

Allowable group capacity = $\dfrac{24\,461}{2.5} = 9784\,\text{kN}$

Allowable group capacity based on single pile capacity = $6 \times 660 = 3960\,\text{kN}$

Design basis is single pile capacity.

Step 8 **Carry out analysis of pile cap**
Assume that pile cap is rigid. Assume 500 mm backfill on top of pile cap. Assume a surcharge of $5\,\text{kN/m}^2$ on backfill with no eccentricity.
It is always advisable to use the table as presented.

W = weight of pile cap
+ weight of backfill on pile cap
+ weight of surcharge on backfill

$= 5.4\,\text{m} \times 3.6\,\text{m} \times 0.9\,\text{m} \times 24\,\text{kN/m}^3$
$+ 5.4 \times 3.6 \times 0.5\,\text{m} \times 20\,\text{kN/m}^3$
$+ 5.4 \times 3.6 \times 5\,\text{kN/m}^2$

$= 712\,\text{kN}$

Maximum service load on pile without wind = 665 kN

Maximum service load on pile with wind = 771 kN

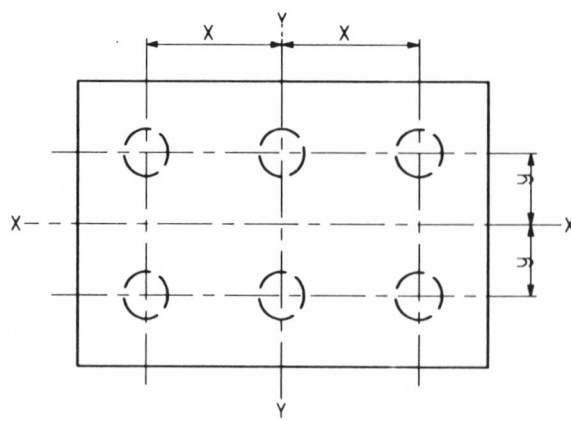

SK 7/37 Calculations of pile group stiffness.

Analysis of loads on pile cap.

Load case	N	M_x	M_y	H_x	H_y	e_x	e_y	e_{hx}	e_{hy}	h	P or P_u	M_x^*	M_y^*	M_{xx}	M_{yy}	T
LC_1	3090	0	184	46	0	0	0	0	0	0.9	3802	0	0	0	225.4	0
LC_3	3090	0	808	202	0	0	0	0	0	0.9	3802	0	0	0	989.8	0
LC_3	3090	448	184	46	112	0	0	0	0	0.9	3802	0	0	548.8	225.4	0
LC_4	1610	0	736	184	0	0	0	0	0	0.9	2322	0	0	0	901.6	0
LC_4	1610	448	112	28	112	0	0	0	0	0.9	2322	0	0	548.8	137.2	0
LC_5	4622	0	272	68	0	0	0	0	0	0.9	5619	0	0	0	333.2	0
LC_6	3708	0	969.6	242.4	0	0	0	0	0	0.9	4562	0	0	0	1187.6	0
LC_6	3708	537.6	220.8	55.2	134.4	0	0	0	0	0.9	4562	0	0	658.6	270.5	0
LC_7	2254	0	1030.4	257.6	0	0	0	0	0	0.9	3251	0	0	0	1262.2	0
LC_7	2254	627.2	156.8	39.2	156.8	0	0	0	0	0.9	3251	0	0	768.3	192.1	0

$M_{xx} = M_x + Ne_y + H_y h + M_x^*$ $M_{yy} = M_y + Ne_x + H_x h + M_y^*$

$T = H_x e_{hy} + H_y e_{hx}$ $P = N + W$ $P_u = N + 1.4W \text{ (or } 1.2W)$

Loads on pile.

Load case	P or P_u	H_x	H_y	M_{xx}	M_{yy}	T	Q_{max}	Q_{min}	H or H_{pu}	M_p or M_{pu}	δ (mm)
LC_1	3802	46	—	—	225.4	—	665	602	7.67	19.0	0.9
LC_3	3802	202	—	—	989.8	—	771	496	33.67	83.5	4.0
LC_3	3802	46	112	548.8	225.4	—	767	501	20.18	50.0	2.4
LC_4	2322	184	—	—	901.6	—	512	262	30.67	76.1	3.7
LC_4	2322	28	112	548.8	137.2	—	508	266	19.24	47.7	2.3
LC_5	5619	68	—	—	333.2	—	983	890	11.33	28.1	1.4
LC_6	4562	242.4	—	—	1187.6	—	925	595	40.40	100.2	4.8
LC_6	4562	55.2	134.4	658.6	270.5	—	920	601	24.21	60.0	2.9
LC_7	3251	257.6	—	—	1262.2	—	717	367	42.93	106.5	5.2
LC_7	3251	39.2	156.8	768.3	192.1	—	711	373	26.94	66.8	3.2

$I_{xx} = \Sigma y^2 = 4.86\,\text{m}^2 \qquad I_{yy} = \Sigma x^2 = 12.96\,\text{m}^2 \qquad I_{zz} = I_{xx} + I_{yy} = 17.82\,\text{m}^2$

$$Q_{max} = \frac{P}{R} + \frac{M_{xx}y}{I_{xx}} + \frac{M_{yy}x}{I_{yy}} \qquad Q_{min} = \frac{P}{R} - \frac{M_{xx}y}{I_{xx}} - \frac{M_{yy}x}{I_{yy}}$$

$$H = \frac{\sqrt{(H_x^2 + H_y^2)}}{R} \qquad R = \text{no. of piles} = 6$$

$M_p = $ bending moment in pile $= 2.48H$ (see Step 3) $x = 1.8\,\text{m}$ $y = 0.9\,\text{m}$

$\delta = $ horizontal displacement at top of pile $= 0.12H$ mm (see Step 3)

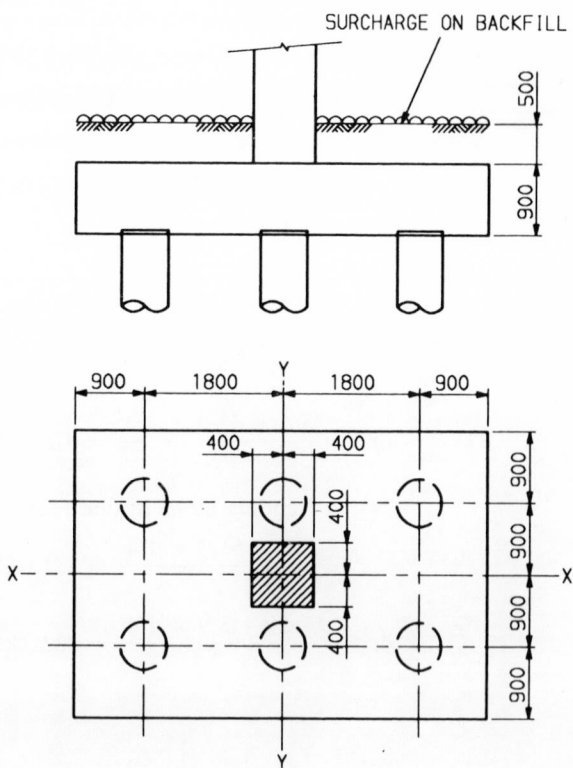

SK 7/38 General arrangement of
pile cap and piles.

Allowable service load on pile without wind = 660 kN   OK

Allowable service load on pile with wind = 660 × 1.25 = 825 kN   OK

Bending moment and shear force in pile cap

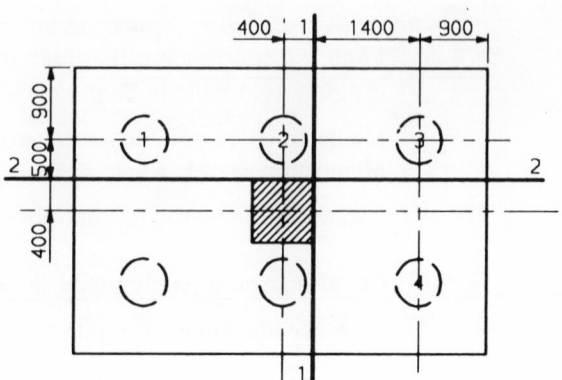

SK 7/39 Critical sections for
calculation of bending moment in
pile cap.

Sections 1−1 and 2−2 are taken at the face of column.
Assume column size = 800 mm × 800 mm

Dead load of pile cap + surcharge + backfill = $0.9 \times 24 + 0.5 \times 20 + 5 =$
$36.6\,kN/m^2$

Applying load factors for different load cases:

$1.4 \times 36.6 = 51.2\,kN/m^2$
$1.2 \times 36.6 = 43.9\,kN/m^2$

M'_{11} = bending moment due to dead load of pile cap etc. on section 1−1

$$= \frac{3.6 \times 51.2 \times 2.3^2}{2} = 487.5\,kNm$$

or $\quad = \dfrac{3.6 \times 43.9 \times 2.3^2}{2} = 418.0\,kNm$

M'_{22} = Bending moment due to dead load of pile cap etc. on section 2−2

$$= \frac{5.4 \times 51.2 \times 1.4^2}{2} = 271.0\,kNm$$

or $\quad = \dfrac{5.4 \times 43.9 \times 1.4^2}{2} = 232.3\,kNm$

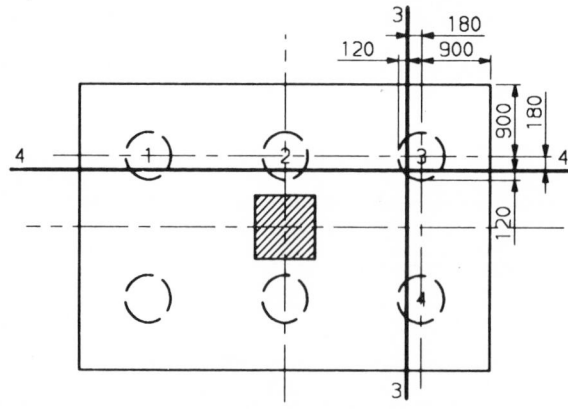

SK 7/40 Critical sections for shear.

Step 9 *Determine cover to reinforcement*
From soil test reports, the total SO_3 is 0.75%. This means it is Class 3
exposure (see table in Step 9 of Section 7.6).

Minimum cover on blinding concrete = 50 mm
Minimum cover elsewhere = 90 mm

Assume 90 mm cover for pile cap everywhere.

Step 10 *Calculate area of reinforcement in pile cap*

M = bending moment in pile cap   as found in Step 8.

$M_{11} = 2264.9\,kNm$ from table in Step 8.

Bending moments and shear in pile cap.

Load case	Q_1	Q_2	Q_3	Q_4	M'_{11}	M'_{22}	M''_{11}	M''_{22}	M_{11}	M_{22}	V'_{33}	V'_{44}	V''_{33}	V''_{44}	V_{33}	V_{44}
LC_5	890	937	983	983	-487.5	-271.0	2752.4	1405	2264.9	1134.0	-199.1	-298.6	1966	2810	1766.9	2511.4
LC_6	595	760	925	925	-418.0	-232.3	2590.0	1140	2172.0	907.7	-170.7	-256.0	1850	2280	1679.3	2024.0
LC_6	844	882	920	676	-418.0	-232.3	2234.4	1323	1816.4	1090.7	-170.7	-256.0	1596	2646	1425.3	2390.0
LC_7	367	542	717	717	-487.5	-271.0	2007.6	813	1520.1	542.0	-199.1	-298.6	1434	1626	1234.9	1327.4
LC_7	657	684	711	427	-487.5	-271.0	1593.2	1026	1105.7	755.0	-199.1	-298.6	1138	2052	938.9	1753.4

Q_1, Q_2, Q_3 and Q_4 are pile reactions

$M''_{11} = 1.4 (Q_3 + Q_4)$ $M''_{22} = 0.5 (Q_1 + Q_2 + Q_3)$ $M_{11} = M'_{11} + M''_{11}$

$V''_{33} = Q_3 + Q_4$ $V'''_{44} = Q_1 + Q_2 + Q_3$ $V_{33} = V'_{33} + V'''_{33}$

M'_{11}, M'_{22}, V'_{33} and V'_{44} are bending moments and shears in pile cap due to dead load of pile cap + surcharge

M''_{11}, M''_{22}, V''_{33} and V''_{44} are bending moments and shears in pile cap due to pile reaction

M_{11}, M_{22}, V_{33} and V_{44} are combined bending moments and shears in pile cap

$\phi = 600\,mm$ $\phi/5 = 120\,mm$ $\phi = $ diameter of pile

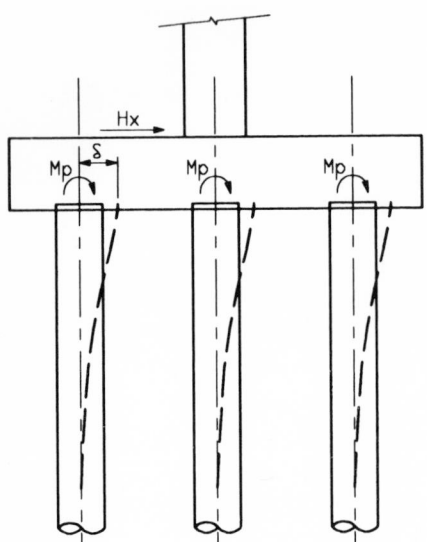

SK 7/41 Moments in pile and pile cap due to pile fixity.

For this load case, pile fixity moment = 19.0 kNm per pile.

Pile fixity moment on pile cap is opposite in sign to moment M_{11} and may be ignored.

Assume 20 mm diameter reinforcement.

d_x = 900 − 90 (cover) − 10 (half bar dia.) = 800 mm   b = 3.6 m

f_{cu} = 30 N/mm² for concrete in pile cap

$$K = \frac{M_{11}}{f_{cu}bd^2} = \frac{2264.9 \times 10^6}{30 \times 3600 \times 800^2} = 0.033$$

$$z = d\left[0.5 + \sqrt{\left(0.25 - \frac{K}{0.9}\right)}\right]$$

$$= 0.96d \leq 0.95d = 760 \text{ mm}$$

$$A_{st} = \frac{M_{11}}{0.87f_yz} = \frac{2264.9 \times 10^6}{0.87 \times 460 \times 760} = 7447 \text{ mm}^2$$

Assume f_y = 460 N/mm² for HT reinforcement

Area of 20 mm dia. bar = 314 mm²   24 × 314 = 7536 mm²

Use 24 no. 20 mm diameter bars equally spaced (approximate spacing 150 mm) in the x−x direction.

M_{22} = 1134 kNm   from table in Step 8.

Ignore the effect of pile fixity moments.
Assume 12 mm diameter reinforcement.

d_y = 900 − 90(cover − 20(bar dia.) − 6(half bar) = 784 mm

$$K = \frac{M_{22}}{f_{cu}bd^2} = \frac{1134 \times 10^6}{30 \times 5400 \times 784^2} = 0.011$$

$z = 0.95d$ by inspection
$\quad = 0.95 \times 784 = 745\,mm$

$$A_{st} = \frac{M_{22}}{0.87f_y z} = \frac{1134 \times 10^6}{0.87 \times 460 \times 745} = 3803\,mm^2$$

Area of 12 mm dia. bar $= 113\,mm^2$ $34 \times 113 = 3842\,mm^2$

Use 34 no. 12 mm diameter bars equally spaced (approximate spacing 155 mm) in the $y-y$ direction.
(See also Step 22 for minimum reinforcement.)

All bars are high tensile reinforcement to be placed at bottom of pile cap. There is no requirement for bars on top of pile cap.

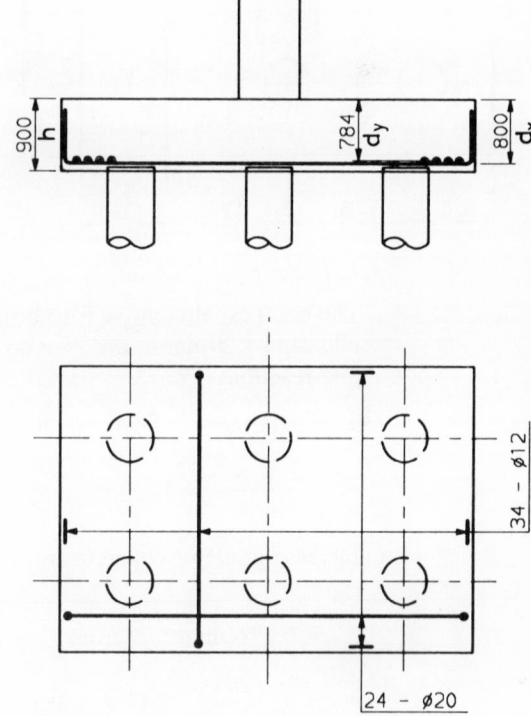

SK 7/42 Pile cap reinforcement.

Step 11   *Check shear stress in pile cap*

$V_{33} =$ shear on critical section $3-3$
$\qquad = 1766.9\,kN$ (see table in Step 8)

$a_v = 2700 - 400$ (half column) $- 1080 = 1220\,mm$

$1.5d_x = 1.5 \times 800 = 1200 \,\text{mm}$

$a_v > 1.5d_x$ hence no enhancement of shear stress is allowed

$$v = \frac{V}{bd} = \frac{1766.9 \times 10^3}{3600 \times 800} = 0.61 \,\text{N/mm}^2$$

$$p = \frac{100A_s}{bd} = \frac{100 \times 7536}{3600 \times 800} = 0.26\%$$

$v_c = 0.425 \,\text{N/mm}^2 < 0.61 \,\text{N/mm}^2$ from Fig. 11.3

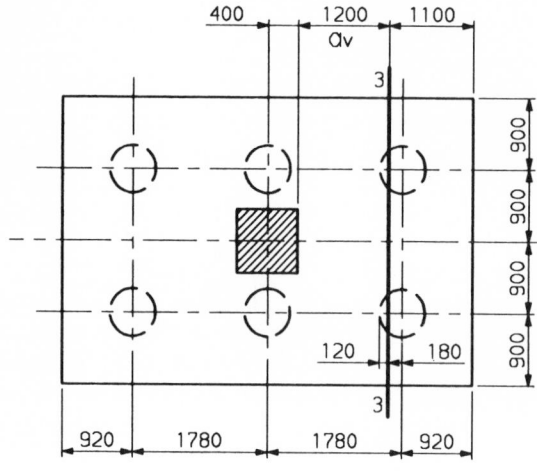

SK 7/43 Critical shear plane in pile cap.

The cheapest alternative is to bring the outer piles in towards the centre of pile cap by 20 mm in the $x-x$ direction only. This has very little effect on pile reactions.

$a_v = 1200 \,\text{mm}$ $1.5d_x = 1200 \,\text{mm}$

$$\frac{2d}{a_v} = \frac{2 \times 800}{1200} = 1.333$$

Increase grade of concrete from $f_{cu} = 30 \,\text{N/mm}^2$ to $f_{cu} = 40 \,\text{N/mm}^2$ in pile cap.

$v_{c1} = 0.47 \,\text{N/mm}^2$ from Figs 11.2 to 11.5

$$v_{c2} = v_{c1}\left(\frac{2d}{a_v}\right) = 0.47 \times 1.333$$

$$= 0.63 \,\text{N/mm}^2 > 0.61 \,\text{N/mm}^2 \quad \text{OK}$$

$V_{44} = $ shear on critical section $4-4$
     $= 2511.4 \,\text{kN}$ (see table in Step 8).

$a_v = 1800 - 1200 + 120 - 400 \,\text{(half column)} = 320 \,\text{mm}$

$1.5d_y = 1.5 \times 784 = 1176 \,\text{mm} > a_v$

$$\frac{2d_y}{a_v} = \frac{2 \times 784}{320} = 4.9$$

$$p = \frac{100A_{sc}}{bd} = \frac{100 \times 3482}{5400 \times 784} = 0.08\%$$

(See Step 22 for minimum percentage of reinforcement.)

$v_{c1} = 0.40 \, \text{N/mm}^2$ for $f_{cu} = 40 \, \text{N/mm}^2$

$v_{c2} = 0.40 \times 4.9 = 1.96 \, \text{N/mm}^2$

$$v_c = \frac{V}{bd} = \frac{2511 \times 10^3}{5400 \times 784}$$

$$= 0.59 \, \text{N/mm}^2 < 1.96 \, \text{N/mm}^2 \quad \text{OK}$$

Step 12   Check punching shear stress in pile cap

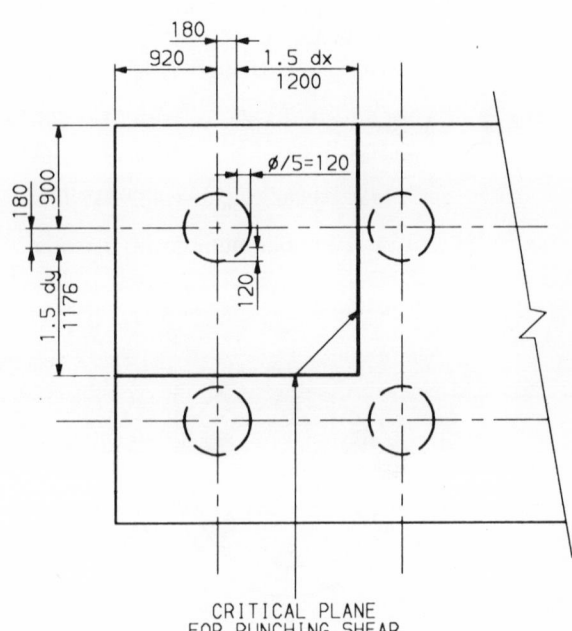

SK 7/44 Critical planes for
punching shear of piles in pile cap.

CRITICAL PLANE
FOR PUNCHING SHEAR

U_1 = perimeter of column = 2 (800 + 800) = 3200 mm

Since pile spacing is not greater than 3 times diameter of pile, then
punching shear stress at critical perimeter for column need not be checked.

U_2 = perimeter on punching shear critical plane for pile load
= 2300 + 2256 = 4556 mm

Ultimate maximum column load, N = 4622 kN   from table in Step 8.

Ultimate maximum pile load, Q = 983 kN

Column punching shear stress $= \dfrac{N}{U_1 d} = \dfrac{4622 \times 10^3}{3200 \times 0.5 \times (800 + 784)}$

$= 1.82 \, \text{N/mm}^2 < 0.8\sqrt{f_{cu}}$ or $5 \, \text{N/mm}^2$ OK

Punching shear stress at perimeter of pile $= \dfrac{983 \times 10^3}{\pi \times 600 \times 800}$

$= 0.65 \, \text{N/mm}^2 < 0.8\sqrt{f_{cu}}$ OK

Pile punching shear stress $= \dfrac{Q}{U_2 d} = \dfrac{983 \times 10^3}{4556 \times 0.5 \, (800 + 784)}$

$= 0.27 \, \text{N/mm}^2$

Minimum v_c for Grade $40 \, \text{N/mm}^2$ concrete $= 0.40 \, \text{N/mm}^2$ OK

Step 13 **Check area of reinforcement in pile**
Unsupported length of pile, l_o, is assumed negligible.
Assume $l_e/h < 10$.
The pile is treated as a short column. From tables in Step 8,

$Q_{max} = 983 \, \text{kN}$ with $M = 28.1 \, \text{kNm}$
$Q_{min} = 367 \, \text{kN}$ with $M = 106.5 \, \text{kNm}$

Max. shear, $V_{max} = 42.93 \, \text{kN}$

Assume minimum cover is 75 mm.

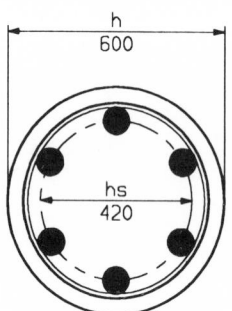

h
600

hs
420

SK 7/45 Pile reinforcement.

Allowing for links and bar diameter, assume $h_s = 420 \, \text{mm}$.

$\dfrac{h_s}{h} = \dfrac{420}{600} = 0.70 = k$

$f_{cu} = 30 \, \text{N/mm}^2$ $e = \dfrac{M}{N} = \dfrac{28.1}{983} = 0.029 \, \text{m}$

$\dfrac{e}{R} = \dfrac{0.029}{0.3} = 0.095$

$\dfrac{Q_{max}}{h^2} = \dfrac{983 \times 10^3}{600 \times 600} = 2.73 \, \text{N/mm}^2$

From Table 11.19, it is observed that minimum reinforcement may be used.
Use minimum reinforcement.

For the second load case,

$$\frac{Q_{min}}{h^2} = \frac{367 \times 10^3}{600 \times 600} = 1 \, N/mm^2$$

$$\frac{e}{R} = 1$$

Again use minimum reinforcement.

Step 14   Check stresses in prestressed concrete piles
Not required.

Step 15   Check shear capacity of RC pile
No shear check is necessary if $M_{pu}/N_u \leq 0.60h$.

$$\frac{M_{pu}}{N_u} = \frac{106.5 \times 10^6}{367 \times 10^3} = 290 \, mm$$

$$0.60h = 0.60 \times 600 = 360 \, mm$$

No shear check is necessary.

$$\frac{H_{pu}}{0.75A_c} = \frac{42.93 \times 10^3}{0.75 \times \pi \times 600^2/4}$$

$$= 0.20 \, N/mm^2 < 0.8\sqrt{f_{cu}} \quad OK$$

Step 16   Check shear capacity of prestressed pile
Not required.

Step 17   Check minimum reinforcement in RC pile

$$\frac{100A_{sc}}{A_c} \geq 0.4$$

$$A_{sc} = \frac{A_c \times 0.4}{100}$$

$$= \frac{\pi \times 300^2 \times 0.4}{100}$$

$$= 1131 \, mm^2$$

Use 6 no. 16 mm dia. HT bars (1206 mm^2).

Step 18   Check minimum prestress in prestressed pile
Not required.

Step 19   Maximum reinforcement in pile
Not required.

Step 20   *Containment of reinforcement in pile*

Minimum dia. of links = $0.25 \times$ bar dia. = $4\,mm \geq 6\,mm$

Maximum spacing of links = $12 \times$ smallest dia. of bar = $12 \times 16 = 192\,mm$

Use 6 mm dia. links at 175 mm centres.

Step 21   *Links in prestressed piles*
Not required.

Step 22   *Minimum tension reinforcement in pile cap*

$A_s \geq 0.0013bh$ in both directions

Minimum reinforcement in the $x-x$ direction = $0.0013 \times 3600 \times 900 = 4212\,mm^2$

Provided $7536\,mm^2$ (see Step 10).

Minimum reinforcement in the $y-y$ direction = $0.0013 \times 5400 \times 900 = 6318\,mm^2$

Area of 16 mm dia. bar = $201\,mm^2$ $32 \times 201 = 6432\,mm^2$

Area required = $3842\,mm^2$ from Step 10

Use 32 no. 16 mm dia. bars equally spaced (approximate spacing 170 mm) in the $y-y$ direction.

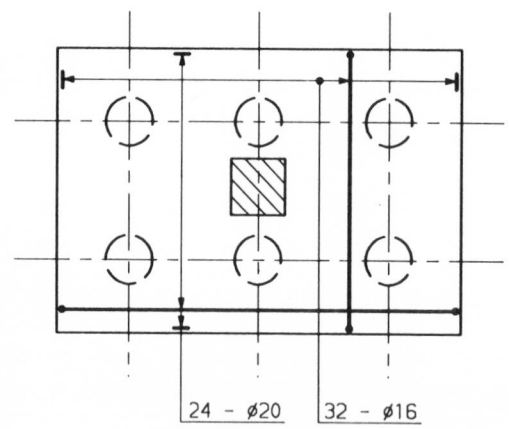

24 – ⌀20   32 – ⌀16

SK 7/46 Pile cap reinforcement revised to suit minimum reinforcement.

Step 23   *Curtailment of bars in pile cap*
Minimum anchorage at ends of bars is $12 \times$ dia. of bar.

$12 \times 20 = 240\,mm$
$12 \times 16 = 192\,mm$

Provide a minimum 250 mm bent up length of pile bottom reinforcement. *Check full anchorage bond length of the main tension bars.*

$f_{cu} = 40 \, \text{N/mm}^2$

Reinforcement used is Type 2 deformed bars.
From Table 3.29 of BS 8110: Part 1: 1985,[1]

tension anchorage length $= 32\phi = 32 \times 20 = 640 \, \text{mm}$

More than 640 mm length of bar is available beyond section 1−1 in Step 8.

Step 24   Spacing of bars in pile cap

$$\text{Maximum percentage of reinforcement} = p = \frac{100 A_s}{bd}$$

$$= \frac{100 \times 7536}{3600 \times 800} = 0.26\%$$

Maximum allowed clear spacing for p less 0.3% is $3d$ or 750 mm, whichever is less.
Spacing of bars adopted is 150 mm.

Step 25   Early thermal cracking

If it is felt necessary to limit early thermal cracking of concrete in pile cap then minimum reinforcement on sides and top of pile cap should be provided based on method of calculation shown in Chapter 2.

Step 26   Assessment of crack width in flexure

Normally the calculations in Step 24 will deem to satisfy the crack width limitations of BS 8110: Part 1: 1985.[1]

If calculations are necessary to prove the limitations of crack width due to flexure in pile cap then methods shown in Chapter 3 should be followed.

Step 27   Connection of pile to pile cap

From Step 17, 16 mm HT Type 2 deformed bars are used.
From Table 3.29 of BS 8110,

full anchorage bond length $= 32\phi; 32 \times 16 = 512 \, \text{mm}$

The bars from the pile will project 600 mm into pile cap. (See general recommendations for design of connections in Chapter 10.)

7.8   FIGURES FOR CHAPTER 7

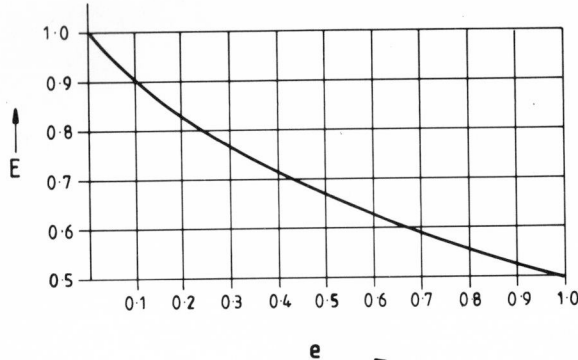

Fig. 7.1 Determination of pile efficiency.

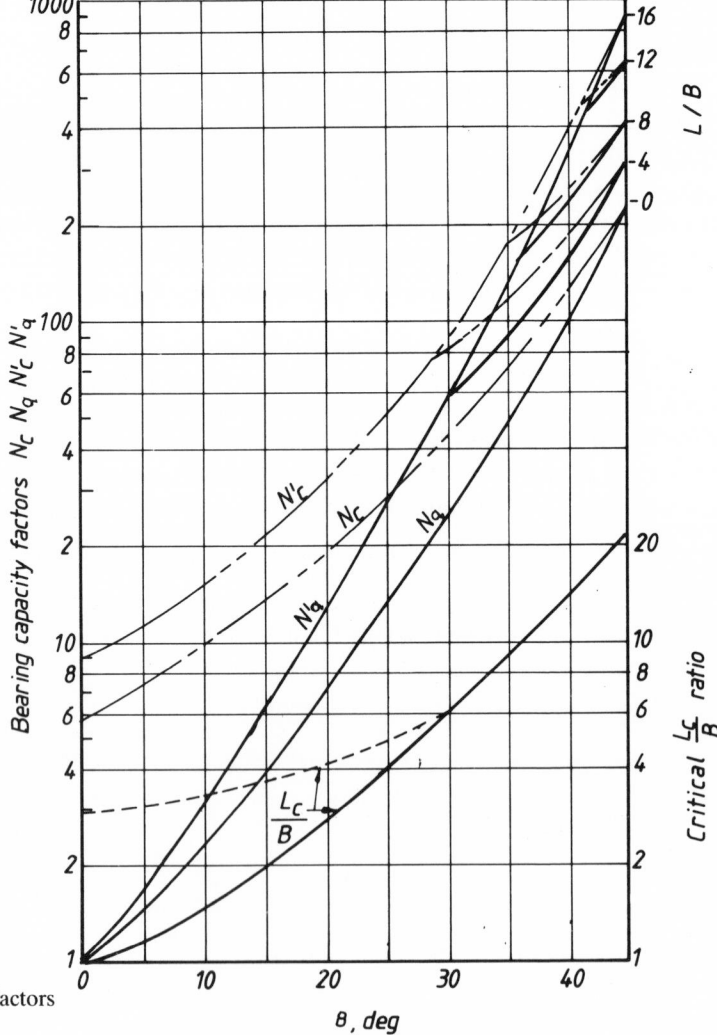

Fig. 7.2 Bearing capacity factors for deep foundations.

Chapter 8
Design of Walls

8.0 NOTATION

a_u	Deflection due to slenderness of wall
a_{cj}	Distances from compression face to centroid of layers of concrete in compression
a_{tj}	Distances from compression face to centroid of layers of tensile reinforcement
A	Area bounded by median line of wall in closed cell
A_c	Net area of concrete in a section of wall
A_C	Centroid of compression in a wall section
A_h	Area of steel in shear reinforcement placed horizontally in in-plane direction
A_T	Centroid of tensile steel in a wall section
A_v	Area of steel in shear reinforcement placed vertically
A_{sc}	Area of steel in compression in a section of wall
A_{si}	Total area of steel in tension in a wall section for in-plane bending
A_{so}	Total area of steel in tension in a wall section for out-of-plane bending
A_{stj}	Layers of tensile steel reinforcement in wall for stress analysis
A'_{stj}	Layers of compressive steel reinforcement in wall for stress analysis
b	Actual width of flange of a shear wall
b	Unit width of wall for out-of-plane bending
b_e	Effective width of flange of a shear wall
B	Plan length of wall for the computation of moment of inertia
c	Coefficient to determine torsional stiffness of a rectangular section
C	Torsional stiffness of a rectangular section
d	Effective depth from compression face to centroid of tensile steel
d_i	Effective depth of tensile steel in wall for in-plane bending
d_o	Effective depth of tensile steel in wall for out-of-plane bending
e	Eccentricity of load on wall section for in-plane bending
e_a	Slenderness coefficient of slender braced plain wall
e_x	Resultant eccentricity of all loads at right angles to plane of wall
$e_{x,1}$	Resultant eccentricity of loads at top of wall
$e_{x,2}$	Resultant eccentricity of loads at bottom of wall
E	Modulus of elasticity
f_c	Stress in concrete compression
f_y	Characteristic yield strength of reinforcement
f_{cu}	Characteristic cube strength of concrete at 28 days
f_{st}	Tensile stress in steel reinforcement

G	Modulus of rigidity
h	Thickness of wall
h_f	Thickness of flange of a shear wall section
h_w	Thickness of web of a shear wall section
H	Height of wall
H_e	Effective height of wall
H_o	Clear height of wall
J	Torsional stiffness of a closed cell structure
K	Factor to determine additional moment due to slenderness
L	Length of wall in in-plane direction
m	Modular ratio E_s/E_c
M	Applied bending moment on a concrete section
M'	Modified applied moment to account for axial load
M_I	In-plane applied bending moment in a wall section
M_{OH}	Out-of-plane bending moment in wall about horizontal plane
M_{OV}	Out-of-plane bending moment in wall about vertical plane
M_{add}	Additional bending moment in out-of-plane direction due to slenderness
n_w	Total design ultimate axial load on a wall
N	Axial load
N_{OV}	In-plane axial load due to out-of-plane loading on wall panel
p_i	Percentage of tensile steel for in-plane bending of wall
p_o	Percentage of tensile steel for out-of-plane bending of wall
q	Shear flow in components of a closed cell (kN/m)
Q_I	In-plane shear flow due to torsion in a closed cell
R	Restraint factor
s	Median length of wall
S_h	Spacing of horizontal shear reinforcement to resist in-plane shear
S_v	Spacing of vertical shear reinforcement to resist in-plane shear
T	Torsion (kNm)
v_i	Shear stress in concrete wall section due to V_i
v_{ci}	Design concrete shear stress in wall section for in-plane bending
v_{co}	Design concrete shear stress in wall section for out-of-plane bending
v_{oh}	Shear stress in concrete wall section due to V_{OH}
v'_{ci}	Modified design concrete shear stress for in-plane bending
v'_{co}	Modified design concrete shear stress for out-of-plane bending
V_i	Combined in-plane flexural shear and torsional shear
V_I	In-plane shear force in a wall section
V_{si}	Shear resistance of shear reinforcement for in-plane shear
V_{so}	Shear resistance of shear reinforcement for out-of-plane shear
V'_{ci}	Available concrete shear strength for in-plane bending after allowing for V_{OH}
V'_{co}	Available concrete shear strength for out-of-plane bending after allowing for V_i
V_{OH}	Out-of-plane shear about horizontal plane
V_{OV}	Out-of-plane shear about vertical plane
x	Depth of neutral axis from compression face
X_i	Shear flow in the components of a closed cell (kN/m)
z	Depth of lever arm

β	Coefficient to determine effective height of wall
β	Factor for determination of deflection due to slenderness of wall
θ	Rate of twist (radians per metre length of member)
ψ	Factor to determine effective width of flange of shear wall

8.1 ANALYSIS OF WALLS

8.1.1 Walls and properties of walls

8.1.1.1 Definitions

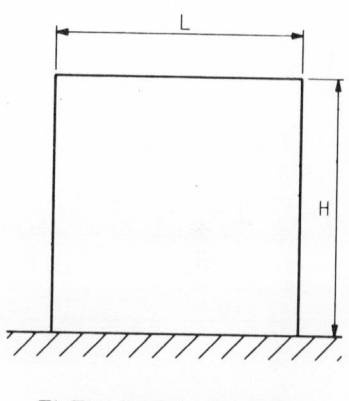

ELEVATION OF WALL

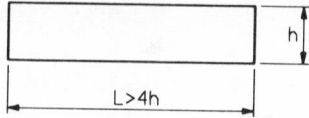

PLAN OF WALL

SK 8/1 Plan and elevation of concrete wall.

Wall is a vertical load-bearing member whose length exceeds four times its thickness.

Unbraced wall is designed to carry lateral loads (horizontal loads) in addition to vertical loads.

Braced wall does not carry any lateral loads (horizontal loads). All horizontal loads are carried by principal structural bracings or lateral supports.

Reinforced wall contains at least the minimum quantities of reinforcement.

Plain wall contains either no reinforcement or less than the minimum quantity of reinforcement.

Stocky wall is where the effective height (H_e) divided by the thickness (h) does not exceed 15 for a braced wall and 10 for an unbraced wall.

Slender wall is a wall other than a stocky wall.

8.1.1.2 Effective heights

8.1.1.2.1 Reinforced wall − monolithic construction

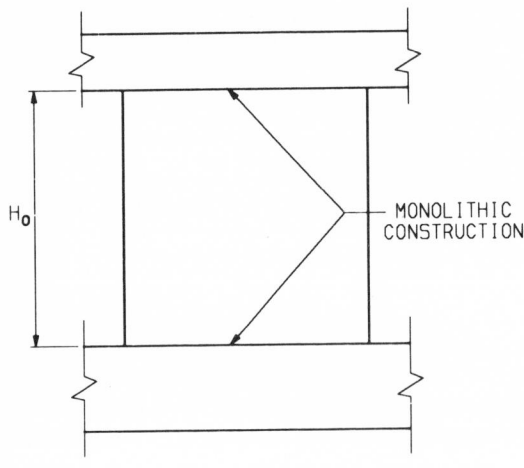

MONOLITHIC
CONSTRUCTION

H_o

SK 8/2 Wall monolithically constructed with slab and foundation.

$$H_e = \beta H_o$$

where H_o = clear height of wall.

Values of β for braced walls.

End condition at top	End condition at bottom	
	1	2
1	0.75	0.80
2	0.80	0.85

Values of β for unbraced walls.

End condition at top	End condition at bottom	
	1	2
1	1.2	1.3
2	1.3	1.5

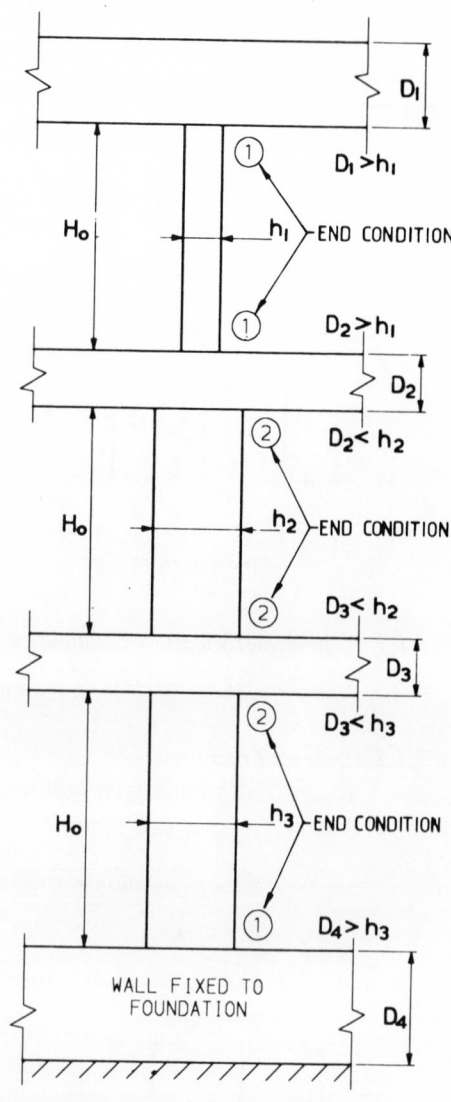

SK 8/3 Wall/slab construction showing end conditions.

<u>WALLS MONOLITHIC WITH SLAB OR FOUNDATION</u>

8.1.1.2.2 *Reinforced wall − simply supported construction*

$H_e = 0.75H_o$ for braced wall where lateral support resists lateral movement and rotation

$H_e = H_o$ for braced walls where lateral supports resist lateral movement

$H_e = 1.5H_o$ for unbraced wall with a roof slab or a floor slab at top

$H_e = 2.0H_o$ for unbraced wall with other forms of construction at top

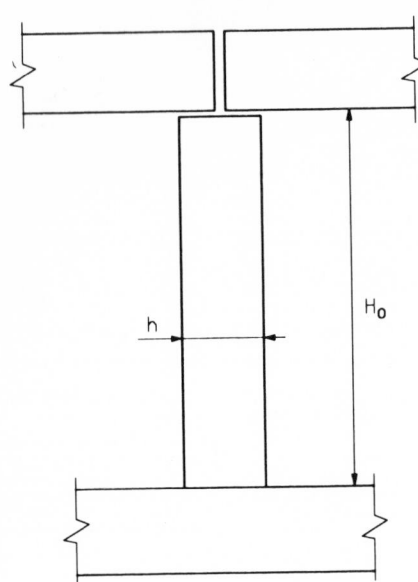

SK 8/4 Slab simply supported on wall.

8.1.1.2.3   Reinforced wall – cantilever construction

$H_e = 2.0H_o$ for moment connection at foundation

8.1.1.2.4   Braced plain wall

With translation and rotation restraint at any lateral support:

$H_e = 0.75H_o$

With translation restraint only at any lateral support:

$H_e = H_o$

Cantilever construction:

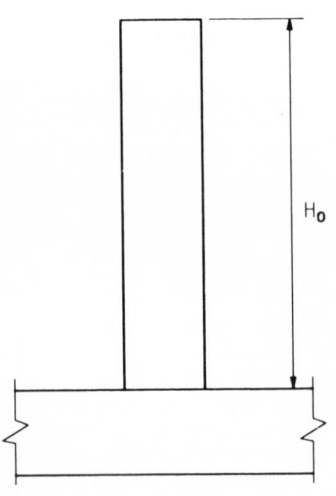

SK 8/5 Cantilever wall.

$H_e = 2H_o$ for rotational and lateral restraint at foundation

8.1.1.2.5   *Unbraced plain wall*

> *Supporting a roof or a floor slab*:
>
> $H_e = 1.5H_o$
>
> *For other walls with lateral restraints*:
>
> $H_e = 2.0H_o$
>
> *For cantilever plain wall*:
>
> $H_e = 3.0H_o$

8.1.1.3   *Effective width of flanges for in-plane bending*

The effective width is width of wall perpendicular to direction of horizontal loading which is considered as effective as compression flange, and also vertical reinforcement provided in this width acts in tension as in tension flange. These factors for effective width are based on the recommendations in BS 5400: Part 5.[3]

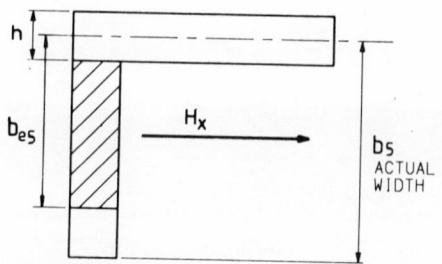

SK 8/6 Effective width of flange on plan of wall arrangement.

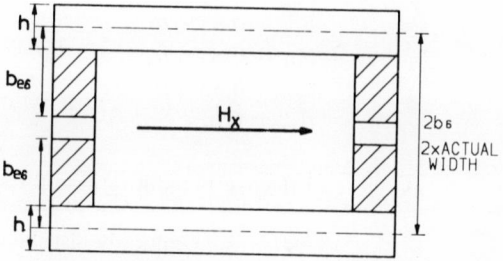

SK 8/7 Effective width of flanges of a closed cell on plan.

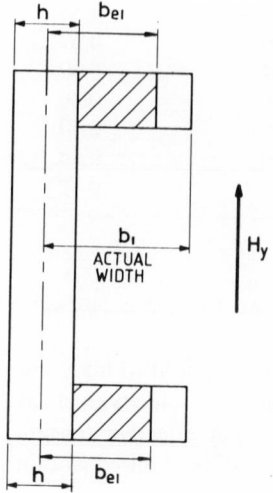

SK 8/8 Effective width of flanges of a channel shaped shear wall on plan.

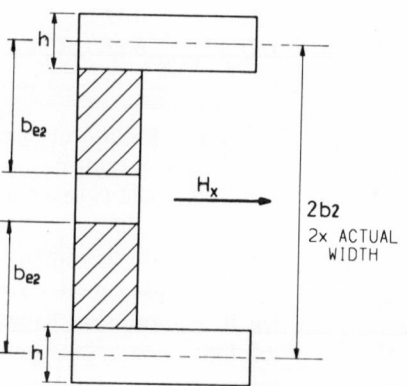

SK 8/9 Effective width of flanges of a channel shaped shear wall on plan.

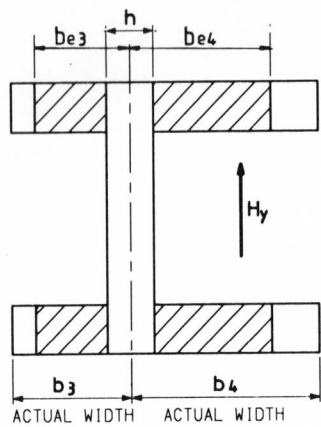

SK 8/10 Effective width of flanges of an I-shaped shear wall on plan.

From diagrams of typical shear wall sections:

$b_{e1} = 0.85\,\psi\,b_1$
$b_{e2} = \psi\,b_2$
$b_{e3} = 0.85\,\psi\,b_3$
$b_{e4} = 0.85\,\psi\,b_4$
$b_{e5} = 0.85\,\psi\,b_5$
$b_{e6} = \psi\,b_6$

Effective breadth ratio ψ for shear walls (see BS 5400: Part 5[3]).

b/H	Uniformly distributed loading		Point loading at top	
	Cantilever wall	Continuous wall	Cantilever wall	Continuous wall
0	1.0	1.0	1.0	1.0
0.05	0.82	0.77	0.91	0.84
0.10	0.68	0.58	0.80	0.67
0.20	0.52	0.41	0.67	0.49
0.40	0.35	0.24	0.49	0.30
0.60	0.27	0.15	0.38	0.19
0.80	0.21	0.12	0.30	0.14
1.00	0.18	0.11	0.24	0.12

b = actual width of flange
H = height of cantilever walls, or
  = half height between monolithic horizontal restraints for continuous walls

Note: The flange width limitations by use of a factor ψ are required to take into account shear lag effects. For ultimate limit state analysis, effects of shear lag in compression flange are sometimes ignored, but effective tension reinforcement in flange for in-plane bending should be limited within effective flange width as given by above expressions.

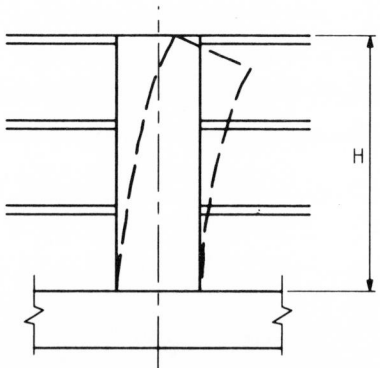

SK 8/11 Cantilever shear wall.

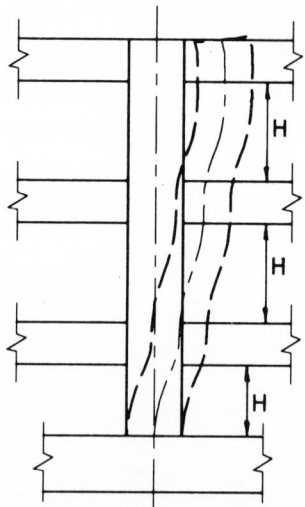

SK 8/12 Continuous shear wall.

8.1.1.4 Moment of inertia and shear area

The moment of inertia and shear area to be used for the computation of deflections of a cantilever shear wall structure and also for input to a computer program with a view to finding the interaction with other walls and frame structures could follow the typical suggestions given below.

Type 1 shear wall

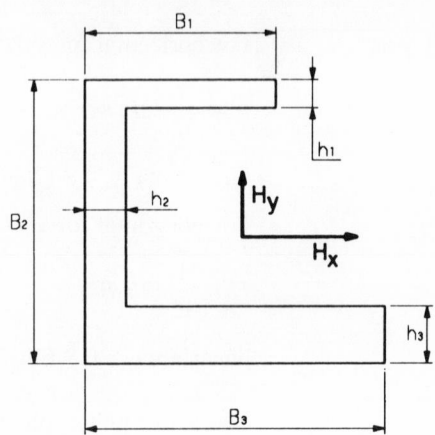

SK 8/13 Type 1 shear wall.

for horizontal force H_x,

$$I_y = \frac{1}{12}h_1B_1^3 + \frac{1}{12}h_3B_3^3$$

Shear area $= 0.8 (B_1h_1 + B_3h_3)$

For horizontal force H_y,

$$I_x = \frac{1}{12}h_2B_2^3$$

Shear area $= 0.8B_2h_2$

Type 2 shear wall

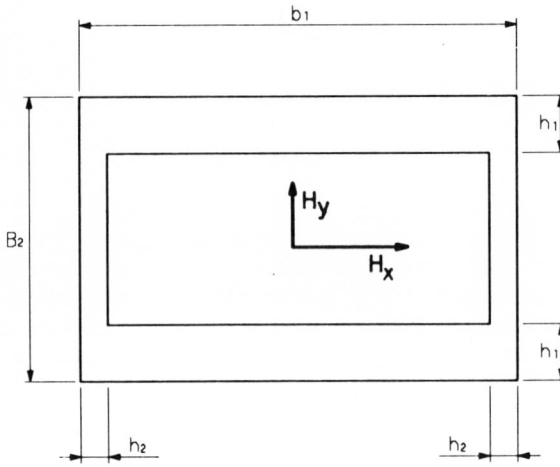

SK 8/14 Type 2 shear wall.

For horizontal force H_x,

$$I_y = \frac{1}{12}(2h_1B_1^3)$$

Shear area $= 0.8 (2h_1B_1)$

For horizontal force H_y,

$$I_x = \frac{1}{12}(2h_2B_2^3)$$

Shear area $= 0.8 (2h_2B_2)$

The above philosophy may be applied to any shape and size of shear wall layout in a building. The stiffness of walls lying parallel to the direction of loading may only be included in the computation.

Note: The flanges of the shear walls have been ignored, as in T-beams in building frames, because the horizontal loads are generally of a reversible nature and concrete in alternate flanges goes into tension. Considering cracked section moment of inertia including effective width of compression flanges does not produce too dissimilar results.

The out-of-plane stiffness of walls may be ignored in the global 3-D frame analysis.

8.1.1.5 Torsional stiffness

8.1.1.5.1 Open cell shear wall

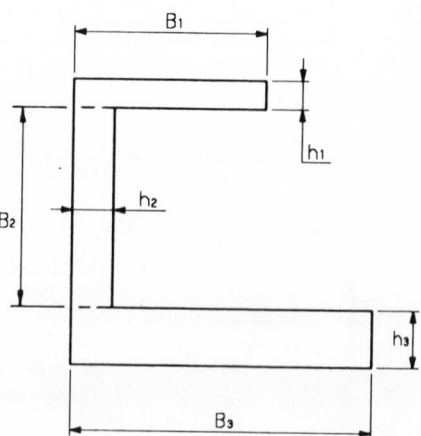

SK 8/15 Open cell shear wall.

The torsional stiffness of individual wall elements should be added. The torsional stiffness of the open cell as a whole is

$$C = c_1 h_1^3 B_1 + c_2 h_2^3 B_2 + c_3 h_3^3 B_3$$

Values of coefficient c.

B/h	1	1.5	2	3	5	10
c	0.14	0.20	0.23	0.26	0.29	0.31

Note: In a global 3-D model each wall of the open cell shear wall may be modelled separately as vertical stiffness elements. The property of each wall will then include the individual torsional stiffness expressed as $C = ch^3 B$.

8.1.1.5.2 Single closed cell shear wall

Torsional stiffness,

$$J = \frac{4A^2}{\Sigma(B/h)} = \frac{4A^2}{\dfrac{2B_1}{h_1} + \dfrac{2B_2}{h_2}}$$

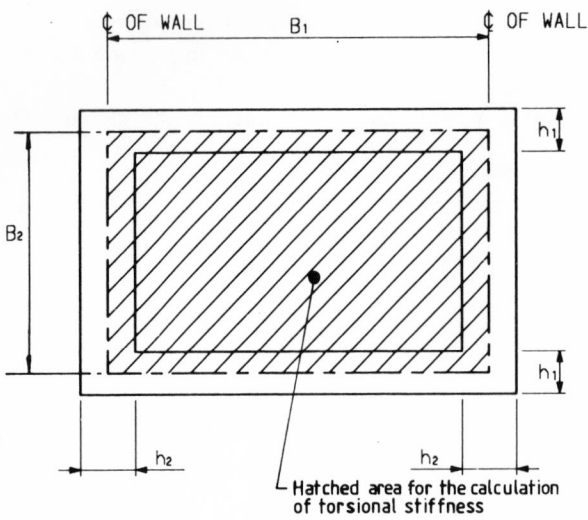

Hatched area for the calculation of torsional stiffness

SK 8/16 Single closed cell shear wall.

$$A = B_1 B_2 \quad \text{(area bounded by median line)}$$

$$T = 2Aq$$

where T = torsion applied
         q = shear flow (kN/m).

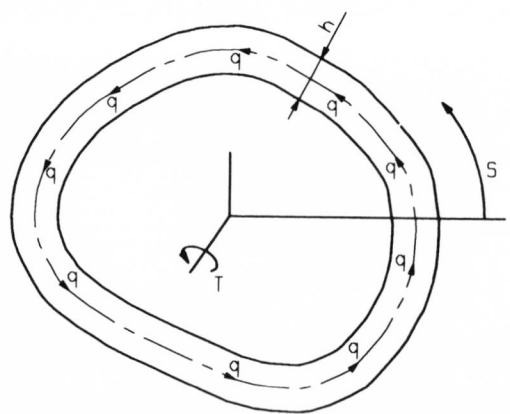

SK 8/17 Closed irregular cell section subject to torsion.

The general formula for any single closed cell is given by:

$$J = \frac{4A^2}{\displaystyle\oint \frac{ds}{h}}$$

where A = area bounded by median line of wall thickness
         h = thickness of wall
         s = median length of wall.

$$J = \frac{T}{G\theta}$$

where T = torque applied = $2Aq$
G = modulus of rigidity = $E/2(1 + \mu)$
E = modulus of elasticity
θ = rate of twist in radians per metre length = $(q/2AG) \int (ds/t)$
q = shear flow (kN/m).

8.1.1.5.3 Multi-cell closed shear wall

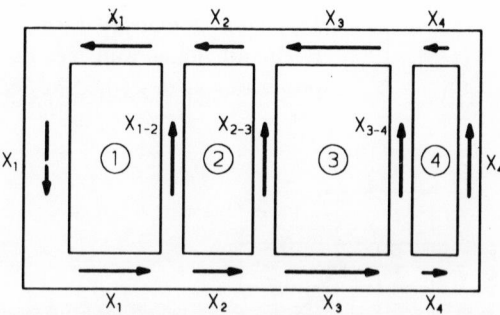

SK 8/18 Closed multiple-cell
section subject to torsion.

General equation for unit twist of one cell:

$$\theta_i = \frac{1}{2A_iG}\left(X_i \int_i \frac{ds}{t} - X_{i-1} \int_{i-1,i} \frac{ds}{t} - X_{i+1} \int_{i,i+1} \frac{ds}{t}\right)$$

$$p_i = \int_i \frac{ds}{t} \qquad p_{i-1,i} = \int_{i-1,i} \frac{ds}{t} \qquad p_{i,i+1} = \int_{i,i+1} \frac{ds}{t}$$

For compatibility, assume $\theta_i = \theta$

$\therefore \quad -X_{i-1}p_{i-1,i} + X_ip_i - X_{i+1}p_{i,i+1} = 2A_iG\theta$

Assume $X_i' = X_i/2G\theta$

$\therefore \quad -X_{i-1}'p_{i-1,i} + X_i'p_i - X_{i+1}'p_{i,i+1} = A_i$

From this general equation:

$$
\begin{array}{llll}
X_1'p_1 & -\ X_2'p_{1,2} & & =\ A_1 \\
-\ X_1'p_{1,2} & +\ X_2'p_2 & -\ X_3'p_{2,3} & =\ A_2 \\
\end{array}
$$

$\cdots\cdots\cdots\cdots\cdots\cdots\cdots\cdots\cdots\cdots\cdots\cdots\cdots\cdots\cdots\cdots$

$$
\begin{array}{llll}
-\ X_{n-2}'p_{n-2,n-1} & +\ X_{n-1}'p_{n-1} & -\ X_n'p_{n-1,n} & =\ A_{n-1} \\
& -\ X_{n-1}'p_{n-1,n} & +\ X_n'p_n & =\ A_n \\
\end{array}
$$

Solving for the unknowns in the above matrix gives values of X_1' to X_n'.

$$T = 4G\theta\sum_1^n A_iX_i'$$

When T is known, θ can be found.

$$J = \frac{T}{G\theta} = 4\sum_{1}^{n} A_i X_i'$$

Shear flow, $X_i = 2G\theta X_i' = \left(\frac{2T}{J}\right) X_i'$

$$X_{i,\,i+1} = X_i - X_{i+1}$$

8.1.2   Modelling for structural analysis

8.1.2.1   Global analysis for in-plane forces

Modelling as individual walls
Each individual wall can be modelled as a vertical beam element with properties as described in Section 8.1.1.4 and 8.1.1.5.

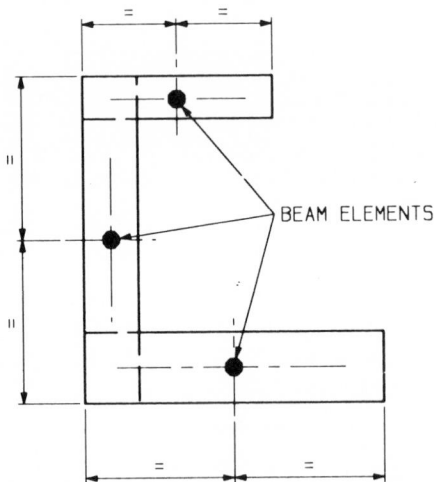

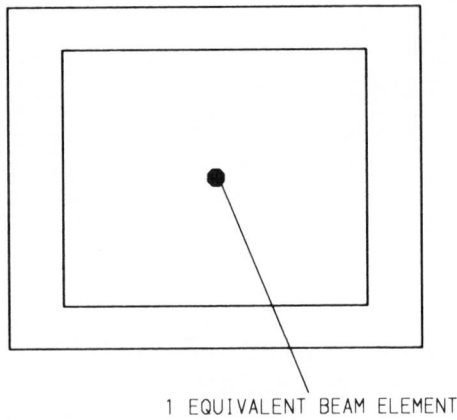

SK 8/20 A closed cell converted to one equivalent beam element.

SK 8/19 Walls of a shear wall system converted to equivalent beam elements.

Modelling as a combined unit
A set of walls can be combined to be represented by one beam element. In this case the property of this beam element will be the summation of properties for individual walls. The representative beam element may be located at the CG of the wall configuration.

In the case of closed cell shear wall structure the equivalent torsional stiffness will not be the sum of the individual torsional stiffnesses of the walls. The equivalent torsional stiffness will be found as per Section 8.1.1.5.2. When a closed cell shear wall structure is modelled as individual wall elements, then the torsional stiffness parameters for these individual wall elements will be considered as negligible. A separate single beam

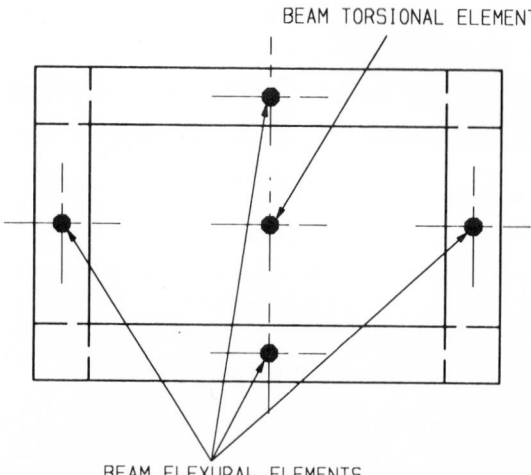

BEAM TORSIONAL ELEMENT

BEAM FLEXURAL ELEMENTS

SK 8/21 Cell converted to a combination of flexural and torsional elements.

element must be modelled to represent the torsional stiffness of the closed cell system. This beam element will have no bending or shear stiffness but only torsional stiffness. This element may be placed at the CG of the closed cell structure. This separate torsional beam element will be connected by rigid offsets with the individual wall bending elements.

The design of walls should be carried out on an individual wall basis. The determination of individual wall moments and shears from the representative single beam element will be carried out by using the relative bending and shear stiffnesses of individual walls.

8.1.2.2 Local analysis for out-of-plane forces

Out-of-plane forces on the wall may be due to the following:

- Eccentric dead and live load.
- Wind pressure on wall panel.
- Earthquake wall mass excitation.
- Earth pressure on wall face.
- Water pressure on wall face.
- Thermal gradient across wall thickness.

Local analyses should be carried out using appropriate boundary conditions. Published tables may be used to find out-of-plane bending moments and shears. Combined bending moment triads using the Wood–Armer principle should be used to find the reinforcement requirement.

For out-of-plane local analyses, follow the general guidelines in Section 3.1.

For a complicated wall geometry, wall panels in the out-of-plane direction may be modelled using hypothetical grillage elements using for solution a grillage suite of a computer software.

8.2   STEP-BY-STEP DESIGN PROCEDURE FOR WALLS

Step 1   *Find properties of wall system*
Find moment of inertia and shear area (follow Section 8.1.1.4).

Step 2   *Find torsional stiffness of wall system*
Follow Section 8.1.1.5.

Step 3   *Carry out modelling for analysis*
Follow Sections 8.1.2.1 and 8.1.2.2.

Step 4   *Carry out global analysis*

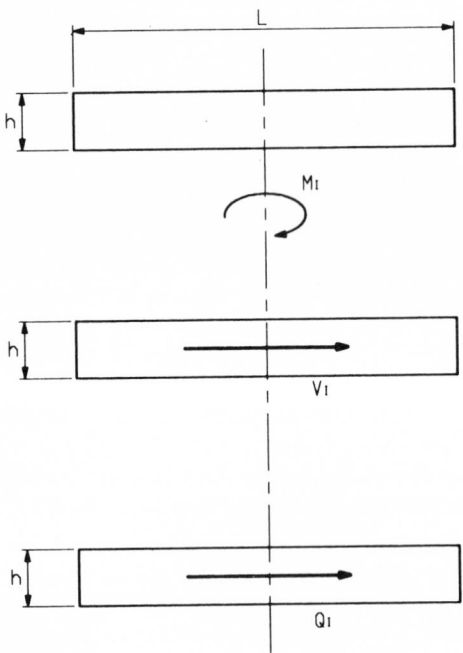

SK 8/22 Description of in-plane forces in a wall.

Find in-plane forces in walls. After analysis the following internal forces should be available for each individual wall section in the system.

M_I = in-plane bending moments
V_I = in-plane shear force
Q_I = in-plane shear flow due to torsion
N = axial load

Step 5   *Carry out local analysis*
Find out-of-plane forces in walls (follow Section 8.1.2.2).
After analysis the following internal forces should be available for each individual wall panel in the system.

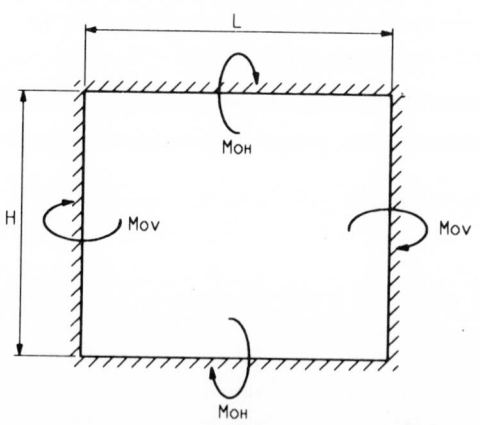

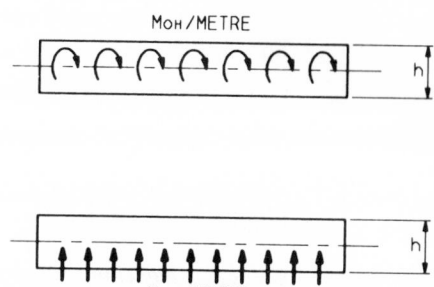

SK 8/24 Out-of-plane internal forces in a wall section.

SK 8/23 Elevation of a wall panel showing out-of-plane moments.

M_{OH} = out-of-plane bending moment about horizontal plane
M_{OV} = out-of-plane bending moment about vertical plane
V_{OH} = out-of-plane shear about horizontal plane
V_{OV} = out-of-plane shear about vertical plane

Step 6   Carry out combination of loading

This should preferably be carried out in a tabular fashion for different load cases. The load combinations should be generally as follows:

$$LC_1 = 1.4DL + 1.6LL + 1.4EP + 1.4WP$$
$$LC_2 = 1.0DL + 1.4EP + 1.4WP$$
$$LC_3 = 1.4DL + 1.4WL + 1.4EP + 1.4WP$$
$$LC_4 = 1.0DL + 1.4WL + 1.4EP + 1.4WP$$
$$LC_5 = 1.2DL + 1.2LL + 1.2WL + 1.2EP + 1.2WP$$

Note: Load combinations LC_2 and LC_4 should be considered only when dead and live load have beneficial effects.

where DL = dead load
  LL = live or superimposed load
  WL = wind load or earthquake load
  WP = water pressure
  EP = earth pressure.

Step 7   Check slenderness of wall

Determine type of wall: braced, unbraced, plain or reinforced.
Find effective height (follow Section 8.1.1.2.1)

$$H_e = \beta H_o$$

Check slenderness ratio H_e/h.
For braced reinforced wall with $<1\%$ reinforcement, limit of $H_e/h \le 40$.

For braced reinforced wall with $\geq 1\%$ reinforcement, limit of $H_e/h \leq 45$. For unbraced reinforced wall and plain wall, limit of $H_e/h \leq 30$.

If $H_e/h \leq 15$ (braced) or 10 (unbraced), then design as a *stocky wall*. Otherwise, design as *slender wall*.

Step 8   ***Find effective width of flanges for reinforced wall***
Follow Section 8.1.1.3.

Step 9   ***Find additional out-of-plane moments about horizontal plane***

   (1) Moments due to minimum eccentricity of $h/20$ or 20 mm of direct loads from beams and slabs simply supported on wall.
   (2) Moments due to slenderness of wall.

   For $H_e/h > 15$ (braced) or >10 (unbraced):

Note: Wall braced or unbraced in the transverse direction only to be considered for additional moments.

   Deflection due to slenderness of wall, $a_u = \beta \, K \, h$

   Assume $K = 1$ for conservatism.

   $$\beta = \frac{1}{2000} \left(\frac{H_e}{h} \right)^2$$

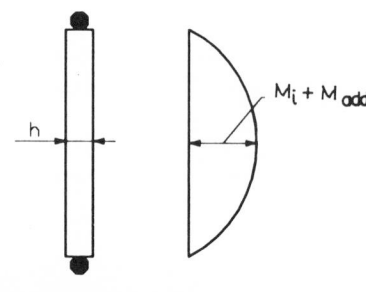

BRACED WALL

BOTH ENDS OF WALL FREE TO ROTATE
BUT RESTRAINED IN POSITION
M_i = Initial wall moment

SK 8/25 Wall additional moment.

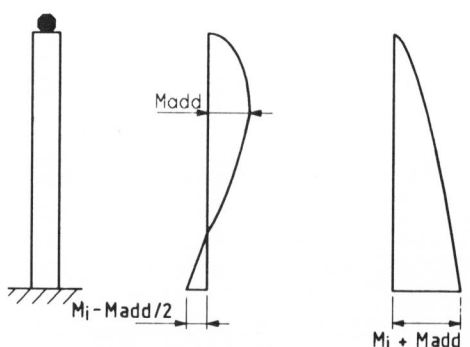

SK 8/26 Wall additional moments.

Additional moment due to slenderness, $M_{add} = Na_u$

where N = direct ultimate load on wall.

Combine this additional moment, M_{add}, with any other out-of-plane moments obtained from analysis using Figure 3.20 or Figure 3.21 of BS 8110: Part 1: 1985.[1]

Note: These additional moments should be doubled if the wall has only one central layer of reinforcement.

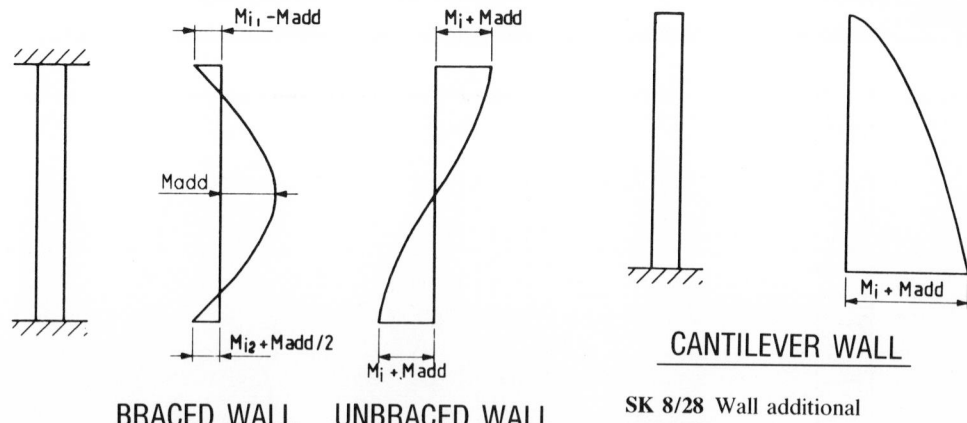

$M_{i1} - M_{add}$ $M_i + M_{add}$

M_{add}

$M_{i2} + M_{add}/2$

$M_i + M_{add}$

BRACED WALL   **UNBRACED WALL**

$M_i + M_{add}$

CANTILEVER WALL

SK 8/28 Wall additional moments.

BOTH ENDS OF WALL RESTRAINED TO ROTATE
M_i = Initial Moments in the wall from analysis.

SK 8/27 Wall additional moments.

These out-of-plane bending moments and shears are about a horizontal plane.

Step 10   ***Design stocky braced reinforced wall with approximately symmetrical arrangement of slabs***
Spans of slab on either side of wall within ± 15% and slab subjected to uniform load.

$$n_w \leq 0.35 f_{cu} A_c + 0.67 A_{sc} f_y$$

where n_w = total design ultimate axial load on wall.

Step 11   ***Determine cover to reinforcement***
Determine cover to reinforcement as per Tables 11.6 and 11.7.

Step 12   ***Design of reinforced wall − rigorous method***
Using the effective flange widths found in Step 8, find by elastic analysis the stresses in the concrete and steel due to in-plane bending moment and axial load only.

(1) Assume initially 0.40% area of steel in wall distributed uniformly in two layers on two faces.
(2) Assume a value of x for depth of neutral axis from compression face.
(3) Divide compression zone into layers of concrete with depths d_{c1}, d_{c2}, d_{c3}, etc. and find centres of these layers from compression face a_{c1}, a_{c2}, a_{c3}, etc.
(4) Conveniently group bars in tension zone and find area of groups A_{st1},

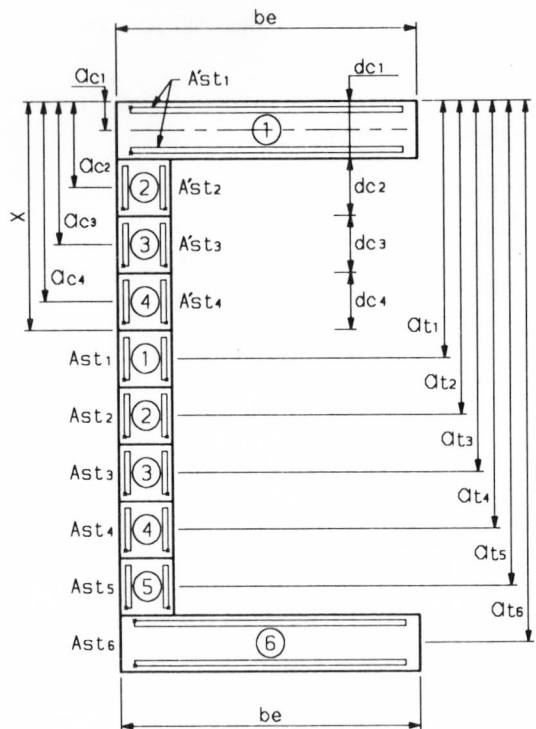

SK 8/29 A typical example of analysis of a shear wall.

A_{st2}, A_{st3}, etc. and also distances of these groups of bars from compression face, i.e. a_{t1}, a_{t2}, a_{t3}, etc.

(5) Find the following:

$$A_T = \frac{\Sigma A_{st} a_t (a_t - x)}{\Sigma A_{st}(a_t - x)}$$

$$S_1 = A_{c1} + (m - 1)A'_{st1}$$

where A_{c1} = area of concrete in the layer 1 of concrete in compression zone

A'_{st1} = area of compressive reinforcement in layer 1 of concrete in compression zone

m = modular ratio = E_s/E_c.

$$A_c = \frac{\Sigma(x - a_c)a_c S}{\Sigma(x - a_c)S}$$

$$\bar{x} = \frac{m\Sigma A_{st} a_t + \Sigma S a_c}{m\Sigma A_{st} + \Sigma S}$$

$$e = \frac{M}{N} = \frac{\text{in-plane bending moment}}{\text{axial compression}}$$

$$f_c = \frac{Nx(e + A_T - \bar{x})}{(A_T - A_c)\,\Sigma\,(x - a_c)S}$$

$$f_{st} = \frac{(d - x)}{\Sigma(a_t - x)A_{st}}\left[\frac{f_c}{x}\Sigma(x - a_c)S - N\right]$$

(6) Finally check:

$$x = \frac{d}{1 + \dfrac{f_{st}}{mf_c}}$$

If x is different from the assumed value then repeat the exercise with a new assumed x until convergence is reached.

(7) Find final stresses f_c and f_{st} after convergence. If f_{st} is greater than $0.87f_y$, then increase area of steel by proportion $f_{st}/0.87f_y$. If f_c is greater than $0.45f_{cu}$, then increase thickness of wall.

(8) Find revised f_c and f_{st} with increased reinforcement. There will be no need to carry out the iteration to find x with increased reinforcement.

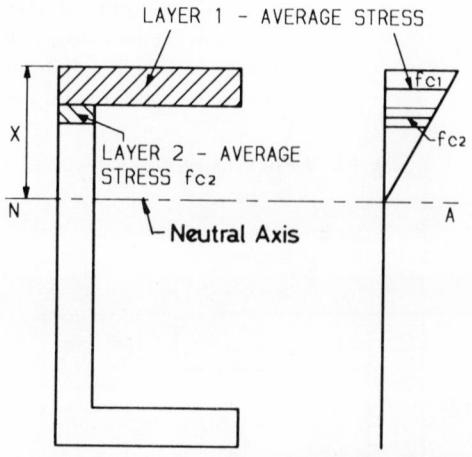

SK 8/30 Elastic stress analysis of a shear wall.

(9) Draw stress diagram for in-plane bending moment and direct axial load. Divide wall into unit lengths. Over each unit length convert the average compressive stress in compression zone into a direct load by multiplying with the area of the unit length of wall. This compression force acts in combination with the out-of-plane bending moment in that length of wall. Design reinforcement for out-of-plane bending additional to that already provided in that unit length using Tables 11.8 to 11.17 − design tables for rectangular columns.

(10) In the tension zone of the wall subject to in-plane bending moment and axial load only, assume the concrete as unstressed. Find reinforcement required for out-of-plane bending moment as in an RC beam following Step 10 of Section 2.3. Add this reinforcement to reinforcement already provided for in-plane bending moment.

If reinforcement provided for in-plane bending is not fully stressed to the ultimate limit of $0.87f_y$, then the residual capacity of this reinforcement may be used to withstand out-of-plane bending moment.

Average tensile strain in the tensile flange may be found and converted to an average tensile force in the flange for computation of reduced shear stress for out-of-plane bending. Conservatively ignore concrete shear resistance in tension flange.

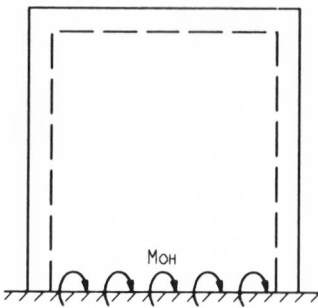

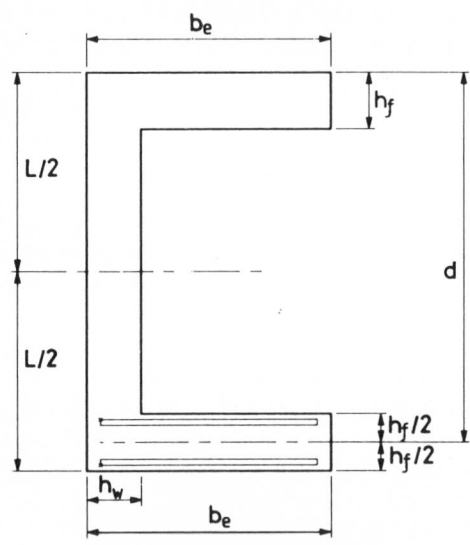

SK 8/31 Out-of-plane bending of a panel of a wall.

Note: Each panel of wall should be checked for global loads in both orthogonal directions separately if these loads are not acting simultaneously. The worst reinforcement from either of the two orthogonal loads will be used. The out-of-plane bending moments for combination with in-plane bending moments are about the horizontal plane.

Step 13   Design of reinforced wall − simple method

SK 8/32 Analysis of a shear wall against in-plane bending.

Flanged wall

$$M' = M + N \left(\frac{L}{2} - \frac{h_f}{2} \right)$$

where M = in-plane bending moment
N = axial load.

$$K = \frac{M'}{f_{cu}bd^2} \leq 0.156$$

$$z = d\left[0.5 + \sqrt{\left(0.25 - \frac{K}{0.9}\right)}\right] \le 0.95d$$

$$x = \frac{d - z}{0.45}$$

$$A_s = \left(\frac{M'}{0.87f_{yz}}\right) - \left(\frac{N}{0.87f_y}\right)$$

This reinforcement will be provided in effective width of flange in two layers as shown. The web of the flanged wall will have the minimum reinforcement unless dictated by out-of-plane bending moments or reinforcement requirement as part of tension flange for other direction of orthogonal load.

If $x > h_f$, then follow Step 11 of Section 2.3.

The out-of-plane bending about a horizontal plane on either the wall flange or the wall web may be due to the following:

(1) Out-of-plane framing action with supported slab.
(2) Slenderness of wall.
(3) Eccentric loads from beam or slab or any other structure on wall.
(4) Coacting horizontal loads on wall panel due to wind, earthquake, water pressure or earth pressure.
(5) Thermal gradient across wall thickness.

The reinforcement for the out-of-plane bending moment about a horizontal plane will be calculated as follows.

In the compression zone of concrete wall due to in-plane bending moment, assume that the concrete has already reached the ultimate stage

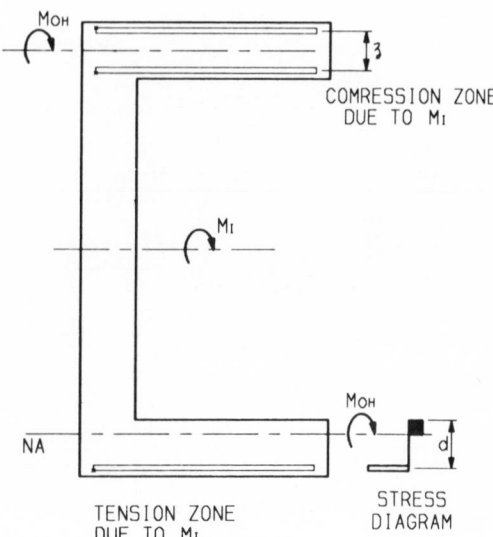

SK 8/33 Analysis of a shear wall against in-plane and out-of-plane bending.

and cannot take any more load. Hence, the bending moment will be resisted by equal amounts of compressive and tensile steel with a lever arm equal to the distance between the two layers of steel.

In the tension zone of concrete wall due to in-plane bending moment, assume that the concrete is unstressed and use the beam theory to find reinforcement due to out-of-plane bending moment.

The reinforcement required due to out-of-plane bending moment will be added to the reinforcement found for in-plane bending.

Step 14   Design of reinforced wall — short and squat cantilever — deep-beam approach

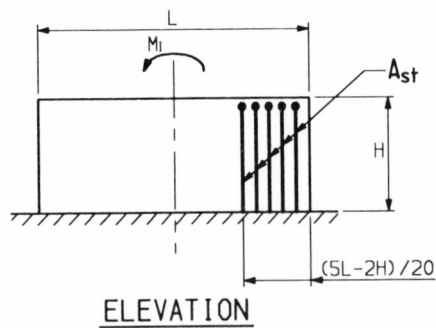

ELEVATION

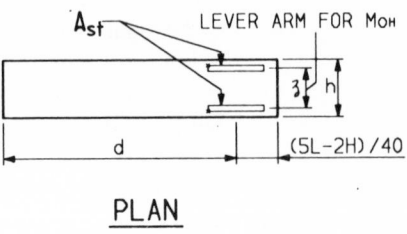

PLAN

SK 8/34 Design of a shear wall by deep-beam method.

This approach may be used for walls with total height less than or equal to their length. For in-plane bending consider the wall as a deep-beam and follow the deep-beam theory of stress distribution.

For horizontal loading to resist in-plane bending moment

$$z = \frac{6H}{5} \quad \text{when } \frac{H}{L} \le 0.5$$

$$= \frac{2(H + L)}{5} \quad \text{when } 0.5 < \frac{H}{L} \le 1$$

Tension reinforcement to be distributed over a length of wall equal to

$$\frac{5L - 2H}{20}$$

$$d = L - \left(\frac{5L - 2H}{40}\right)$$

$$M' = M + N \left(\frac{L}{2} - \frac{5L - 2H}{40}\right)$$

$$A_s = \left(\frac{M'}{0.87 f_y z}\right) - \left(\frac{N}{0.87 f_y}\right)$$

Note: The flexural strain in concrete is very small in a short and squat cantilever wall and for all practical purposes may be ignored when designing for the transverse out-of-plane bending moment. Use the normal beam theory to find reinforcement for the transverse out-of-plane bending moment.

Add this additional reinforcement for transverse out-of-plane bending moment to reinforcement already found for in-plane bending moment. The out-of-plane bending moment in this context is about the horizontal plane.

The flanges of the wall, if present, either in tension or compression may be ignored if this deep-beam approach is used. The shear strains in a wall with the aspect ratios of a deep beam may be high and a conservative approach taking the shear-lag effect would be to ignore the flanges.

Step 15 Check shear

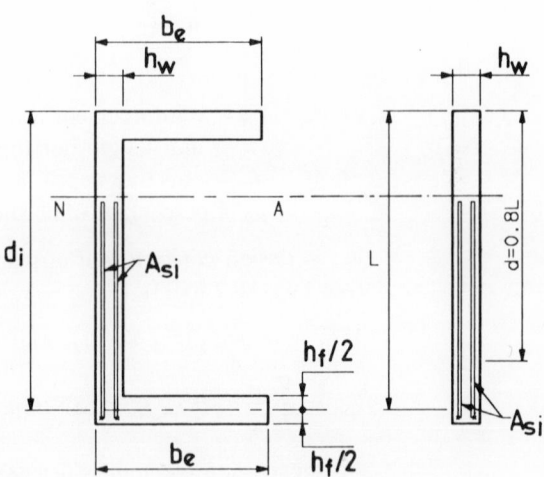

SK 8/35 Design for in-plane shear.

In-plane bending

$$p_i = \frac{100 A_{si}}{h_w d_i}$$

where p_i = percentage of tensile reinforcement in in-plane direction
A_{si} = reinforcement available to resist in-plane bending
d_i = effective depth as shown.

$$v_i = \frac{V_i}{h_w d_i} \leq 0.8\sqrt{f_{cu}} \leq 5\,\text{N/mm}^2$$

where v_i = shear stress due to in-plane bending and torsional shear flow in wall
V_i = combined in-plane shear
= flexural shear + torsional shear
= $V_1 + Q_1$ (see Step 4).

v_{ci} = design concrete in-plane shear stress depending on p_i and f_{cu} (see Figs 11.2 to 11.5)

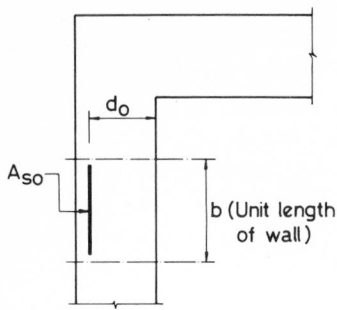

SK 8/36 Design for out-of-plane shear.

Out-of-plane bending

$$p_o = \frac{100A_{so}}{bd_o}$$

where A_{so} = reinforcement available to resist out-of-plane bending
b = unit length horizontally
d_o = effective depth in transverse out-of-plane direction
p_o = percentage of tensile reinforcement in out-of-plane direction.

v_{co} = design concrete out-of-plane shear stress depending on p_o and f_{cu} (see Figs 11.2 to 11.5)

V_{OH} = out-of-plane shear about a horizontal plane over a unit width b.

$$v_{oh} = \frac{V_{OH}}{bd_o} \leq 0.8\sqrt{f_{cu}} \leq 5\,\text{N/mm}^2$$

No shear reinforcement is necessary if the following equation is satisfied:

$$\frac{v_i}{v_{ci}} + \frac{v_{oh}}{v_{co}} \leq 1$$

Note: In Step 12 and Step 13 the wall is designed as flanged beams for in-plane loading. For out-of-plane shear in the flanges which acts together with the in-plane loading, the check should be carried out separately for compression and tension flange. For compression flange the enhancement of design shear stress due to axial load may be allowed based on average compressive stress. For tension flange the concrete may be conservatively ignored and the shear force will be totally carried by shear reinforcement. Alternatively, average tensile strain in concrete may be found and the shear stress reduction formula may be used.

Step 16   Calculate shear reinforcement

Note: Increase or decrease of design concrete shear stress due to presence of axial load may be allowed following the formula on page 160.

SK 8/37 In-plane shear
reinforcement in walls.

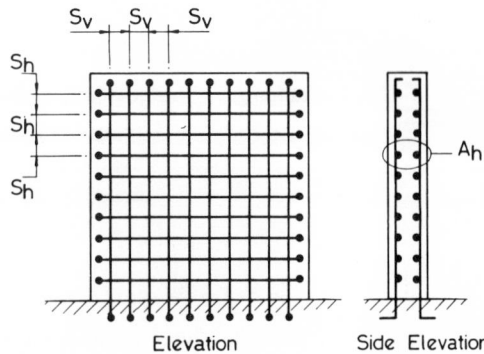

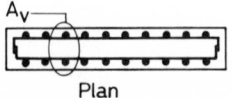

Plan   Elevation   Side Elevation

Case 1: $v_{oh} < v_{co}$

$$v'_{ci} = \left(1 - \frac{v_{oh}}{v_{co}}\right)v_{ci}$$

where v'_{ci} = available concrete shear strength in in-plane direction for use
with shear reinforcement.

$$V'_{ci} = v'_{ci}h_w d_i$$

Provide shear reinforcement in in-plane direction for a shear force equal to
$(V_i - V'_{ci})$ and check:

$$(V_i - V'_{ci}) \leq V_{si}$$

$$V_{si} = \frac{0.87 A_h f_y d_i}{S_h}$$

where V_{si} = shear resistance of horizontal bars in wall for in-plane shear
      S_h = spacing of horizontal bar in wall
      A_h = area of horizontal shear reinforcement
      f_y = characteristic yield strength of reinforcement.

Note: Provide equal amount of vertical shear reinforcement with horizontal
shear reinforcement.

$$\frac{A_v}{S_v} = \frac{A_h}{S_h}$$

where A_v = area of vertical shear reinforcement
      S_v = spacing of vertical shear reinforcement.

In this Case 1, no shear reinforcement is required for out-of-plane flexure.
Provide the shear reinforcement for in-plane shear in addition to other
bars required for in-plane and out-of-plane bending moments.

Case 2: $v_{oh} > v_{co}$

$$v'_{ci} = \frac{v_{ci}\, v_i}{v_{oh} + v_i}$$

$$v'_{co} = \frac{v_{co}\, v_{oh}}{v_{oh} + v_i}$$

where v'_{ci} = available concrete shear stress strength in in-plane direction for use with shear reinforcement

v'_{co} = available concrete shear stress strength in out-of-plane direction for use with shear reinforcement.

$$V'_{ci} = v'_{ci}\, hd_i$$

$$V'_{co} = v'_{co}\, bd_o$$

Provide shear reinforcement in in-plane direction, as in Case 1, which satisfies

$$V_{si} \geq (V_i - V'_{ci})$$

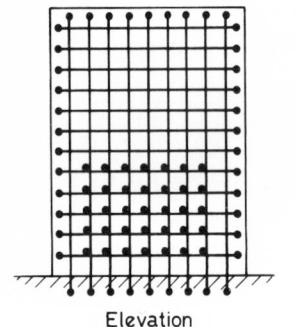
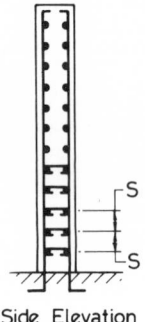
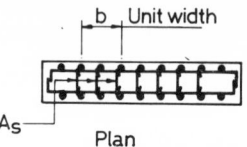

SK 8/38 Out-of-plane shear reinforcement in walls.

Elevation   Side Elevation A_s Plan

For shear reinforcement in out-of-plane horizontal direction, use links through thickness of wall.

For out-of-plane horizontal directional shear, resistance from links, V_{so}, for a unit length b is given by

$$V_{so} = \frac{0.87 f_y A_s d}{S}$$

where A_s = area of links over a unit width b
S = vertical spacing of links.

Check $V_{OH} - V'_{co} \leq V_{so}$

Note: If considerable ductility is required of a shear wall, as in seismic design, the whole shear force should be carried by reinforcement and the shear capacity of concrete may be ignored if the shear capacity of concrete is exceeded.

Step 17   ***Check-out-of-plane bending about vertical plane***
After local analysis of wall panel the bending moments, direct loads and shears about the vertical plane in the panel are obtained.

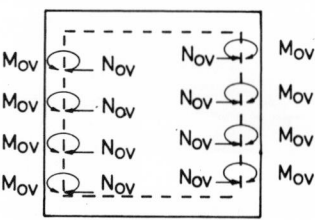

Elevation of Wall Panel

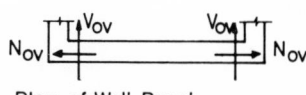

SK 8/39 Internal forces in wall
panel due to out-of-plane loading.

Plan of Wall Panel

Design horizontal reinforcement for flexure of wall panel about a vertical
plane. The procedure is the same as in Step 4 of Section 3.3.

Check shear stress and reinforcement for shear as in Step 6 of Section 3.3.

Step 18   Design of plain (not adequately reinforced) walls

(A) Stocky braced plain wall

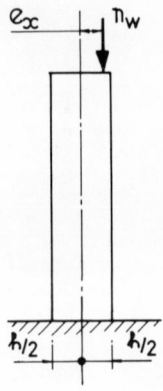

SK 8/40 Eccentric loading on wall
in out-of-plane direction.

Check $n_w \leq 0.3 \, (h - 2e_x) f_{cu}$

where n_w = maximum ultimate axial load per unit length on wall

e_x = resultant eccentricity of all loads at right angles to plane of
wall (minimum value of e_x is $h/20$).

(B) Slender braced plain wall

Check $n_w \leq 0.3 \, (h - 2e_x) f_{cu}$

and $n_w \leq 0.3 \, (h - 1.2e_x - 2e_a) f_{cu}$

$$e_a = \frac{H_e^2}{2500h}$$

where H_e = effective height (as per Section 8.1.1.2).

(C) Unbraced plain wall

Check $\quad n_w \leq 0.3\ (h - 2e_{x,1})f_{cu}$

and $\quad n_w \leq 0.3\ [h - 2\ (e_{x,2} + e_a)]f_{cu}$

where $\quad e_{x,1}$ = resultant eccentricity of loads at top of wall

$\quad\quad e_{x,2}$ = resultant eccentricity of loads at bottom of wall.

Step 19 **Shear check of plain walls**

Check $\quad V \leq 0.25n_w$

where $\quad V$ = in-plane ultimate shear force per unit length.

Check $\quad v = \dfrac{V}{hb} \leq 0.45\,\text{N/mm}^2$

where $\quad v$ = shear stress

$\quad\quad b$ = unit length (mm).

Note: A plain wall subjected to in-plane shear should satisfy at least one of the above checks.

Step 20 **Check minimum reinforcement**

Minimum compression (vertical) reinforcement in reinforced wall = 0.4% $(f_y = 460\,\text{N/mm}^2)$ of gross cross section

Minimum horizontal tension reinforcement to withstand out-of-plane loads = 0.13% $(f_y = 460\,\text{N/mm}^2)$ of gross cross-section on each face

Minimum anti-crack reinforcement = 0.25% $(f_y = 460\,\text{N/mm}^2)$ of gross cross-section

Step 21 **Check maximum reinforcement**

Maximum vertical reinforcement in wall = 4% of gross cross-section

Step 22 **Check containment of wall reinforcement**

For vertical compression reinforcement in walls up to 2% of gross cross-sectional area, use the following minimum horizontal bars:

0.25% of gross concrete area

Horizontal bar diameter should be greater than or equal to $\frac{1}{4}$ size of vertical bars but not less than 6 mm diameter.

For vertical compression reinforcement in walls greater than 2% of gross cross-sectional area, use links through the thickness of wall.

Dia. of links $\geq \frac{1}{4}$ dia. of vertical bars or 6 mm, whichever is greater

Spacing of links $\geq 2h$ in horizontal and vertical direction

Spacing of links in vertical direction should not be more than 16 times vertical bar diameter.

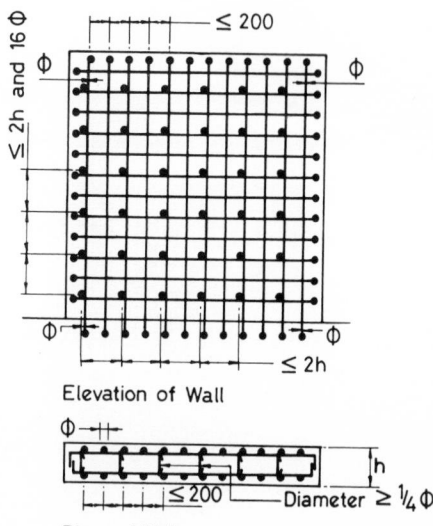

Elevation of Wall

Plan of Wall

SK 8/41 Detailing rules for walls.

Any vertical compression bar not enclosed by link should be within 200 mm of a restrained bar.

Step 23   Check early thermal cracking
See Step 14 of Section 3.3.

Step 24   Clear spacing of bars in tension
Follow Step 13 of Section 3.3.

Step 25   Connections
See Chapter 10.

8.3 WORKED EXAMPLE

Example 8.1   Reinforced concrete cell
Design the walls of a reinforced concrete cell which forms part of the horizontal stability system of a building.

Step 1   Find properties of wall system
Follow Section 8.1.1.4.

Stiffness in y-direction
Divide the wall cell system into six beam elements located at the centroid of each wall.

Equivalent beam elements 1, 2, 3 and 4 contribute to stiffness in y-direction.

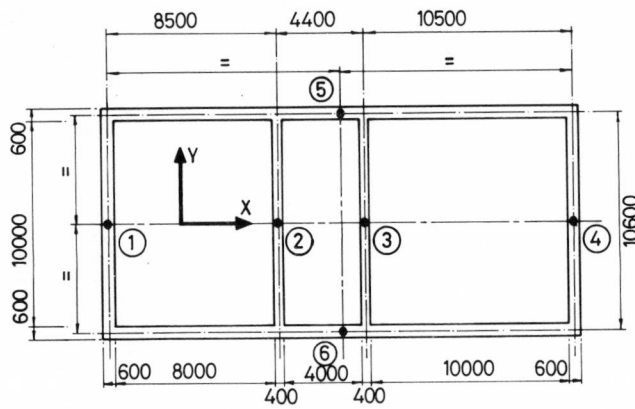

SK 8/42 Location of flexural vertical beam elements.

Equivalent beam elements 1 and 4 (ignoring flanges)

$$I_x = \frac{1}{12} \times 600 \times 11\,200^3 = 70.25\,\text{m}^4$$

Shear area $= 0.8 \times 600 \times 11\,200 = 5.38\,\text{m}^2$

Equivalent beam elements 2 and 3 (ignoring flanges)

$$I_x = \frac{1}{12} \times 400 \times 11\,200^3 = 46.84\,\text{m}^4$$

Shear area $= 0.8 \times 400 \times 11\,200 = 3.58\,\text{m}^2$

Note: The moments of inertia and shear area of equivalent beam elements 1, 2, 3, and 4 about the *y*-axis will be ignored in the analysis.

Stiffness in x-direction
Equivalent beam elements 5 and 6 contribute to stiffness in *x*-direction.

Equivalent beam elements 5 and 6 (ignoring flanges)

$$I_y = \frac{1}{12} \times 600 \times 24\,000^3 = 691.2\,\text{m}^4$$

Shear area $= 0.8 \times 600 \times 24\,000 = 11.52\,\text{m}^2$

Note: The moments of inertia and shear area of equivalent beam elements 5 and 6 about the *x*-axis will be ignored in the analysis.

Step 2 *Find torsional stiffness of wall system*
Equivalent torsional rigidity element of the closed cell structure may be located at the centroid of the cell.

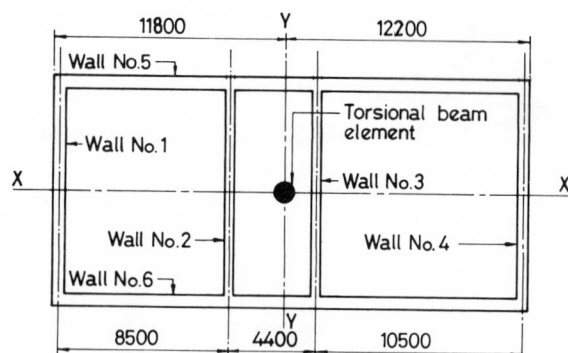

SK 8/43 Location of torsional vertical beam element.

By taking moments of areas about the left-hand edge of the cell,

$$x = \frac{\begin{array}{c}0.6 \times 10 \times 0.3 + 0.4 \times 10 \times 8.8 + 0.4 \times 10 \times 13.2 + 0.6 \times 10 \times 23.7 \\ + \ 2 \times 0.6 \times 24 \times 12\end{array}}{6 + 4 + 4 + 6 + 2 \times 14.4}$$

$$= 11.8 \, \text{m}$$

Areas of cells on centreline are as follows:

$A_1 = 10.6 \times 8.5 = 90.1 \, \text{m}^2$
$A_2 = 10.6 \times 4.4 = 46.6 \, \text{m}^2$
$A_3 = 10.6 \times 10.5 = 111.3 \, \text{m}^2$

$$p_1 = \sum \left(\frac{\text{length of each arm of cell 1}}{\text{thickness of arm}} \right)$$

$$= \left(\frac{2 \times 8.5}{0.6} \right) + \left(\frac{10.6}{0.6} \right) + \left(\frac{10.6}{0.4} \right)$$

$$= 72.5$$

Similarly,

$$p_2 = \left(\frac{2 \times 4.4}{0.6} \right) + \left(\frac{2 \times 10.6}{0.4} \right)$$

$$= 67.7$$

$$p_3 = \left(\frac{2 \times 10.5}{0.6} \right) + \left(\frac{10.6}{0.6} \right) + \left(\frac{10.6}{0.4} \right)$$

$$= 79.2$$

$$p_{1,2} = \frac{10.6}{0.4} = 26.5$$

$$p_{2,3} = \frac{10.6}{0.4} = 26.5$$

Substituting $n = 3$ in the general equations in Section 8.1.1.5.3.

$$X_1'p_1 - X_2'p_{1,2} + 0 = A_1$$
$$-X_1'p_{1,2} + X_2'p_2 - X_3'p_{2,3} = A_2$$
$$0 - X_2'p_{2,3} + X_3'p_3 = A_3$$

or
$$72.5X_1' - 26.5X_2' + 0 = 90.1$$
$$-26.5X_1' + 67.7X_2' - 26.5X_3' = 46.6$$
$$0 - 26.5X_2' + 79.2X_3' = 111.3$$

Solving these equations:

$$X_1' = 2.11\,\text{m}^2 \qquad X_2' = 2.38\,\text{m}^2 \qquad X_3' = 2.2\,\text{m}^2$$

$$J = 4 \sum_1^{n=3} (A_i X_i')$$

$$= 4(A_1 X_1' + A_2 X_2' + A_3 X_3')$$
$$= 4(90.1 \times 2.11 + 46.6 \times 2.38 + 111.3 \times 2.2)$$
$$= 2183.5\,\text{m}^4$$

It is always useful to check at this stage the torsional rigidity of the outer cell, ignoring the internal dividing walls. This gives confidence in the numerical accuracy of the analysis.
For single outer cell:

$$J = \frac{4A^2}{\Sigma(B/h)}$$

$$= \frac{4 \times (23.4 \times 10.6)^2}{(2 \times 23.4/0.6) + (2 \times 10.6/0.6)}$$

$$= 2171\,\text{m}^4$$

This value is very close to the multiple cell rigidity.

Note: The torsional beam element to be used in the analysis will have negligible moments of inertia and shear area.

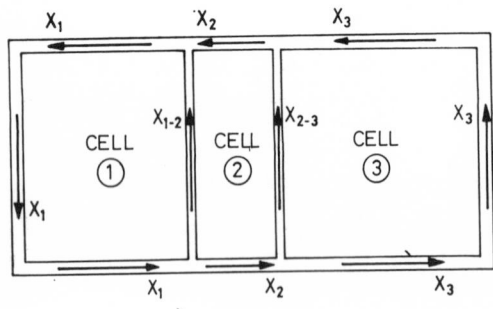

SK 8/44 Torsional shear flow diagram.

$$X_1 = \left(\frac{2T}{J}\right)X_1' = 2.18 \times 10^{-3}T\,\text{kN/m}$$

$$X_2 = \left(\frac{2T}{J}\right)X_2' = 2.18 \times 10^{-3}T\,\text{kN/m}$$

$$X_3 = \left(\frac{2T}{J}\right)X_3' = 2.02 \times 10^{-3} T \, \text{kN/m}$$

Step 3   Carry out modelling for analysis
Follow recommendations in Section 8.1.2.1 and 8.1.2.2.

Step 4   Carry out global analysis
The results of the analysis for different loadings are as follows:

Global torsion
$= +50\,000$ kNm (clockwise) for horizontal load in y-direction
$= +40\,000$ kNm (clockwise) for horizontal load in x-direction

Wall no.	Load case	N (kN)	M_1 (kNm)	V_1 (kN)	Local shear flow (kN/m)
1	DL	3180	—		
	LL	1325	—		
	WL(y)	—	+20 000	+1700	+97
2	DL	980	—		
	LL	615	—		
	WL(y)	—	+13 350	+1200	−12
3	DL	980	—		
	LL	615	—		
	WL(y)	—	+13 350	+1200	+8
4	DL	3180	—		
	LL	1325	—		
	WL(y)	—	+20 000	+1700	−101
5	DL	7200	—		
	LL	3125	—		
	WL(x)	—	+35 000	+3000	+87
6	DL	7200	—		
	LL	3125	—		
	WL(x)	—	+35 000	+3000	−87

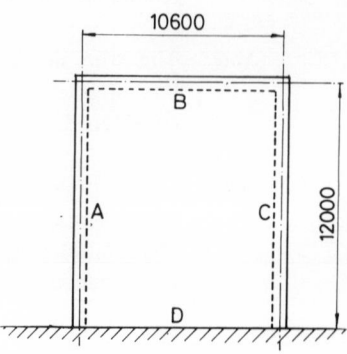

SK 8/45 Elevation − Wall 1.

Step 5 ***Carry out local analysis***

Find out-of-plane internal forces in wall panels (follow Section 8.1.2.2). After analysis the following internal forces are reported:

Wall no.	Line	Load case	M_{OH} (kNm/m)	V_{OH} (kN/m)	M_{OV} (kNm/m)	V_{OV} (kN/m)
1	A	DL	—	—	—	—
		LL	—	—	—	—
		WL	—	—	28	26
	B	DL	20	3	—	—
		LL	10	1	—	—
		WL	32	30	—	—
	C	DL	—	—	—	—
		LL	—	—	—	—
		WL	—	—	28	26
	D	DL	10	3	—	—
		LL	5	1	—	—
		WL	32	30	—	—

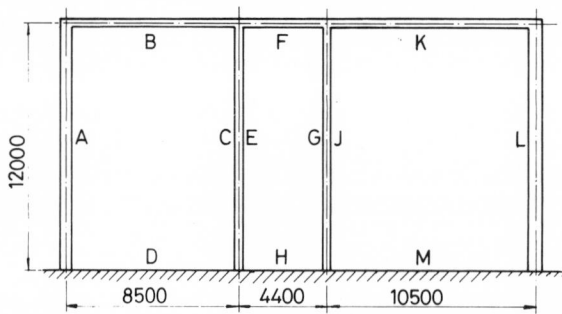

SK 8/46 Elevation – Wall 5.

Wall 1 only will be designed as an example.
Wall 5 panels are shown in sketch to illustrate the location of lines where results should be available for out-of-plane bending.

Note: The example shown uses only one value of bending moment and shear per line of interest. In practice, more values along the line will have to be considered.

Step 6 ***Carry out combination of loading***

Most computer programs used for the analysis will automatically carry out the combination according to principles described in Step 6 of Section 8.2. Reproduced below is the result of one combination of Wall 1.

Load case $LC_3 = 1.4DL + 1.4WL$

Wall 1 subject to WL (y-direction)
In-plane forces (see Step 4 in Section 8.2):

$N = 4452\,kN$
$M_I = 28\,000\,kNm$
$V_I = 2380\,kN$
$Q_I = 1440\,kN\ (97 \times 10.6 \times 1.4)$

Out-of-plane forces (see Step 5 in Section 8.2):

At line D (WL in y-direction),

$M_{OH} = 14\,kNm/m$ (dead load $\times$ 1.4)
$V_{OH} = 42\,kN/m$ (DL $\times$ 1.4)

On flanges (part of Wall 5 and 6),

$M_{OH} = 39\,kNm/m$ (WL in y-direction)
$V_{OH} = 36\,kN/m$ (WL in y-direction)

Step 7   *Check slenderness of wall*
Follow Section 8.1.1.2.1.

Type of wall = unbraced, reinforced in the in-plane direction

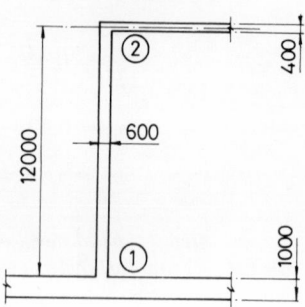

SK 8/47 Section through Wall 1.

$f_{cu} = 30\,N/mm^2$ $f_y = 460\,N/mm^2$

$H_e = \beta\,H_o$

H_o = clear height = 12.0 m

Monolithic construction at top and bottom of wall.
Assume thickness of slab at top is 400 mm.
End conditions are 1 at bottom and 2 at top.

$\therefore\ \ \beta = 1.3$

$H_e = 1.3 \times 12\,000 = 15\,600\,mm$

$\dfrac{H_e}{h} = \dfrac{15\,600}{600} = 26 > 10 < 30$ (limit for unbraced reinforced wall)

Design as slender wall.

Step 8   Find effective width of flanges

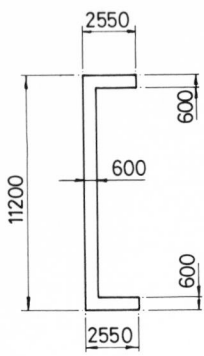

SK 8/48 Plan of Wall 1 showing effective flange widths.

Follow Section 8.1.1.3.

Assume the shear wall behaves as a cantilever.

$$b = \frac{8500}{2} = 4250 \qquad H = 12\,000$$

$$\frac{b}{H} = 0.35$$

$\psi = 0.53$ for loading at top of wall

$b_e = \psi\, b = 0.53 \times 4250 = 2250\,\text{mm}$

Step 9   Find additional out-of-plane moments

Wall is assumed braced in the out-of-plane direction.

$a_u = \beta\, Kh, \; H_e = \beta\, H_o = 0.8 \times 12\,000 = 9600$

Assume $K = 1$ for conservatism.

$$\beta = \frac{1}{2000}\left(\frac{H_e}{h}\right)^2 = \frac{256}{2000} = 0.128$$

$a_u = 0.128h = 76.8\,\text{mm}$

$$\begin{aligned}
M_{\text{add}} &= Na_u \quad \text{(out-of-plane)}\\
&= 4452 \times 0.0768\\
&= 342\,\text{kNm}\\
&= \frac{342}{10.6} = 32.3\,\text{kNm/m}
\end{aligned}$$

Step 10   Design stocky braced reinforced wall

Not applicable.

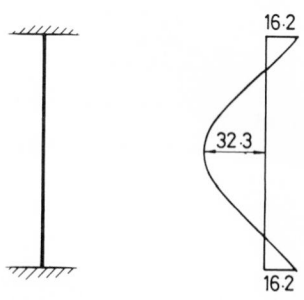

SK 8/49 Moments due to
slenderness.

Out-of-Plane M_{add}

Step 11   *Determine cover to reinforcement*
Maximum size of aggregate = 20 mm
Condition of exposure = mild
Grade of concrete = C30
Minimum cement content = 275 kg/m³
Nominal cover = 25 mm

Step 12   *Design of reinforced wall − rigorous method*

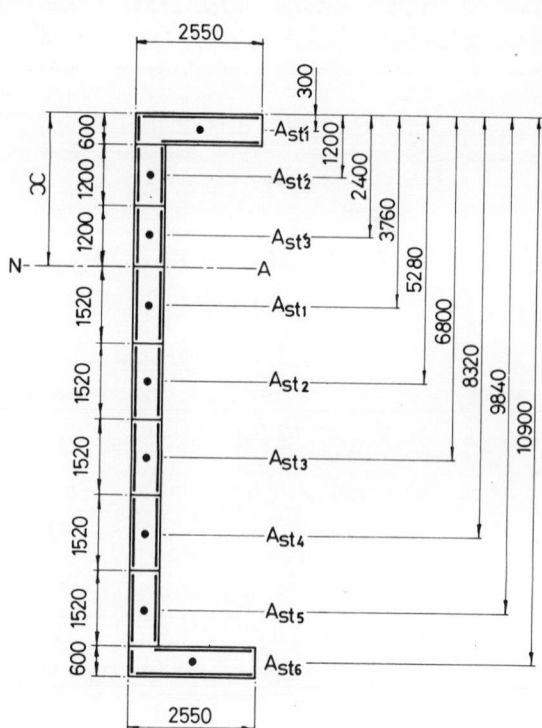

SK 8/50 Elastic stress analysis of
Wall 1.

(1) Assume 0.40% reinforcement in wall.

$$\text{Reinforcement per metre length} = \frac{600 \times 1000 \times 0.40}{100} = 2400 \, \text{mm}^2/\text{m}$$

Use $1200 \, \text{mm}^2$ on each face per metre length.

(2) Assume a value of x for depth of neutral axis from compression face.

$$x = 3000 \, \text{mm} \quad \text{assumed.}$$

(3) Divide compression zone into convenient layers of concrete.
(4) Divide tension zone into convenient layers of steel.
(5) Find the following in completing the table:

$$S = A_c + (m - 1)A'_{st}$$

$$C_1 = (x - a_c)S$$
$$C_2 = (x - a_c)a_c S$$
$$C_3 = (a_t - x)A_{st}$$
$$C_4 = (a_t - x)a_t A_{st}$$

Number	A_c $(\times 10^6)$	A'_{st}	S $(\times 10^6)$	a_c	A_{st}	a_t	C_1 $(\times 10^9)$	C_2 $(\times 10^{12})$	C_3 $(\times 10^6)$	C_4 $(\times 10^{10})$
1	1.53	6120	1.62	300	3648	3760	4.374	1.312	2.772	1.042
2	0.72	2880	0.76	1200	3648	5280	1.368	1.642	8.317	4.392
3	0.72	2880	0.76	2400	3648	6800	0.456	1.094	13.862	9.426
4	—	—	—	—	3648	8320	—	—	19.407	16.147
5	—	—	—	—	3648	9840	—	—	24.952	24.553
6	—	—	—	—	6120	10 900	—	—	48.348	52.699
Totals	2.97	11 880	3.14		24 360		6.198	4.048	117.658	108.259

$$A_c = \frac{\Sigma C_2}{\Sigma C_1} = \frac{4.048 \times 10^{12}}{6.198 \times 10^9} = 653 \, \text{mm}$$

$$A_T = \frac{\Sigma C_4}{\Sigma C_3} = 9201 \, \text{mm}$$

$$\bar{x} = \frac{m\Sigma A_{st}a_t + \Sigma Sa_c}{m\Sigma A_{st} + \Sigma S} = 1735 \, \text{mm}$$

$$e = \frac{M}{N} = \frac{28\,000 \times 10^3}{4452} = 6289 \, \text{mm}$$

$$f_c = \frac{Nx(e + A_T - \bar{x})}{(A_T - A_c)\Sigma (x - a_c)S}$$

$$= \frac{4452 \times 10^3 \times 3000 \times (6289 + 9201 - 1735)}{(9201 - 653) \times 6.198 \times 10^9}$$

$$= 3.467 \, \text{N/mm}^2$$

$$f_{st} = \left[\frac{d - x}{\Sigma\ (a_t - x)A_{st}}\right]\left[\left(\frac{f_c}{x}\right) \Sigma\ (x - a_c)S - N\right]$$

$$= \left(\frac{11\,100 - 3000}{117.658 \times 10^6}\right) \times \left[\left(\frac{3.467}{3000}\right) \times 6.198 \times 10^9 - 4452 \times 10^3\right]$$

$$= 186.6\,\text{N/mm}^2$$

(6) Check $x = \dfrac{d}{1 + \dfrac{f_{st}}{mf_c}} = 2419\,\text{mm}$

Second approximation for x is halfway between first approximation and the check result.

Assume $x = 2700\,\text{mm}$

After carrying out the same tabular exercise as before it is found that:

$f_c = 3.66\,\text{N/mm}^2$
$f_{st} = 182.4\,\text{N/mm}^2$

Check $x = 2570\,\text{mm}$.

No further iteration is necessary.

Check reinforcement in compression flange due to out-of-plane bending

Average compressive stress in flange $= \dfrac{1}{2}(3.7 + 2.8) = 3.25\,\text{N/mm}^2$

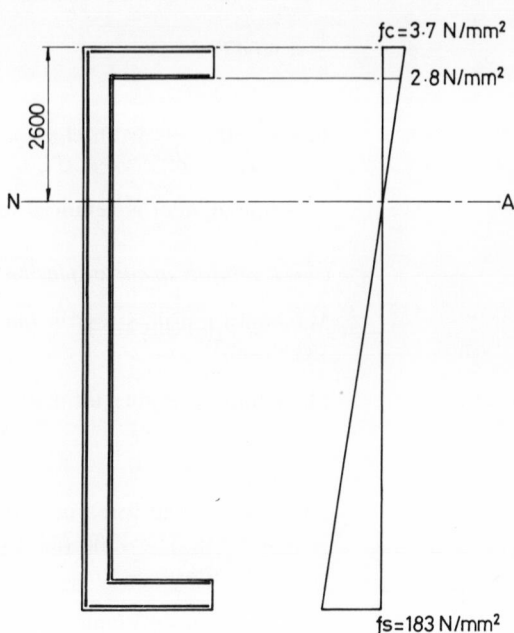

SK 8/51 Elastic analysis – stress
diagram.

Average compressive stress in reinforcement in flange $= 3.25 \times 15$
for $m = 15$
$= 48.75\,\text{N/mm}^2$

Over a unit length of wall,

compressive force, $N = 3.25 \times 600 = 1950\,\text{kN}$

Out-of-plane bending moment due to $DL + WL(y)$ + additional moment
due to slenderness

$$= 39 + \frac{32.3}{2} \quad \text{(see Step 9)}$$

$$= 55.2\,\text{kNm/m}$$

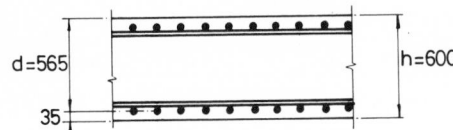

d=565

35

h=600

SK 8/52 Section through wall for
out-of-plane analysis.

Material strengths chosen:

$f_{cu} = 30\,\text{N/mm}^2 \qquad f_y = 460\,\text{N/mm}^2$

$k = \dfrac{d}{h} = \dfrac{565}{600} = 0.95 \quad e = \dfrac{M}{N} = 0.028\,\text{m}$

See Table 11.8.

$$\frac{e}{h} = 0.047$$

For $p = 0.4$, $\dfrac{N}{bh}$ from chart is $13.53 > 3.25\,\text{N/mm}^2$

$\therefore$ Nominal steel is required as per chart.

Check reinforcement in tension flange for out-of-plane bending

Maximum tensile stress in bar due to in-plane bending moment $=$
$183\,\text{N/mm}^2$

Maximum allowable ultimate tensile stress in bars $= 0.87 f_y$
$= 0.87 \times 460$
$= 400\,\text{N/mm}^2$

Available tensile force in bars per metre length of wall per face of wall
$= (400 - 183) \times 1200$ (area on each face)
$= 260.4\,\text{kN/m}$

Maximum out-of-plane bending moment $= 55.2\,\text{kNm/m}$

$$K = \frac{M}{f_{cu}bd^2}$$

$$= \frac{55.2 \times 10^6}{30 \times 1000 \times 565^2}$$

$$= 5.76 \times 10^{-3}$$

$$z = d\left[0.5 + \sqrt{\left(0.25 - \frac{K}{0.9}\right)}\right] \leq 0.95d$$

$$= 0.95d = 537 \, mm$$

Required tensile force in bars $= \dfrac{M}{z} = \dfrac{55.2 \times 10^3}{537} = 102.8 \, kN/m$

This is less than 260.4 kN/m available. Hence, no additional reinforcement is required in tension flange.

Step 13   **Design of reinforced wall – simple method**
Not required. The design principle is exactly similar to beam design and has not been illustrated.

Step 14   **Design of short and squat cantilever wall – deep beam approach**
Not required, because $H/L > 1$.

Step 15   **Check shear**

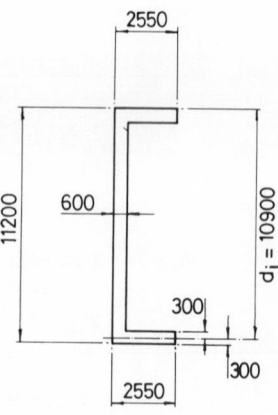

SK 8/53 Shear check of Wall 1.

A_{si} = available tension reinforcement below neutral axis in web ignoring flange
$$= (11.2 - 2.6) \times 2400$$
$$= 20\,640 \, m^2$$

$d_i = 11200 - 300 = 10\,900 \, mm$ (approx.)

$$p_i = \frac{100A_{si}}{h_w d_i} = \frac{100 \times 20\,640}{600 \times 10\,900} = 0.32\%$$

$$f_{cu} = 30\,\text{N/mm}^2$$

From Figs 11.2 to 11.5,

$$v_{ci} = 0.47\,\text{N/mm}^2$$

$$V_i = \text{combined in-plane shear}$$
$$= V_I + Q_I = 2380 + 1440 = 3820\,\text{kN}$$

$$v_i = \frac{V_i}{h_w d_i} = 0.58\,\text{N/mm}^2$$

$$A_{so} = \text{available tension reinforcement for out-of-plane bending}$$
$$= 1200\,\text{mm}^2/\text{m}\quad(\text{each face})$$

$$d_o = \text{effective depth in out-of-plane direction} = 565\,\text{mm}$$

$$p_o = \frac{100 A_{so}}{b d_o} = \frac{100 \times 1200}{1000 \times 565} = 0.21\%$$

From Figs 11.2 to 11.5,

$$v_{co} = 0.4\,\text{N/mm}^2$$

$$V_{OH} = \text{out-of-plane shear coacting with } V_i$$
$$= 4.2\,\text{kN/mm}\quad\text{on the web}$$

$$v_{oh} = \frac{V_{OH}}{b d_o} = 0.007\,\text{N/mm}^2$$

$$\frac{v_i}{v_{ci}} + \frac{v_{oh}}{v_{co}} = \left(\frac{0.58}{0.47}\right) + \left(\frac{0.007}{0.4}\right) = 1.25 > 1$$

Shear reinforcement is necessary for in-plane shear.

Note: Increase of design concrete shear stress due to presence of axial load has been ignored in these calculations but may be allowed as per formula on page 160.

Step 16 *Calculate shear reinforcement*

Case 1: $v_{oh} < v_{co}$

$$v_i' = \left(1 - \frac{v_{oh}}{v_{co}}\right) v_{ci}$$
$$= \left(1 - \frac{0.007}{0.4}\right) \times 0.47$$
$$= 0.46\,\text{N/mm}^2$$

$$V_i' = v_i' h_w d_i = 3008.4\,\text{kN}$$

$$V_i - V_i' = 3820 - 3008.4 = 811.6\,\text{kN}$$

Shear reinforcement is required to resist 811.6 kN.

$$\frac{A_h}{S_h} = \frac{V_{si}}{0.87 f_y d_i}$$

$$= \frac{811.6 \times 10^3}{0.87 \times 460 \times 10\,900}$$

$$= 0.19$$

If $S_h = 300$, then $A_h = 300 \times 0.19 = 57\,\text{mm}^2$ which is $29\,\text{mm}^2$ of horizontal bar on each face at $300\,\text{mm}$ centres, or, $97\,\text{mm}^2$ per metre on each face.

$$\frac{A_v}{S_v} = \frac{A_h}{S_h} = 0.19 \qquad \text{for } f_y = 460\,\text{N/mm}^2$$

Vertical shear reinforcement additional to vertical bars provided for bending is required if available vertical bars have no residual capacity.
In the web $2400\,\text{mm}^2/\text{m}$ vertical bars are available at a maximum average stress level of, say, $160\,\text{N/mm}^2$ (see Step 12). Hence residual capacity available in vertical bars in web $= 0.87 \times 460 - 160 = 240\,\text{N/mm}^2$

Modified A_v/S_v to take into account the residual capacity

$$= 0.19 \times 0.87 \times \frac{460}{240} = 0.32$$

A_v for shear required per metre length of wall $= 320\,\text{mm}^2$ (modified)

Available vertical bars $= 2400\,\text{mm}^2/\text{m}$ in web

Hence no additional vertical bars are necessary to resist shear in web. No shear reinforcement is required in out-of-plane direction.

Step 17 *Check out-of-plane bending about vertical plane*

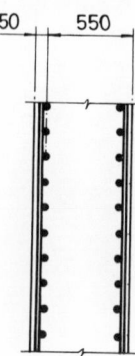

SK 8/54 Out-of-plane bending about vertical plane.

$$M_{OV} = 1.4 \times 28 \qquad \text{(see Step 5)}$$
$$= 39.2\,\text{kNm/m}$$

$$V_{OV} = 1.4 \times 26$$
$$= 36.4\,\text{kN/m}$$

$$K = \frac{M}{f_{cu}bd^2} = \frac{39.2 \times 10^6}{30 \times 1000 \times 550^2} = 4.3 \times 10^{-3}$$

$$z = d\left[0.5 + \sqrt{\left(0.25 - \frac{K}{0.9}\right)}\right] \le 0.95d$$

$$= 0.95d = 522\,\text{mm}$$

$$A_s = \frac{M}{0.87f_yz} = \frac{39.2 \times 10^6}{0.87 \times 460 \times 522} = 187.6\,\text{mm}^2$$

Add to this reinforcement the horizontal reinforcement required in Step 16 for in-plane shear.

Total horizontal reinforcement required on each face (assuming the load WL is reversible in direction)
$$= 187.6 + 97 = 284.6\,\text{mm}^2/\text{m}$$

$$v_{ov} = \frac{V_{ov}}{bd} = \frac{36.4 \times 10^3}{1000 \times 550} = 0.07\,\text{N/mm}^2$$

Shear stress is negligible.

Step 18 *Design of plain walls*
Not required.

Step 19 *Shear check of plain walls*
Not required.

Step 20 *Check minimum reinforcement*
Minimum compression vertical reinforcement in wall = 0.4%
($f_y = 460\,\text{N/mm}^2$)
This has been provided.

Minimum horizontal tension reinforcement on each face
$= 0.13\%$ ($f_y = 460\,\text{N/mm}^2$)

$$= 0.13 \times 1000 \times \frac{600}{100}$$

$$= 780\,\text{mm}^2/\text{m}\quad\text{on each face}$$

This amount is greater than horizontal reinforcement found in Step 17. This reinforcement will be adopted.
Minimum anti-crack reinforcement is 0.25% in both directions on each face. This has been provided.

Step 21 *Check maximum reinforcement*
Not required.

Step 22 *Check containment of wall reinforcement*
Vertical reinforcement is less than 2% of gross concrete area.

Hence requirement is to provide horizontal reinforcement equal to 0.2% of gross cross-sectional area. This is provided.

Vertical bar diameter = 20 mm

Horizontal bar diameter = 10 mm $> \dfrac{1}{4}$ (20 mm)

Step 23   Check early thermal cracking

Crack width limitation = 0.3 mm

(see Step 14 of Section 3.3).
Assume $R = 0.8$ at base.

$T_1 = 32°C$

$\varepsilon_r = 0.8 T_1 \alpha R$
    $= 0.8 \times 32 \times 12 \times 10^{-6} \times 0.8 = 2.46 \times 10^{-4}$

Check horizontal bars for vertical cracks

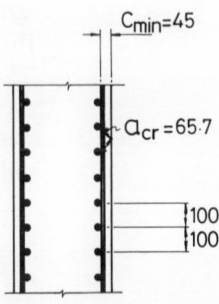

$C_{min}=45$

$a_{cr}=65.7$

100
100

SK 8/55 Crack width for
horizontal bars 10 mm @ 100 c/c.

Assume 10 mm diameter bars at 100 mm centres (785 mm²/m).

$a_{cr} = 65.7$ mm $(1.414 \times 50 - 5 = 65.7)$

take $x = h/2$

$$W_{max} = \frac{3 a_{cr}\, \varepsilon_r}{1 + \dfrac{2(a_{cr} - C_{min})}{h - x}}$$

$$= \frac{3 \times 65.7 \times 2.46 \times 10^{-4}}{1 + \dfrac{2(65.7 - 45)}{300}}$$

$$= 0.04 \text{ mm} < 0.3 \text{ mm}   \text{OK}$$

Check vertical bars for horizontal cracks
Assume 20 mm diameter bars at 250 mm centres vertically (1256 mm²/m each face).

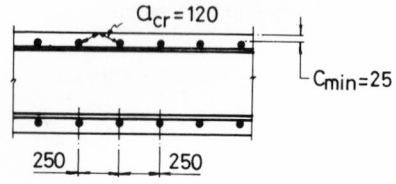

SK 8/57 Sketch to find a_{cr}.

SK 8/56 Crack width for vertical
bars 20 mm @ 250 c/c.

$$a_{cr} = 120 \, \text{mm}$$

$$W_{max} = \frac{3 \times 120 \times 2.46 \times 10^{-4}}{1 + \dfrac{2(120 - 25)}{300}}$$

$$= 0.05 \, \text{mm} < 0.3 \, \text{mm} \quad \text{OK}$$

Step 24 **Clear spacings of bars in tension**
Reinforcement provided is 20 mm diameter at 250 mm centres both faces
vertically and 10 mm diameter at 100 mm centres both faces horizontally.
These spacings satisfy the requirements according to Step 13 of Section
3.3.

Step 25 **Connections**
Follow Chapter 10.

Chapter 9
Design of Flat Slabs

9.0 NOTATION

A	Area of column or area of effective column head
A_{st}	Area of steel in tension
b_e	Effective width of slab for transfer of moment to edge column
C_x	Size of a rectangular column in x-direction
C_y	Size of a rectangular column in y-direction
d	Effective depth of tensile reinforcement
d_h	Depth of column head
f_y	Characteristic yield strength of reinforcement
f_{cu}	Characteristic cube strength of concrete at 28 days
h_c	Effective diameter of column or effective column head
G_k	Characteristic dead load
l_c	Dimension of column in direction of l_h
l_h	Effective dimension of column head
l_x	Shorter span framing onto columns
l_y	Longer span framing onto columns
l_{ho}	Actual dimension of column head
$l_{h,max}$	Maximum dimension of column head taking 45° dispersion
l_1	Centre-to-centre of column in direction of span being considered
l_2	Centre-to-centre of column perpendicular to direction of span being considered
M'	Design limit moment at $h_c/2$
M_t	Moment transferred to column by frame analysis
$M_{t,max}$	Limiting moment between flat slab and edge column
n	Total ultimate load per unit area on flat slab
Q_k	Characteristic live load
V_t	Calculated shear from analysis
V_{eff}	Effective shear at column/slab interface
W_k	Characteristic wind loading
x	Length of side of a perimeter parallel to axis of bending

9.1 DEFINITIONS

Flat slab is a reinforced concrete slab supported by columns with, or without, drops. The columns may be with, or without, column heads.

Drop is a local thickening of the slab in the region of the column.

403

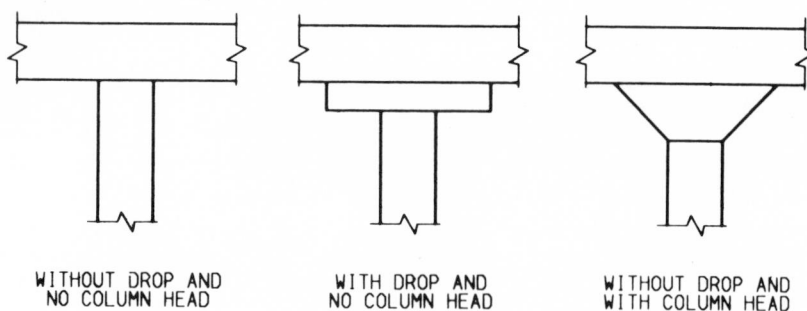

WITHOUT DROP AND
NO COLUMN HEAD

WITH DROP AND
NO COLUMN HEAD

WITHOUT DROP AND
WITH COLUMN HEAD

SK 9/1 Flat slab – section. **SK 9/2** Flat slab – section. **SK 9/3** Flat slab – section.

Column head is a local enlargement of the column at the junction with the slab.

9.2 ANALYSIS OF FLAT SLABS

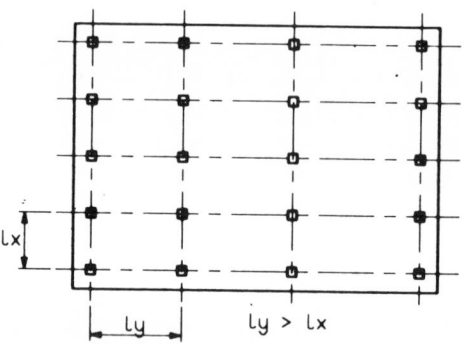

SK 9/4 Typical plan of flat slab.

Flat slabs are usually supported by a rectangular arrangement of columns. The analysis may be carried out by an equivalent frame method or by the use of a finite element computer code. When using the equivalent frame method the ratio of the longer to the shorter span should not exceed 2. The analysis for uniformly distributed vertical load may be carried out by using Tables 9.1 to 9.6.

The properties of the flat slab for analysis are similar to those already discussed for solid slabs in Chapter 3.

9.2.1 Effective dimension of column head

l_h = effective dimension of head
l_{ho} = actual dimension of head
$l_{h,max} = l_c + 2(d_h - 40)$
l_c = dimension of column in direction of l_h
d_h = depth of head
l_h = is taken as the lesser of l_{ho} or $l_{h,max}$

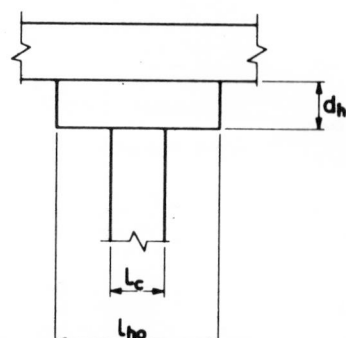

SK 9/5 Flat slab – definitions.

Note: This means that the maximum dimension is limited by a 45° dispersion of column up to 40 mm below the slab.

9.2.2   Effective diameter of a column head

$h_c = (4A/\pi)^{\frac{1}{2}} \leq 0.25 l_x$
h_c = effective diameter of column or column head
A = area of column or area of effective column head as defined by l_h
l_x = shortest span framing onto column

h_c should not be taken greater than one-quarter of shortest span of slab framing into column.

9.2.3   Drops

Drops will be effective in the analysis if the smaller dimension of the drop is at least one-third of the smallest span of surrounding panels.
For the checking of punching shear, this limitation does not apply.

9.2.4   Load combinations for analysis

$LC_1 = 1.4G_k + 1.6Q_k$ on all spans
$LC_2 = 1.4G_k + 1.6Q_k$ on alternate spans and other spans loaded with
$\quad\quad\quad 1.0G_k$

where G_k = characteristic dead load
$\quad\quad\quad Q_k$ = characteristic live load.

9.2.5   Effective width of slab for analysis

For vertical loading assume full width of panel between columns for frame analysis.
For horizontal loading as a frame assume stiffness of half width of panel.

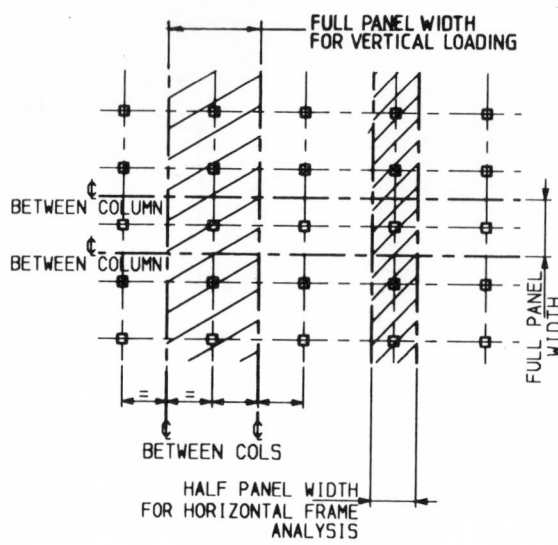

SK 9/6 Plan of flat slab showing
panel widths for analysis.

The analysis should be carried out using a computer program or a moment distribution method. The analysis may also be carried out for uniformly distributed vertical loads using Tables 9.1 to 9.6.

The analysis may be carried out using Table 3.13 of BS 8110: Part 1: 1985[1] provided the lateral stability is not dependent on the slab−column connection and loading on the flat slab for the design is based on a single load case, i.e. LC_1, the ratio of Q_k/G_k does not exceed 1.25, Q_k does not exceed $5\,kN/m^2$, and there are at least three rows of panels.

9.3 DESIGN OF FLAT SLABS

The design may be based on the negative moment at $h_c/2$ from the centreline of the column. But this negative moment will have to be modified if the sum of the positive design moment and the average negative design moment is less than the following expression:

$$M' = \left(\frac{nl_2}{8}\right)\left(l_1 - \frac{2h_c}{3}\right)^2$$

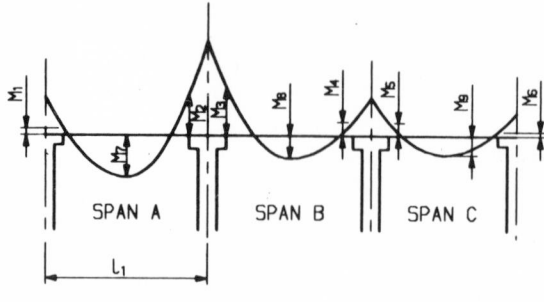

SK 9/7 Negative moment
limitation for flat slabs − section.

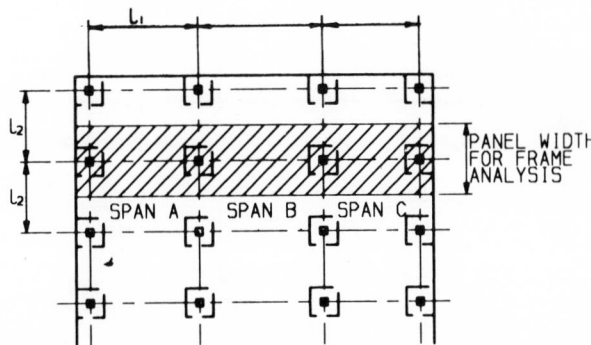

SK 9/8 Typical plan of flat slab – negative moment limitation.

where l_1 = centre-to-centre of column in direction of span being considered

l_2 = centre-to-centre of column perpendicular to direction of span being considered

n = total ultimate load on slab (kN/m^2).

To give an example:

For Span A

$0.5\,(M_1 + M_2) + M_7 \geq M'$

For Span B

$0.5(M_3 + M_4) + M_7 \geq M'$

Increase negative moments M_1, M_2, M_3, etc. until these conditions are satisfied.

9.3.1 Division of panels

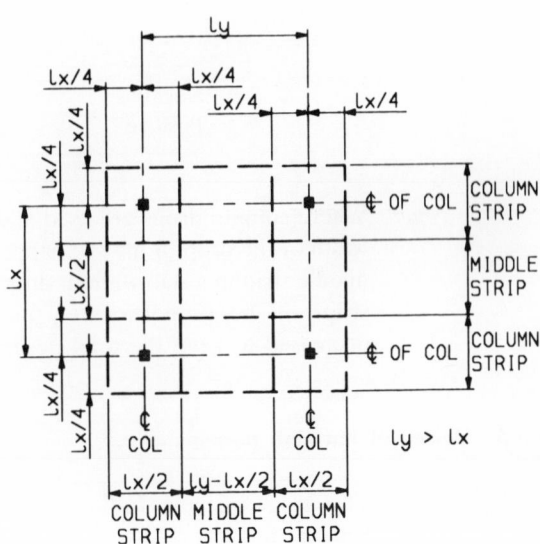

SK 9/9 Flat slab – division of strips.

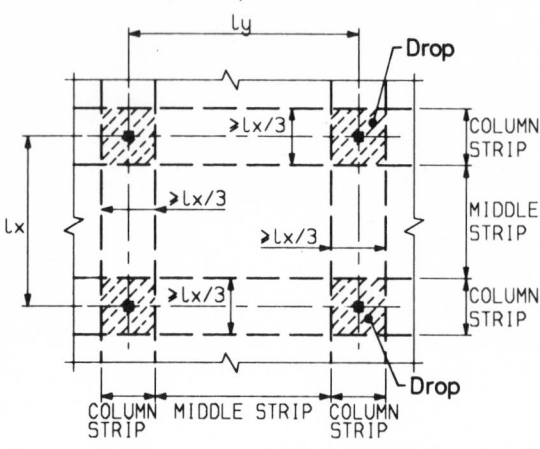

PLAN OF SLAB WITH DROP

IGNORE DROP IF DROP WIDTH < Lx/3

SK 9/10 Flat slab − division of strips.

Panels are divided into column strips and middle strips as shown.

For slab without drop the column strip is $l_x/4$ wide on either side of the centreline of column, where l_x is the shorter span.

For slab with drop the column strip is the size of the drop. Ignore drop if the size of the drop is less than $l_x/3$.

9.3.2   Division of moments between columns and middle strips

The moments obtained from analysis of frames should be divided as follows (these percentages are for slabs without drops):

	Column strip	Middle strip
Negative	75%	25%
Positive	55%	45%

Note: Where column drops are used and column strips are determined from the width of the drop, it may so happen that the middle strip is bigger than the middle strip in a slab without drop. In that case the moments in the middle strip will be proportionately increased and those in the column strip decreased to keep the total positive and negative moment unchanged.

9.3.3   Design of flat slab panels

The design is similar to the design of slabs and the worked examples are in Chapter 3.

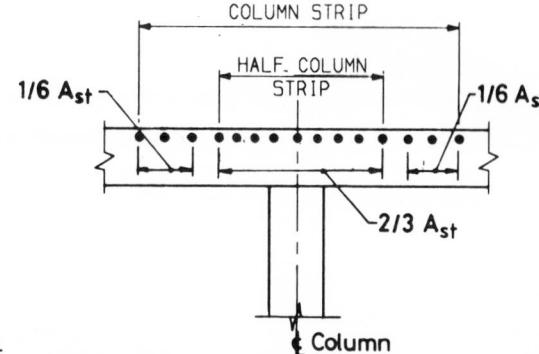

SK 9/11 Detailing of reinforcement in flat slabs.

Internal panels and edge panels
Two-thirds of the negative support reinforcement in the column strip should be placed in half the width of the column strip centred over the column.

9.3.4   Moment connection to edge column

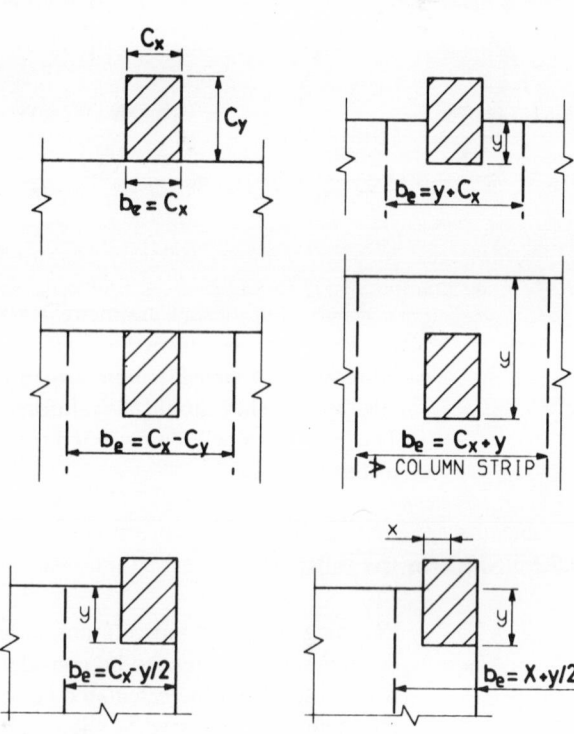

SK 9/12 Effective width of slab for moment connection to edge column.

See sketches above to find effective width of slab b_e for transfer of moment between flat slab and edge column. This moment should be limited to

$$M_{t,max} = 0.15b_e d^2 f_{cu}$$

where d = effective depth of top reinforcement in column strip.

The moment $M_{t,max}$ should not be less than half the design moment from an equivalent frame analysis or 70% of the design moment from a grillage or finite element analysis. The structural arrangement may be changed if $M_{t,max}$ does not satisfy the above condition.

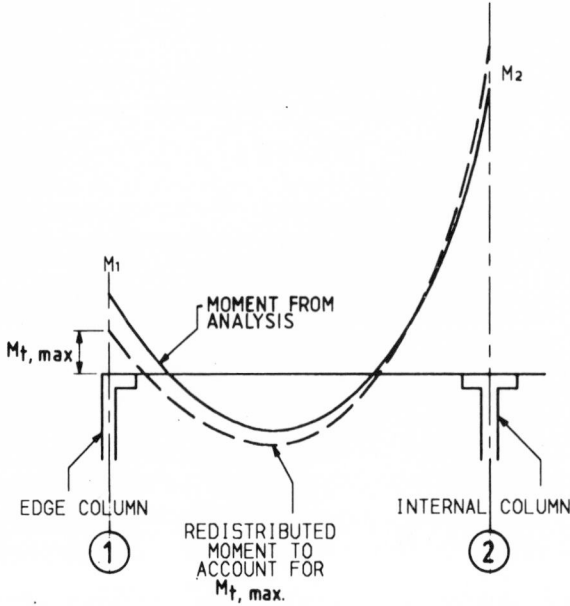

SK 9/13 Insufficient moment transfer capacity at edge column.

Where the design moment is larger than $M_{t,max}$, redistribution of moment may be carried out to reduce the design moment to $M_{t,max}$. Otherwise, to transfer moments in excess of $M_{t,max}$ to edge column, the edge of the slab should be reinforced by an edge beam or an edge strip. The edge beam will be designed to carry the additional moment by torsion to the column.

9.3.5 Shear in flat slabs

Punching shear around columns should be checked according to Step 7 of Section 3.3. The shear to be considered for the punching shear calculation is increased from the calculated column shear by an amount dependent on the moment transferred to the column by frame action.

For internal column connections,

$$V_{eff} = V_t\left(1 + \frac{1.5M_t}{V_t x}\right)$$

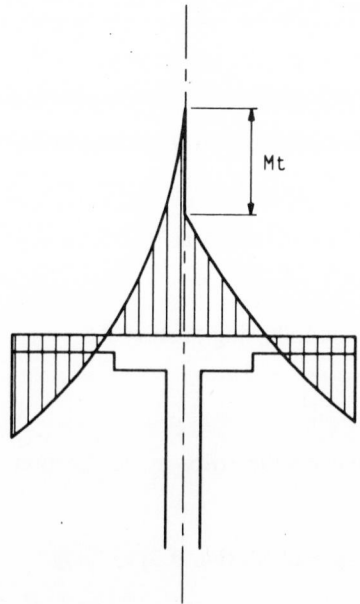

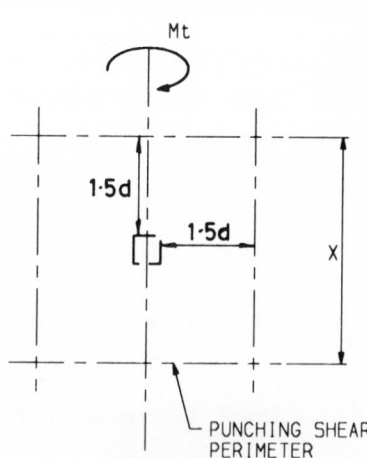

SK 9/14 Moment diagram at an internal column of a flat slab.

SK 9/15 Definition of dimension x.

where V_t = calculated shear from analysis

M_t = moment transferred to column by frame analysis

x = length of side of perimeter considered parallel to axis of bending.

Alternatively,

$V_{eff} = 1.15V_t$ for simplicity

For corner column connections,

$V_{eff} = 1.25V_t$

For edge column connections,

$V_{eff} = 1.25V_t$ for bending about axis parallel to free edge

$V_{eff} = V_t\left(1.25 + \dfrac{1.5M_t}{V_t x}\right)$ for bending about axis perpendicular to free edge

Alternatively,

$V_{eff} = 1.4V_t$

The moment M_t may be reduced by 30% where the equivalent frame analysis is used and both load cases LC_1 and LC_2 have been considered. The shear reinforcement will be calculated according to Step 7 of Section 3.3.

9.4   STEP-BY-STEP DESIGN PROCEDURE FOR FLAT SLABS

Step 1   Carry out analysis as in Section 9.2.

Step 2   Find moment connection to edge column as per Section 9.3.4 and redistribute moments if necessary.

Step 3   Draw bending moment diagrams and calculate moments at $h_c/2$ following Section 9.3.

Step 4   Check limitation of negative design moments following Section 9.3.

Step 5   Carry out division of panels as in Section 9.3.1.

Step 6   Divide moments between column strips and middle strips as per Section 9.3.2.

Step 7   Determine cover to reinforcement (see Step 3 of Section 3.3).

Step 8   Carry out design for flexure as per Step 4 of Section 3.3.

Step 9   Distribute reinforcement as per Section 9.3.3.

Step 10   **Check punching shear stress**
Follow Step 7 of Section 3.3.

Step 11   **Check span/effective depth ratio**
Follow Step 11 of Section 3.3 for slabs with drops. For slabs without drops follow the same step but multiply l_e/d from Table 11.3 by 0.9.

Step 12   **Curtailment of bars**
Follow Step 12 of Section 3.3.

Step 13   **Spacing of bars**
Follow Step 13 of Section 3.3.

Step 14   **Check early thermal cracking**
Follow Step 14 of Section 3.3.

Step 15   **Calculate minimum reinforcement**
Follow Step 9 and Step 15 of Section 3.3.

Step 16   **Calculate flexural crack width**
Follow Step 16 of Section 3.3.

Step 17   **Design of connections**
Follow Chapter 11.

9.5   WORKED EXAMPLE

Example 9.1   Flat slab construction for a sports hall

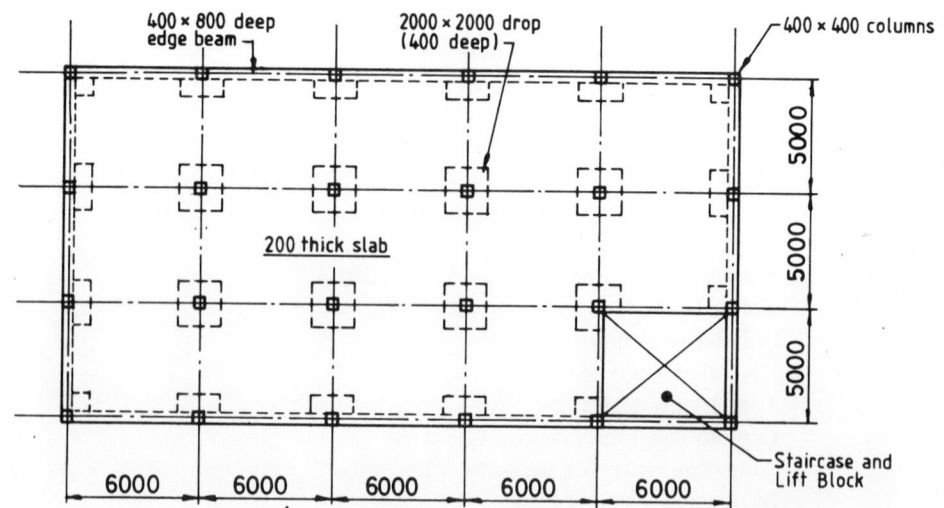

SK 9/16 Plan on first floor.

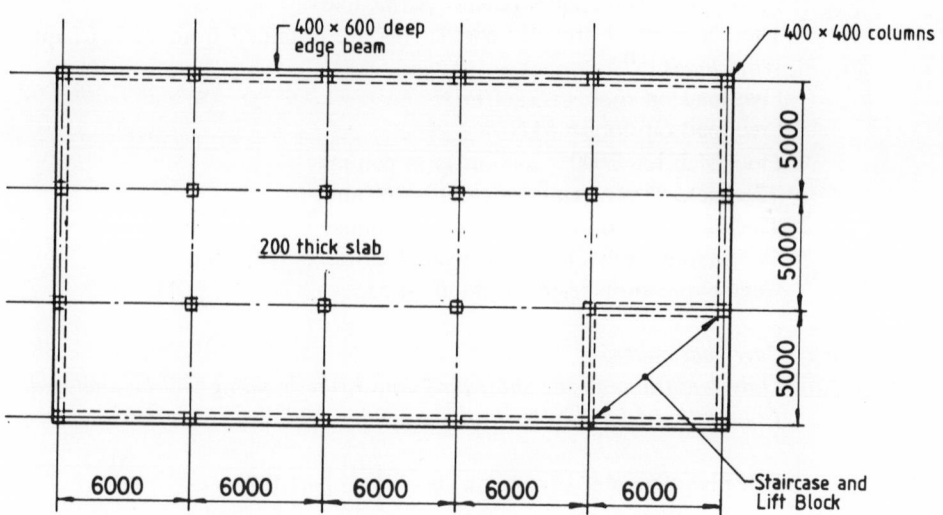

SK 9/17 Plan on roof.

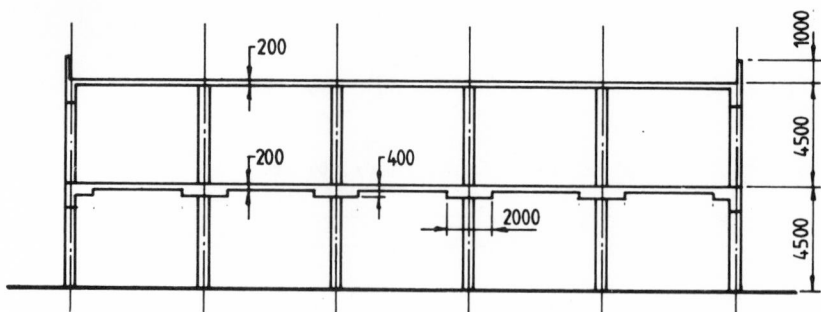

SK 9/18 Section through building.

Two-storey building plan size: $15\,\text{m} \times 30\,\text{m}$
Column grid: $5\,\text{m} \times 6\,\text{m}$
Column size: 400×400
Height of building $= 10\,\text{m}$ overall
Topography factor, $S_1 = 1.0$
Ground roughness factor, $S_2 = 0.95$
Statistical factor for wind, $S_3 = 1.0$
Basic wind speed $= 42\,\text{m/s} = V$
Design wind speed $= S_1 S_2 S_3\ V = 40\,\text{m/s}$
$q = kV_s^2 = 1\,\text{kN/m}^2$
$C_{pe} = +0.7$ and -0.3
$C_{pi} = -0.3$ (four faces equally permeable)
The above wind pressure coefficients are obtained from CP3: Chapter V
Wind loads.[14]
Live load on roof $= 1.5\,\text{kN/m}^2$
Live load on floor $= 5\,\text{kN/m}^2$
Floor slab has 2000×2000 drop at columns
Thickness of roof and floor slab $= 200\,\text{mm}$
Thickness at drop of floor slab $= 400\,\text{mm}$
Continuous perimeter edge beam 400 wide $\times$ 800 deep
Centre-to-centre height of floor $= 4.5\,\text{m}$

Step 1   *Carry out analysis*
Only one frame in the short direction of the building will be analysed.
Column head has not been used.

Effective diameter of column, $h_c = \left(\dfrac{4A}{\pi}\right)^{\frac{1}{2}} \leq 0.25 l_x$

$A = 400 \times 400 = 160\,000\,\text{mm}^2$

$\therefore\ \ h_c = \left(\dfrac{4 \times 160\,000}{\pi}\right)^{\frac{1}{2}}$

$\qquad = 451\,\text{mm} < 0.25 l_x$

$\quad l_x = 5000\,\text{mm}$

$0.25 l_x = 1250\,\text{mm}$

Drop of 2000 mm in floor slab is greater than $l_x/3 = 1667$ mm. Drop will be effective in the distribution of moment.

Loading
Frames in short direction are 6 m apart.

Roof slab

G_k = characteristic dead load = $0.2 \times 25 = 5$ kN/m^2

$Q_k = 1.5$ kN/m^2

$LC_1 = 1.4G_k + 1.6Q_k = 9.4$ kN/m^2 = 56.4 kN/m

$LC_2 = 9.4$ kN/m^2 and 5 kN/m^2 on alternate spans
or $LC_2 = 56.4$ kN/m and 30 kN/m on alternate spans

Floor slab

G_k = 5 kN/m^2 at slab without drop
 = 10 kN/m^2 at slab with drop (area 2 m $\times$ 2 m)
 = 30 kN/m or $30 + 5 \times 2 = 40$ kN/m

Q_k = 5 kN/m^2
 = 30 kN/m

$LC_1 = 1.4G_k + 1.6Q_k$

LC_2 = alternate spans loaded with LC_1 and dead load only

Columns

Horizontal load on columns is due to wind load at the rate of 1 kN/m^2 which is equivalent to 6 kN/m on the column. The wind loading analysis will be carried out separately and combined later with the vertical loading because the stiffness of the slab to resist horizontal loading is half of that to resist vertical loading.
Load cases with wind load W_k are as follows:

$LC_3 = 1.4G_k + 1.4W_k$
$LC_4 = 1.2G_k + 1.2Q_k + 1.2W_k$

Frame analysis using a computer software
E = Young's modulus = 28×10^6 kN/m^2
12 joints 14 members
Joints 1, 4, 7 and 10 rigidly fixed.
Column size 400×400
Slab size 6000×200 (deep)
Load cases:
B_1 – dead load
B_2 to B_7 – live loads on members 9 to 14 respectively
B_8 – wind load

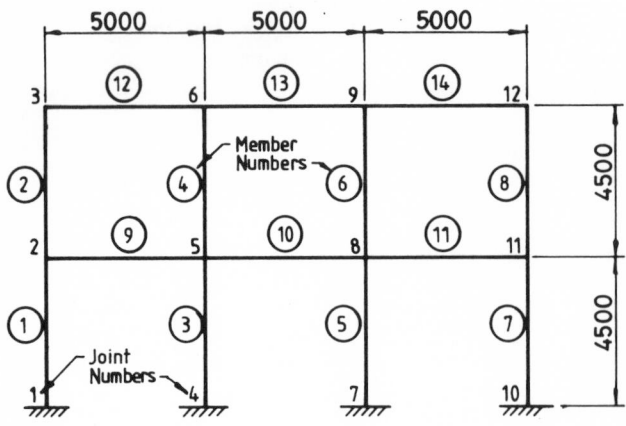

SK 9/19 Frame diagram for analysis.

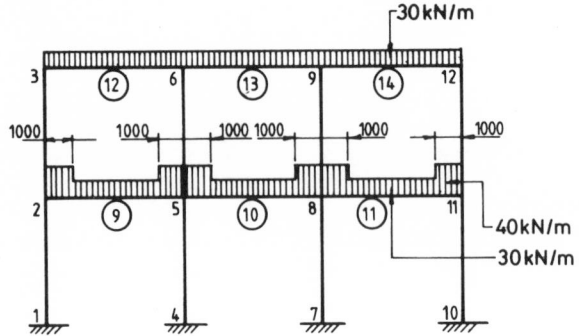

SK 9/20 Dead load on frame (B_1).

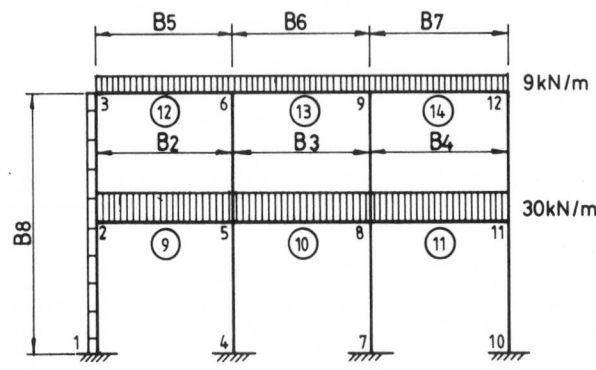

SK 9/21 Basic live loads B_2 to B_8.

Combinations:
$$C_1 = 1.4B_1 + 1.6(B_2 + B_3 + B_4 + B_5 + B_6 + B_7)$$
$$C_2 = 1.4B_1 + 1.6(B_2 + B_4 + B_5 + B_7)$$
$$C_3 = 1.4B_1 + 1.6(B_3 + B_6)$$
$$C_4 = 1.4B_1 + 1.6(B_2 + B_3 + B_4)$$

Output from analysis
Envelope of load cases (vertical loads)
Elastic analysis − no redistribution

Floor slab	Member 9		
Joint	Maximum BM	Shear	Combination
2	130.80	228.5	C_2
5	215.5	258.2	C_4
Midspan	131.2	−	C_2

Floor slab	Member 10		
Joint	Maximum BM	Shear	Combination
5	199.9	239.0	C_4
8	199.9	239.0	C_4
Midspan	112.5	−	C_3

Roof slab	Member 12		
Joint	Maximum BM	Shear	Combination
3	67.8	130.0	C_2
6	129.0	153.5	C_1
Midspan	82.0	−	C_2

Roof slab	Member 13		
Joint	Maximum BM	Shear	Combination
6	121.3	141.0	C_1
9	121.3	141.0	C_1
Midspan	60.4	−	C_3

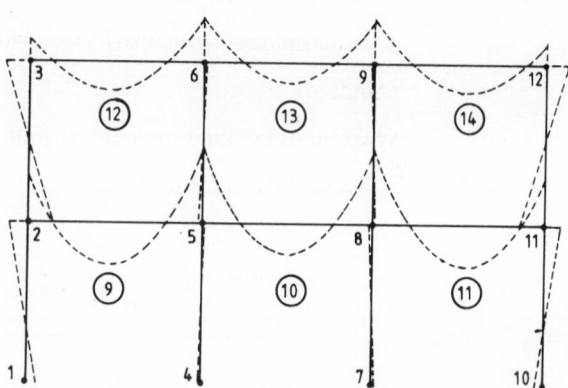

SK 9/22 Combination C_1 − bending moment diagram.

Envelope of load cases (vertical + horizontal loads) (The analysis of horizontal load is carried out with half stiffness of slab)
Elastic analysis − no redistribution
Combinations:
$$C_5 = 1.4B_1 + 1.4B_8$$
$$C_6 = 1.2B_1 + 1.2(B_2 + B_3 + B_4 + B_5 + B_6 + B_7) + 1.2B_8$$

Floor slab	Member 9		
Joint	Maximum BM	Shear	Combination
2	121.9	185.9	C_6
5	191.4	215.2	C_6
Midspan	95.8	—	C_6

Floor slab	Member 10		
Joint	Maximum BM	Shear	Combination
5	176.0	198.6	C_6
8	176.0	198.6	C_6
Midspan	71.4	—	C_6

Roof slab	Member 12		
Joint	Maximum BM	Shear	Combination
3	57.0	108	C_6
6	112.2	128.9	C_6
Midspan	67.1	—	C_6

Roof slab	Member 13		
Joint	Maximum BM	Shear	Combination
6	108.1	119.8	C_6
9	108.1	119.8	C_6
Midspan	45.4	—	C_6

Carry out redistribution of moment:

Maximum bending moment at joint $5 = 215.5\,\text{kNm}$

Assume 20% redistribution.

Set plastic moment capacity at joint $5 = 0.8 \times 215.5 = 172.4\,\text{kNm}$

Similarly

maximum bending moment at joint $6 = 129\,\text{kNm}$

Assume 20% redistribution.

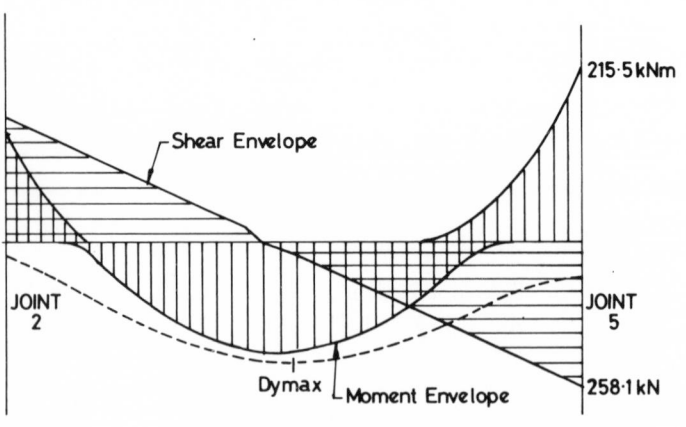

SK 9/23 Shear and moment envelope for member 9.

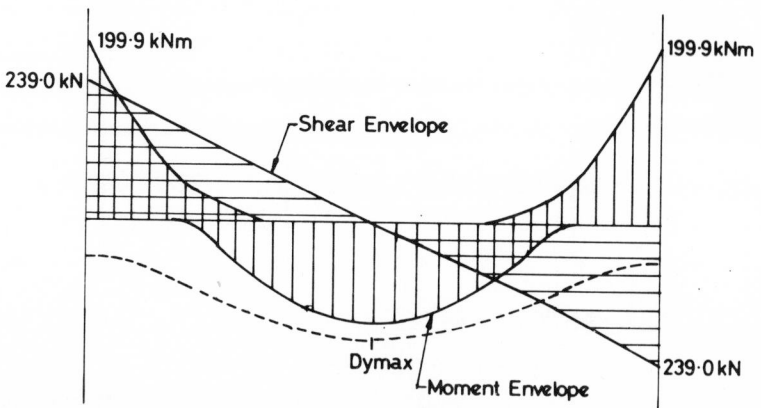

SK 9/24 Shear and moment envelope for member 10.

Set plastic moment capacity at joint $6 = 0.8 \times 129 = 103.2 \, \text{kNm}$

The following steps of reanalysis of frame are carried out:

Step 1: For one combination at a time increase live load on span until plastic moment is reached at a joint in a member. Plastic moment capacity of members on first floor is 172.4 kNm and member on roof is 103.2 kNm.
Step 2: Release joint where plastic moment is reached and increase loading until plastic moment capacity is reached at another joint.
Step 3: Progressively release joints and increase live load until full complement of live load is on structure.
Step 4: Find cumulative effect of all incremental live load on structure.

The following tables become useful if a non-linear finite element computer package is not available.

Frame types:

F_1 = no member end releases
F_2 = member in F_1 nos 12 and 14 ends released at joints 6 and 9
F_3 = member in F_2 nos 9 and 11 ends released at joints 5 and 8
F_4 = member in F_3 no. 13 ends released at joints 6 and 9
F_5 = member in F_4 no. 10 ends released at joints 5 and 8

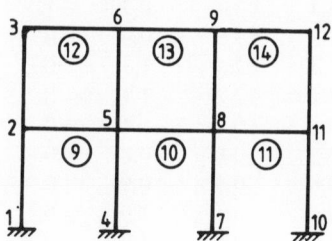

SK 9/25 Frame type F_1.

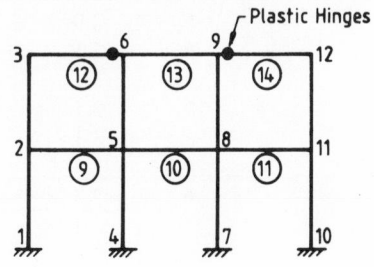

SK 9/26 Frame type F_2.

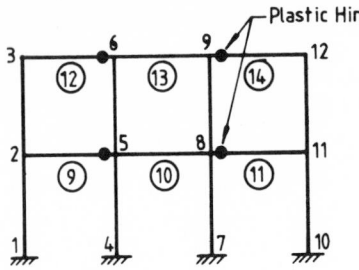

SK 9/27 Frame type F_3.

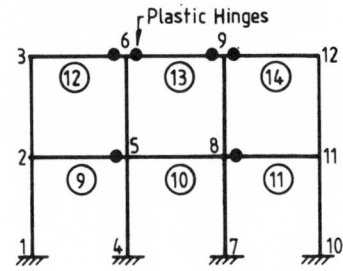

SK 9/28 Frame type F_4.

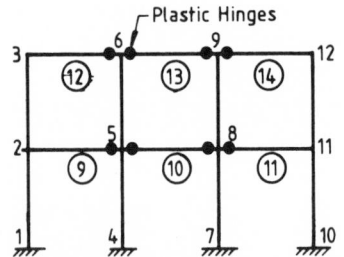

SK 9/29 Frame type F_5.

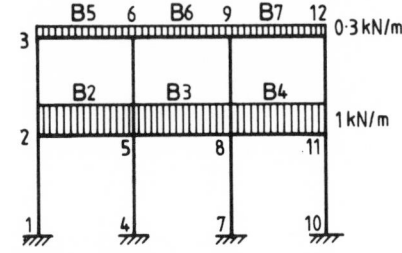

SK 9/30 Unit live load on frame.

Combination $C_7 = 1.4 \times$ dead load   or $1.4B_1$

The method is illustrated for combinations C_1' and C_2' only.

$$C_1' = 1.6(B_2' + B_3' + B_4' + B_5' + B_6' + B_7')$$
$$C_2' = 1.6(B_2' + B_4' + B_5' + B_7')$$

$B_2', B_3', B_4' = 1 \, \text{kN/m}$
$B_5', B_6', B_7' = 0.3 \, \text{kN/m}$

Combination	Frame type	Member end bending moments (kNm)							
		Member 9		Member 10		Member 12		Member 13	
		Joint 2	Joint 5	Joint 5	Joint 8	Joint 3	Joint 6	Joint 6	Joint 9
C_7	F_1	62.7	102.5	95.5	95.5	44.7	97.5	91.6	91.6
C_1'	F_1	2.0	3.7	3.4	3.4	0.7	1.1	1.0	1.0
C_1'	F_2	2.1	3.6	3.5	3.5	0.9	0	0.5	0.5
C_1'	F_3	3.1	0	2.4	2.4	1.1	0	0.8	0.8
C_1'	F_4	3.1	0	2.2	2.2	1.1	0	0	0
C_1'	F_5	3.1	0	0	0	1.1	0	0	0
C_2'	F_1	2.3	2.7	0.8	0.8	0.8	0.8	0.2	0.2
C_2'	F_2	2.3	2.6	0.8	0.8	0.9	0	-0.2	-0.2
C_2'	F_3	3.1	0	0	0	1.0	0	0	0
C_2'	F_4	3.1	0	0	0	1.0	0	0	0
C_2'	F_5	3.1	0	0	0	1.0	0	0	0

$C'_1 = 1$ unit of live load combination in combination C_1 i.e. $C'_1 = 1$ kN/m of B_2, B_3 and B_4 and 9/30 kN/m of B_5, B_6 and B_7.

Full compliment of B_2, B_3 and B_4 is 30 kN/m and of B_5, B_6 and B_7 is 9 kN/m.

Plastic moment at joint 5 is fixed at 172 kNm
Plastic moment at joint 6 is fixed at 103.2 kNm
Dead load moment at joint 6 = 97.5 kNm

Each unit of combination C_1 produces 1.1 kNm at joint 6 for frame type F_1. Therefore

units of live load in combination C_1 required to form first plastic hinges at joint 6 and joint 9 in members 12 and 14

$$= \frac{103.2 - 97.5}{1.1} = 5 \text{ units, say}$$

Frame type F_2 has joints released at joints 6 and 9 for members 12 and 14. After 5 units of combination C_1 the bending moments at joints are as follows:

Frame type F_1

Member 9	Joint 2	$62.7 + 5 \times 2.0 = 72.7$ kNm	
	Joint 5	$102.5 + 5 \times 3.7 = 121.0$ kNm	
Member 10	Joint 5	$95.5 + 5 \times 3.4 = 112.5$ kNm	
Member 12	Joint 3	$44.7 + 5 \times 0.7 = 48.2$ kNm	
	Joint 6	$97.5 + 5 \times 1.1 = 103$ kNm	*plastic
Member 13	Joint 6	$91.6 + 5 \times 1.0 = 96.6$ kNm	

Units of live load in combination C_1 to form second plastic hinges at joints 5 and 8 in members 9 and 11

$$= \frac{172 - 121}{3.6} = 14 \text{ units of combination } C_1$$

Total number of units of C_1 to cause plastic hinges at joints 5 and 8 in members 9 and 11 is 19.
After 19 units of combination C_1, the bending moments at joints are as follows:

Frame type F_2

Member 9	Joint 2	$72.7 + 14 \times 2.1 = 102.1$ kNm	
	Joint 5	$121 + 14 \times 3.6 = 171.4$ kNm	*plastic
Member 10	Joint 5	$112.5 + 14 \times 3.5 = 161.5$ kNm	
Member 12	Joint 3	$48.2 + 14 \times 0.9 = 60.8$ kNm	
	Joint 6	plastic $= 103$ kNm	*plastic
Member 13	Joint 6	$96.6 + 14 \times 0.5 = 103.6$ kNm	*plastic

Joint 5 of member 9 and joint 6 of member 13 have gone plastic simultaneously at 19 units of combination C_1. Therefore *frame type F_3 is not considered*.

After 24 units of combination C_1, the bending moments at joints are as follows:

Frame type F_4

Member 9	Joint 2	$102.1 + 5 \times 3.1 = 117.6\,kNm$	
	Joint 5	plastic $= 171.4\,kNm$	*plastic
Member 10	Joint 5	$161.5 + 5 \times 2.2 = 172.5\,kNm$	*plastic
Member 12	Joint 3	$60.8 + 5 \times 1.1 = 66.3\,kNm$	
	Joint 6	plastic $= 103\,kNm$	*plastic
Member 13	Joint 6	plastic $= 103.6\,kNm$	*plastic

Frame type F_5
After 30 units of combination C_1, the bending moments at joints are as follows:

Member 9	Joint 2	$117.6 + 6 \times 3.1 = 136.2\,kNm$
	Joint 5	$= 171.4\,kNm$
Member 10	Joint 5	$= 172.5\,kNm$
Member 12	Joint 3	$66.3 + 6 \times 1.1 = 72.9\,kNm$
	Joint 6	$= 103.0\,kNm$
Member 13	Joint 6	$= 103.6\,kNm$

Formula for calculating midspan bending moment and shear

$C_7 + 5$ units of C'_1 $(F_1) + 14$ units of C'_1 $(F_2) + 5$ units of C'_1 $(F_4) + 6$ units of C'_1 (F_5)

Member 9 — combination C_1 (20% redistribution)

Midspan moment $= 56.3 + 2.2 \times 5 + 2.2 \times 14 + 5 \times 3.5 + 6 \times 3.5 = 136.6\,kNm$

End shear, joint $2 = 111.1 + 3.7 \times 5 + 3.7 \times 14 + 4.6 \times 5 + 4.6 \times 6 = 232.0\,kN$

End shear, joint $5 = 126.9 + 4.3 \times 5 + 4.3 \times 14 + 3.4 \times 5 + 3.4 \times 6 = 246.0\,kN$

Similarly

Member 10 — combination C_1 (20% redistribution)

Midspan moment $= 115.7\,kNm$
End shear $= 239\,kN$

Member 12 — combination C_1 (20% redistribution)

Midspan moment $= 90.9\,kNm$
End shear $3 = 134.9\,kN$
End shear $6 = 147.1\,kN$

Member 13 — combination C_1 (20% redistribution)

Midspan moment $= 72.7\,kNm$
End shear $6 = 141\,kN$

From computer output:

Member midspan moments (kNm) and shears (kN) for various frame types under unit loading.

Combination	Frame type	Member 9			Member 10			Member 12			Member 13		
		Midspan moment	Shear		Midspan moment	Shear		Midspan moment	Shear		Midspan moment	Shear	
			2	5		5	8		3	6		6	9
C_7	F_1	56.3	111.1	126.9	42.7	119.0	119.0	61.5	94.4	115.6	39.7	105.0	105.0
C_1'	F_1	2.2	3.7	4.3	1.6	4.0	4.0	0.6	1.1	1.3	0.5	1.2	1.2
C_1'	F_2	2.2	3.7	4.3	1.5	4.0	4.0	1.1	1.4	1.0	1.0	1.2	1.2
C_1'	F_3	3.5	4.6	3.4	2.6	4.0	4.0	1.0	1.4	1.0	0.7	1.2	1.2
C_1'	F_4	3.5	4.6	3.4	2.8	4.0	4.0	1.0	1.4	1.0	1.5	1.2	1.2
C_1'	F_5	3.5	4.6	3.4	5.0	4.0	4.0	1.0	1.4	1.0	1.5	1.2	1.2

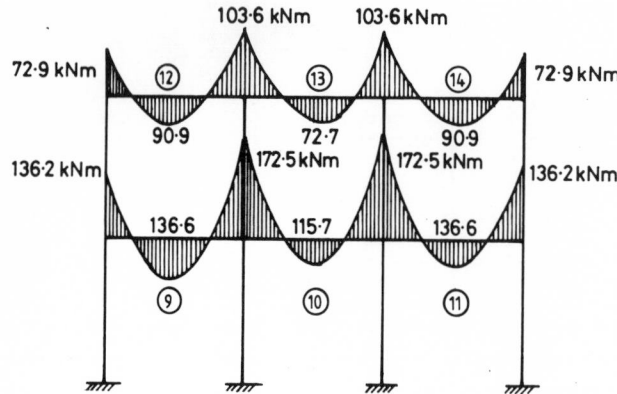

SK 9/31 Bending moment diagram combination C_1 (20% redistribution).

Note: Only one combination C_1 has been fully analysed to demonstrate the procedure for redistribution of moments in a frame structure. In practice all combinations of loads should be similarly processed to get an envelope of moments and shears. For all combinations of loads the plastic hinges will form at the same moment, i.e. 172 kNm at first floor level and 103.2 kNm at roof level.

Step 2   Check moment connection to edge column

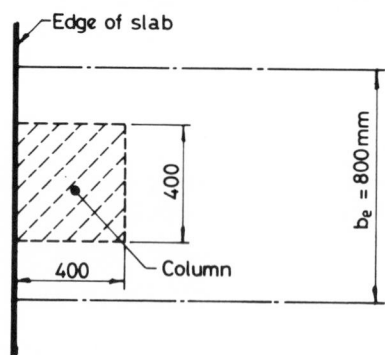

SK 9/32 Effective width of slab for moment transfer.

$M_{t, \text{max}} = 0.15 b_e d^2 f_{cu}$

$b_e = C_x + C_y = 400 + 400 = 800 \, \text{mm}$

$d = 175 \, \text{mm}$ assumed

$f_{cu} = 40 \, \text{N/mm}^2$

$M_{t, \text{max}} = 0.15 \times 800 \times 175^2 \times 40$
$\qquad\qquad = 147 \, \text{kNm} > 136.2 \, \text{kNm}$ (member 9, joint 2)

The column slab connection at the edge can transfer the applied moment and no further redistribution is necessary. It is conservatively assumed in this analysis that the depth of the slab at the column is 200 mm, ignoring the drop. The moment $M_{t, \text{max}}$ is greater than the design moment obtained from an equivalent frame analysis.

Step 3 ***Find bending moments at $h_c/2$ and at edge of drop***

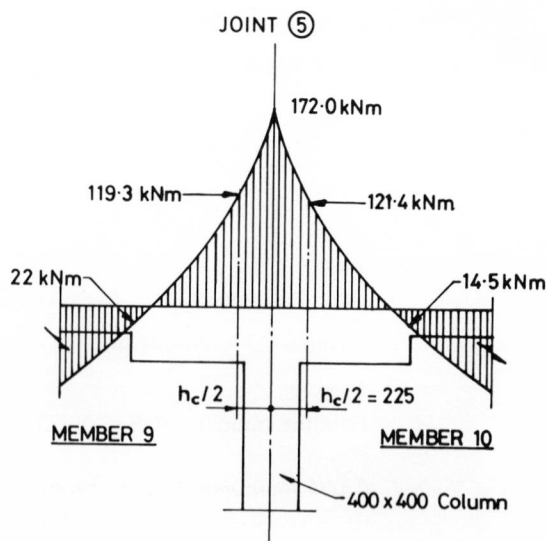

JOINT ⑤

172·0 kNm

119·3 kNm—

—121·4 kNm

22 kNm—

—14·5 kNm

$h_c/2$

$h_c/2 = 225$

MEMBER 9

MEMBER 10

400 x 400 Column

SK 9/33 Bending moments at
critical points − combination C_1
(20% redistribution).

Member 9 − combination C_1 (20% redistribution)

Joint 5:
Bending moment = 172 kNm
Shear = 246 kN
Dead load: 1.4 × 40 = 56 kN/m   near support
Live load: 1.6 × 30 = 48 kN/m   near support
$h_c/2$ = 0.225 m

Bending moment at $h_c/2 = 172 - 246 \times 0.225 + \dfrac{(56 + 48) \times 0.225^2}{2}$

$= 119.3$ kNm   (top tension)

Edge of drop = 1000 mm from centreline of column

Bending moment at edge of drop $= 172 - 246 \times 1 + \dfrac{(56 + 48) \times 1^2}{2}$

$= -22$ kNm   (bottom tension)

Joint 2, similarly:

Bending moment at $h_c/2 = 136.2 - 232 \times 0.225 + \dfrac{(56 + 48) \times 0.225^2}{2}$

$= 86.6$ kNm   (top tension)

Bending moment at edge of drop $= -43.8$ kNm   (bottom tension)

Member 10 − combination C_1 (20% redistribution)

Joint 5:

Bending moment at $h_c/2 = 121.4\,\text{kNm}$ (top tension)

Bending moment at edge of drop $= -14.5\,\text{kNm}$ (bottom tension)

Member 12 – combination C_1 (20% redistribution)

Joint 3:

Bending moment at $h_c/2$

$$= 72.9 - 134.9 \times 0.225 + \frac{(1.4 \times 30 + 1.6 \times 9) \times 0.225^2}{2} = 44.0\,\text{kNm}$$

Joint 6:

Bending moment at $h_c/2 = 71.3\,\text{kNm}$ (top tension)

Member 13 – combination C_1 (20% redistribution)

Joint 6:

Bending moment at $h_c/2 = 73.3\,\text{kNm}$ (top tension)

Step 4 Check limitation of negative design moment

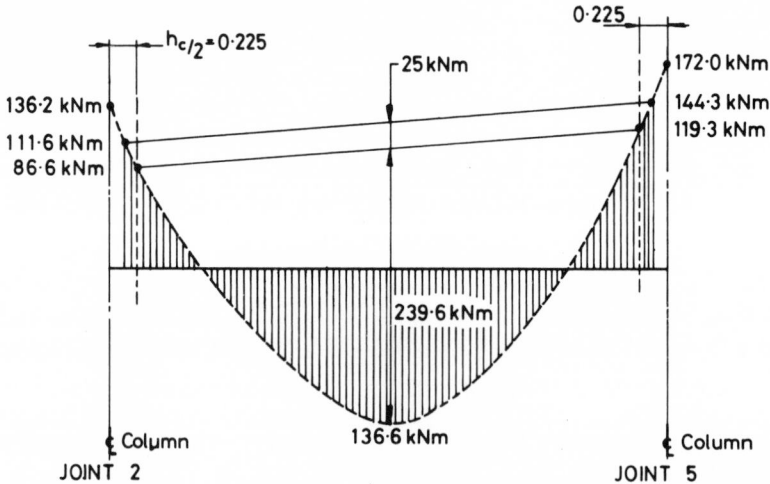

SK 9/34 Member 9 – combination C_1 (20% redistribution): limitation of negative design moment.

$$M' = \left(\frac{nl_2}{8}\right)\left(l_1 - \frac{2h_c}{3}\right)^2$$

where $n =$ loading per unit area on slab.

Average n on span on first floor $= 1.4 \times 5 + 1.6 \times 5 = 15\,\text{kN/m}^2$

$l_1 = 5.0\,\text{m}$ $l_2 = 6.0\,\text{m}$ $h_c = 0.225\,\text{m}$

$$M' = \left(\frac{15 \times 6.0}{8}\right)\left(5.0 - \frac{2 \times 0.225}{3}\right)^2$$

$$= 264.6\,\text{kNm} \text{(floor slab)}$$

Average n on span on roof $= 1.4 \times 5 + 1.6 \times 1.5 = 9.4\,\text{kN/m}^2$

$$M' = \left(\frac{9.4 \times 6.0}{8}\right)\left(5.0 - \frac{2 \times 0.225}{3}\right)^2$$

$\qquad = 165.8\,\text{kNm}$ (roof slab)

Check negative moment limitation

Member 9
Joint 2 at $h_c/2 = M_2 = 86.6\,\text{kNm}$ (see Step 3)
Joint 5 at $h_c/2 = M_5 = 119.3\,\text{kNm}$
Midspan moment $= 136.6\,\text{kNm}$ (see Step 1)

Average of M_2 and M_5 plus midspan moment
$= 0.5\,(86.6 + 119.3) + 136.6$
$= 239.6\,\text{kNm} < M' = 264.6\,\text{kNm}$

The negative moments will have to be increased by $(264.6 - 239.6) = 25.0\,\text{kNm}$

Revised negative moments:
Joint 2: $86.6 + 25.0 = 111.6\,\text{kNm}$
Joint 5: $119.3 + 25.0 = 144.3\,\text{kNm}$

Member 10
Joint 5 at $h_c/2 = M_5 = 121.4\,\text{kNm}$
Midspan moment $= 115.7\,\text{kNm}$
Average of negative and positive $= 121.4 + 115.7 = 237.1\,\text{kNm} < M' = 264.6\,\text{kNm}$

The negative moments will have to be increased by $(264.6 - 237.1) = 27.5\,\text{kNm}$

Revised negative moments:
Joint 5: $121.4 + 27.5 = 148.9\,\text{kNm}$

Member 12
Joint 3 at $h_c/2 = M_3 = 44.0\,\text{kNm}$
Joint 6 at $h_c/2 = M_6 = 71.3\,\text{kNm}$
Midspan moment $= 90.9\,\text{kNm}$

Average of negative and positive $= 0.5\,(44.0 + 71.3) + 90.9 = 148.55\,\text{kNm} < M' = 165.8\,\text{kNm}$

The negative moments will have to be increased by $(165.8 - 148.6) = 17.2\,\text{kNm}$

Revised negative moments:
Joint 3: $44.0 + 17.2 = 61.2\,\text{kNm}$
Joint 6: $71.3 + 17.2 = 88.5\,\text{kNm}$

Member 13
Joint 6 at $h_c/2 = M_6 = 73.3\,\text{kNm}$
Midspan moment $= 72.7\,\text{kNm}$

Average of negative and positive $= 73.3 + 72.7 = 146 \, \text{kNm} < M' = 165.8 \, \text{kNm}$

The negative moments will have to be increased by $(165.8 - 146) = 19.8 \, \text{kNm}$

Revised negative moments:
Joint 6: $73.3 + 19.8 = 93.1 \, \text{kNm}$

Step 5   *Carry out division of panels*

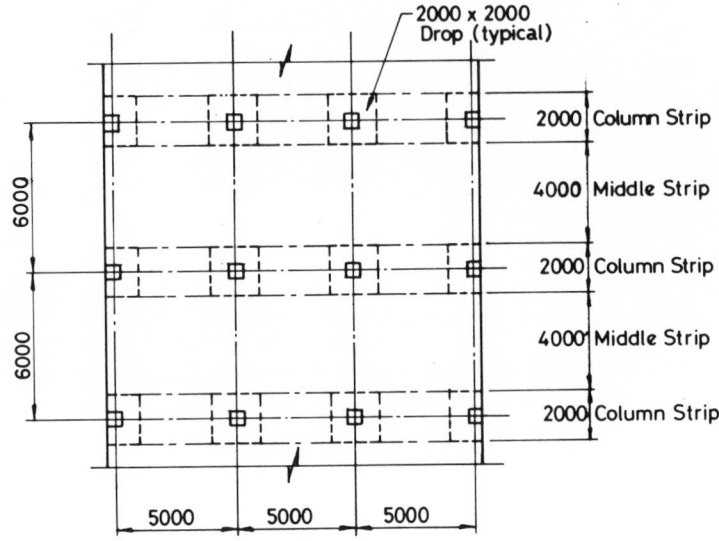

SK 9/35 Plan of floor slab — division of strips.

First floor slab

Size of drop $= 2000 \, \text{mm}$

$$\frac{l_x}{3} = \frac{5000}{3} = 1667 \, \text{mm} < 2000 \, \text{mm}$$

Column strip $= 2000 \, \text{mm}$

With drop middle strip $= 5000 - 2000 = 3000 \, \text{mm}$ $(y-y)$
and $= 6000 - 2000 = 4000 \, \text{mm}$ $(x-x)$

Middle strip in a slab without drop $= 6000 - l_x/2 = 6000 - 2500 = 3500 \, \text{mm}$

Proportion of middle strip with drop and without drop $= \dfrac{4000}{3500} = 1.14$

Roof slab

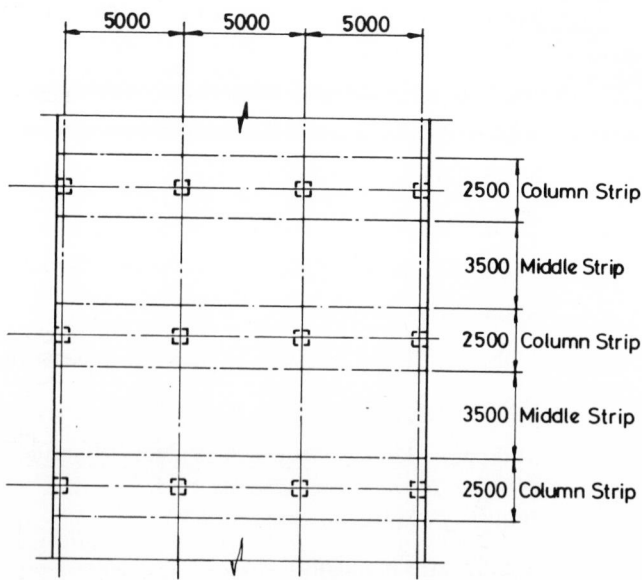

SK 9/36 Plan of roof slab − division of strips.

$$\text{Column strip} = \frac{l_x}{2} = \frac{5000}{2} = 2500\,\text{mm}$$

$$\begin{aligned}\text{Middle strip} &= 6000 - 2500 = 3500\,\text{mm}\quad(x-x)\\ &= 5000 - 2500 = 2500\,\text{mm}\quad(y-y)\end{aligned}$$

Step 6   **Divide moments between column strip and middle strip**
For slabs without drops,

Negative moments − 75% column strip
                                    25% middle strip
Positive moments − 55% column strip
                                    45% middle strip

Floor slab: design moments

Member 9: negative moments

Joint 2:   111.6 kNm   (see Step 4)
Joint 5:   144.3 kNm   (see Step 4)

Middle strip moments

Joint 2:   111.6 × 0.25 × 1.14   (see Step 4) = 31.8 kNm   (top tension)

Edge of drop = 43.8 × 0.25 × 1.14 = 12.5 kNm   (top tension)

Joint 5:   144.3 × 0.25 × 1.14 = 41.1 kNm   (top tension)

Edge of drop = 22.0 × 0.25 × 1.14 = 6.3 kNm   (top tension)

Column strip moments

Joint 2: 111.6 − 31.8 = 79.8 kNm (top tension)
Edge of drop: 43.8 − 12.5 = 31.3 kNm (top tension)
Joint 5: 144.3 − 41.1 = 103.2 kNm (top tension)
Edge of drop: 22 − 6.3 = 15.7 kNm (top tension)

Member 9: positive moments

Design midspan moment = 136.6 kNm (bottom tension)

Middle strip moments

Midspan: 136.6 × 0.45 × 1.14 = 70.1 kNm (bottom tension)

Column strip moments

Midspan: 136.6 − 70.1 = 66.5 kNm (bottom tension)
Member 10: negative moments
Joint 5: 148.9 kNm (see Step 4)

Middle strip moments

Joint 5: 148.9 × 0.25 × 1.14 = 42.4 kNm (top tension)
Edge of drop: 14.5 × 0.25 × 1.14 = 4.1 kNm (top tension)

Column strip moments

Joint 5: 148.9 − 42.4 = 106.3 kNm (top tension)
Edge of drop: 14.5 − 4.1 = 10.4 kNm (top tension)

Member 10: positive moments

Design midspan moment = 115.7 kNm (bottom tension)

Middle strip moments

Midspan: 115.7 × 0.45 × 1.14 = 59.4 kNm (bottom tension)

Column strip moments

Midspan: 115.7 − 59.4 = 56.3 kNm (bottom tension)

Note: Similarly calculate moments in column strips and middle strips in roof slab.

Step 7 **Determine cover to reinforcement**
See Step 3 of Section 3.3.

Step 8 **Design for flexure**
See Step 4 of Section 3.3.
The increased slab thickness at drops may be considered for the determination of reinforcement provided all reinforcement is properly anchored. Check reinforcement also at edge of drop.

Note: In this example the reinforcement is found for the flat slab spanning in the short direction only. Exactly the same method of analysis and design should be used to find the reinforcement in the long direction.

Step 9 *Detailing of reinforcement*
Two-thirds of the negative support reinforcement in the column strip
should be placed in half the width of the column strip centred over the
column.

Step 10 *Calculate punching shear and shear stress*

Punching shear at floor slab
Check joint 5.

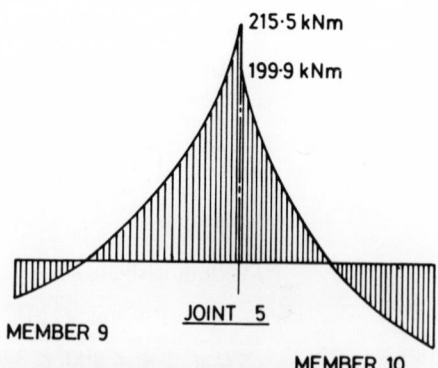

SK 9/37 Moment transfer to
column for punching shear
calculation.

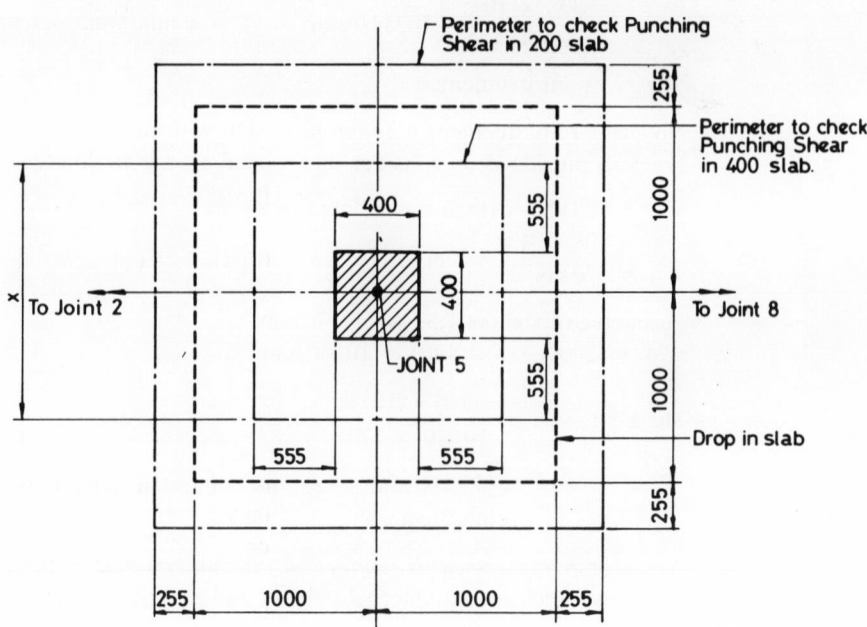

SK 9/38 Punching shear perimeters — plan of floor slab.

Use results of elastic analysis of frame before redistribution.

Maximum column moment at
joint 5 = 215.5 − 199.9 = 15.6 kNm
$\qquad M_t = 15.6\,\text{kNm}$

A 30% reduction is allowed if frame analysis is carried out.

$\therefore \quad M_t = 0.7 \times 15.6 = 10.9\,\text{kNm}$

$V_t = 258.2 + 239.0 = 497.2\,\text{kN}$

Punching shear perimeter at 1.5d from face of column.

$d = 400 - 30 = 370\,\text{mm}$
$1.5d = 1.5 \times 370 = 555\,\text{mm}$
$x = 400 + 2 \times 555 = 1510\,\text{mm} = 1.510\,\text{m}$

$$V_{\text{eff}} = V_t\left(1 + \frac{1.5M_t}{V_t x}\right) = 497.2\left(1 + \frac{1.5 \times 10.9}{497.2 \times 1.510}\right)$$
$$= 508\,\text{kN}$$

Maximum shear stress at column perimeter ($U_o = 4 \times 400$)

$$= \frac{V_{\text{eff}}}{U_o d} = \frac{508 \times 10^3}{4 \times 400 \times 370} = 0.86\,\text{N/m}^2 < 5\,\text{N/m}^2 \quad \text{OK}$$

Shear stress, $v = \dfrac{V_{\text{eff}}}{Ud}$

$U = 4 \times 1510 = 6040$

$\therefore \quad v = \dfrac{508 \times 10^3}{6040 \times 370} = 0.23\,\text{N/mm}^2 < v_c$ for minimum percentage of

reinforcement

No shear reinforcement is required in slab with drop.
For slab outside drop consider that loaded area is perimeter of drop.

$V_{\text{eff}} = 508\,\text{kN} -$ (load on the area of drop)
$\qquad = 508 - 4 \times 22$
$\qquad = 420\,\text{kN}$

Perimeter of slab at 1.5d ($d = 170\,\text{mm}$)
$= 4 \times (2000 + 3 \times 170) = 10040\,\text{mm}$

Shear stress, $v = \dfrac{420 \times 10^3}{10040 \times 170}$

$\qquad = 0.25\,\text{N/mm}^2 < v_c$ for minimum percentage of
reinforcement

No shear reinforcement is required at internal columns of floor slab.
Similarly check for an external column and a corner column.

Rules for calculation of perimeters of external and corner columns

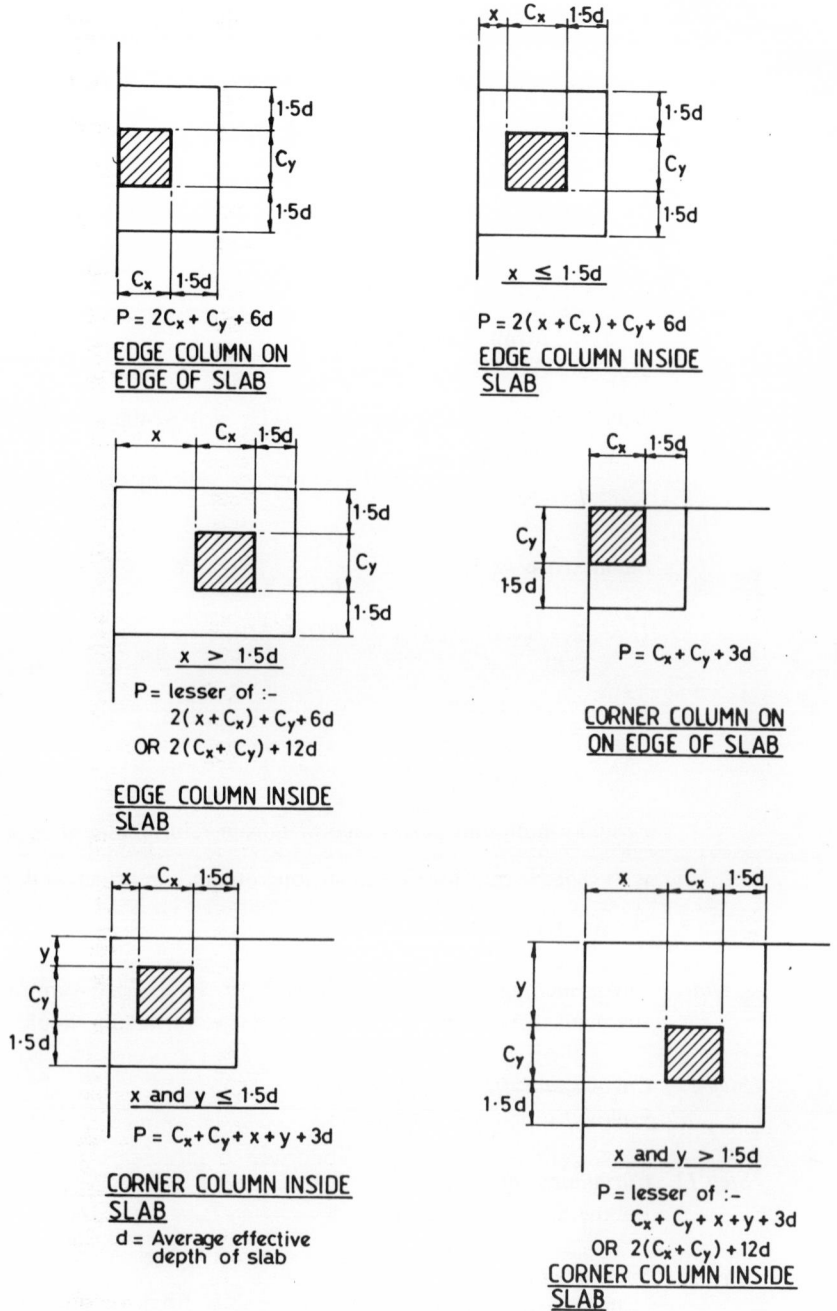

EDGE COLUMN ON EDGE OF SLAB

$P = 2C_x + C_y + 6d$

EDGE COLUMN INSIDE SLAB

$x \leq 1.5d$

$P = 2(x + C_x) + C_y + 6d$

EDGE COLUMN INSIDE SLAB

$x > 1.5d$

P = lesser of :–
  $2(x + C_x) + C_y + 6d$
OR $2(C_x + C_y) + 12d$

CORNER COLUMN ON ON EDGE OF SLAB

$P = C_x + C_y + 3d$

CORNER COLUMN INSIDE SLAB

x and $y \leq 1.5d$

$P = C_x + C_y + x + y + 3d$

d = Average effective depth of slab

CORNER COLUMN INSIDE SLAB

x and $y > 1.5d$

P = lesser of :–
  $C_x + C_y + x + y + 3d$
OR $2(C_x + C_y) + 12d$

SK 9/39 Punching shear perimeters for flat slab.

The illustrations show the different column configurations with respect to a free edge and the corresponding perimeters for the calculation of punching shear stresses.

When the column face is more than $1.5d$ away from a free edge of slab, then there are two alternative perimeters possible as illustrated. Take the least of these two alternatives for the calculation of punching shear stress.

Punching shear at roof slab
Check joint 3 – edge column.

$V_t = 130\,kN$

The frame action considered is in the $x-x$ direction as explained in Section 9.3.5.

$V_{eff} = 1.25V_t = 1.25 \times 130 = 162.5\,kN$

$$d = 170\,mm \quad \text{assumed}$$
$$1.5d = 1.5 \times 170 = 255\,mm$$

Shear stress at column perimeter ($U_o = 3 \times 400 = 1200$)

$$= \frac{V_{eff}}{U_o d} = \frac{162.5 \times 10^3}{1200 \times 170} = 0.8\,N/mm^2 < 5\,N/mm^2 \quad OK$$

Shear stress at $1.5d = v = \dfrac{V_{eff}}{Ud}$

$$U = [2 \times (400 + 255)] + (400 + 510)$$
$$= 2220$$

$$\therefore \quad v = \frac{162.5 \times 10^3}{2220 \times 170}$$
$$= 0.43\,N/mm^2$$

Assume minimum percentage of tensile reinforcement in slab.

$v_c = 0.48\,N/mm^2$ for Grade 40 concrete and an effective depth of 170 mm.

No shear reinforcement is necessary.

Note: The punching shear check should also be carried out for the flat slab spanning in the long direction and the worst result should be used.

Step 11 **Check span/effective depth ratio**
Follow Step 11 of Section 3.3

Step 12 **Curtailment of bars**
Follow Step 12 of Section 3.3.

Step 13 **Spacing of bars**
Follow Step 13 of Section 3.3.

Step 14 **Check early thermal cracking**
Follow Step 14 of Section 3.3.

Step 15 **Calculate minimum reinforcement**
Follow Step 9 and Step 15 of Section 3.3.

Step 16 *Calculate flexural crack width*
Follow Step 16 of Section 3.3.

Step 17 *Design of connections*
Follow Chapter 11

9.6 TABLES AND GRAPHS FOR CHAPTER 9

How to use Tables 9.1 to 9.6

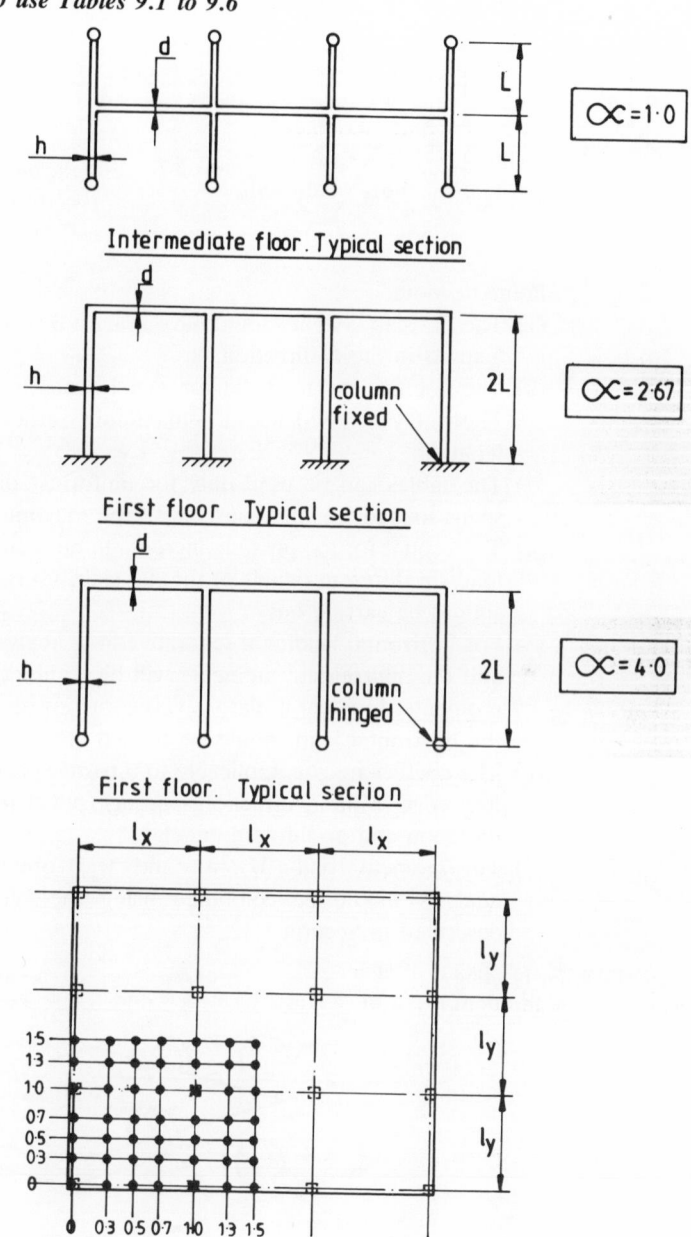

Intermediate floor. Typical section

First floor . Typical section

First floor. Typical section

Plan showing points for which coefficients are in tables 9·1 to 9·6

SK 9/40 Sketches to be used in conjunction with Tables 9.1–9.6.

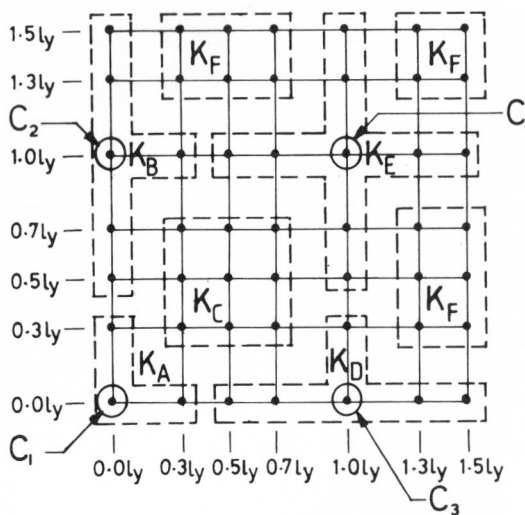

SK 9/41 Zones of stiffness correction factors to be applied to points of interest.

Points to note:

(1) The flat slab system should have at least 3 spans in the l_x direction and 3 spans in the l_y direction.

(2) The coefficients are valid for equal spans in the l_x and l_y direction. They may be used up to a maximum variation of 20% in the span lengths.

(3) The tables can be used only for uniformly distributed loads with all spans loaded simultaneously with the maximum load.

(4) To account for the possible increase in moment due to variation of live loads in different panels of the flat slab, no redistribution of moments should be carried out.

(5) For horizontal loading a separate frame analysis should be carried out and the appropriate moments will be combined with the vertical load moments. In general, flat slab construction should be fully braced and the horizontal load should be carried entirely by a shear-wall system.

(6) The coefficients are applicable to a corner panel, an edge panel with a free edge in the l_x direction, an edge panel with a free edge in the l_y direction and an internal panel.

(7) The moment triad (M_x, M_y and M_{xy}) obtained by this method of analysis should be combined using the Wood−Armer method as described in Section 1.12.

Sign convention for bending moments (+ ve in directions shown)

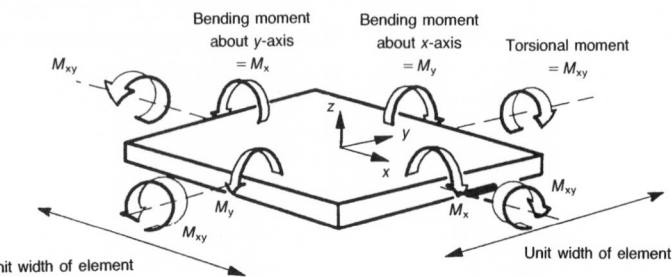

Note: A positive moment denotes hogging. This sign convention is opposite to the Wood−Armer convention. Reverse the signs of the moments before carrying out Wood−Armer combination as per Section 1.12.

Step-by-step analysis procedure
Step 1: Determine value of α (see SK 9/40 and find l_x/l_y).
Step 2: Assume d is the thickness of slab and h is the dimension of a side of a square column.
Step 3: From available L, as shown in SK 9/40, determine $S = \alpha d^3 L/h^4$.
Step 4: Select a point of interest from SK 9/40 or SK 9/41 where the moments have to be found. Corresponding to the zone of influence, find appropriate stiffness connection factor K depending on S from Graphs 9.1 to 9.18.
Step 5: Find moment coefficients from Tables 9.1 to 9.3 corresponding to l_x/l_y and the location of the point of interest.
Step 6: Find the ultimate uniformly distributed load on the flat slab = n kN/m^2.
Step 7: Find moment triads:

$$M_x = nC_xK_xl_y^2 \text{ kNm/m}$$
$$M_y = nC_yK_yl_y^2 \text{ kNm/m}$$
$$M_{xy} = nC_{xy}K_{xy}l_y^2 \text{ kNm/m}$$

Step 8: Carry out Wood–Armer combination.
Step 9: Find column reactions corresponding to l_x/l_y (see SK 9/41 for column locations).
Step 10: The moments obtained using these coefficients are in kNm/m. Find the effective width b_e as in Section 9.3.4. Multiply the moment obtained by analysis at edge and corner columns with the effective width b_e to find the slab-to-column connection moment. This transfer moment should be less than $M_{t,max}$ as defined in Section 9.3.4.

Reaction at column $C_1 = nC_1l_xl_y$
Reaction at column $C_2 = nC_2l_xl_y$, etc.

Note: If in the zone of K_A (as shown in SK 9/41) the point of interest is located, then to find M_x use stiffness correction factor K_{AX} corresponding to S and l_x/l_y as in Graph 19.1. Similarly if the point of interest lies in the zone of K_F then to find M_x use K_{FX} as in Graph 9.16.

The benefits of using these tables and graphs are that the analysis can be done very quickly and the necessity of carrying out the two analyses for the two orthogonal directions may be avoided. These tables can also be used for the analyses of raft foundation where the loading n may be assumed to be uniformly distributed over an inverted flat slab. The total loads from a structure will be assumed uniformly distributed at the underside of the raft.
Step 11: Calculate total column moments using Table 9.7 and divide the total moment between the columns at the junction depending on their relative stiffness. The stiffer the column, the more moment it will carry. The stiffness of a column may be calculated as I/l where I is the moment of inertia and l is the effective height.

Table 9.1 Bending moment coefficients for design of flat slabs $(l_x/l_y = 1.0)$.

Location of point of interest	$0l_x$	$0.3l_x$	$0.5l_x$	$0.7l_x$	$1.0l_x$	$1.3l_x$	$1.5l_x$	Moment coefficient, C
1.5l_y	−0.01184	−0.05370	−0.05697	−0.02496	+0.02633	−0.00554	−0.02230	C_x
	−0.04987	−0.03300	−0.02794	−0.03301	−0.04358	−0.02844	−0.02230	C_y
	0.00541	0.00296	0.00100	0.00400	0.00159	0.00380	0.00113	C_{xy}
1.3l_y	−0.01048	−0.05739	−0.06249	−0.02620	+0.04690	−0.00757	−0.02844	C_x
	−0.02776	−0.01667	−0.01322	−0.01265	−0.01954	−0.00757	−0.00554	C_y
	0.02030	0.01103	0.00338	0.01476	0.00744	0.01420	0.00380	C_{xy}
1.0l_y	+0.05482	−0.07014	−0.07560	−0.03647	+0.15656	−0.01954	−0.04358	C_x
	+0.11571	+0.02704	+0.01521	−0.04247	+0.15656	+0.04690	+0.02633	C_y
	0.03189	0.00619	0.00159	0.00819	0.04291	0.00744	0.00159	C_{xy}
0.7l_y	−0.01049	−0.05749	−0.06357	−0.02941	+0.04247	−0.01265	−0.03301	C_x
	−0.04132	−0.03218	−0.02993	−0.02941	−0.03647	−0.02620	−0.02496	C_y
	0.02524	0.01214	0.00395	0.01790	0.00819	0.01476	0.00400	C_{xy}
0.5l_y	−0.00157	−0.05245	−0.05718	−0.02993	+0.01521	−0.01322	−0.02794	C_x
	−0.07603	−0.06063	−0.05718	−0.06357	−0.07560	−0.06249	−0.05697	C_y
	0.00562	0.00262	0.00113	0.00395	0.00159	0.00338	0.00100	C_{xy}
0.3l_y	−0.01059	−0.05423	−0.06063	−0.03218	+0.02704	−0.01667	−0.03300	C_x
	−0.06515	−0.05423	−0.05245	−0.05749	−0.07014	−0.05739	−0.05370	C_y
	0.01419	0.00744	0.00262	0.01214	0.00619	0.01103	0.00296	C_{xy}
0l_y	+0.06370	−0.06515	−0.07603	−0.04132	+0.11571	−0.02776	−0.04987	C_x
	+0.06370	−0.01059	−0.01157	−0.01049	+0.05482	−0.01048	−0.01184	C_y
	0.01277	0.01419	0.00562	0.02524	0.03189	0.02030	0.00541	C_{xy}

$C_1 = 0.194355 \quad C_2 = 0.45153 \quad C_3 = 0.45153 \quad C_4 = 1.15263$

Table 9.2 Bending moment coefficients for design of flat slabs ($l_x/l_y = 1.2$).

Location of point of interest	$0l_x$	$0.3l_x$	$0.5l_x$	$0.7l_x$	$1.0l_x$	$1.3l_x$	$1.5l_x$	Moment coefficient, C
$1.5l_y$	-0.01657	-0.07980	-0.08586	-0.03473	+0.05340	-0.00898	-0.03784	C_x
	-0.05758	-0.03557	-0.02828	-0.03295	-0.04754	-0.02534	-0.01780	C_y
	0.00608	0.00347	0.00108	0.00474	0.00236	0.00466	0.00123	C_{xy}
$1.3l_y$	-0.01414	-0.08348	-0.09097	-0.03764	-0.07768	-0.01256	-0.04371	C_x
	-0.03235	-0.02082	-0.01716	-0.01332	-0.01930	-0.00593	-0.00543	C_y
	0.02211	0.01211	0.00337	0.01624	0.01089	0.01604	0.00383	C_{xy}
$1.0l_y$	+0.07262	-0.09512	-0.10176	-0.05003	+0.19583	-0.02621	-0.05675	C_x
	+0.12706	+0.01542	+0.00164	-0.03408	+0.18274	+0.03907	+0.01480	C_y
	0.04229	0.00573	0.00158	0.00776	0.05087	0.00707	0.00145	C_{xy}
$0.7l_y$	-0.01399	-0.08230	-0.09065	-0.04080	+0.07160	-0.01733	-0.04690	C_x
	-0.04465	-0.03575	-0.03378	-0.03032	-0.03656	-0.02569	-0.02623	C_y
	0.02831	0.01360	0.00421	0.02085	0.01213	0.01747	0.00432	C_{xy}
$0.5l_y$	-0.01604	-0.07664	-0.08390	-0.03994	+0.03919	-0.01588	-0.04109	C_x
	-0.08182	-0.06117	-0.05595	-0.05809	-0.08095	-0.06046	-0.05311	C_y
	0.00710	0.00316	0.00148	0.00502	0.00236	0.00431	0.00115	C_{xy}
$0.3l_y$	-0.01365	-0.07796	-0.08693	-0.04249	+0.05316	-0.02017	-0.04558	C_x
	-0.06907	-0.05432	-0.05111	-0.05730	-0.07482	-0.05576	-0.05069	C_y
	0.01457	0.00963	0.00299	0.01417	0.00935	0.01330	0.00331	C_{xy}
$0l_y$	+0.09004	-0.08934	-0.10261	-0.05444	+0.15200	-0.03380	-0.06265	C_x
	+0.06476	-0.01134	-0.01169	-0.01106	+0.05945	-0.01081	-0.01176	C_y
	0.01871	0.01902	0.00641	0.03097	0.03657	0.02540	0.00656	C_{xy}

$C_1 = 0.195075$ $C_2 = 0.44811$ $C_3 = 0.453375$ $C_4 = 1.15344$

Table 9.3 Bending moment coefficients for design of flat slabs ($l_x/l_y = 1.4$).

Location of point of interest	$0l_x$	$0.3l_x$	$0.5l_x$	$0.7l_x$	$1.0l_x$	$1.3l_x$	$1.5l_x$	Moment coefficient, C
$1.5l_y$	-0.02136	-0.10961	-0.11941	-0.04705	$+0.08741$	-0.01469	-0.05713	C_x
	-0.06491	-0.03952	-0.03062	-0.03282	-0.05018	-0.02241	-0.01520	C_y
	0.00639	0.00369	0.00100	0.00498	0.00320	0.01037	0.00113	C_{xy}
$1.3l_y$	-0.01752	-0.11318	-0.12362	-0.05114	$+0.11466$	-0.01923	-0.06200	C_x
	-0.03625	-0.02623	-0.02280	-0.01445	-0.01742	-0.00486	-0.00687	C_y
	0.02248	0.01222	0.00296	0.01615	0.01479	0.01609	0.00335	C_{xy}
$1.0l_y$	$+0.09384$	-0.12354	-0.13173	-0.06459	$+0.23834$	-0.03332	-0.07210	C_x
	$+0.13729$	$+0.00397$	-0.01129	$+0.02551$	$+0.20889$	$+0.03142$	$+0.00431$	C_y
	0.05628	0.00527	0.00170	0.00717	0.05755	0.00649	0.00125	C_{xy}
$0.7l_y$	-0.01723	-0.11078	-0.12186	-0.05410	$+0.10672$	-0.02342	-0.06343	C_x
	-0.04721	-0.03997	-0.03854	-0.03142	-0.03513	-0.02532	-0.02823	C_y
	0.03003	0.01418	0.00411	0.02227	0.01662	0.01867	0.00422	C_{xy}
$0.5l_y$	-0.02056	-0.10476	-0.11551	-0.05103	$+0.07003$	-0.02048	-0.05780	C_x
	-0.08747	-0.06227	-0.05539	-0.06281	-0.08560	-0.05789	-0.04970	C_y
	0.00842	0.00355	0.00170	0.00575	0.00318	0.00483	0.00116	C_{xy}
$0.3l_y$	-0.01606	-0.10580	-0.11820	-0.05479	$+0.08586$	-0.02520	-0.06155	C_x
	-0.07295	-0.05462	-0.04976	-0.05667	-0.07905	-0.05361	-0.04740	C_y
	0.01377	0.01071	0.00382	0.01612	0.01305	0.01473	0.00343	C_{xy}
$0l_y$	$+0.12167$	-0.11753	-0.13360	-0.06868	$+0.19234$	-0.04028	-0.07779	C_x
	$+0.06593$	-0.01189	-0.01158	-0.01135	$+0.06438$	-0.01080	-0.01135	C_y
	0.03124	0.02358	0.00741	0.03596	0.04237	0.02947	0.00739	C_{xy}

$C_1 = 0.19683$ $C_2 = 0.445545$ $C_3 = 0.45477$ $C_4 = 1.1529$

Table 9.4 Bending moment coefficients for design of flat slabs ($l_x/l_y = 1.6$).

Location of point of interest	$0l_x$	$0.3l_x$	$0.5l_x$	$0.7l_x$	$1.0l_x$	$1.3l_x$	$1.5l_x$	Moment coefficient, C
$1.5l_y$	-0.02596	-0.14272	-0.15690	-0.06172	+0.12785	-0.02244	-0.07938	C_x
	-0.07131	-0.04461	-0.03496	-0.03274	-0.05106	-0.01984	-0.01475	C_y
	0.00638	0.00367	0.00083	0.00484	0.00411	0.00478	0.00091	C_{xy}
$1.3l_y$	-0.02028	-0.14612	-0.16010	-0.06652	+0.15746	-0.02746	-0.08306	C_x
	-0.03924	-0.03265	-0.02981	-0.01601	-0.01375	-0.00439	-0.00969	C_y
	0.02163	0.01167	0.00236	0.01496	0.01912	0.01489	0.00261	C_{xy}
$1.0l_y$	+0.11858	-0.15531	-0.16570	-0.08020	+0.28462	-0.04116	-0.09017	C_x
	+0.14626	-0.00727	-0.02353	+0.01704	+0.23486	+0.02416	-0.00511	C_y
	0.07490	0.00493	0.00211	0.00649	0.06314	0.00578	0.00102	C_{xy}
$0.7l_y$	-0.01988	-0.14273	-0.15713	-0.06914	+0.14768	-0.03091	-0.08271	C_x
	-0.04901	-0.04471	-0.04392	-0.03274	-0.03231	-0.02516	-0.03077	C_y
	0.03044	0.01397	0.00373	0.02243	0.02162	0.01871	0.00386	C_{xy}
$0.5l_y$	-0.02479	-0.13655	-0.15162	-0.06494	+0.10749	-0.02697	-0.07773	C_x
	-0.09266	-0.06388	-0.05566	-0.06185	-0.08934	-0.05511	-0.04713	C_y
	0.00954	0.00383	0.00182	0.00602	0.00408	0.00504	0.00110	C_{xy}
$0.3l_y$	-0.01737	-0.13751	-0.15419	-0.06902	+0.12493	-0.03182	-0.08087	C_x
	-0.07649	-0.05520	-0.04868	-0.05576	-0.08263	-0.05125	-0.04433	C_y
	0.01189	0.01251	0.00267	0.01759	0.01727	0.01559	0.00346	C_{xy}
$0l_y$	+0.15859	-0.14962	-0.16914	-0.08415	+0.23712	-0.04757	-0.09590	C_x
	+0.06699	-0.01231	-0.01139	-0.01145	+0.06950	-0.01061	-0.01082	C_y
	0.05131	0.02810	0.00830	0.04046	0.04891	0.03267	0.00799	C_{xy}

$C_1 = 0.199\,125$ $C_2 = 0.443\,385$ $C_3 = 0.45576$ $C_4 = 1.15173$

Table 9.5 Bending moment coefficients for design of flat slabs ($l_x/l_y = 1.8$).

Location of point of interest	$0l_x$	$0.3l_x$	$0.5l_x$	$0.7l_x$	$1.0l_x$	$1.3l_x$	$1.5l_x$	Moment coefficient, C
$1.5l_y$	−0.030 10	−0.178 90	−0.197 96	−0.078 54	+0.174 36	−0.032 03	−0.104 20	C_x
	−0.076 52	−0.050 63	−0.041 05	−0.032 79	−0.050 01	−0.017 74	−0.016 33	C_y
	0.006 08	0.003 50	0.000 63	0.004 42	0.005 11	0.004 31	0.000 66	C_{xy}
$1.3l_y$	−0.022 10	−0.182 09	−0.200 23	−0.083 64	+0.205 82	−0.037 50	−0.106 77	C_x
	−0.041 20	−0.039 86	−0.037 92	−0.017 91	−0.008 29	−0.004 47	−0.013 68	C_y
	0.019 76	0.010 78	0.001 71	0.013 11	0.023 88	0.012 93	0.001 81	C_{xy}
$1.0l_y$	+0.146 87	−0.190 28	−0.203 70	−0.096 95	+0.335 05	−0.049 97	−0.111 26	C_x
	+0.154 02	−0.018 30	−0.035 10	+0.008 88	+0.260 64	+0.017 36	−0.013 59	C_y
	0.099 11	0.004 89	0.002 47	0.006 31	0.067 83	0.004 98	0.000 81	C_{xy}
$0.7l_y$	−0.021 83	−0.178 00	−0.196 42	−0.085 82	+0.194 37	−0.039 82	−0.104 88	C_x
	−0.050 07	−0.049 86	−0.049 70	−0.034 25	−0.028 18	−0.025 23	−0.033 68	C_y
	0.029 65	0.013 13	0.003 44	0.021 59	0.027 08	0.017 92	0.003 37	C_{xy}
$0.5l_y$	−0.028 40	−0.171 82	−0.191 96	−0.080 93	+0.151 36	−0.035 31	−0.100 74	C_x
	−0.097 18	−0.065 94	−0.056 75	−0.060 72	−0.092 04	−0.052 32	−0.045 52	C_y
	0.010 47	0.004 03	0.001 84	0.005 89	0.005 16	0.005 03	0.000 99	C_{xy}
$0.3l_y$	−0.017 16	−0.172 85	−0.194 64	−0.085 11	+0.170 18	−0.040 05	−0.103 47	C_x
	−0.079 44	−0.056 07	−0.048 04	−0.546 80	−0.085 44	−0.048 87	−0.041 77	C_y
	0.009 08	0.014 58	0.003 96	0.019 06	0.021 96	0.016 15	0.003 50	C_{xy}
$0l_y$	+0.200 72	−0.185 39	−0.209 19	−0.100 93	+0.286 59	−0.055 94	−0.117 31	C_x
	+0.067 85	−0.012 73	−0.011 21	−0.011 43	+0.074 73	−0.010 29	−0.010 27	C_y
	0.080 02	0.032 92	0.009 14	0.044 72	0.055 97	0.035 22	0.008 44	C_{xy}

$C_1 = 0.201 78$ $C_2 = 0.441 495$ $C_3 = 0.456 39$ $C_4 = 1.150 38$

Table 9.6 Bending moment coefficients for design of flat slabs ($l_x/l_y = 2.0$).

Location of point of interest	$0l_x$	$0.3l_x$	$0.5l_x$	$0.7l_x$	$1.0l_x$	$1.3l_x$	$1.5l_x$	Moment coefficient, C
$1.5l_y$	−0.03352	−0.21798	−0.24238	−0.09727	+0.22671	−0.04331	−0.13147	C_x
	−0.08045	−0.05733	−0.04855	−0.03300	−0.04700	−0.01614	−0.01960	C_y
	0.00553	0.00326	0.00044	0.00386	0.00621	0.00368	0.00041	C_{xy}
$1.3l_y$	−0.02266	−0.22093	−0.24390	−0.10237	+0.25960	−0.04825	−0.13312	C_x
	−0.04210	−0.04764	−0.04682	−0.02007	−0.00105	−0.00502	−0.01861	C_y
	0.01707	0.00981	0.00113	0.01098	0.02901	0.01064	0.00107	C_{xy}
$1.0l_y$	+0.17869	−0.22830	−0.24572	−0.11492	+0.38991	−0.05993	−0.13551	C_x
	+0.16068	−0.02909	−0.04607	+0.00116	+0.28634	−0.01105	−0.02132	C_y
	0.12976	0.00537	0.00275	0.00633	0.07181	0.00436	0.00102	C_{xy}
$0.7l_y$	−0.02187	−0.21636	−0.23968	−0.10405	+0.24672	−0.05017	−0.13005	C_x
	−0.05039	−0.05533	−0.05573	−0.03591	−0.02278	−0.02554	−0.03688	C_y
	0.02777	0.01179	0.00331	0.02000	0.03292	0.01656	0.00283	C_{xy}
$0.5l_y$	−0.03105	−0.21043	−0.23636	−0.09883	+0.20147	−0.04540	−0.12677	C_x
	−0.10089	−0.06837	−0.05857	−0.05954	−0.09363	−0.04968	−0.04482	C_y
	0.01121	0.00418	0.00178	0.00588	0.00635	0.00488	0.00090	C_{xy}
$0.3l_y$	−0.01505	−0.21163	−0.23936	−0.10299	+0.22144	−0.04992	−0.12932	C_x
	−0.08166	−0.05719	−0.04790	−0.05350	−0.08741	−0.04657	−0.03984	C_y
	0.00549	0.01703	0.00446	0.02071	0.02705	0.01663	0.00361	C_{xy}
$0l_y$	+0.24789	−0.22465	−0.25366	−0.11907	+0.34090	−0.06557	−0.14215	C_x
	+0.06845	−0.01316	−0.01109	−0.01131	+0.08006	−0.00991	−0.00979	C_y
	0.11841	0.03847	0.01002	0.04900	0.06341	0.03730	0.00882	C_{xy}

$C_1 = 0.204\,615$ $C_2 = 0.439\,74$ $C_3 = 0.456\,66$ $C_4 = 1.149\,03$

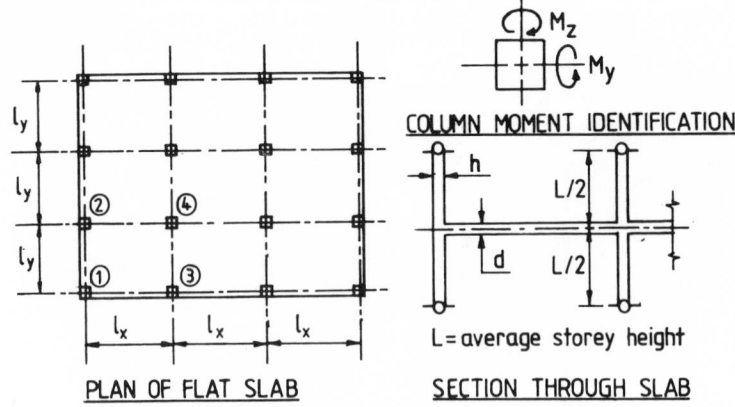

SK 9/42 Column number identification.

Table 9.7 Bending moment coefficient for design of columns in flat slab construction.

l_x/l_y	1	1.2	1.4	1.6	1.8	2	Moment coefficients
Column no.							
1	0.011 75	0.013 45	0.015 26	0.017 17	0.019 16	0.021 23	C_{cy}
1	−0.011 75	−0.017 92	−0.025 94	−0.036 03	−0.048 36	−0.063 07	C_{cz}
2	−0.002 44	−0.002 33	−0.002 24	−0.002 18	−0.002 16	−0.002 18	C_{cy}
2	−0.022 80	−0.032 71	−0.045 25	−0.060 73	−0.079 39	−0.101 47	C_{cz}
3	0.022 80	0.027 65	0.033 01	−0.038 85	0.045 13	0.051 84	C_{cy}
3	0.002 44	0.004 41	0.007 06	0.010 40	0.014 45	0.019 19	C_{cz}
4	−0.003 76	−0.003 93	−0.004 18	−0.004 54	−0.005 06	−0.005 71	C_{cy}
4	0.003 76	0.006 59	0.010 28	0.014 84	0.020 31	0.026 67	C_{cz}

Column total moment $= C_c \times R \times l_y^3 \times \eta$

where η = load per unit area on slab.
See Graphs 9.19−9.26 for factors R_y and R_z.

Note: Divide total moment in column to top and bottom column in proportion to their stiffness.

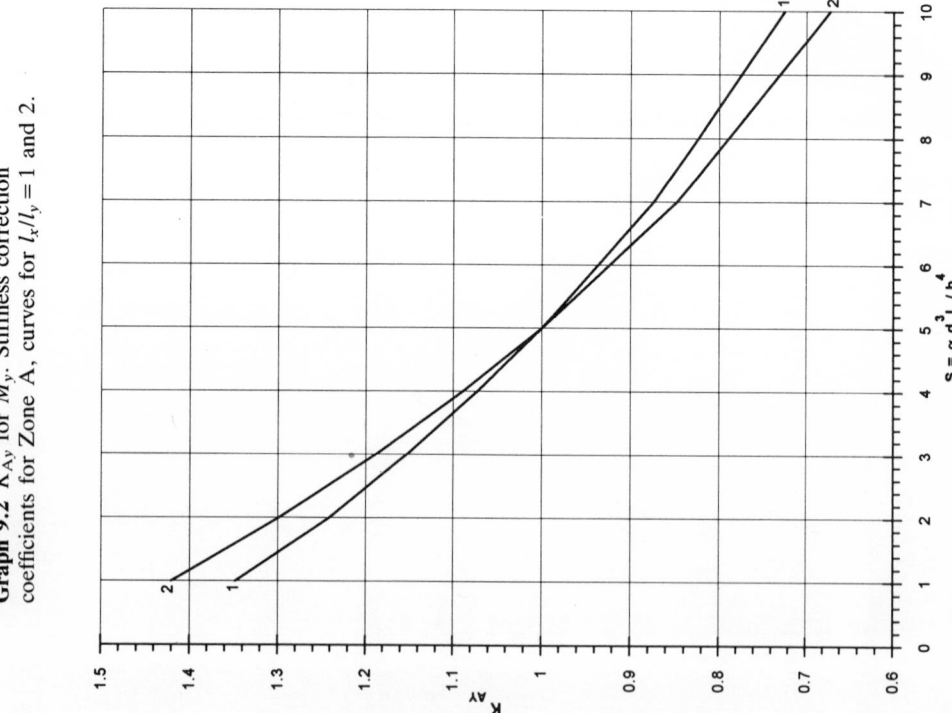

Graph 9.2 K_{Ay} for M_y. Stiffness correction coefficients for Zone A, curves for $l_x/l_y = 1$ and 2.

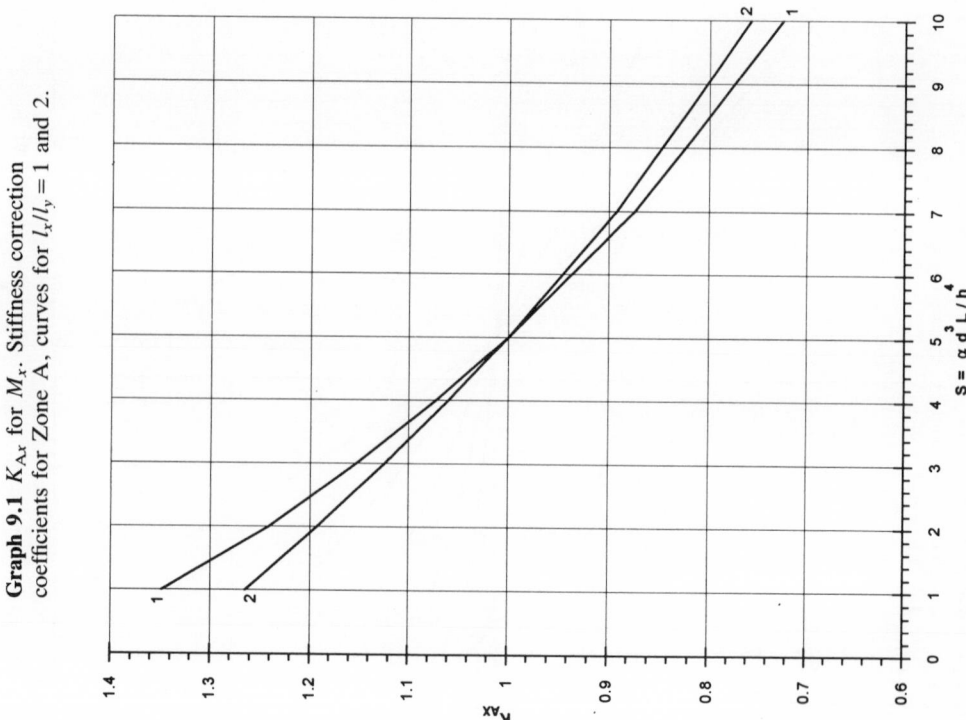

Graph 9.1 K_{Ax} for M_x. Stiffness correction coefficients for Zone A, curves for $l_x/l_y = 1$ and 2.

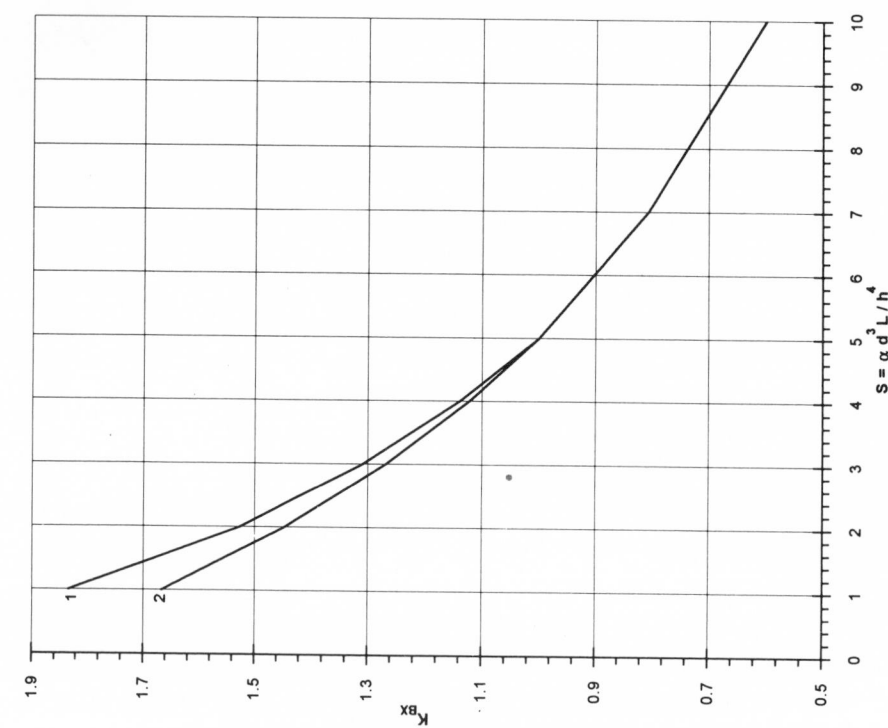

Graph 9.4 K_{Bx} for M_x. Stiffness correction coefficients for Zone B, curves for $l_x/l_y = 1$ and 2.

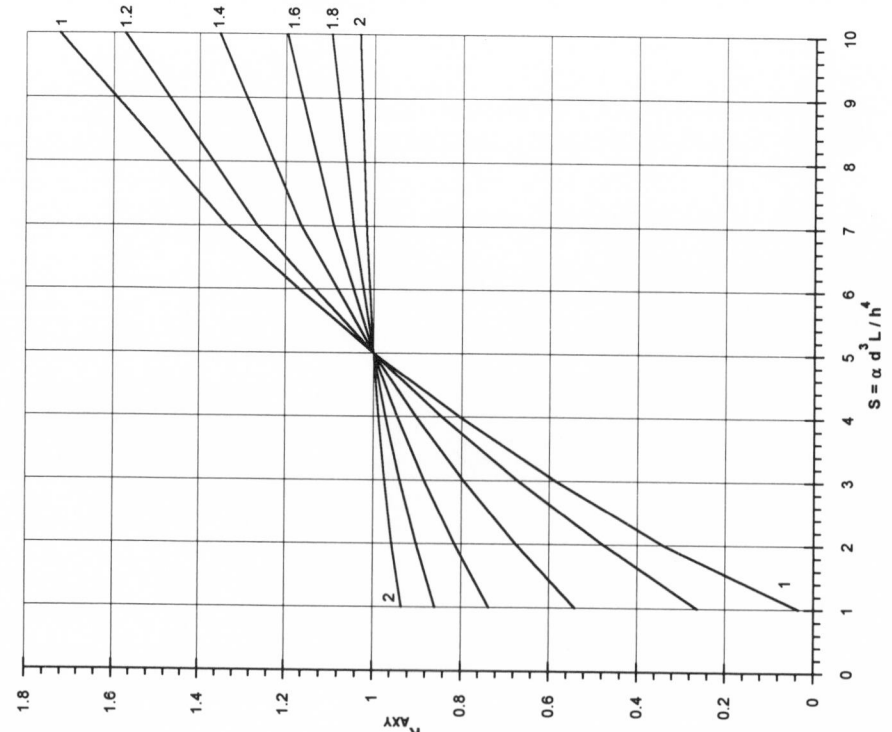

Graph 9.3 K_{Axy} for M_{xy}. Stiffness correction coefficients for Zone A, curves for $l_x/l_y = 1$, 1.2, 1.4, 1.6, 1.8 and 2.

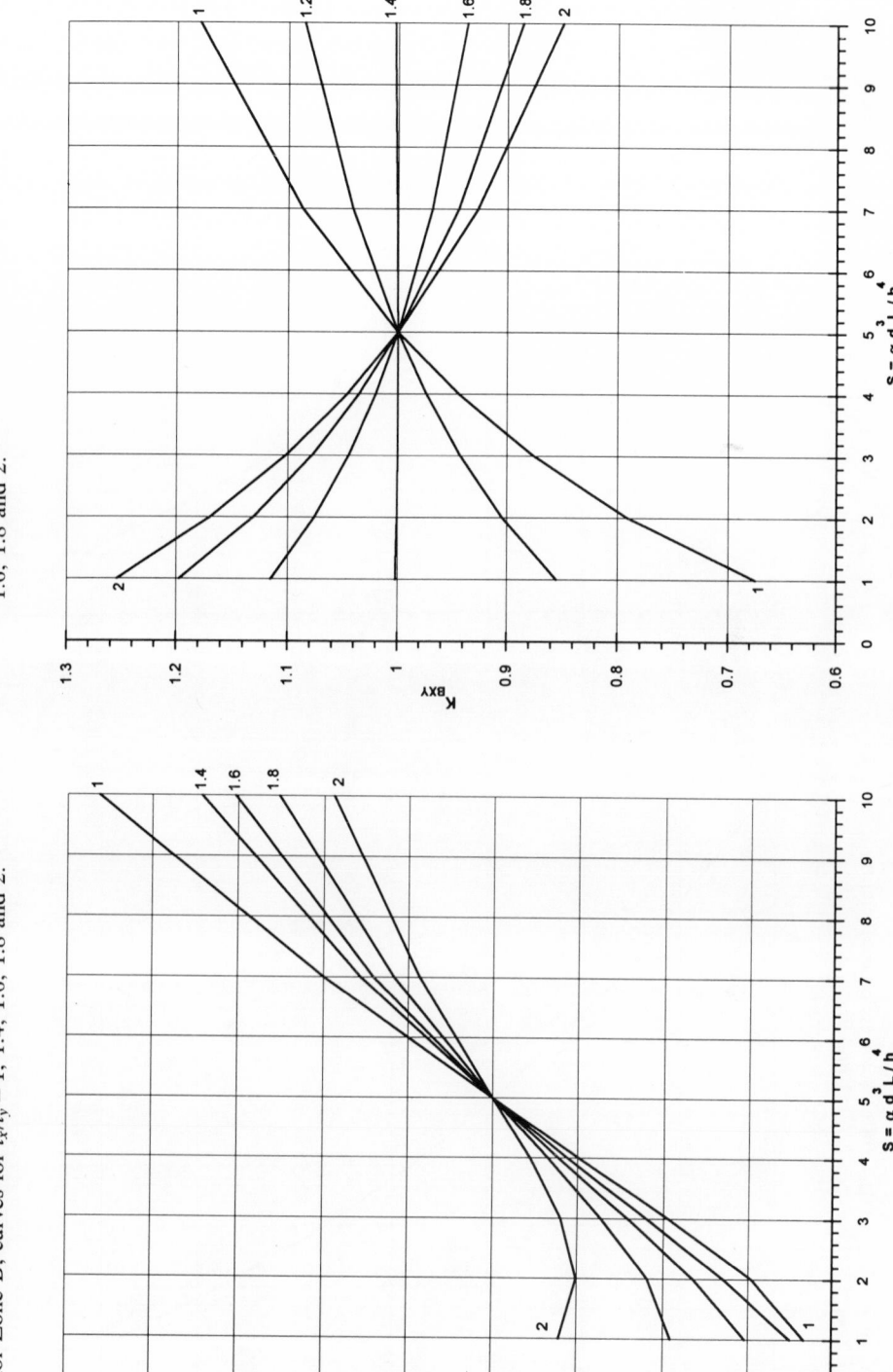

Graph 9.6 K_{Bxy} for M_{xy}. Stiffness correction coefficients for Zone B, curves for $l_x/l_y = 1$, 1.2, 1.4, 1.6, 1.8 and 2.

Graph 9.5 K_{By} for M_y. Stiffness correction coefficients for Zone B, curves for $l_x/l_y = 1$, 1, 1.4, 1.6, 1.8 and 2.

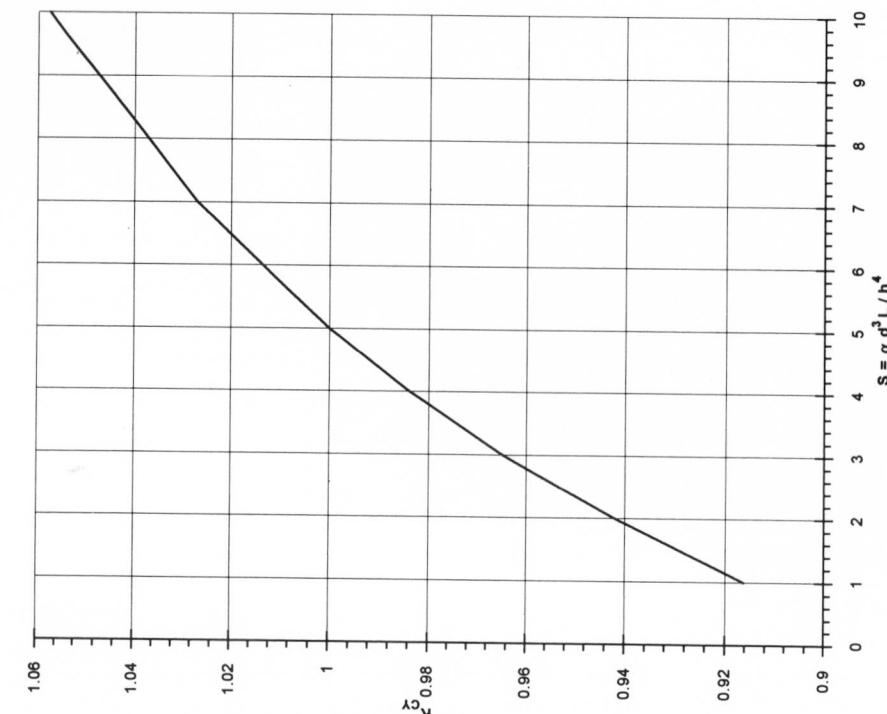

Graph 9.8 K_{Cy} for M_y. Stiffness correction coefficients for Zone C, curves for $l_x/l_y = 1$ and 2.

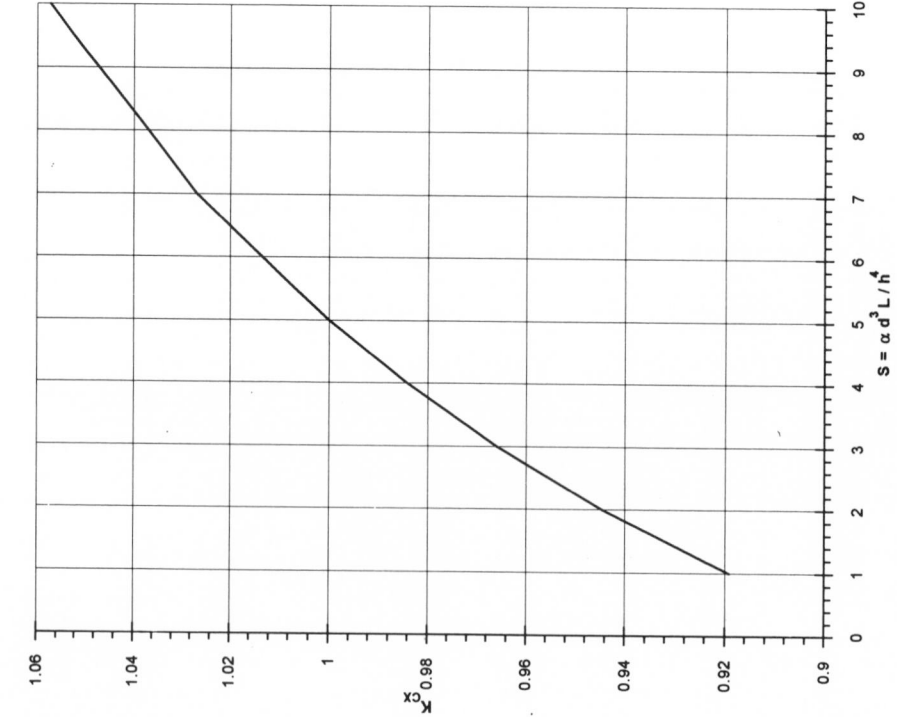

Graph 9.7 K_{Cx} for M_x. Stiffness correction coefficients for Zone C, curves for $l_x/l_y = 1$ and 2.

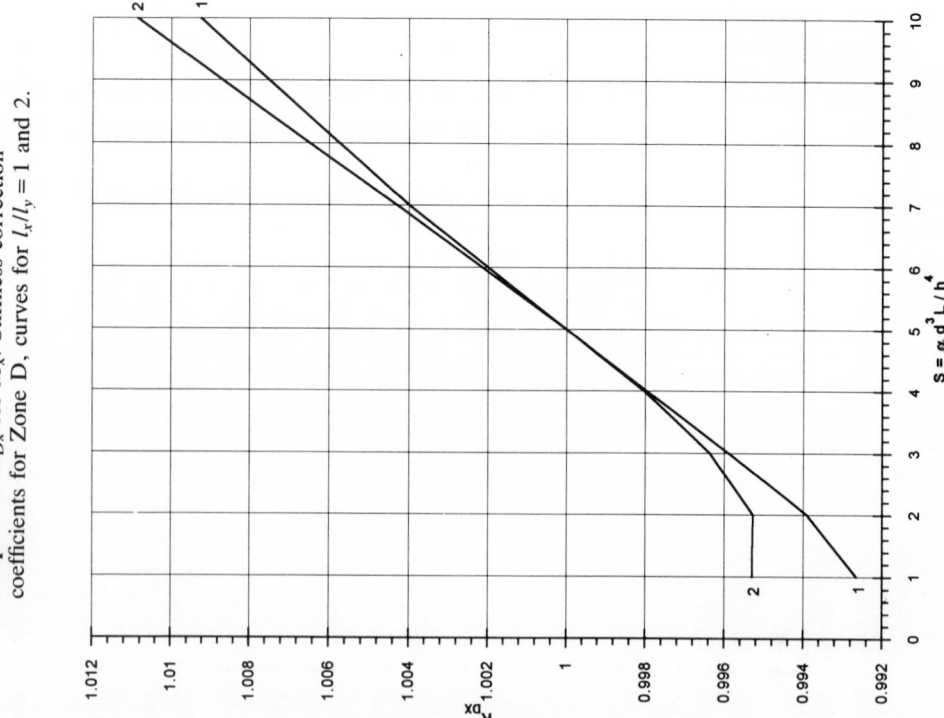

Graph 9.10 K_{Dx} for M_x. Stiffness correction coefficients for Zone D, curves for $l_x/l_y = 1$ and 2.

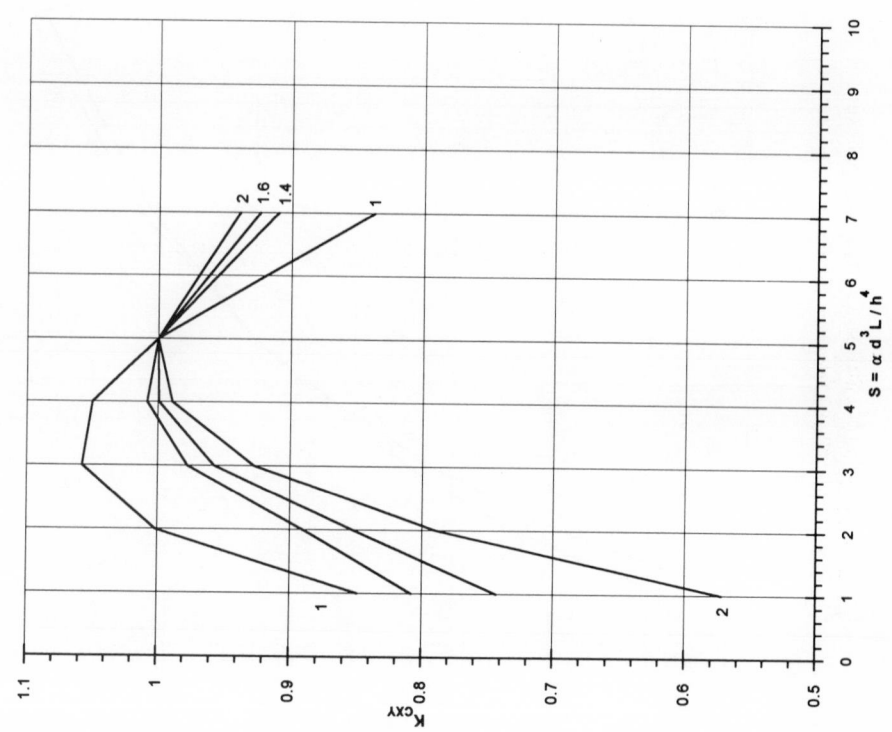

Graph 9.9 K_{Cxy} for M_{xy}. Stiffness correction coefficients for Zone C, curves for $l_x/l_y = 1$, 1.4, 1.6 and 2.

Graph 9.12 K_{Dxy} for M_{xy}. Stiffness correction coefficients for Zone D, curves for $l_x/l_y = 1$, 1.2, 1.6 and 2.

Graph 9.11 K_{Dy} for M_y. Stiffness correction coefficients for Zone D, curves for $l_x/l_y = 1$ and 2.

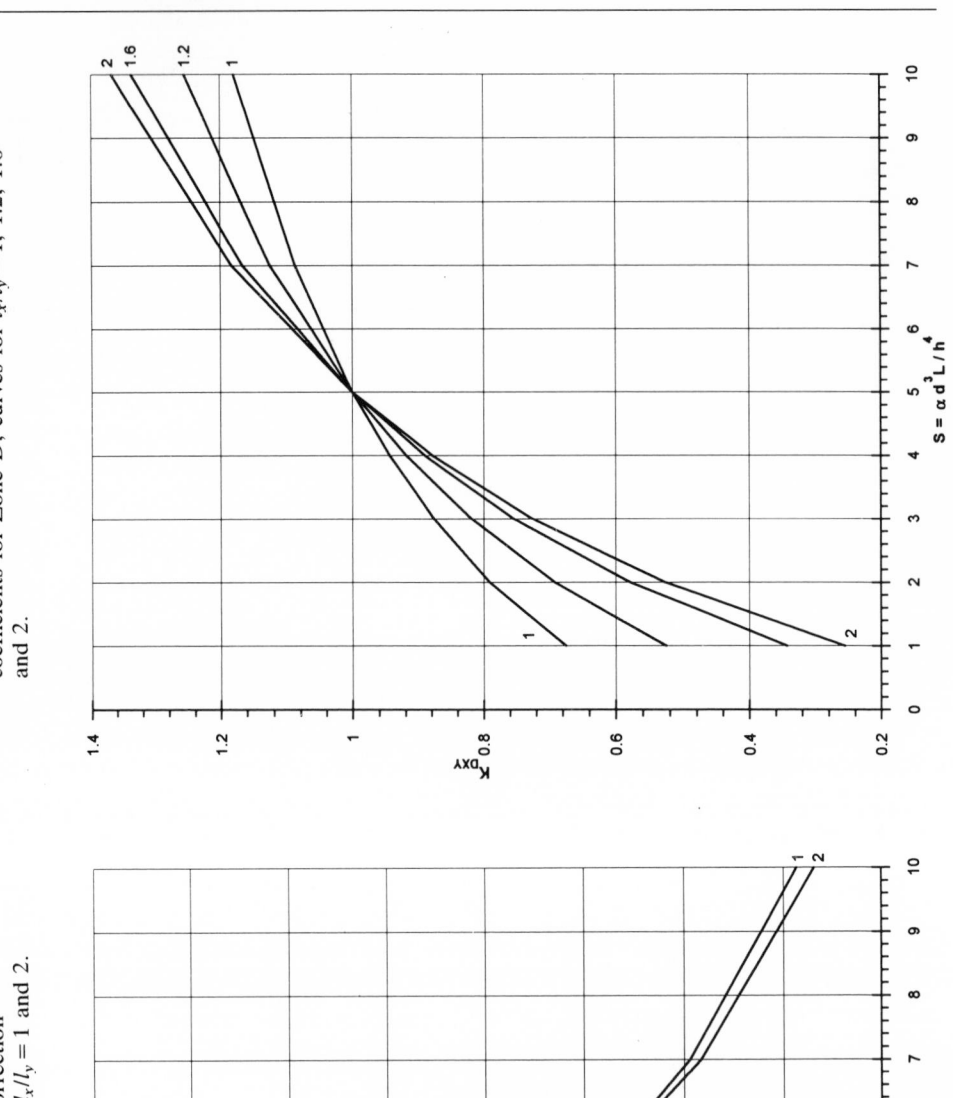

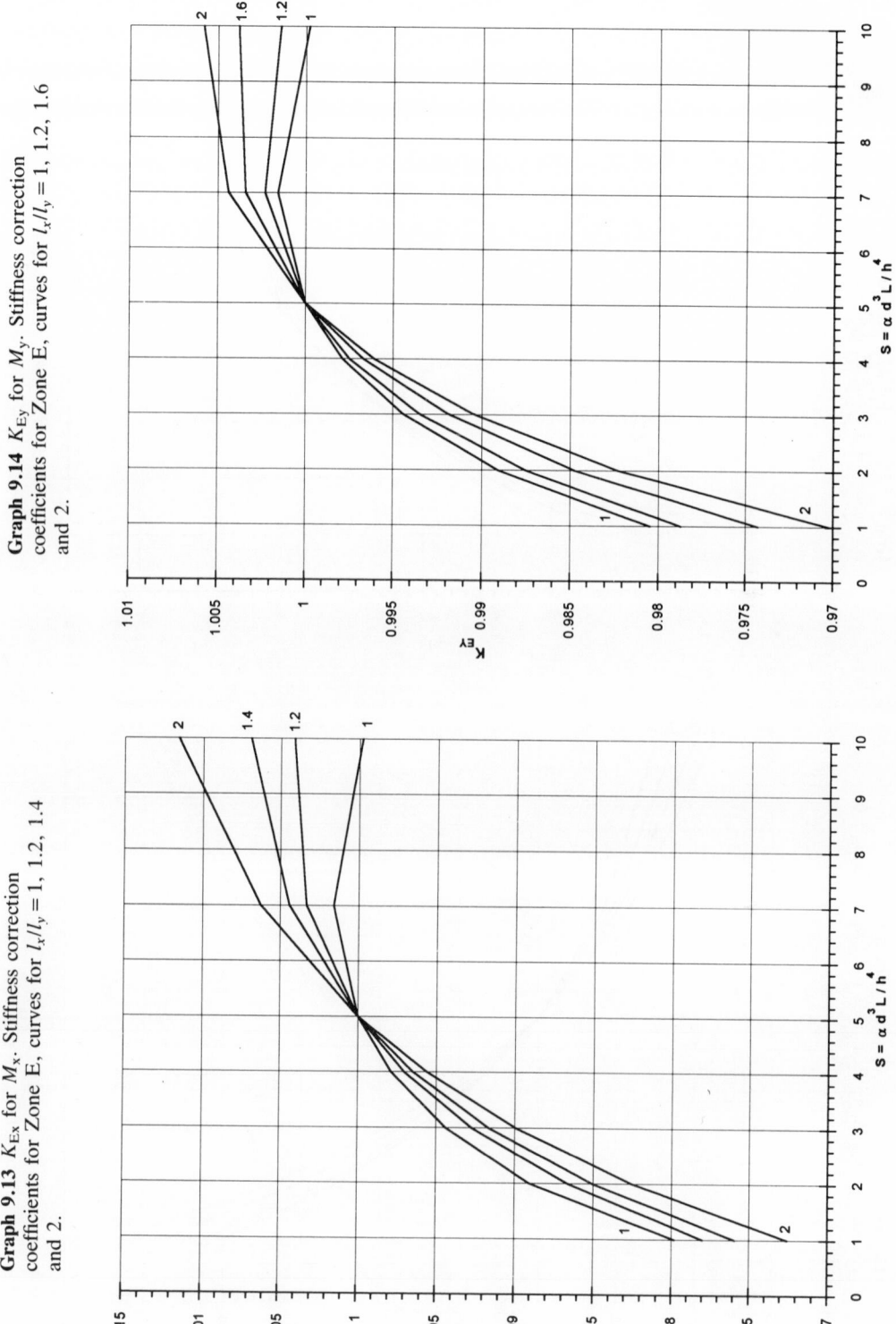

Graph 9.14 K_{Ey} for M_y. Stiffness correction coefficients for Zone E, curves for $l_x/l_y = 1, 1.2, 1.6$ and 2.

Graph 9.13 K_{Ex} for M_x. Stiffness correction coefficients for Zone E, curves for $l_x/l_y = 1, 1.2, 1.4$ and 2.

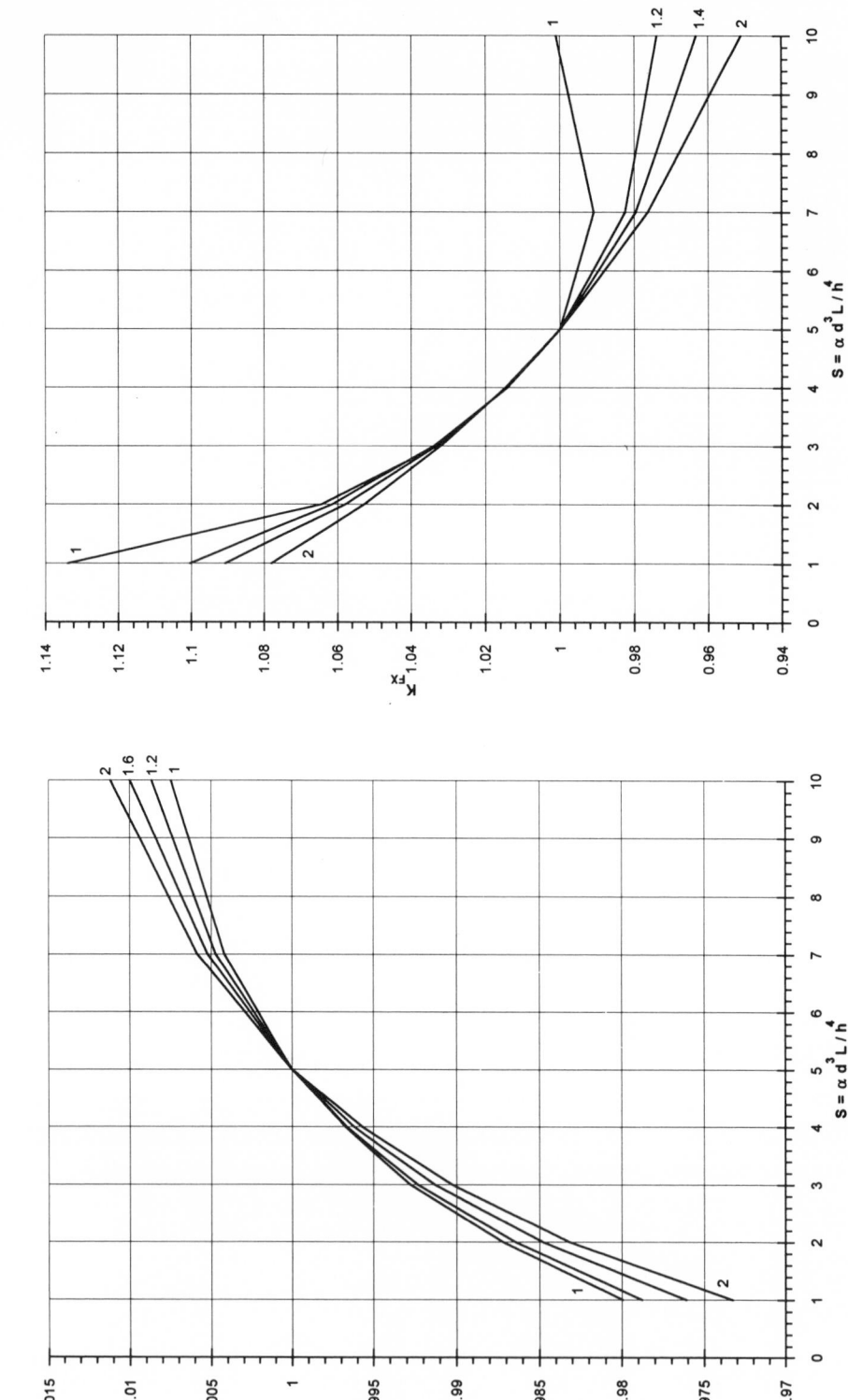

Graph 9.16 K_{Fx} for M_x. Stiffness correction coefficients for Zone F, curves for $l_x/l_y = 1$, 1.2, 1.4 and 2.

Graph 9.15 K_{Exy} for M_{xy}. Stiffness correction coefficients for Zone E, curves for $l_x/l_y = 1$, 1.2, 1.6 and 2.

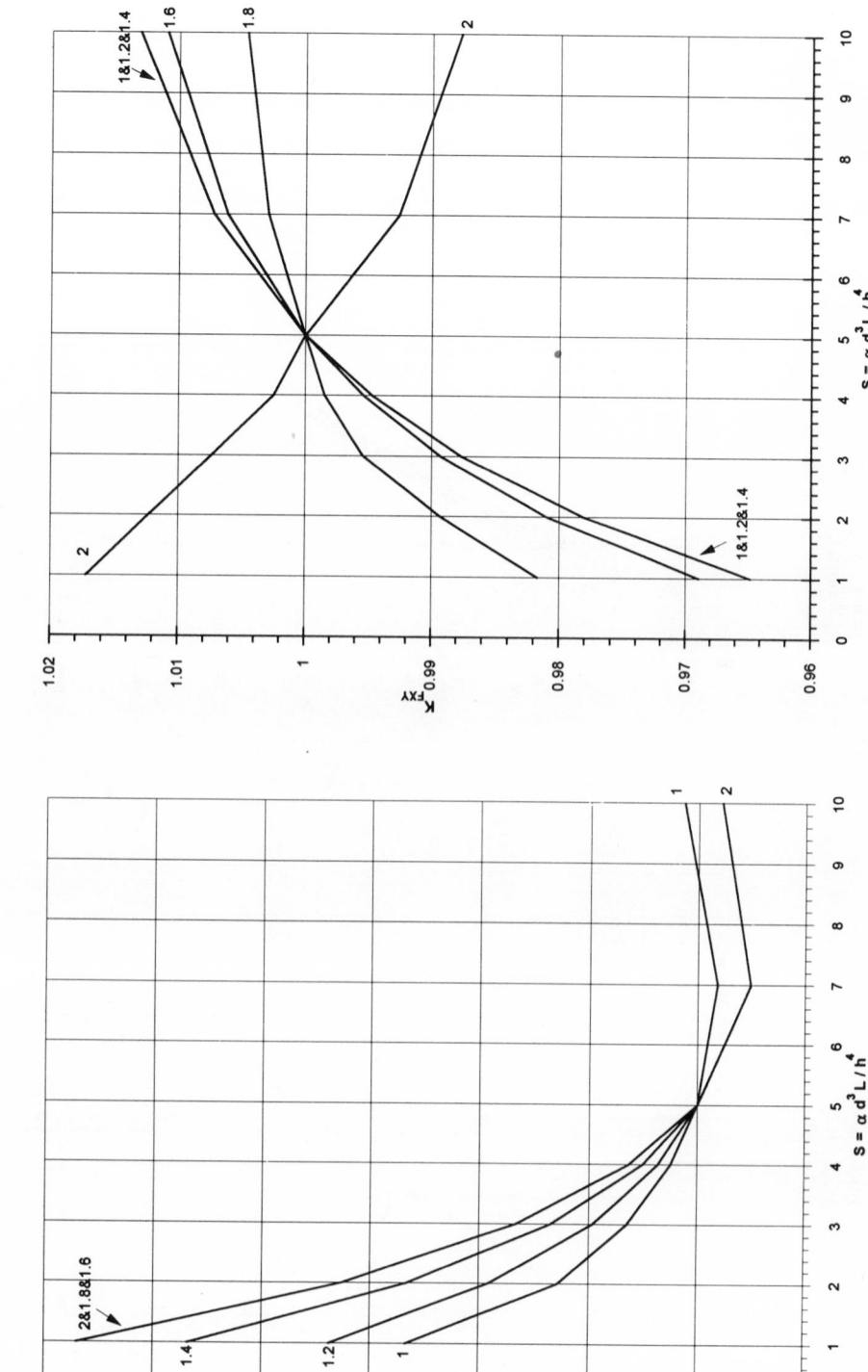

Graph 9.18 K_{Fxy} for M_{xy}. Stiffness correction coefficients for Zone F, curves for $l_x/l_y = 1$, 1.2, 1.4, 1.6, 1.8 and 2.

Graph 9.17 K_{Fy} for M_y. Stiffness correction coefficients for Zone F, curves for $l_x/l_y = 1$, 1.2, 1.4, 1.6, 1.8 and 2.

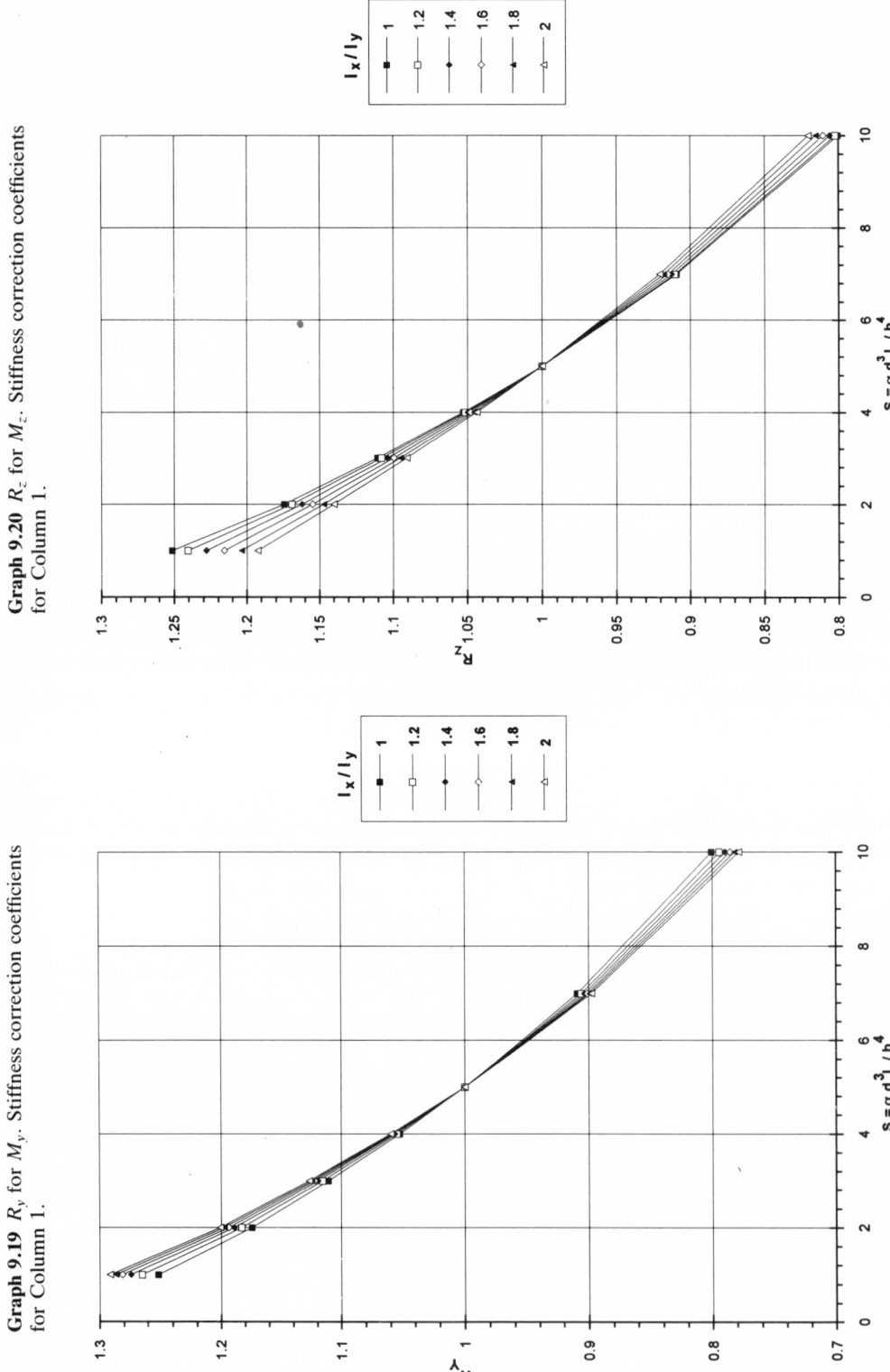

Graph 9.20 R_z for M_z. Stiffness correction coefficients for Column 1.

Graph 9.19 R_y for M_y. Stiffness correction coefficients for Column 1.

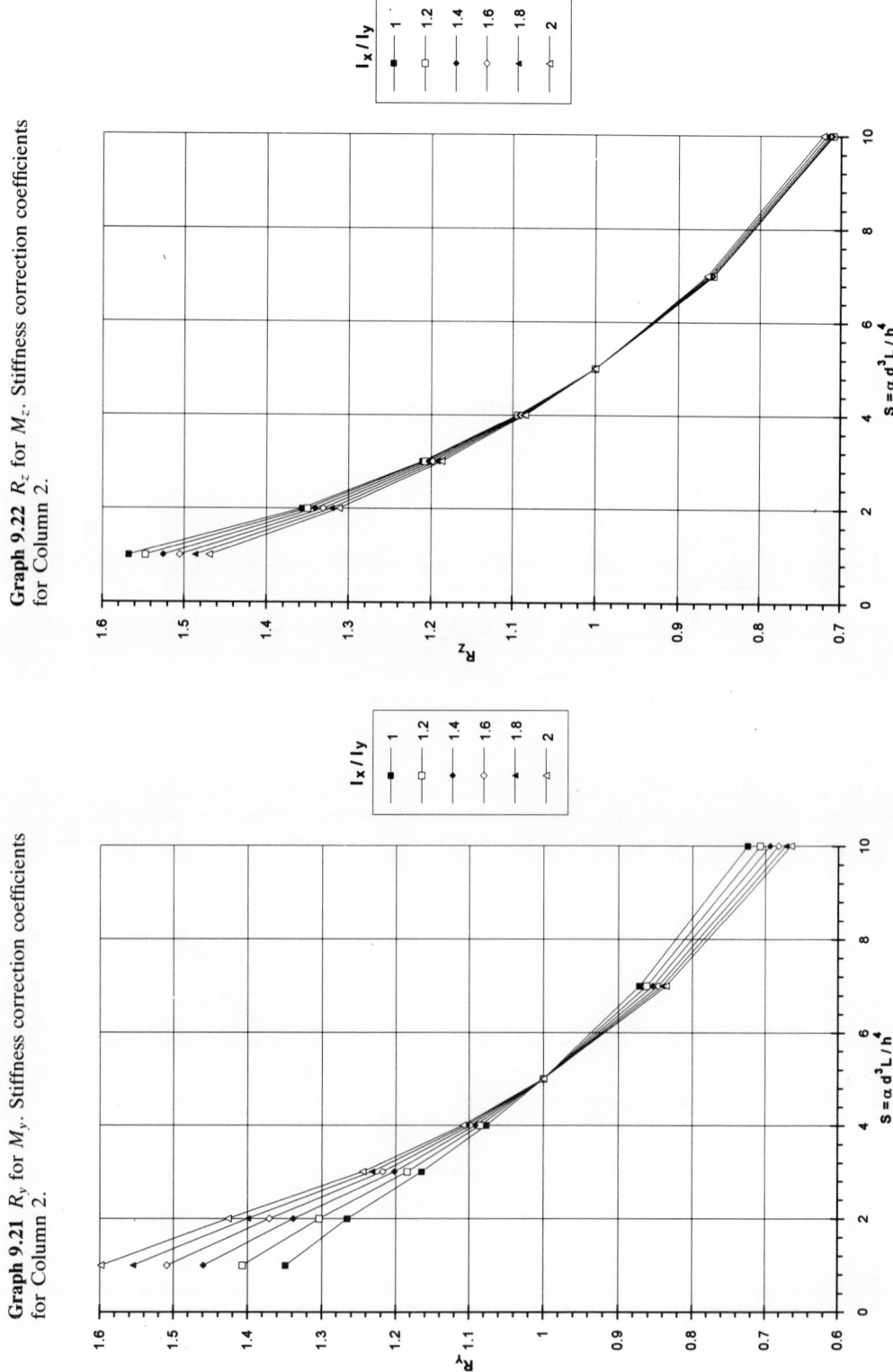

Graph 9.22 R_z for M_z. Stiffness correction coefficients for Column 2.

Graph 9.21 R_y for M_y. Stiffness correction coefficients for Column 2.

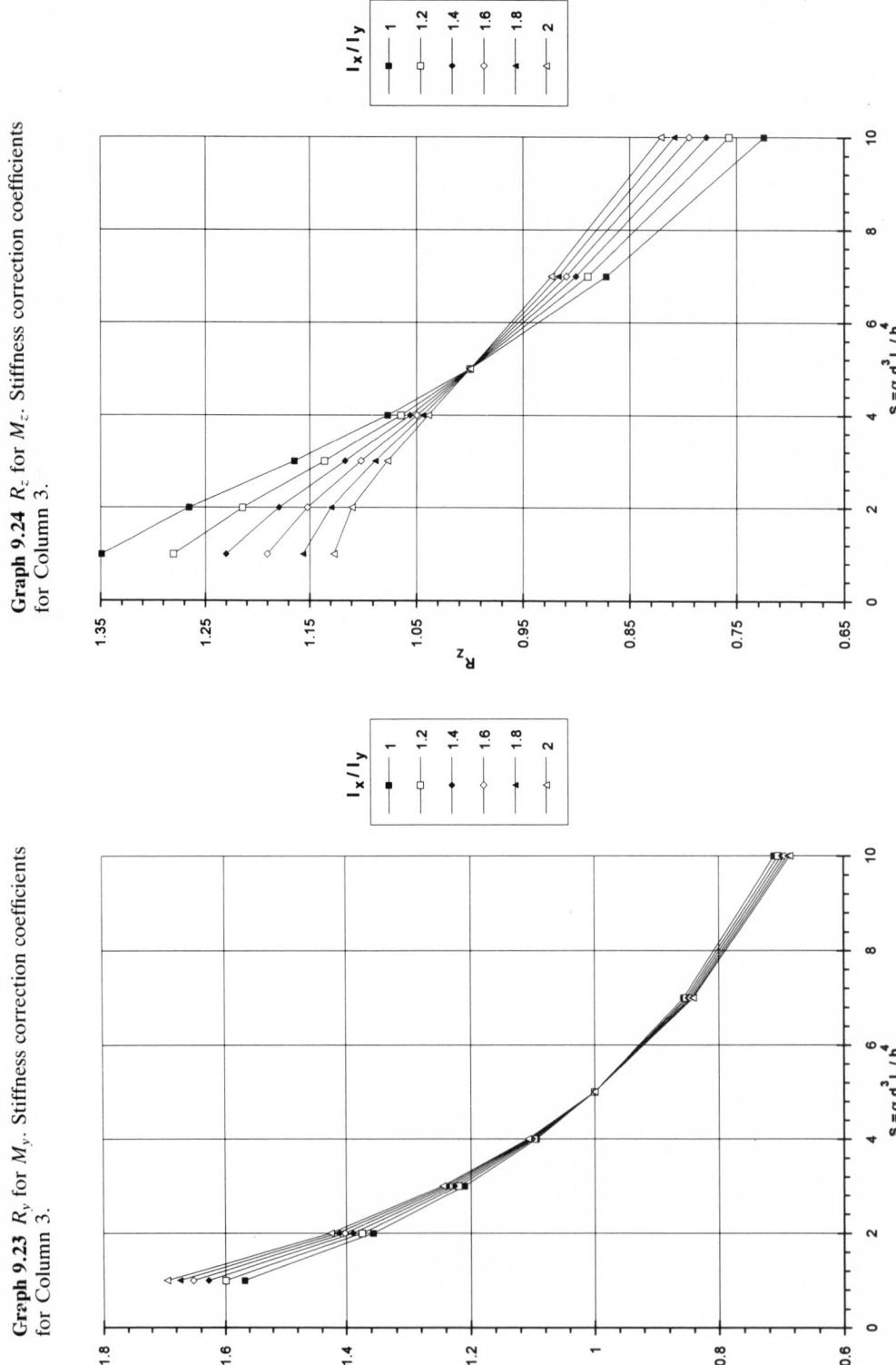

Graph 9.24 R_z for M_z. Stiffness correction coefficients for Column 3.

Graph 9.23 R_y for M_y. Stiffness correction coefficients for Column 3.

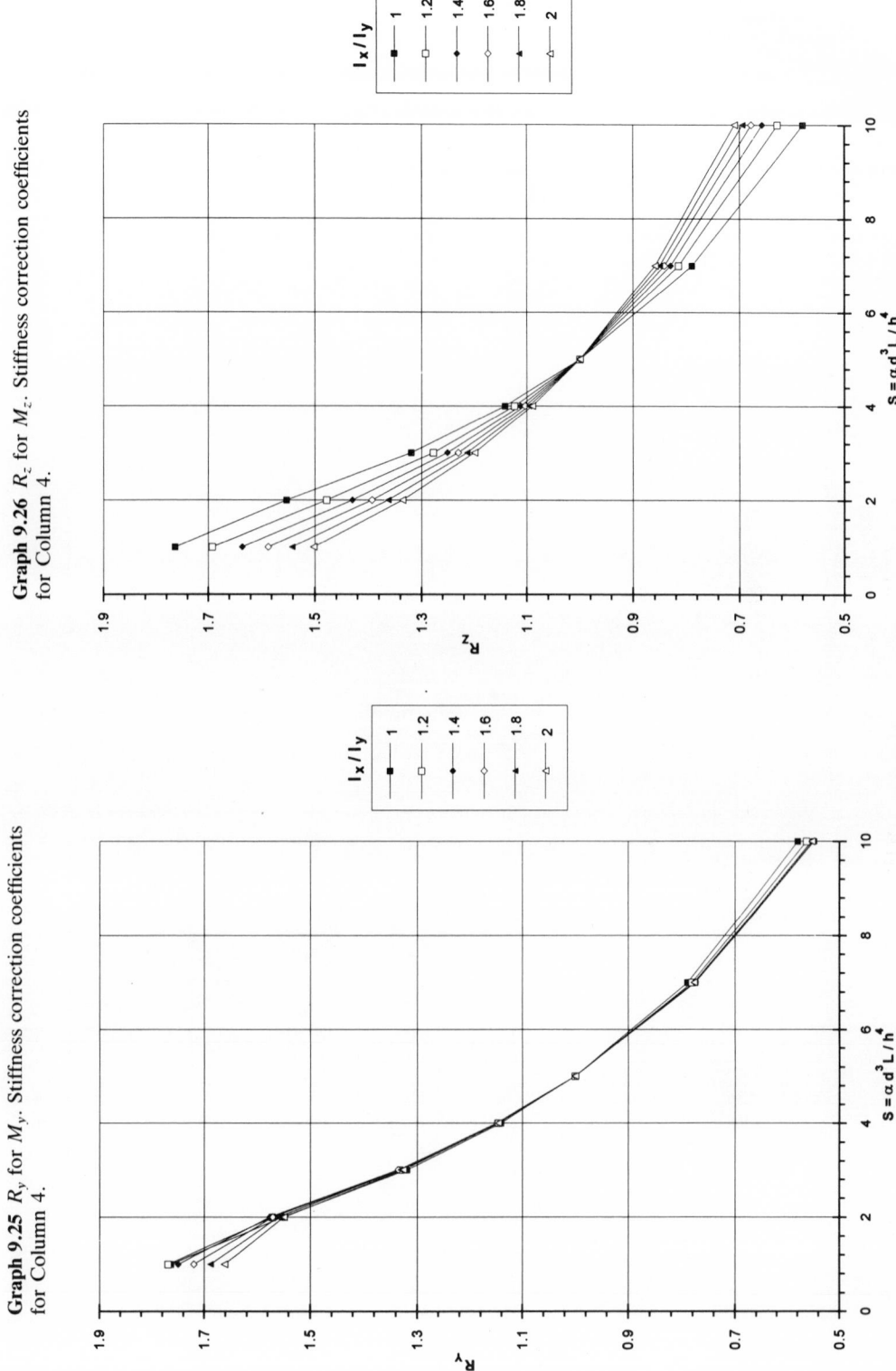

Graph 9.26 R_z for M_z. Stiffness correction coefficients for Column 4.

Graph 9.25 R_y for M_y. Stiffness correction coefficients for Column 4.

Chapter 10

Design of Connections

10.0 NOTATION

a	Distance between two rows of bars resisting bending moment
A_{st}	Area of steel in tension
C	Embedment of pile into pile cap
D	Diameter or width of pile
f_b	Ultimate anchorage bond stress
f_y	Characteristic yield strength of reinforcement
f_{bu}	Design ultimate bond stress
f_{cu}	Characteristic cube strength of concrete at 28 days
F_s	Ultimate force in a bar or a group of bars
F_t	Tie force (kN)
H	Design ultimate horizontal load on pile
K_t	Coefficient to determine transmission length of prestressing tendons
l	Anchorage bond length
l_r	Greatest distance between vertical load-bearing elements in direction of a tie
l_s	Floor to ceiling height (m)
l_t	Transmission length of prestressing tendons
M	Applied bending moment on concrete section of pile
n	Number of reinforcement bars in each row
n_o	Number of stories in a building
O	Perimeter of a bar of reinforcement in tension
r	Internal radius of a bend in a bar
v_c	Design concrete shear stress
V	Shear force in concrete section
z	Depth of lever arm
β	Coefficient to determine design ultimate bond stress
ϕ	Diameter of bar
ϕ_e	Diameter of one bar or equivalent diameter of a group of bars

10.1 INTRODUCTION

To make a complete building or structure, the elements described and designed in the previous chapters will have to be connected together and also tied together to give horizontal stability. This chapter describes the principles of design of these connections and the ties.

There are basically two types of connections: *rigid* and *free*. The rigid connections will have full moment and other internal force transfer capability. The connections classed as free do not offer resistance to rotation to members at the connection. These connections should be capable of transferring shear and axial loads.

10.2 CONTENTS: TYPE OF CONNECTIONS

The following connections have been described in this chapter:

(1) Requirement of building ties as per Codes of Practice.
(2) Pile-to-foundation/pile cap connection.
(3) Column-to-foundation connection.
(4) Wall-to-foundation connection.
(5) Column-to-column connection.
(6) Wall-to-wall connection.
(7) Column-to-beam connection.
(8) Wall-to-beam connection.
(9) Wall-to-slab connection.
(10) Column-to-wall connection.
(11) Slab-to-beam connection.

The theory of anchorage and bond length requirements is described initially.

10.3 ANCHORAGE AND BOND

Local bond stress is dependent on shear, i.e. the rate of change of bending moment at any section.

$$\text{Local bond stress} = \frac{V}{z\Sigma o}$$

where V = shear at section
z = lever arm of bending moment
Σo = summation of perimeter of bars in tension.

The local bond stress at ultimate state need not be checked provided there is adequate anchorage of the bars in tension on both sides of a section. The ultimate anchorage bond stress is assumed constant over the anchorage length of a bar.

$$f_b = \frac{F_s}{\pi\phi_e l}$$

where f_b = ultimate anchorage bond stress
F_s = ultimate force in bar or group of bars bundled together
ϕ_e = diameter of one bar or equivalent diameter of a bar, the area of which equals the total area of the bundle of bars
l = anchorage bond length.

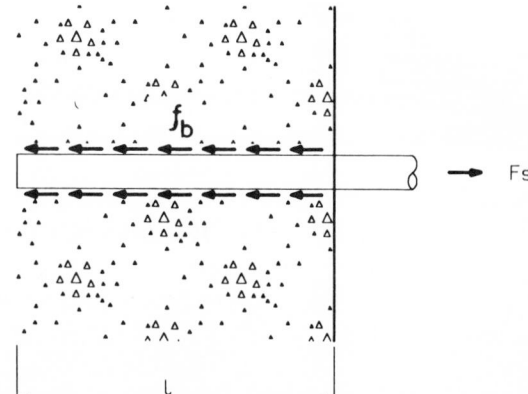

SK 10/1 Development of bond
stress in concrete.

The design ultimate bond stress depends on the characteristic strength of
the concrete and is given by the following formula:

$$f_{bu} = \beta \sqrt{f_{cu}}$$

Values of β.

	In tension	In compression
Plain bars	0.28	0.35
Type 1: deformed	0.40	0.50
Type 2: deformed	0.50	0.63
Fabric	0.65	0.81

A partial safety factor $\gamma_m = 1.4$ is included.
The *fabric reinforcement* should be according to clause 3.12.8.5 of BS 8110:
Part 1: 1985.[1]

10.3.1 Basic rules of anchorage and laps

Anchorage of links
The link is to pass round another bar of equal or greater dimension
through an angle of 90° and continue for a minimum length of 8 × diameter,
or through an angle of 180° and continue for a minimum length of
4 × diameter.

Anchorage of column starter bars
Column starter bars in compression need not be checked for anchorage.
They should be taken down to the level of the bottom layer of reinforcement
in the foundation.

Column starter bars in tension should be checked for anchorage.

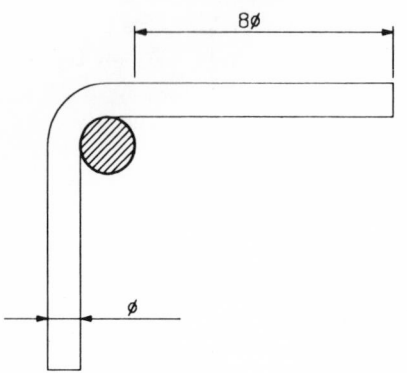

SK 10/2 A 90° bend.

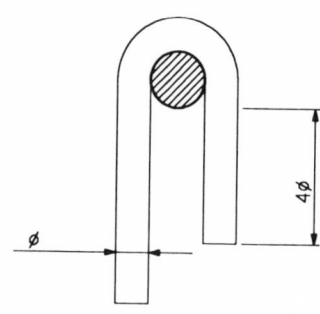

SK 10/3 A 180° bend.

Laps and joints

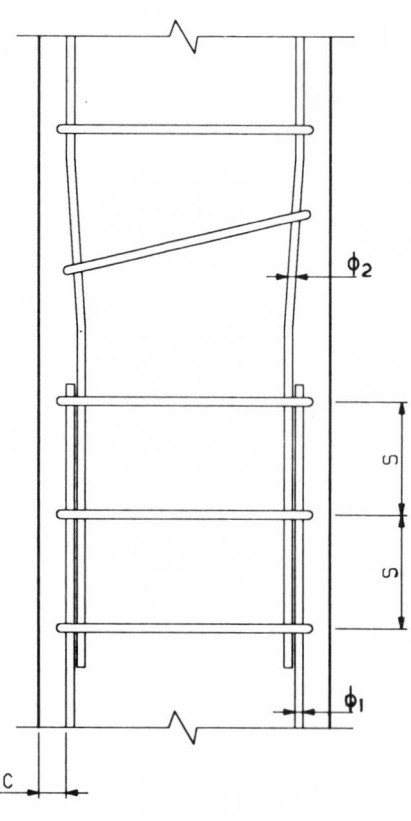

$S \leqslant 200$ IF
ϕ_1 AND $\phi_2 > 20$
AND $C < 1.5\phi_1$ OR, $1.5\phi_2$

SK 10/4 Detailing rules at lap of
column bars.

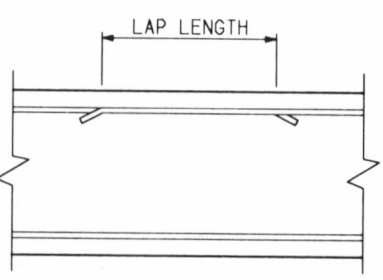

SK 10/5 Lapping of bars.

Laps and joints should be staggered. Welded joints should not be used for cyclic loading. Minimum lap length is 15 times bar size or 300 mm. Links to be used at laps of bars in beams and columns at a maximum spacing of 200 mm where both bars at a lap exceed 20 mm diameter and the cover is less than 1.5 times the diameter of the smaller bar.

10.3.2 Design of tension laps

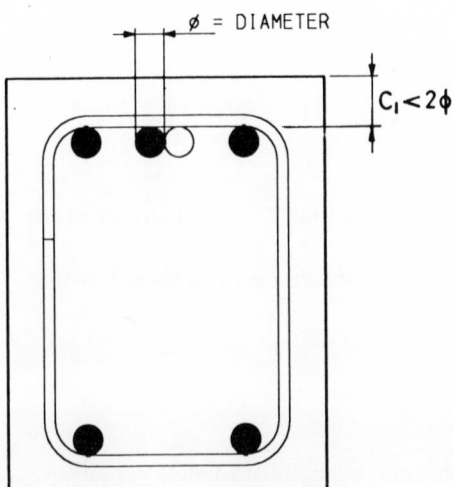

SK 10/6 Case 1 — anchorage length.

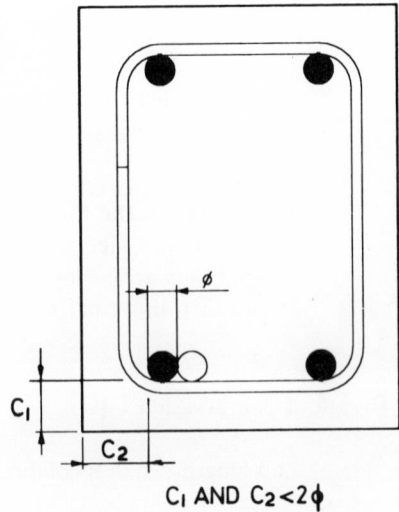

SK 10/7 Case 2 — anchorage length.

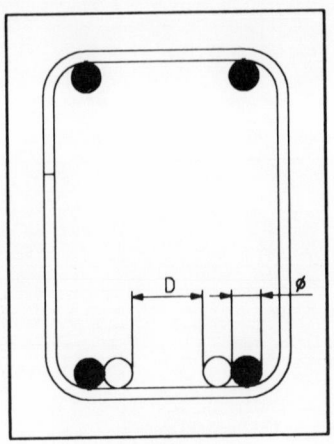

SK 10/8 Case 3 — anchorage length.

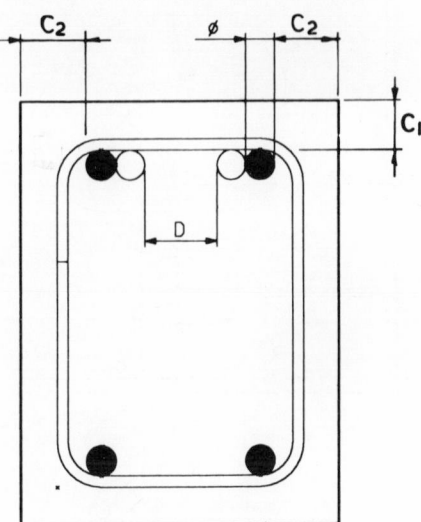

SK 10/9 Case 4 — anchorage length.

Lap length = tension anchorage length normally
        = 1.4 × tension anchorage length for Cases 1, 2 and 3
        = 2 × tension anchorage length for Case 4

Case 1
Bars lapped are at the top of a section and cover is less than 2 times the size of lapped bars.

Case 2
Bars lapped are at a corner of a section and cover to either face is less than 2 times the size of lapped bar.

Case 3
The distance between adjacent laps is less than 75 mm or 6 times bar diameter, whichever is the greater.

Case 4
Corner bars at the top of a section with less than 2 times diameter of bar cover to either face.
Lapped bars at the top of a section with distance between them less than 75 mm or 6 times bar diameter.

10.3.3   Design of compression laps

Lap length = 1.25 × compression anchorage length of smaller bar at lap

Effective anchorage length of a hook or a bend

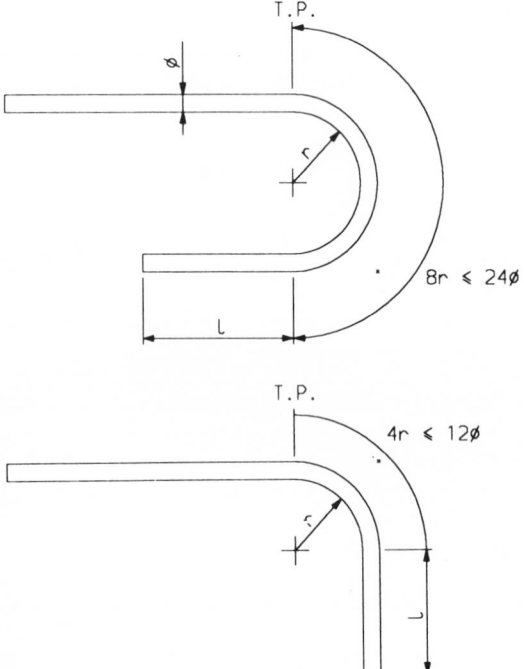

SK 10/10 Effective anchorage bond lengths.

180° hook

Effective anchorage length $= 8r(\leq 24\phi) + l - 4\phi$

or $=$ actual length of bar from tangent point
whichever is larger

90° bend

Effective anchorage length $= 4r(\leq 12\phi) + l - 4\phi$

or $=$ actual length of bar from tangent point
whichever is larger

10.3.4   Curtailment and anchorage of bars

Minimum anchorage length $= d$ or 12ϕ

In tension zone of a flexural member, take a bar to:

● a full tension anchorage length beyond a point where it is not required,
  or
● a point where the shear capacity of the section is twice the shear force at
  the point, or
● a point where the available bars continuing beyond provide a moment
  of resistance twice the bending moment at the point.

The curtailment of bars should be staggered.

Anchorage of bars at a simply supported end

(1) An effective anchorage length of 12ϕ beyond centreline of support.
    Hook or bend should not begin before centreline. Effective anchorage
    lengths of a hook or a bend may be considered.

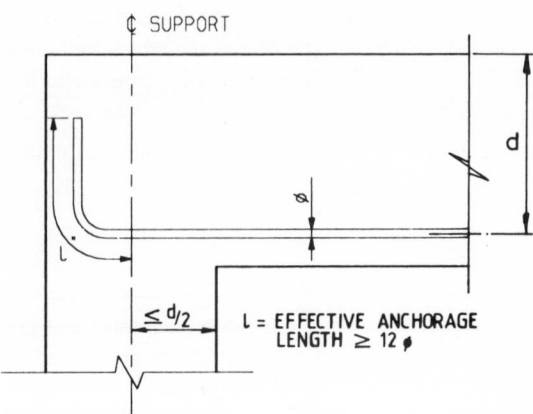

SK 10/11 Required effective
anchorage at simply supported end
where centreline of support is less
than or equal to $d/2$ away from
face of support.

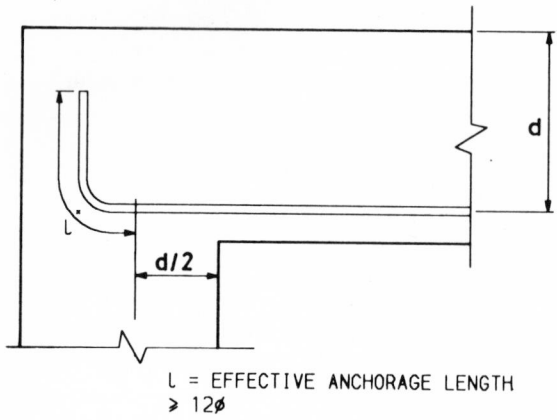

L = EFFECTIVE ANCHORAGE LENGTH
≥ 12∅

SK 10/12 Required effective anchorage at simply supported end where centreline of support is more than $d/2$ away from face of support.

(2) An effective anchorage length of 12φ beyond $d/2$ from face of support. Hook or bend should not begin before $d/2$ from face of support. Effective anchorage lengths of a hook or a bend may be considered.

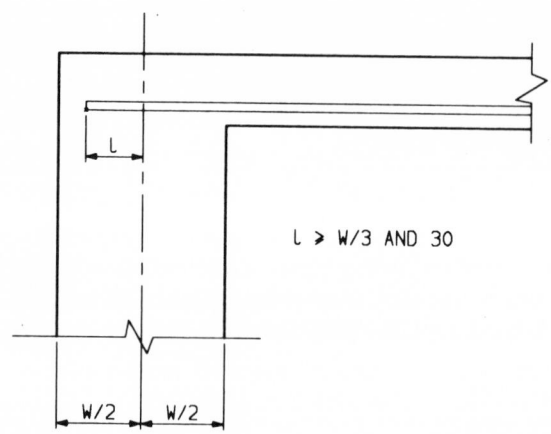

L ≥ W/3 AND 30

SK 10/13 Required effective anchorage at end of slab where shear stress is less than $0.5V_c$.

(3) For slabs, if shear stress at face of support is less than half v_c, then project a straight length of bar beyond centreline of support equal to one third of support width or 30 mm, whichever is greater.

Anchorage bond lengths in multiples of bar sizes for Type 2 deformed bars ($f_y = 460 \, \text{N/mm}^2$).

	Grades of concrete – f_{cu} (N/mm^2)					
	C25	C30	C35	C40	C45	C50
Tension anchorage and lap	40	37	34	32	30	28
Compression anchorage	32	29	27	26	24	22
Compression lap	40	37	34	32	30	28

Anchorage bond lengths in multiples of bar sizes for plain grade $250\,\mathrm{N/mm^2}$ bars

	Grades of concrete – f_{cu} $(\mathrm{N/mm^2})$					
	C25	C30	C35	C40	C45	C50
Tension anchorage and lap	39	36	33	31	29	27
Compression anchorage	32	29	27	25	23	22
Compression lap	39	36	33	31	29	27

Note: The tension anchorage bond lengths will be multiplied by either 1.4 or 2.0, depending on the location of the bar as described in Section 10.3.2.

10.4 BUILDING TIES

The following ties will be considered:

- Peripheral ties.
- Internal ties.
- Horizontal column and wall ties.
- Vertical ties.

Ties are continuous fully anchored and properly lapped welded or mechanically connected tension reinforcement.

The reinforcement required to act as continuous ties is additional to other designed reinforcement. Available excess design reinforcement if properly tied and continuous and capable of carrying the prescribed tie forces may be considered.

10.4.1 Peripheral ties

A continuous tie should be provided at each floor level and roof level within 1.2 m of edge of building or within perimeter wall. This tie should be capable of resisting a tensile force equal to $F_t\mathrm{kN}$.

$F_t = 20 + 4n_o$ or   60   whichever is less

where n_o = number of storeys in the structure.

Required area of steel for peripheral tie $= A_{st} = \dfrac{F_t}{0.87f_y}$

This means that the maximum area of steel for peripheral tie at each floor and roof level is given by:

$\dfrac{60 \times 10^3}{0.87 \times 460} = 150\,\mathrm{mm^2}$

or   two 10 mm dia. bars $(f_y = 460\,\mathrm{N/mm^2})$ fully lapped and anchored.

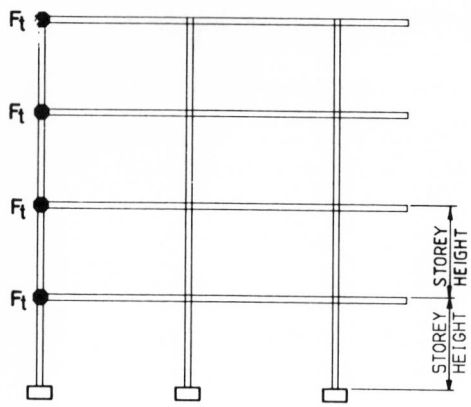

SK 10/14 Typical frame elevation showing tie forces.

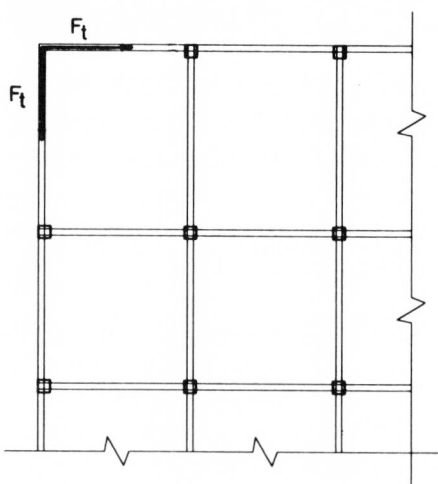

SK 10/15 Typical floor plan showing ties required.

10.4.2   Internal ties

These ties are at floor and roof levels in two orthogonal directions and anchored to peripheral ties or columns or perimeter walls.

The spacings of these ties will not be greater than $1.5l_r$, where l_r is greatest

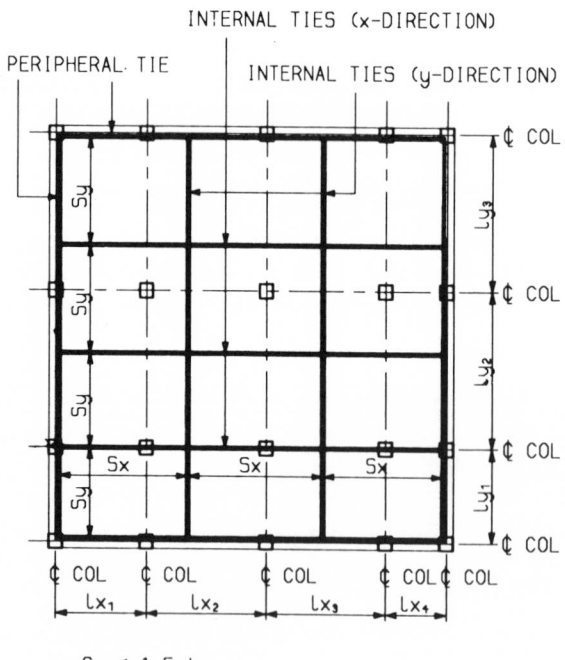

$$S_x \leqslant 1.5 \, L_x$$
$$S_y \leqslant 1.5 \, L_y$$

SK 10/16 Typical floor plan showing ties required.

distance between centres of vertical load-bearing elements in direction of tie.

The ties should be capable of resisting a tensile force equal to the greater of:

$$0.0267(g_k + q_k)lF_t \qquad \text{or} \qquad F_t$$

where $(g_k + q_k)$ is the sum of the average characteristic dead and imposed floor loads (kN/m^2).

10.4.3   Horizontal column and wall ties

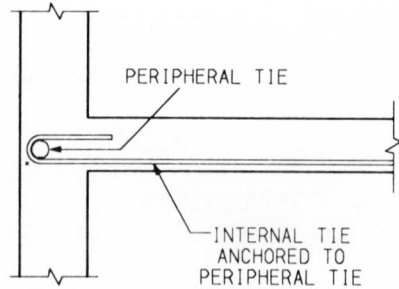

SK 10/17 Anchorage of ties.

If peripheral tie is located in wall, then internal ties should be anchored to peripheral tie. No other wall tie is required.

Each external column should be tied back horizontally at each floor or roof level. The tie force will be the greater of (a) or (b) below.

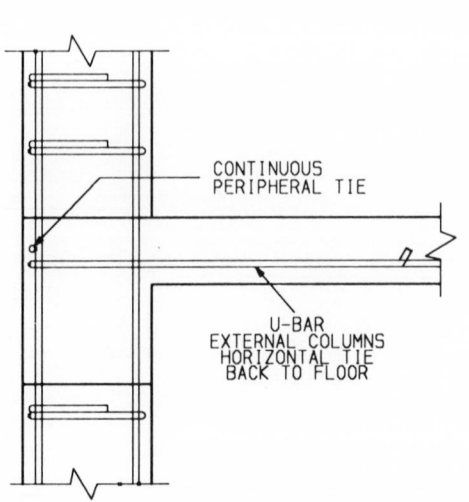

SK 10/18 External column − elevation showing tie back.

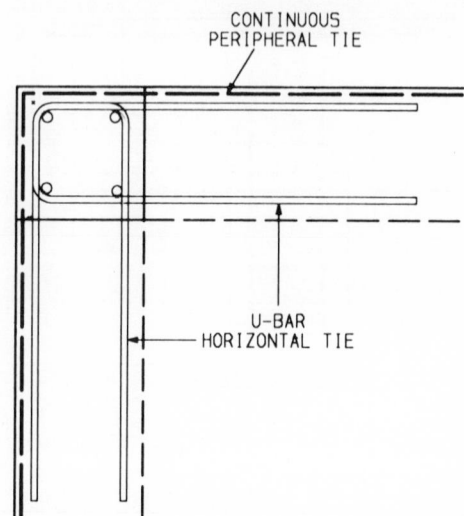

SK 10/19 Corner column − plan view showing tie back.

(a) $2F_t$ or $(l_s/2.5) F_t$ if less, where l_s = floor-to-ceiling height (m)
(b) 3% of total design ultimate axial load carried by column

If peripheral tie is not located in wall, then every metre of wall should be tied back at each floor or roof level. The tie force will be either (a) or (b), as above.

The corner column will have horizontal ties at each floor level or roof level in each of two directions, capable of developing a tie force equal to either (a) or (b), as above.

10.4.4   Vertical ties

Each column and each wall should be tied continuously from the lowest to the highest level. The tie force in tension will be the maximum design ultimate dead and live load imposed on column or wall from any one storey.

10.5   CONNECTIONS

The most commonly occurring structural connections are illustrated in this section with guidance on the preferred detailing methods.

10.5.1   Pile-to-pile cap/foundation raft/ground beam

10.5.1.1   *Bored and cast in-situ concrete pile*

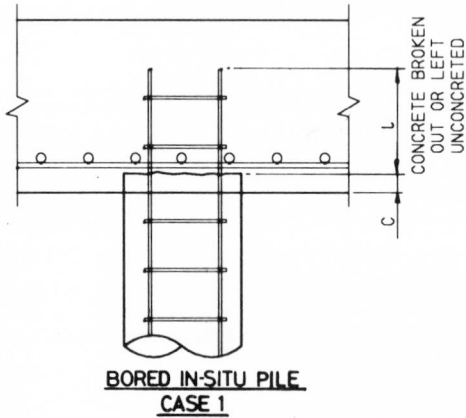

BORED IN-SITU PILE
CASE 1

SK 10/20 Pile-to-foundation connection.

Case 1
Mainly vertical loads.
Small horizontal load.
No bending moment in pile at connection.
No tension loads in pile.

Pile embedment C into pile cap, or raft, or ground beam, up to bottom layer of reinforcement.

Check bearing stress on concrete (pile and pile cap) due to horizontal load.

$$\text{Bearing stress} = \frac{H}{DC} \leq 0.6 f_{cu}$$

where H = design ultimate horizontal load on pile
D = diameter or width of pile
C = embedment of pile into pile cap.

Check anchorage of pile reinforcement.

l = compression anchorage length

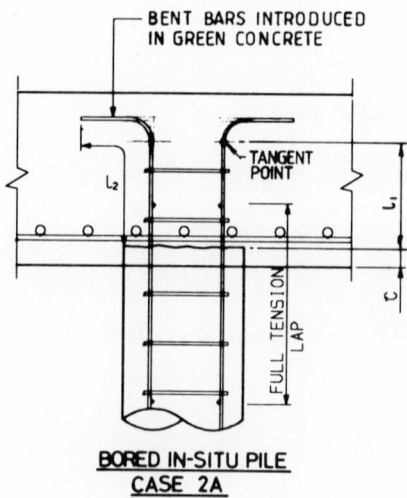

BORED IN-SITU PILE
CASE 2A

SK 10/21 Pile-to-foundation connection.

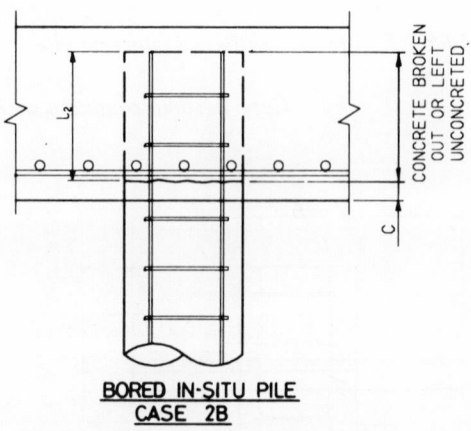

BORED IN-SITU PILE
CASE 2B

SK 10/22 Pile-to-foundation connection.

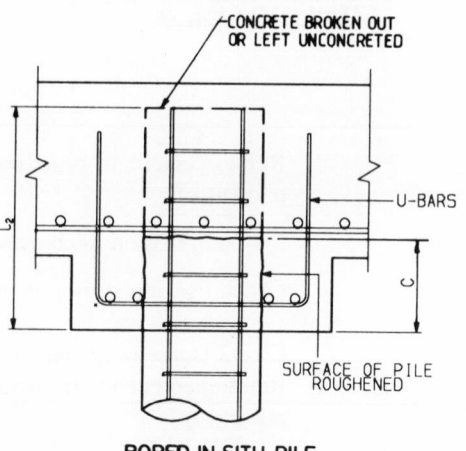

BORED IN-SITU PILE
CASE 2C

SK 10/23 Pile-to-foundation connection.

Case 2
Vertical compression load.
Vertical tension load.
Piles in swelling clay.
Bending moment in pile at connection.
Horizontal load on pile.

Check bearing stress on concrete as in Case 1.
Check anchorage of pile reinforcement.

l_1 = compression anchorage length
l_2 = tension anchorage length (the effective anchorage length of the bend
      may be considered)

If bearing stress is higher than allowed, then embedment may be increased
by adopting solution in Case 2C.

10.5.1.2 *Precast reinforced concrete pile*

Case 1 (same condition as in Section 10.5.1.1, Case 1)

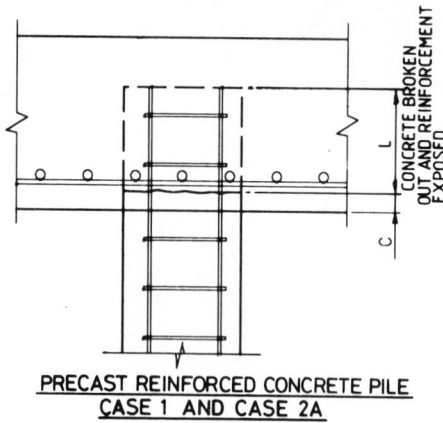

PRECAST REINFORCED CONCRETE PILE
CASE 1 AND CASE 2A

SK 10/24 Pile-to-foundation
connection.

Reinforcement to be exposed by breaking out a length equal to *l* above
pile cut off.

l = compression anchorage length

Check bearing stress, as in Section 10.5.1.1, Case 1.

Case 2 (same condition as in Section 10.5.1.1, Case 2)
Reinforcement to be exposed by breaking out a length equal to *l* above
pile cut off.

l = tension anchorage length

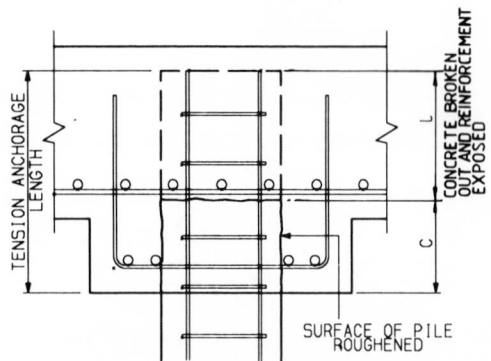

SK 10/25 Pile-to-foundation
connection.

<u>PRECAST REINFORCED CONCRETE PILE</u>
<u>CASE 2B</u>

Check bearing stress, as in Section 10.5.1.1, Case 2.

If bearing stress is higher than allowed then embedment C may be increased by adopting solution in Case 2B.

10.5.1.3 Precast prestressed concrete pile

Case 1 (same condition as in Section 10.5.1.1, Case 1)

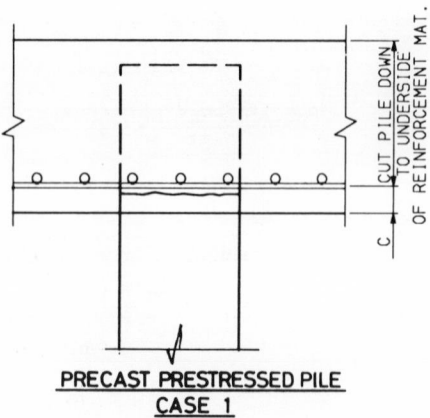

SK 10/26 Pile-to-foundation
connection.

<u>PRECAST PRESTRESSED PILE</u>
<u>CASE 1</u>

Check bearing stress, as in Section 10.5.1.1, Case 1.

Case 2 (same condition as in Section 10.5.1.1, Case 2)

l = transmission length of prestressing tendons

$$= \frac{K_t \phi}{\sqrt{f_{cu}}}$$

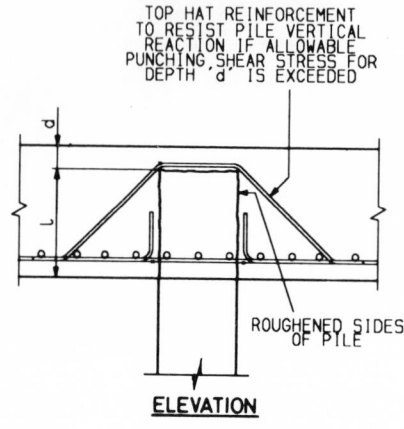

TOP HAT REINFORCEMENT
TO RESIST PILE VERTICAL
REACTION IF ALLOWABLE
PUNCHING SHEAR STRESS FOR
DEPTH 'd' IS EXCEEDED

ROUGHENED SIDES
OF PILE

ELEVATION

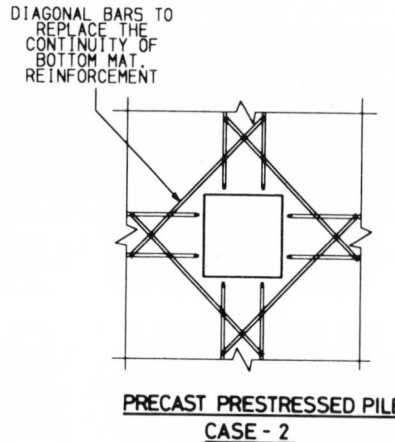

DIAGONAL BARS TO
REPLACE THE
CONTINUITY OF
BOTTOM MAT
REINFORCEMENT

PRECAST PRESTRESSED PILE
CASE - 2

SK 10/27 Pile-to-foundation connection.

where ϕ = nominal diameter of tendon

K_t = 600 for plain or indented wire

= 400 for crimped wire with wave height not less than 0.15 mm

= 240 for 7-wire strand or super-strand

= 360 for 7-wire drawn strand.

The top-hat reinforcement detailing at pile connection may be adopted for large vertical load and significant bending moment in pile.

10.5.1.4 Steel H-pile or steel tubular pile

Case 1 (same conditions as in Section 10.5.1.1, Case 1)
Check bearing stress, as in Section 10.5.1.1 Case 1, using width of flange or depth of section, whichever is smaller.
Check bearing stress on concrete on top of mild steel plate using maximum ultimate vertical load on pile.

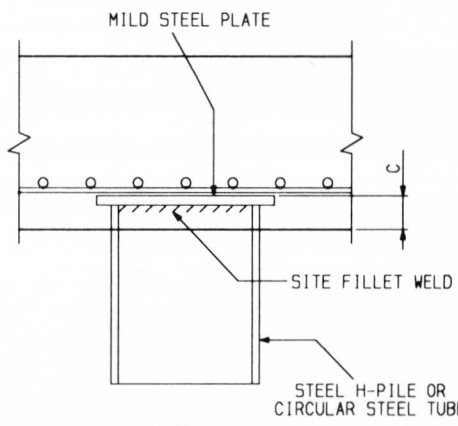

SK 10/28 Connection of steel pile to foundation.

CASE 1

Case 2 (same conditions as in Section 10.5.1.1, Case 2)
Check bearing stress, as in Section 10.5.1.1 Case 1, using width of flange or depth of section, whichever is smaller.
Check bearing stress on concrete on top of mild steel plate using maximum ultimate vertical load on pile.

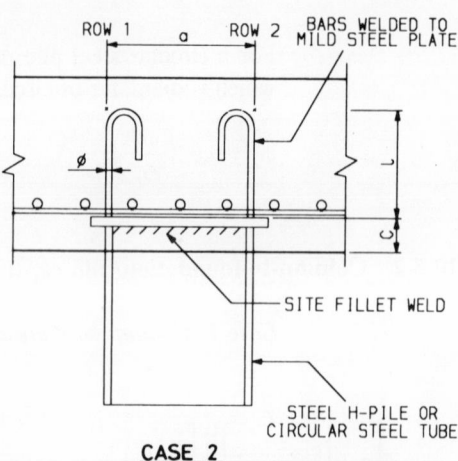

SK 10/29 Moment connection of steel pile to foundation.

CASE 2

l = tension anchorage length of the type of bar used (the effective anchorage length of a hook or a bend may be used)

$$\phi = \text{diameter of bar} = 1.21\left(\frac{M}{af_yn}\right)^{\frac{1}{2}}$$

M = maximum ultimate bending moment in pile at connection

a = distance between bars perpendicular to axis of rotation or moment, or distance between two rows.

n = number of bars in each row

Note: During the driving operation it is difficult to control the orientation of the flanges of H-piles to match the axis of bending moment. In practice, the anchoring bars should be used in such a fashion that the bending moment capacity of these anchor bars are equal in both the orthogonal directions. The mild steel plate should be checked for strength to transfer the anchoring tension of the bars to the web of the H-pile.

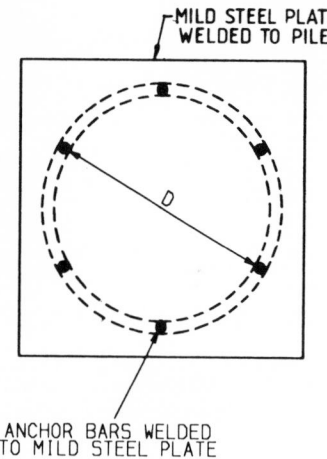

MILD STEEL PLATE
WELDED TO PILE

ANCHOR BARS WELDED
TO MILD STEEL PLATE

SK 10/30 Plan of circular pile showing location of anchor bars for moment connection as in Case 2.

For a circular steel pile using 6 no. anchor bar on a circle of diameter D which is diameter of circular pile, diameter of each anchor bar is given by:

$$\phi = 0.92\left(\frac{M}{Df_y}\right)^{\frac{1}{2}}$$

10.5.2 Column-to-foundation/pile cap/raft

Case 1: Column bars always in compression

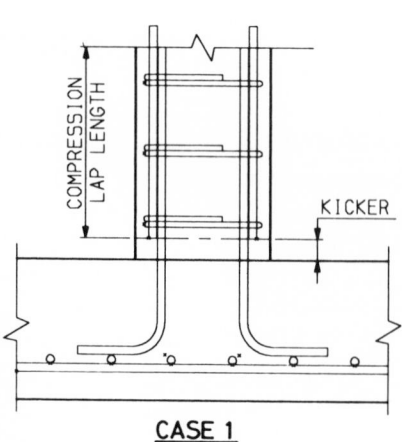

COMPRESSION
LAP LENGTH

KICKER

CASE 1

SK 10/31 Connection of column to foundation — no tension in bars.

No check necessary for anchorage bond length.
The column bars should be taken down to bottom reinforcement mat of foundation.

Case 2: Significant tension in column bars due to foundation fixity bending moment and/or direct axial tension load

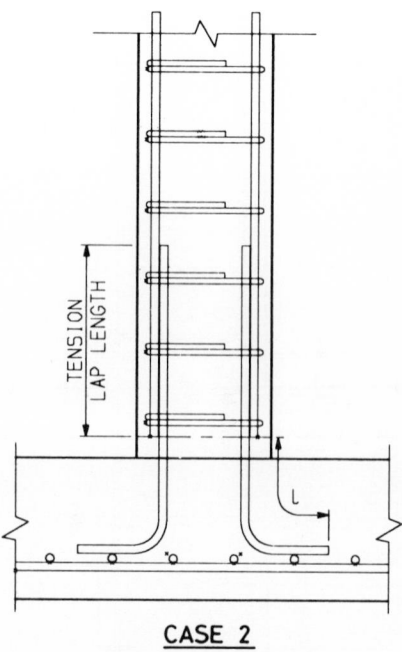

CASE 2

SK 10/32 Connection of column to foundation — significant tension in bars.

l = tension anchorage length (effective anchorage length of a bend may be used)

10.5.3 Wall-to-foundation/pile cap/raft

Case 1 (same condition as in Section 10.5.2, Case 1)
Use same principle as in 10.5.2, Case 1.

Case 2 (same condition as in Section 10.5.2, Case 2)
Use same principle as in Section 10.5.2, Case 2.

10.5.4 Column-to-column connection

Case 1: Column bars always in compression

l = compression lap length

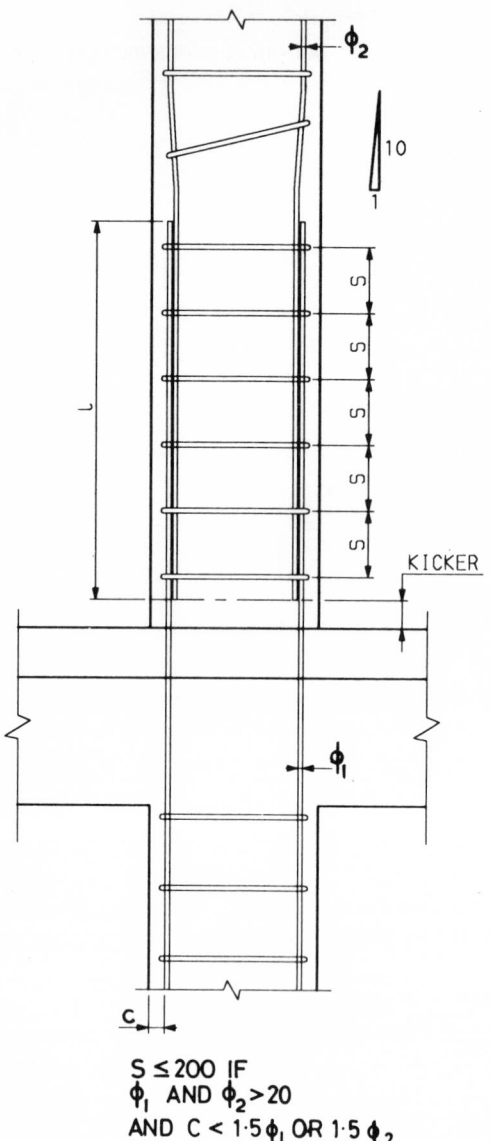

$S \leq 200$ IF
ϕ_1 AND $\phi_2 > 20$
AND $C < 1.5\phi_1$ OR $1.5\phi_2$

SK 10/33 Column-to-column connection.

The links at the lap will be at a maximum spacing of 200 mm if bars lapped are greater than 20 mm in diameter and cover is less than 1.5 times bar diameter.

Case 2: Significant tension in column bars due to bending moment or axial tension

l = tension anchorage length if cover is at least 2 times diameter of lapped bars.

$l = 1.4 \times$ tension anchorage length for corner bars where cover to either face is less than 2 times diameter of lapped bars

$l = 1.4 \times$ tension anchorage length if adjacent laps are less than 75 mm or 6 times bar diameter away

10.5.5 Wall-to-wall connection

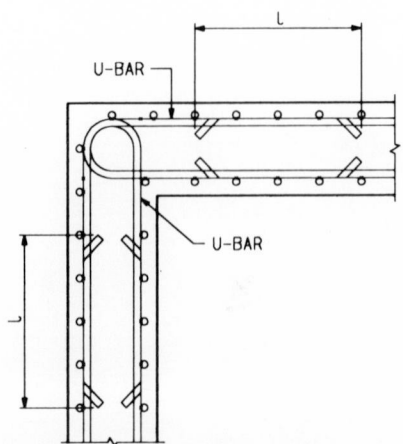

SK 10/34 Type 1 – plan of wall at corner.

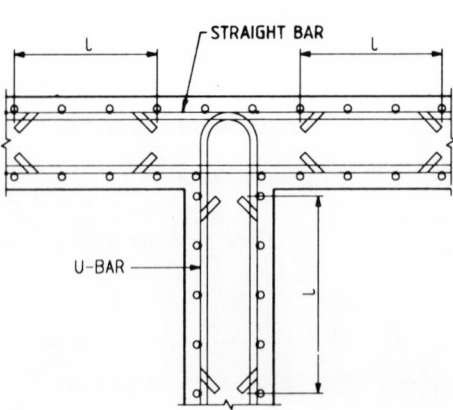

SK 10/35 Type 1 – plan of wall at intersection.

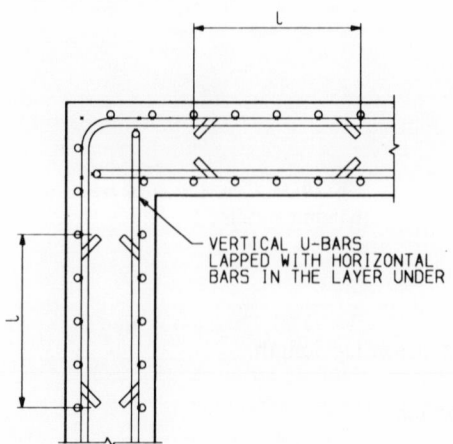

SK 10/36 Type 2 – plan of wall at corner.

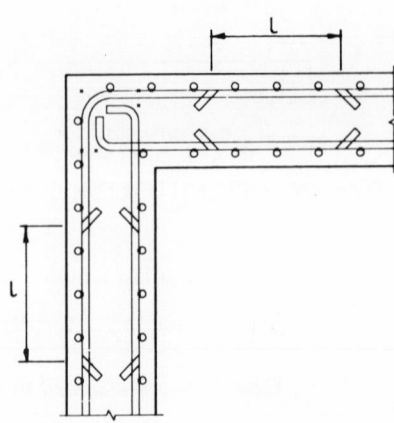

SK 10/37 Type 3 – plan of wall at corner.

Type 1 and *Type 2* connections are efficient for significant reversible bending moment at connection. If horizontal bars are designed to carry significant tension at junction, then lap length is given by:

l = tension anchorage lap length

Type 3 connection may be used where nominal horizontal reinforcement is required to prevent cracking and to contain vertical reinforcement. The lap length is given by:

l = 15 times bar diameter or 300 mm, whichever is greater

Note: If the loading causes the corner of the wall to open up then Type 2 connection becomes most efficient. Moreover, with Type 2 detailing the horizontal bars could be of different diameters at the inside and the outside faces.

10.5.6 Column-to-beam connection

10.5.6.1 External columns

Case 1: Beam assumed simply supported

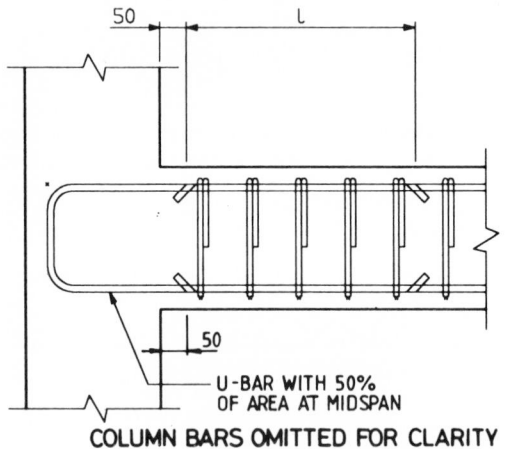

U-BAR WITH 50%
OF AREA AT MIDSPAN

COLUMN BARS OMITTED FOR CLARITY

SK 10/38 Connection of beam to column – Case 1.

l = tension lap length

See Section 10.3.2 for design of tension lap length.

Case 2: Beam assumed fixed to column

l_1 = length required by design calculations $\geq$ 0.15 span $\geq$ tension lap length

r = radius of bend (special radius may be necessary)
Check bearing stress inside bend (see Step 22 of Section 2.3).

l_2 = tension lap length (see Section 10.3.2)

l_3 = tension anchorage length (to be checked if available within depth)

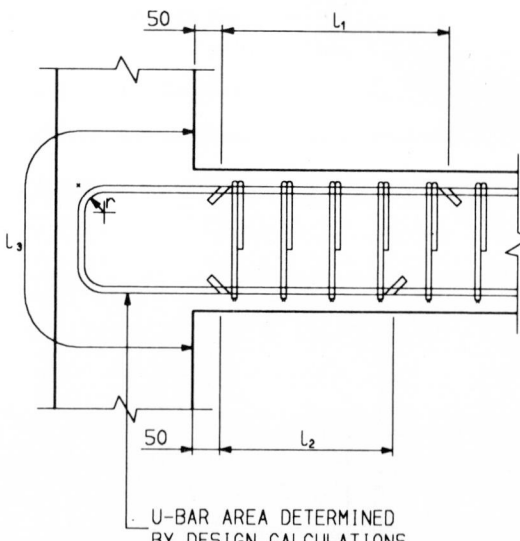

SK 10/39 Connection of beam to
column − Case 2.

U–BAR AREA DETERMINED
BY DESIGN CALCULATIONS

Case 3: Beam assumed fixed to column (no reversal of moment)

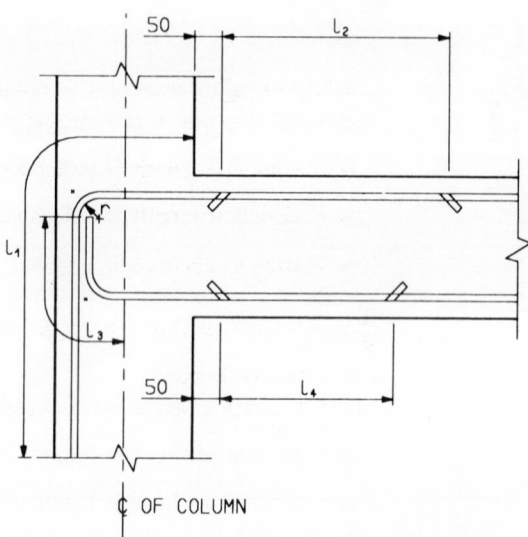

SK 10/40 Connection of beam to
column − Case 3.

₵ OF COLUMN

If l_3 in Case 2 is less than tension anchorage length, use detail in Case 3.

l_1 = tension anchorage length

r = radius of bend
Check bearing stress inside bend (see Step 22 of Section 2.3).

l_2 = length required by design calculations $\geq$ 0.15 span $\geq$ tension lap
length

l_3 = 12 × bar diameter of effective anchorage

l_4 = tension lap length (see Section 10.3.2)

Case 4: Beam assumed fixed to column (ductile connection for reversible moment)

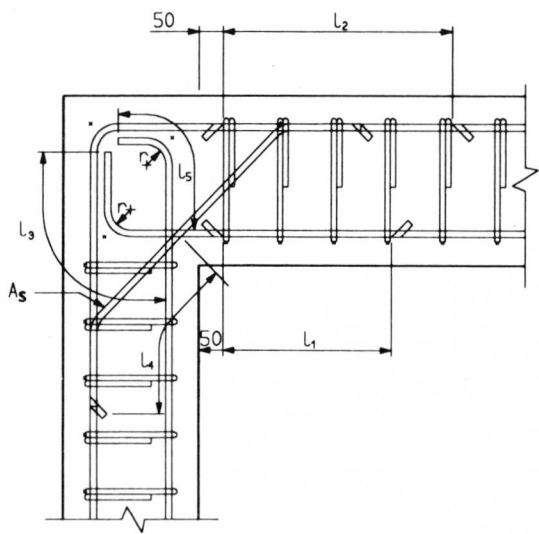

SK 10/41 Ductile column/beam connection – Case 4.

Where bending moments at connection are very large and reversible, e.g. at knee of a portal frame, use detail in Case 4.

l_1 = tension lap length (see Section 10.3.2)

l_2 = designed length ≥ 0.15 span ≥ tension lap length

l_3 = tension anchorage length

May be provided with a hook at end to get full effective anchorage length.

r = radius of bend
Check bearing stress inside bend (see Step 22 of Section 2.3).

l_4 = tension anchorage length

l_5 = tension anchorage length

A_s = same area of steel as beam design bottom steel at column

10.5.6.2 *Internal columns*

Connection uses straight splice bars at intersection.
Splice bars for secondary beam may be placed inside splice bars of main beam.

l_1 = tension lap length (see Section 10.3.2)

l_2 = designed length ≥ tension anchorage length

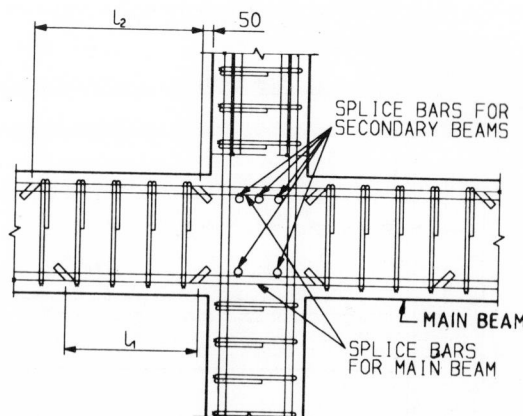

SK 10/42 Internal column −
beams from both orthogonal
directions.

10.5.7   Wall-to-beam connection

The same principles apply as in Section 10.5.6.

10.5.8   Wall-to-slab connection

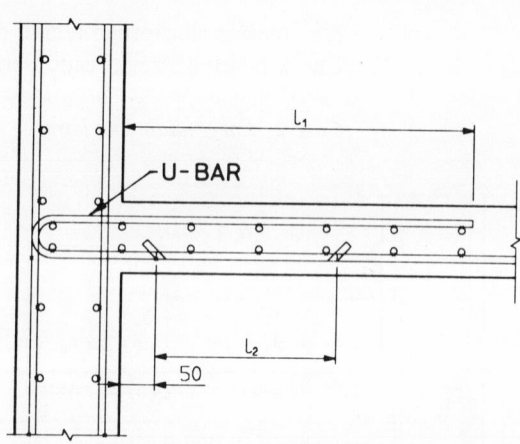

SK 10/43 Connection of slab to
wall. (Slab assumed simply
supported.)

10.5.8.1   *Slab simply supported on wall*

$l_1 = 4 \times$ thickness of slab or 600 mm or 0.1 × span, whichever is the
greatest

U-bars are same diameter as bottom bars.

l_2 = tension lap or 500 mm, whichever is greater

10.5.8.2   *Slab restrained by wall-moment connection*

Case 1: Small diameter bars

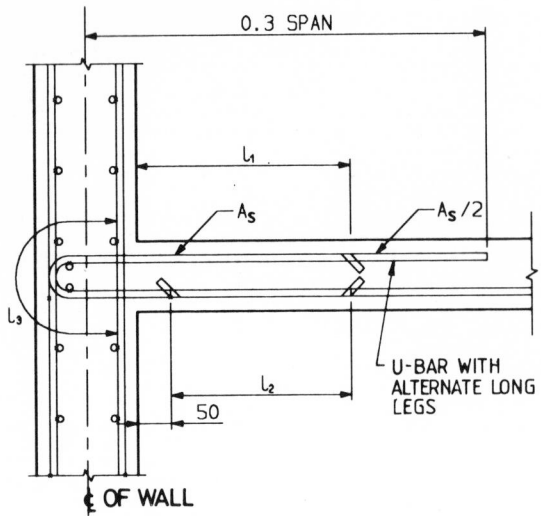

SK 10/44 Connection of slab to wall − Case 1 (Slab rigidly connected to wall.)

l_1 = the greatest of designed length, tension anchorage length, 4 × thickness of slab, 600 mm, or 0.1 × span

l_2 = tension lap length

l_3 = tension anchorage length allowing for bends
Check bearing stress inside bend.

Case 2: Large diameter bars

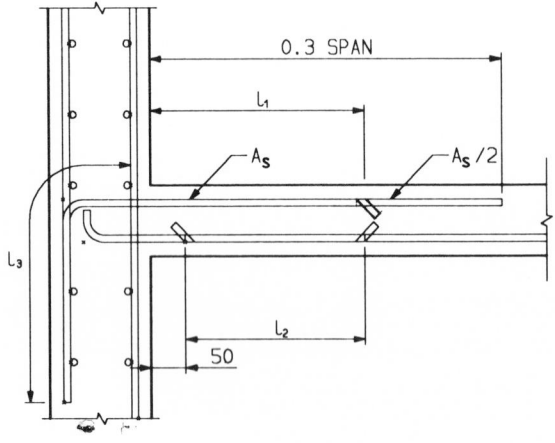

SK 10/45 Connection of slab to wall − Case 2 (Slab rigidly connected to wall.)

l_1 = same as in Case 1

l_2 = tension lap length

l_3 = tension anchorage length allowing for a bend
Check bearing stress inside bend.

Note: Case 2 detail is used when tension anchorage length cannot be accommodated within bend of U-bar in Case 1.

10.5.9   Column-to-wall connection

Case 1: No significant tension in column bars

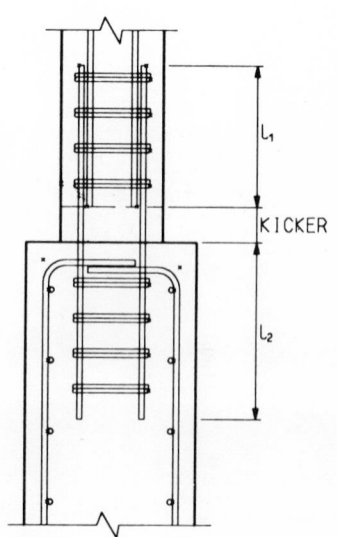

KICKER

SK 10/46 Wall-to-column connection.

l_1 = compression lap length

l_2 = compression anchorage length

Case 2: Significant tension in column bars

l_1 = tension lap length (see Section 10.3.2 for design of lap length)

l_2 = tension anchorage length

10.5.10   Slab-to-beam connection

The same principles apply as in Section 10.5.8.1.

Chapter 11
General Figures, Tables and Charts

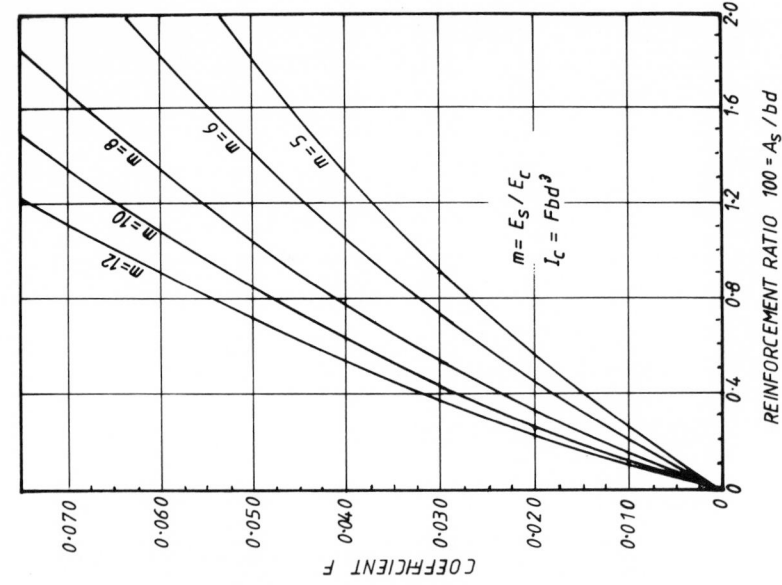

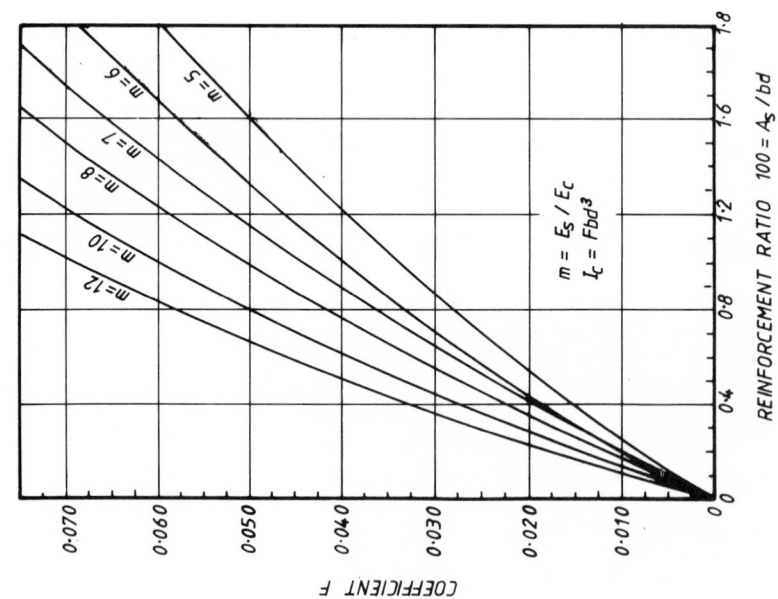

Fig. 11.1 Coefficient for moment of inertia of cracked sections with (a) equal reinforcement on opposite faces and (b) tension reinforcement only.

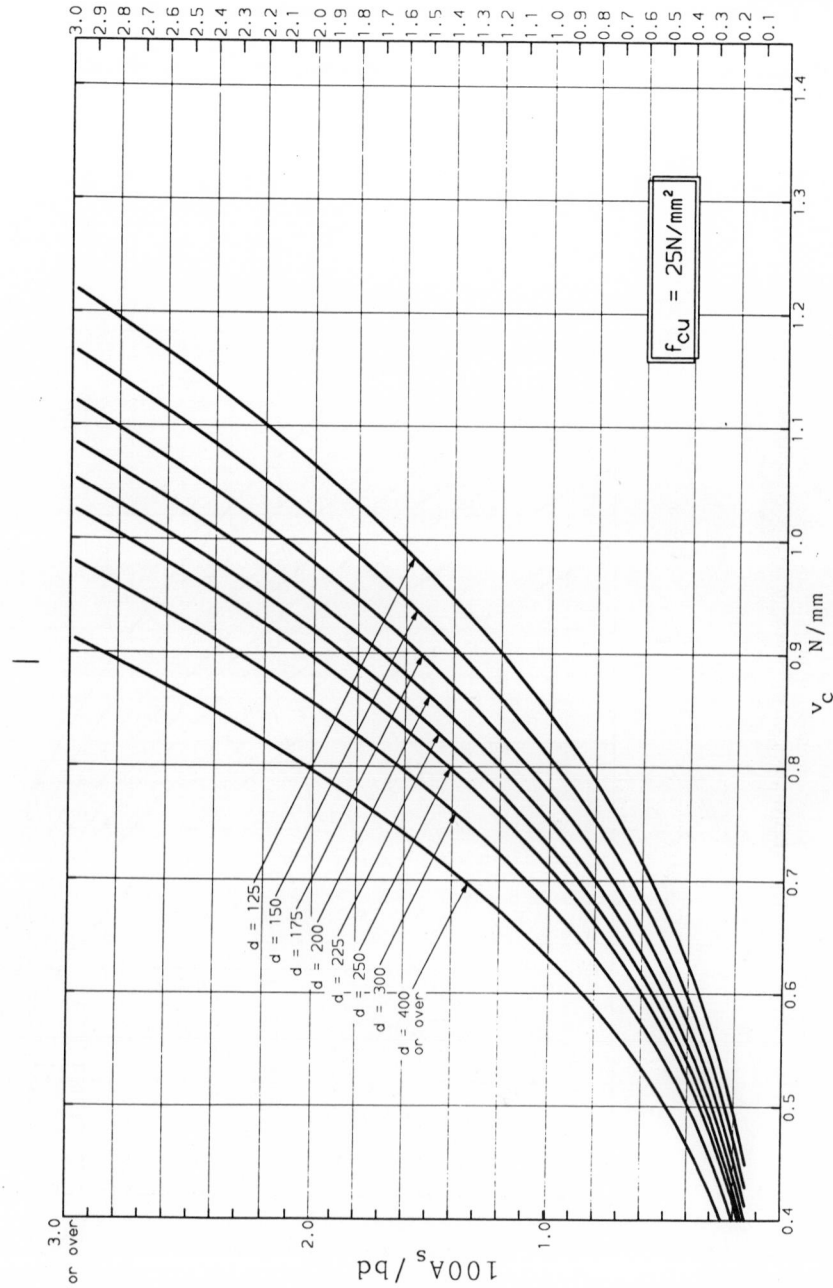

Fig. 11.2 Values of v_c – design concrete shear stress.

Fig. 11.3 Values of v_c — design concrete shear stress.

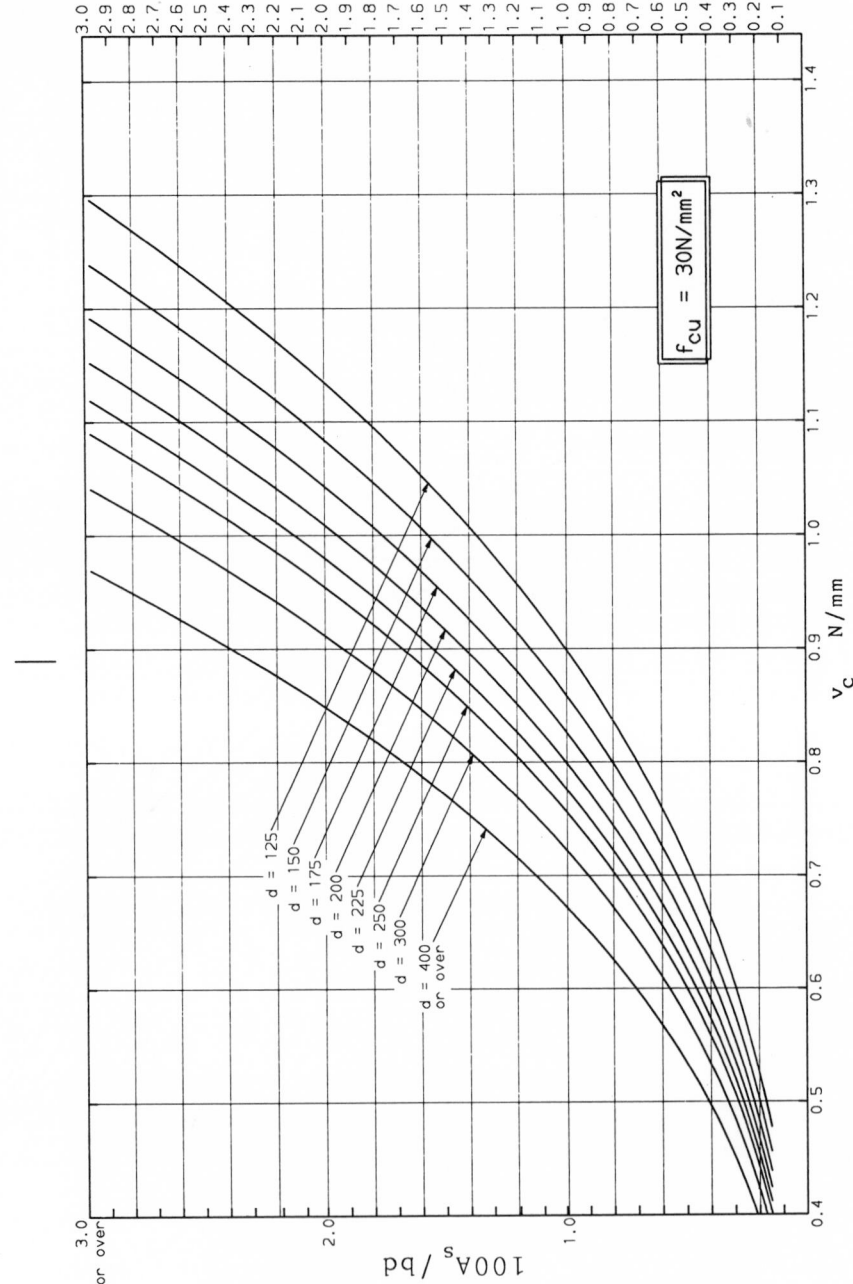

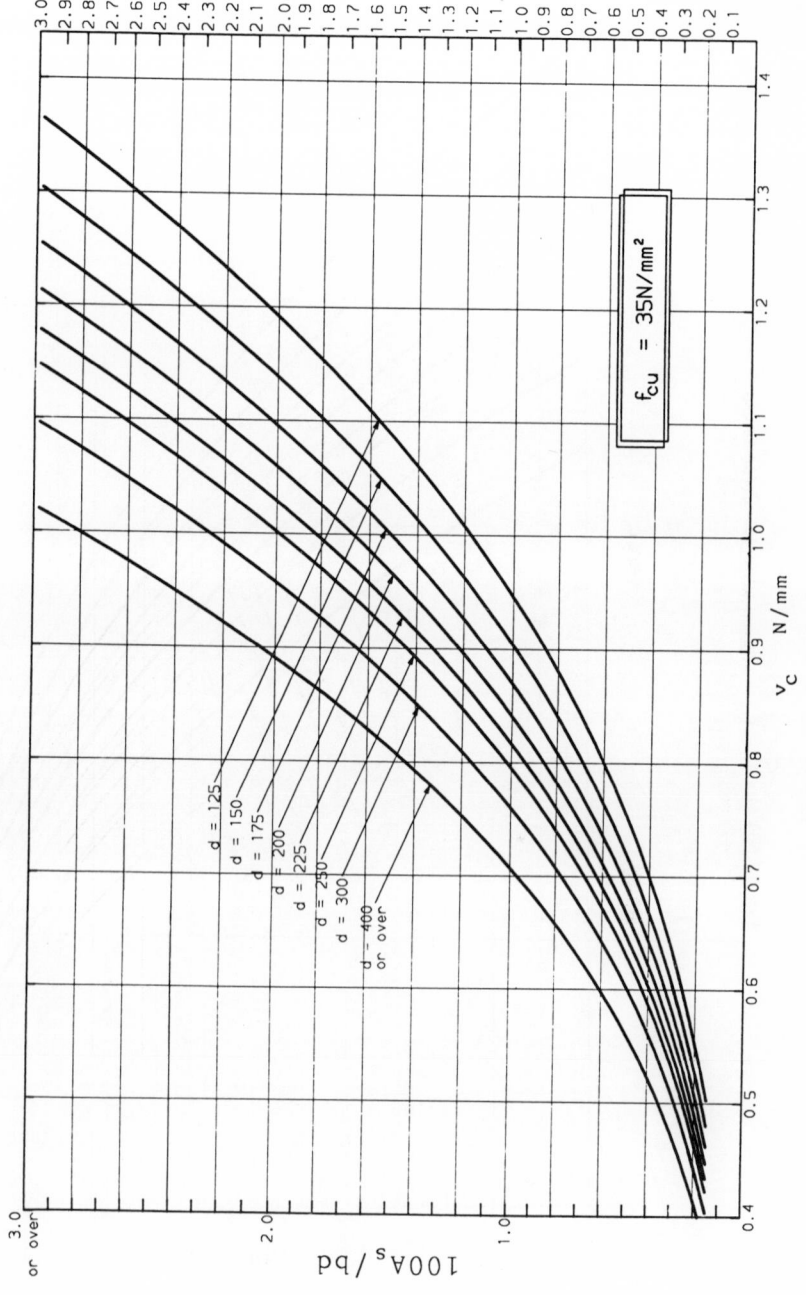

Fig. 11.4 Values of v_c – design concrete shear stress.

Fig. 11.5 Values of v_c – design concrete shear stress.

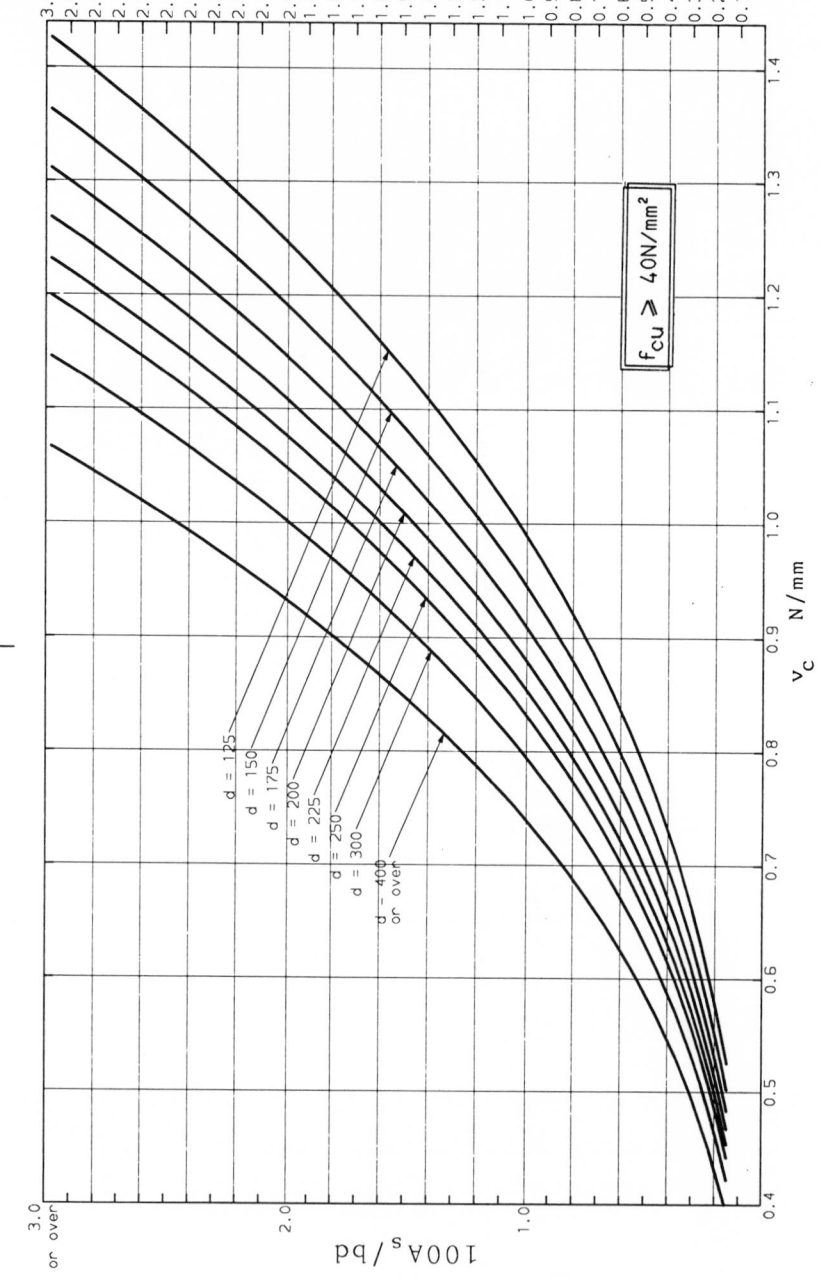

Table 11.1 Area of steel reinforcement for various spacings.

Diameter (mm)	Perimeter (mm)	Area (mm²)	Weight (kg/m)	Area of steel reinforcement for different spacings (mm²/m)							
				50 mm	75 mm	100 mm	125 mm	150 mm	200 mm	250 mm	300 mm
6	18.8	28.3	0.222	566	377	283	226	188	141	113	94
8	25.1	50.3	0.395	1006	670	503	402	335	251	201	167
10	31.4	78.5	0.616	1570	1046	785	628	523	392	314	261
12	37.7	113.1	0.888	2262	1508	1131	904	754	565	452	377
16	50.3	201.1	1.579	4022	2681	2011	1608	1340	1005	804	670
20	62.8	314.2	2.466	6284	4189	3142	2513	2094	1571	1256	1047
25	78.5	490.9	3.854	9818	6544	4909	3926	3272	2454	1963	1636
32	100.5	804.2	6.313	—	10722	8042	6433	5361	4021	3216	2680
40	125.7	1256.6	9.864	—	—	12566	10052	8377	6283	5026	4188
50	157.1	1963.5	15.413	—	—	—	15708	13090	9817	7854	6545

Table 11.2 Sectional properties.

TYPE OF SECTION	AREA	MOMENT OF INERTIA	TORSIONAL CONSTANT
Circle, diameter D	$A = \pi D^2/4$	$I = \pi D^4/64$	$I_z = \pi D^4/32$
Annulus, D_1, D_2	$A = \frac{\pi}{4}[D_1^2 - D_2^2]$	$I = \frac{\pi}{64}[D_1^4 - D_2^4]$	$I_z = \frac{\pi}{32}(D_1^4 - D_2^4)$
Octagon, $0.414h$, $0.414h$, $45°$, h	$A = 0.828h^2$	$I_{xx} = I_{yy} = 0.0547h^4$	$I_z \approx \pi h^4/32$ (Approximately)
Hexagon, $0.577h$, $0.577h$, h	$A = 0.866h^2$	$I_{xx} = I_{yy} = 0.0601h^4$	$I_z \approx \pi h^4/32$ (Approximately)
Rectangle, $h/2$, $h/2$, $b/2$, $b/2$	$A = bh$	$I_{xx} = \frac{1}{12}bh^3$ $I_{yy} = \frac{1}{12}hb^3$	$h \geq b$ $I_z = 0.5Cb^3h$ C = Coefficient in table 2-2
T-section, $B/2$, $B/2$, h_f, h, x, $b/2$, $b/2$	$A = bh + (B-b)h_f$ $x = \dfrac{[bh^2 + (B-b)h_f^2]}{2[bh + (B-b)h_f]}$	$I_{xx} = \frac{1}{12}bh^3 + \frac{1}{12}(B-b)h_f^3 + bh(\frac{h}{2} - x)^2 + (B-b)h_f(x - h_f/2)^2$ $I_{yy} = \frac{1}{12}(h - h_f)b^3 + \frac{1}{12}h_f B^3$	$(h - h_f) \geq b$ $I_z = 0.5CBh_f^3 + 0.5Cb^3(h - h_f)$ C = Coefficient in table 2-2
I-section, $B/2$, $B/2$, h_f, h, h_f, $b/2$, $b/2$	$A = bh + 2h_f(B-b)$	$I_{xx} = \frac{1}{12}bh^3 + \frac{1}{6}(B-b)h_f^3 + 2(B-b)h_f(h/2 - h_f/2)^2$ $I_{yy} = \frac{1}{6}h_f B^3 + \frac{1}{12}(h - 2h_f)b^3$	$(h - 2h_f) \geq b$ $I_z = CBh_f^3 + 0.5Cb^3(h - 2h_f)$ C = Coefficient in table 2-2
Box section, C of wall, $B/2$, $B/2$, B_1, $H/2$, $h/2$, $h/2$, $H/2$, $b/2$, $b/2$	$A = BH - bh$	$I_{xx} = \frac{1}{12}BH^3 - \frac{1}{12}bh^3$ $I_{yy} = \frac{1}{12}HB^3 - \frac{1}{12}hb^3$	$I_z = \dfrac{4A_1^2}{\sum \dfrac{B_1}{t}}$ A_1 = Area of closed cell on C of wall. B_1 = Length of each side of closed cell on C. t = Thickness of each side of closed cell on C.
Channel, B, x, h_f, H, b_w, h_f, e, b, Shear centre	$A = 2Bh_f + (H-2h_f)b_w$ $x = \dfrac{[B^2 h_f + (H-2h_f)b_w^2/2]}{A}$ $e = \dfrac{b^2 h_f(H - h_f)^2}{4I_{xx}}$	$I_{xx} = \frac{1}{12}b_w H^3 + \frac{1}{6}(B - b_w)h_f^3$ $I_{yy} = \frac{1}{6}B^3 h_f + \frac{1}{12}(H - 2h_f)b_w^3 + 2Bh_f(x - B/2)^2 + (H - 2h_f)(x - b_w/2)^2$	$B > h_f$ $(H - 2h_f) > b_w$ $I_z = CBh_f^3 + C(H - 2h_f)b_w^3$ C = Coefficient in table 2-2

Table 11.3 Basic span/effective depth ratios for rectangular and flanged beams.

Support conditions	Rectangular section	Flanged beam with $b_w/b \leq 0.3$
Cantilever	7	5.6
Simply supported	20	16.0
Continuous	26	20.8

Chart 11.4 Modification factor for compression reinforcement.

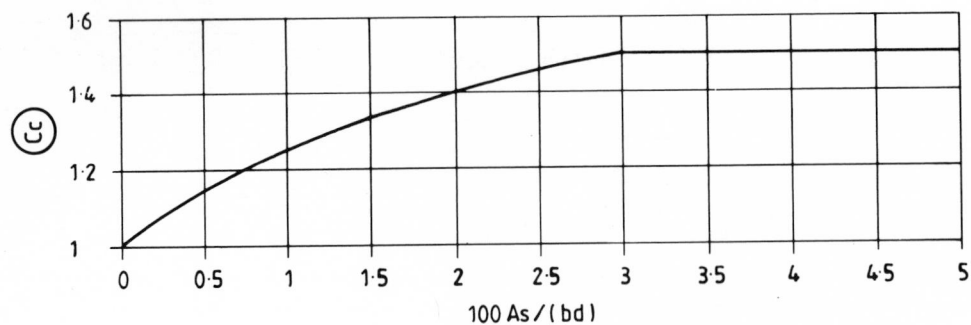

Cc = Modification factor for Compression Reinforcement

Chart 11.5 Modification factor for tension reinforcement.

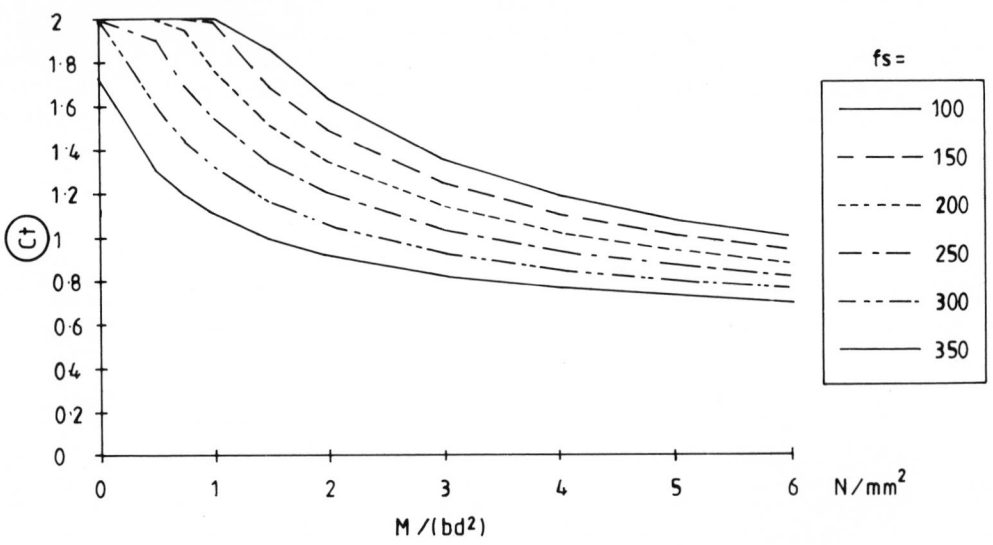

Ct = Modification factor for Tension Reinforcement

Table 11.6 Nominal cover (mm) to all reinforcement including links to meet durability requirements.

Condition of exposure	Lowest grades of concrete (N/mm^2)				
	30	35	40	45	50
Mild: protected against aggressive environment	25	20	20	20	20
Moderate: sheltered from rain and freezing, subject to condensation, or, continuously under water	—	35	30	25	20
Severe: subject to alternate wetting and drying and freezing	—	—	40	30	25
Very severe: subject to sea water spray, deicing salts, etc.	—	—	50	40	30
Extreme: exposed to abrasive action by machinery or vehicles or water-carrying solids with pH ≤ 4.5	—	—	—	60	50

Table 11.7 Nominal cover (mm) to all reinforcement including links to meet specified periods of fire resistance.

Fire resistance (hours)	Beams		Slabs		Columns
	Simply supported	Continuous	Simply supported	Continuous	
0.5	20	20	20	20	20
1.0	20	20	20	20	20
1.5	20	20	25	20	20
2.0	40	30	35	25	25
3.0	60	40	45	35	25
4.0	70	50	55	45	25

Note: cover in excess of 40 mm may require additional measure to reduce risk of spalling

Table 11.8 Rectangular columns – $f_{cu} = 30\,\text{N/mm}^2$, $k = 0.95$.

e/h	p=0.4		p=1.0		p=2.0		p=3.0		p=4.0		p=5.0		p=6.0	
	N/bh	x/h	N/bh	x/h	N/bh	x/h	N/bh	x/h	N/bh	x/h	N/bh	x/h	N/bh	x/h
0.05	13.53	1.045	15.68	1.095	19.27	1.062	22.87	1.104	26.47	******	30.07	******	33.67	******
0.10	12.18	0.944	14.24	1.000	17.54	0.980	20.79	1.022	24.06	1.053	27.34	******	30.61	******
0.15	10.90	0.851	12.90	0.915	16.03	0.910	19.07	0.953	22.08	0.983	25.07	1.078	28.06	******
0.20	9.71	0.767	11.69	0.841	14.66	0.851	17.52	0.895	20.34	0.926	23.13	1.007	25.91	1.098
0.25	8.65	0.694	10.61	0.778	13.46	0.803	16.15	0.847	18.79	0.878	21.41	0.948	24.00	1.025
0.30	7.73	0.633	9.68	0.726	12.40	0.763	14.95	0.808	17.43	0.838	19.89	0.900	22.32	0.966
0.35	6.84	0.567	8.87	0.683	11.48	0.729	13.89	0.775	16.24	0.805	18.55	0.860	20.84	0.918
0.40	5.93	0.492	8.17	0.647	10.67	0.701	12.96	0.747	15.18	0.777	17.37	0.827	19.53	0.877
0.45	5.11	0.424	7.56	0.617	9.96	0.677	12.14	0.723	14.25	0.753	16.32	0.799	18.37	0.844
0.50	4.39	0.364	6.95	0.576	9.33	0.656	11.42	0.702	13.42	0.733	15.39	0.775	17.33	0.815
0.55	3.77	0.313	6.31	0.523	8.78	0.639	10.77	0.685	12.68	0.715	14.55	0.754	16.41	0.791
0.60	3.25	0.270	5.73	0.475	8.29	0.623	10.19	0.669	12.02	0.699	13.80	0.736	15.57	0.771
0.65	2.82	0.234	5.22	0.433	7.85	0.610	9.67	0.656	11.42	0.686	13.13	0.721	14.81	0.753
0.70	2.47	0.204	4.76	0.395	7.45	0.583	9.20	0.644	10.87	0.674	12.51	0.707	14.13	0.737
0.75	2.17	0.180	4.36	0.362	7.03	0.547	8.77	0.633	10.38	0.663	11.95	0.695	13.50	0.723
0.80	1.93	0.160	4.00	0.332	6.59	0.514	8.38	0.624	9.93	0.653	11.44	0.684	12.93	0.711
0.85	1.74	0.144	3.69	0.306	6.19	0.483	8.03	0.615	9.51	0.644	10.97	0.674	12.40	0.700
0.90	1.57	0.130	3.41	0.283	5.83	0.456	7.70	0.607	9.13	0.637	10.53	0.665	11.91	0.690
0.95	1.43	0.119	3.17	0.263	5.50	0.430	7.40	0.596	8.78	0.629	10.13	0.657	11.47	0.681
1.00	1.31	0.111	2.95	0.245	5.19	0.386	7.07	0.533	8.46	0.617	9.76	0.650	11.05	0.673
1.10	1.12	0.101	2.58	0.214	4.66	0.348	6.43	0.487	7.87	0.606	9.09	0.637	10.30	0.666
1.20			2.29	0.190	4.20	0.317	5.87	0.447	7.36	0.566	8.51	0.626	9.65	0.653
1.30			2.05	0.170	3.82	0.290	5.39	0.412	6.83	0.525	8.00	0.617	9.07	0.642
1.40			1.86	0.154	3.49	0.267	4.97	0.382	6.34	0.489	7.55	0.609	8.56	0.632
1.50			1.69	0.140	3.21	0.247	4.61	0.355	5.90	0.457	7.11	0.590	8.10	0.624
1.60			1.55	0.129	2.97	0.229	4.29	0.332	5.51	0.429	6.67	0.553	7.69	0.617
1.70			1.44	0.119	2.76	0.214	4.00	0.311	5.17	0.403	6.28	0.520	7.32	0.611
1.80			1.33	0.114	2.58	0.200	3.75	0.292	4.86	0.380	5.92	0.491	6.93	0.605
1.90			1.24	0.110	2.42	0.188	3.53	0.276	4.58	0.359	5.59	0.464	6.57	0.575
2.00			1.16	0.107	2.27	0.168	3.32	0.247	4.33	0.323	5.30	0.440	6.23	0.544
2.20			1.03	0.103	2.03	0.152	2.98	0.224	3.90	0.294	4.79	0.397	5.65	0.517
2.40					1.83	0.138	2.70	0.204	3.54	0.269	4.36	0.362	5.16	0.469
2.60					1.67	0.127	2.46	0.188	3.24	0.248	4.00	0.332	4.74	0.428
2.80					1.53	0.117	2.26	0.174	2.99	0.229	3.69	0.306	4.38	0.393
3.00					1.41	0.114	2.09	0.161	2.77	0.214	3.42	0.284	4.07	0.364
3.20					1.31	0.112	1.95	0.151	2.58	0.200	3.19	0.265	3.80	0.338
3.40					1.22	0.110	1.82	0.142	2.41	0.187	2.99	0.248	3.56	0.315
3.60					1.14	0.108	1.71	0.133	2.26	0.177	2.81	0.233	3.35	0.295
3.80					1.08	0.107	1.61	0.126	2.13	0.167	2.65	0.220	3.16	0.278
4.00					1.01	0.102	1.52	0.112	2.01		2.50	0.208	2.99	0.262
5.00					0.79		1.19	0.109	1.58	0.131	1.97	0.163	2.35	0.195
6.00							0.97	0.106	1.30	0.115	1.62	0.134	1.94	0.161
7.00							0.82	0.104	1.10	0.112	1.37	0.116	1.65	0.136
8.00							0.72	0.102	0.95	0.110	1.19	0.114	1.43	0.119
10.00							0.57		0.75	0.107	0.94	0.111	1.13	0.114

Table 11.8 (contd) − $f_{cu} = 30\,\text{N/mm}^2$, $k = 0.90$.

e/h	N/bh (p=0.4)	x/h (p=0.4)	N/bh (p=1.0)	x/h (p=1.0)	N/bh (p=2.0)	x/h (p=2.0)	N/bh (p=3.0)	x/h (p=3.0)	N/bh (p=4.0)	x/h (p=4.0)	N/bh (p=5.0)	x/h (p=5.0)	N/bh (p=6.0)	x/h (p=6.0)
0.05	13.47	1.035	15.55	1.076	19.03	1.027	22.59	1.062	26.14	1.089	29.70	......	33.25	......
0.10	12.08	0.932	14.05	0.976	17.25	0.941	20.41	0.976	23.57	1.002	26.73	1.022	29.93	1.038
0.15	10.76	0.835	12.65	0.887	15.65	0.869	18.59	0.905	21.50	0.930	24.41	0.949	27.31	0.964
0.20	9.54	0.748	11.39	0.810	14.22	0.809	16.96	0.846	19.67	0.872	22.36	0.891	25.03	0.905
0.25	8.44	0.673	10.27	0.746	12.97	0.760	15.54	0.798	18.06	0.824	20.56	0.843	23.04	0.858
0.30	7.49	0.610	9.31	0.693	11.88	0.720	14.30	0.760	16.66	0.786	18.99	0.805	21.31	0.819
0.35	6.61	0.548	8.48	0.649	10.94	0.687	13.22	0.727	15.44	0.754	17.63	0.773	19.80	0.787
0.40	5.69	0.472	7.77	0.614	10.12	0.660	12.28	0.701	14.37	0.727	16.43	0.746	18.48	0.760
0.45	4.87	0.403	7.17	0.585	9.41	0.637	11.46	0.678	13.44	0.705	15.38	0.724	17.31	0.738
0.50	4.14	0.343	6.55	0.543	8.79	0.618	10.74	0.659	12.61	0.685	14.46	0.704	16.28	0.719
0.55	3.53	0.292	5.91	0.490	8.25	0.601	10.10	0.642	11.88	0.669	13.63	0.688	15.37	0.702
0.60	3.01	0.250	5.34	0.443	7.76	0.587	9.53	0.628	11.23	0.655	12.90	0.673	14.55	0.688
0.65	2.58	0.218	4.84	0.401	7.33	0.574	9.02	0.615	10.65	0.642	12.24	0.661	13.81	0.675
0.70	2.23	0.195	4.40	0.364	6.95	0.539	8.57	0.604	10.12	0.631	11.64	0.650	13.14	0.664
0.75	1.95	0.177	4.01	0.332	6.50	0.504	8.15	0.595	9.64	0.621	11.10	0.640	12.53	0.654
0.80	1.73	0.164	3.66	0.304	6.08	0.472	7.78	0.586	9.21	0.612	10.60	0.631	11.98	0.645
0.85	1.55	0.154	3.36	0.279	5.69	0.443	7.44	0.578	8.81	0.604	10.15	0.623	11.47	0.637
0.90	1.40	0.146	3.10	0.257	5.34	0.416	7.12	0.565	8.44	0.597	9.74	0.616	11.01	0.630
0.95	1.28	0.139	2.87	0.238	5.02	0.392	6.82	0.537	8.11	0.591	9.35	0.609	10.58	0.623
1.00	1.17	0.134	2.66	0.225	4.73	0.350	6.47	0.496	7.80	0.579	9.00	0.598	10.18	0.612
1.10	1.01	0.126	2.31	0.207	4.22	0.315	5.86	0.443	7.24	0.558	8.37	0.588	9.47	0.602
1.20	0.88	0.120	2.04	0.194	3.80	0.286	5.34	0.405	6.73	0.514	7.82	0.580	8.85	0.593
1.30	0.79	0.116	1.82	0.184	3.45	0.261	4.89	0.373	6.20	0.476	7.33	0.573	8.31	0.586
1.40	0.71	0.112	1.64	0.176	3.15	0.240	4.50	0.345	5.74	0.443	6.91	0.535	7.83	0.580
1.50	0.64	0.110	1.50	0.170	2.89	0.227	4.16	0.320	5.34	0.413	6.45	0.501	7.40	0.574
1.60	0.59	0.107	1.38	0.165	2.66	0.220	3.86	0.298	4.98	0.386	6.04	0.470	7.02	0.550
1.70	0.54	0.105	1.27	0.161	2.46	0.213	3.60	0.279	4.66	0.363	5.67	0.442	6.63	0.519
1.80	0.50	0.104	1.18	0.158	2.29	0.208	3.37	0.262	4.37	0.342	5.34	0.418	6.26	0.491
1.90	0.47	0.103	1.10	0.155	2.14	0.203	3.16	0.247	4.12	0.323	5.04	0.395	5.92	0.466
2.00	0.44	0.101	1.04	0.152	2.01	0.196	2.98	0.229	3.89	0.290	4.77	0.357	5.62	0.421
2.20			0.92	0.148	1.79	0.190	2.66	0.221	3.50	0.263	4.30	0.324	5.08	0.384
2.40			0.83	0.145	1.62	0.186	2.40	0.214	3.17	0.240	3.91	0.297	4.63	0.352
2.60			0.76	0.142	1.47	0.182	2.19	0.209	2.90	0.229	3.58	0.274	4.25	0.325
2.80			0.69	0.140	1.35	0.179	2.01	0.205	2.66	0.224	3.30	0.254	3.92	0.302
3.00			0.64	0.138	1.25	0.176	1.85	0.202	2.46	0.220	3.06	0.236	3.64	0.282
3.20			0.59	0.137	1.16	0.174	1.72	0.199	2.29	0.216	2.85	0.230	3.40	0.264
3.40			0.55	0.136	1.08	0.172	1.61	0.196	2.14	0.213	2.66	0.226	3.18	0.248
3.60			0.52	0.135	1.01	0.171	1.51	0.194	2.00	0.210	2.50	0.223	2.99	0.234
3.80			0.49	0.134	0.96	0.169	1.42	0.192	1.89	0.208	2.35	0.221	2.82	0.230
4.00			0.46	0.133	0.90	0.164	1.34	0.185	1.78	0.200	2.22	0.211	2.66	0.220
5.00			0.36	0.130	0.71	0.161	1.05	0.181	1.40	0.195	1.74	0.205	2.09	0.213
6.00			0.30	0.128	0.58	0.158	0.86	0.178	1.15	0.191	1.43	0.201	1.72	0.209
7.00			0.25	0.126	0.49	0.157	0.73	0.176	0.97	0.189	1.22	0.198	1.46	0.206
8.00			0.22	0.125	0.43	0.155	0.64	0.173	0.85	0.185	1.06	0.194	1.27	0.201
10.00			0.17	0.124	0.34		0.50		0.67		0.84		1.00	

Table 11.9 Rectangular columns – $f_{cu} = 30\,\text{N/mm}^2$, $k = 0.85$.

e/h	p=0.4 N/bh	p=0.4 x/h	p=1.0 N/bh	p=1.0 x/h	p=2.0 N/bh	p=2.0 x/h	p=3.0 N/bh	p=3.0 x/h	p=4.0 N/bh	p=4.0 x/h	p=5.0 N/bh	p=5.0 x/h	p=6.0 N/bh	p=6.0 x/h
0.05	13.41	1.026	15.43	1.057	18.79	1.096	22.23	******	25.73	******	29.23	******	32.73	******
0.10	11.99	0.919	13.87	0.953	16.95	0.991	20.02	1.019	23.09	1.040	26.16	1.056	29.23	1.070
0.15	10.63	0.820	12.40	0.859	15.26	0.901	18.08	0.928	20.88	0.948	23.68	0.963	26.47	0.976
0.20	9.36	0.729	11.08	0.779	13.75	0.826	16.36	0.854	18.93	0.874	21.50	0.889	24.06	0.901
0.25	8.22	0.651	9.91	0.712	12.44	0.765	14.87	0.795	17.25	0.816	19.62	0.831	21.97	0.843
0.30	7.24	0.586	8.92	0.658	11.32	0.716	13.59	0.748	15.81	0.769	18.00	0.785	20.19	0.796
0.35	6.37	0.529	8.07	0.615	10.36	0.676	12.49	0.710	14.56	0.732	16.61	0.747	18.65	0.759
0.40	5.44	0.451	7.36	0.580	9.54	0.644	11.54	0.679	13.49	0.701	15.41	0.717	17.32	0.729
0.45	4.60	0.382	6.75	0.551	8.83	0.618	10.73	0.654	12.56	0.676	14.37	0.692	16.16	0.705
0.50	3.85	0.323	6.13	0.508	8.22	0.596	10.01	0.632	11.75	0.655	13.46	0.672	15.14	0.684
0.55	3.21	0.279	5.49	0.455	7.68	0.578	9.39	0.615	11.03	0.638	12.65	0.654	14.25	0.666
0.60	2.71	0.246	4.93	0.409	7.21	0.563	8.84	0.599	10.40	0.623	11.93	0.639	13.45	0.651
0.65	2.32	0.221	4.44	0.368	6.79	0.549	8.34	0.586	9.83	0.610	11.29	0.626	12.73	0.638
0.70	2.01	0.203	3.99	0.336	6.39	0.530	7.90	0.575	9.32	0.598	10.71	0.615	12.09	0.627
0.75	1.77	0.189	3.59	0.312	5.94	0.492	7.51	0.565	8.87	0.588	10.19	0.605	11.51	0.617
0.80	1.58	0.179	3.25	0.294	5.53	0.459	7.15	0.556	8.45	0.579	9.72	0.596	10.98	0.608
0.85	1.42	0.170	2.97	0.278	5.16	0.428	6.82	0.548	8.07	0.572	9.29	0.588	10.50	0.600
0.90	1.29	0.164	2.72	0.265	4.83	0.400	6.53	0.541	7.73	0.564	8.90	0.581	10.06	0.593
0.95	1.18	0.158	2.51	0.255	4.52	0.375	6.17	0.512	7.41	0.558	8.54	0.575	9.65	0.587
1.00	1.09	0.154	2.33	0.246	4.25	0.353	5.85	0.485	7.12	0.552	8.20	0.569	9.28	0.581
1.10	0.94	0.147	2.04	0.232	3.75	0.328	5.27	0.437	6.60	0.542	7.61	0.559	8.61	0.571
1.20	0.83	0.142	1.81	0.222	3.34	0.310	4.78	0.396	6.05	0.502	7.09	0.550	8.03	0.562
1.30	0.74	0.138	1.62	0.214	3.01	0.296	4.36	0.362	5.56	0.461	6.65	0.543	7.53	0.555
1.40	0.67	0.135	1.47	0.207	2.74	0.284	3.99	0.341	5.13	0.425	6.19	0.513	7.08	0.549
1.50	0.61	0.132	1.34	0.202	2.51	0.275	3.66	0.328	4.76	0.394	5.76	0.478	6.68	0.543
1.60	0.56	0.130	1.24	0.198	2.32	0.268	3.38	0.318	4.43	0.367	5.38	0.446	6.29	0.522
1.70	0.52	0.129	1.15	0.194	2.15	0.262	3.14	0.309	4.13	0.347	5.04	0.418	5.91	0.490
1.80	0.48	0.127	1.07	0.191	2.00	0.257	2.93	0.302	3.86	0.338	4.74	0.393	5.57	0.462
1.90	0.45	0.126	1.00	0.189	1.88	0.252	2.75	0.296	3.62	0.330	4.47	0.370	5.26	0.436
2.00	0.43	0.125	0.94	0.187	1.77	0.248	2.59	0.291	3.40	0.323	4.22	0.350	4.98	0.413
2.20	0.38	0.123	0.84	0.183	1.58	0.242	2.31	0.282	3.05	0.312	3.78	0.337	4.50	0.373
2.40	0.34	0.122	0.76	0.180	1.43	0.237	2.09	0.275	2.76	0.304	3.42	0.327	4.09	0.346
2.60	0.31	0.121	0.69	0.178	1.30	0.233	1.91	0.270	2.51	0.297	3.12	0.319	3.73	0.337
2.80	0.29	0.120	0.63	0.176	1.20	0.230	1.75	0.265	2.31	0.292	2.87	0.313	3.43	0.330
3.00	0.26	0.119	0.59	0.174	1.11	0.227	1.62	0.261	2.14	0.287	2.66	0.307	3.18	0.324
3.20	0.24	0.118	0.54	0.173	1.03	0.224	1.51	0.258	1.99	0.283	2.47	0.303	2.96	0.319
3.40	0.23	0.117	0.51	0.172	0.96	0.222	1.41	0.255	1.86	0.280	2.31	0.299	2.76	0.314
3.60	0.21	0.117	0.48	0.171	0.90	0.221	1.33	0.253	1.75	0.277	2.17	0.295	2.60	0.310
3.80	0.20	0.116	0.45	0.170	0.85	0.219	1.25	0.251	1.65	0.274	2.05	0.292	2.45	0.307
4.00	0.19	0.116	0.42	0.169	0.80	0.218	1.18	0.249	1.56	0.272	1.94	0.290	2.31	0.304
5.00	0.15	0.115	0.33	0.166	0.63	0.213	0.93	0.242	1.23	0.264	1.52	0.280	1.82	0.294
6.00	0.12	0.114	0.27	0.164	0.52	0.210	0.77	0.238	1.01	0.259	1.25	0.275	1.50	0.287
7.00	0.10	0.113	0.23	0.163	0.44	0.208	0.65	0.235	0.86	0.255	1.07	0.271	1.27	0.283
8.00	0.09	0.113	0.20	0.162	0.38	0.206	0.57	0.233	0.75	0.253	0.93	0.268	1.11	0.280
10.00	0.07	0.112	0.16	0.160	0.31	0.204	0.45	0.231	0.59	0.249	0.74	0.264	0.88	0.275

Table 11.9 (contd) $- f_{cu} = 30\,\text{N/mm}^2$, $k = 0.80$.

e/h	p=0.4 N/bh	p=0.4 x/h	p=1.0 N/bh	p=1.0 x/h	p=2.0 N/bh	p=2.0 x/h	p=3.0 N/bh	p=3.0 x/h	p=4.0 N/bh	p=4.0 x/h	p=5.0 N/bh	p=5.0 x/h	p=6.0 N/bh	p=6.0 x/h
0.05	13.37	1.017	15.33	1.039	18.62	1.067	21.92	1.089	25.23	1.107	28.64	******	32.07	******
0.10	11.92	0.908	13.70	0.930	16.66	0.955	19.62	0.974	22.59	0.988	25.55	1.000	28.52	1.009
0.15	10.51	0.804	12.16	0.831	14.85	0.860	17.53	0.878	20.20	0.892	22.86	0.902	25.52	0.910
0.20	9.19	0.710	10.76	0.747	13.25	0.781	15.69	0.802	18.12	0.816	20.53	0.827	22.94	0.835
0.25	8.00	0.629	9.54	0.677	11.87	0.719	14.13	0.742	16.35	0.758	18.56	0.769	20.76	0.777
0.30	6.98	0.562	8.50	0.622	10.71	0.670	12.81	0.695	14.86	0.712	16.90	0.724	18.93	0.733
0.35	6.12	0.508	7.64	0.579	9.73	0.631	11.69	0.659	13.61	0.677	15.49	0.689	17.37	0.698
0.40	5.14	0.430	6.92	0.544	8.91	0.600	10.75	0.629	12.53	0.648	14.29	0.661	16.04	0.671
0.45	4.24	0.366	6.32	0.517	8.21	0.575	9.94	0.606	11.61	0.625	13.26	0.638	14.90	0.648
0.50	3.51	0.315	5.67	0.470	7.60	0.555	9.24	0.586	10.82	0.606	12.37	0.619	13.90	0.629
0.55	2.93	0.277	4.98	0.425	7.08	0.538	8.63	0.570	10.12	0.590	11.58	0.604	13.03	0.614
0.60	2.48	0.249	4.39	0.389	6.63	0.524	8.10	0.556	9.51	0.576	10.89	0.590	12.26	0.600
0.65	2.14	0.229	3.90	0.360	6.23	0.512	7.62	0.544	8.97	0.564	10.28	0.579	11.58	0.589
0.70	1.87	0.214	3.49	0.337	5.78	0.479	7.20	0.534	8.48	0.554	9.73	0.569	10.97	0.579
0.75	1.66	0.202	3.15	0.319	5.31	0.449	6.83	0.525	8.05	0.545	9.24	0.560	10.42	0.570
0.80	1.49	0.194	2.86	0.304	4.88	0.427	6.49	0.517	7.65	0.538	8.79	0.552	9.92	0.563
0.85	1.35	0.186	2.62	0.292	4.51	0.409	6.18	0.510	7.30	0.531	8.39	0.545	9.47	0.556
0.90	1.23	0.181	2.41	0.282	4.18	0.393	5.83	0.483	6.97	0.525	8.02	0.539	9.05	0.549
0.95	1.13	0.176	2.24	0.273	3.90	0.380	5.48	0.460	6.68	0.519	7.68	0.533	8.67	0.544
1.00	1.05	0.172	2.08	0.266	3.65	0.369	5.14	0.445	6.40	0.514	7.37	0.528	8.33	0.539
1.10	0.91	0.166	1.83	0.255	3.22	0.350	4.57	0.420	5.86	0.486	6.82	0.520	7.71	0.530
1.20	0.81	0.161	1.63	0.246	2.89	0.336	4.10	0.401	5.31	0.454	6.35	0.512	7.17	0.523
1.30	0.73	0.158	1.47	0.240	2.61	0.325	3.72	0.386	4.82	0.435	5.88	0.488	6.71	0.517
1.40	0.66	0.155	1.33	0.234	2.38	0.316	3.40	0.374	4.41	0.421	5.42	0.459	6.30	0.511
1.50	0.60	0.152	1.22	0.230	2.19	0.309	3.13	0.364	4.07	0.408	5.00	0.445	5.90	0.499
1.60	0.55	0.150	1.13	0.226	2.03	0.303	2.90	0.356	3.77	0.398	4.64	0.433	5.50	0.463
1.70	0.51	0.149	1.05	0.223	1.88	0.298	2.70	0.349	3.51	0.390	4.32	0.423	5.13	0.451
1.80	0.48	0.147	0.98	0.220	1.76	0.293	2.53	0.343	3.29	0.382	4.05	0.414	4.81	0.441
1.90	0.45	0.146	0.92	0.218	1.65	0.289	2.37	0.338	3.09	0.376	3.80	0.407	4.52	0.433
2.00	0.42	0.145	0.86	0.216	1.56	0.286	2.24	0.334	2.91	0.371	3.59	0.400	4.26	0.426
2.20	0.38	0.143	0.77	0.213	1.39	0.281	2.00	0.326	2.61	0.361	3.22	0.390	3.83	0.413
2.40	0.34	0.142	0.70	0.210	1.26	0.276	1.82	0.321	2.37	0.354	2.92	0.381	3.47	0.404
2.60	0.31	0.141	0.64	0.208	1.15	0.273	1.66	0.316	2.17	0.348	2.67	0.374	3.18	0.396
2.80	0.28	0.140	0.59	0.206	1.06	0.270	1.53	0.312	2.00	0.344	2.46	0.369	2.93	0.390
3.00	0.26	0.139	0.54	0.204	0.98	0.267	1.42	0.309	1.85	0.340	2.28	0.364	2.71	0.384
3.20	0.24	0.138	0.51	0.203	0.92	0.265	1.32	0.306	1.72	0.336	2.13	0.360	2.53	0.380
3.40	0.23	0.138	0.47	0.202	0.86	0.263	1.24	0.303	1.61	0.333	1.99	0.357	2.37	0.376
3.60	0.21	0.137	0.44	0.201	0.81	0.262	1.16	0.301	1.52	0.331	1.87	0.354	2.23	0.373
3.80	0.20	0.137	0.42	0.200	0.76	0.260	1.10	0.299	1.43	0.328	1.77	0.351	2.10	0.370
4.00	0.19	0.136	0.40	0.199	0.72	0.259	1.04	0.298	1.35	0.326	1.67	0.349	1.99	0.367
5.00	0.15	0.135	0.31	0.197	0.57	0.254	0.82	0.292	1.07	0.319	1.32	0.340	1.57	0.358
6.00	0.12	0.134	0.26	0.195	0.47	0.252	0.67	0.288	0.88	0.314	1.09	0.335	1.29	0.352
7.00	0.11	0.133	0.22	0.194	0.40	0.249	0.57	0.285	0.75	0.311	0.93	0.331	1.10	0.347
8.00	0.09	0.133	0.19	0.193	0.35	0.248	0.50	0.283	0.65	0.309	0.81	0.328	0.96	0.344
10.00	0.07	0.132	0.15	0.191	0.27	0.246	0.40	0.281	0.52	0.305	0.64	0.325	0.76	0.340

Table 11.10 Rectangular columns − $f_{cu} = 35\,\text{N/mm}^2$, $k = 0.95$.

e/h	p=0.4 N/bh	p=0.4 x/h	p=1.0 N/bh	p=1.0 x/h	p=2.0 N/bh	p=2.0 x/h	p=3.0 N/bh	p=3.0 x/h	p=4.0 N/bh	p=4.0 x/h	p=5.0 N/bh	p=5.0 x/h	p=6.0 N/bh	p=6.0 x/h
0.05	15.55	1.039	17.70	1.085	21.28	1.047	24.88	******	28.48	******	32.08	******	35.68	******
0.10	13.98	0.938	16.05	0.989	19.37	0.964	22.63	1.088	25.89	1.037	29.16	1.061	32.44	1.081
0.15	12.48	0.843	14.51	0.902	17.67	0.894	20.73	1.006	23.75	0.967	26.76	0.991	29.75	1.010
0.20	11.09	0.757	13.11	0.827	16.14	0.835	19.02	0.936	21.85	0.909	24.66	0.933	27.45	0.951
0.25	9.84	0.682	11.87	0.763	14.78		17.51	0.879	20.17		22.80		25.41	
0.30	8.75	0.619	10.79	0.709	13.59	0.796	16.18	0.831	18.69	0.862	21.16	0.885	23.61	0.903
0.35	7.64	0.543	9.86	0.665	12.55	0.745	15.01	0.791	17.39	0.822	19.72	0.845	22.02	0.863
0.40	6.56	0.466	9.06	0.629	11.65	0.711	13.99	0.758	16.24	0.789	18.45	0.812	20.63	0.830
0.45	5.59	0.397	8.32	0.592	10.86	0.683	13.09	0.730	15.23	0.761	17.32	0.784	19.38	0.801
0.50	4.74	0.337	7.50	0.533	10.16	0.658	12.29	0.706	14.33	0.737	16.32	0.760	18.28	0.777
0.55	4.03	0.286	6.76	0.481	9.55	0.638	11.59	0.685	13.53	0.717	15.42	0.739	17.29	0.757
0.60	3.43	0.244	6.10	0.434	9.00	0.620	10.95	0.668	12.81	0.699	14.62	0.722	16.40	0.739
0.65	2.95	0.209	5.51	0.392	8.52	0.605	10.39	0.652	12.16	0.684	13.89	0.706	15.60	0.723
0.70	2.56	0.182	5.00	0.355	7.93	0.564	9.87	0.639	11.58	0.670	13.23	0.692	14.87	0.710
0.75	2.24	0.159	4.55	0.323	7.40	0.526	9.41	0.627	11.04	0.658	12.64	0.680	14.20	0.697
0.80	1.98	0.141	4.16	0.295	6.91	0.491	8.99	0.616	10.56	0.647	12.09	0.670	13.59	0.686
0.85	1.77	0.126	3.81	0.271	6.47	0.460	8.60	0.607	10.11	0.638	11.59	0.660	13.03	0.677
0.90	1.59	0.114	3.51	0.250	6.06	0.431	8.17	0.580	9.71	0.629	11.12	0.651	12.52	0.668
0.95	1.45	0.107	3.25	0.231	5.69	0.405	7.74	0.550	9.33	0.621	10.70	0.643	12.04	0.660
1.00	1.32	0.101	3.02	0.214	5.36	0.381	7.35	0.522	8.98	0.614	10.30	0.636	11.60	0.652
1.10			2.63	0.187	4.78	0.340	6.65	0.472	8.31	0.591	9.59	0.623	10.81	0.640
1.20			2.32	0.165	4.30	0.306	6.04	0.430	7.63	0.542	8.98	0.613	10.12	0.629
1.30			2.08	0.148	3.89	0.277	5.53	0.393	7.03	0.499	8.42	0.598	9.51	0.620
1.40			1.88	0.133	3.55	0.252	5.08	0.361	6.50	0.462	7.82	0.556	8.98	0.611
1.50			1.71	0.121	3.26	0.232	4.69	0.334	6.04	0.429	7.30	0.519	8.49	0.604
1.60			1.56	0.114	3.01	0.214	4.36	0.310	5.63	0.400	6.83	0.485	7.97	0.567
1.70			1.44	0.109	2.79	0.198	4.06	0.289	5.26	0.374	6.41	0.455	7.50	0.533
1.80			1.34	0.106	2.60	0.185	3.80	0.270	4.94	0.351	6.03	0.428	7.08	0.503
1.90			1.25	0.103	2.44	0.173	3.57	0.253	4.65	0.330	5.69	0.404	6.69	0.476
2.00			1.17	0.100	2.29	0.163	3.36	0.239	4.39	0.312	5.38	0.383	6.34	0.451
2.20					2.04	0.145	3.01	0.214	3.94	0.280	4.85	0.345	5.73	0.408
2.40					1.84	0.131	2.72	0.193	3.57	0.254	4.41	0.313	5.22	0.371
2.60					1.67	0.119	2.48	0.176	3.27	0.232	4.04	0.287	4.79	0.341
2.80					1.53	0.114	2.28	0.162	3.00	0.214	3.72	0.264	4.42	0.314
3.00					1.41	0.111	2.10	0.149	2.78	0.198	3.45	0.245	4.10	0.292
3.20					1.31	0.109	1.95	0.139	2.59	0.184	3.21	0.228	3.83	0.272
3.40					1.22	0.107	1.83	0.130	2.42	0.172	3.00	0.213	3.58	0.255
3.60					1.14	0.105	1.71	0.122	2.27	0.161	2.82	0.201	3.37	0.239
3.80					1.08	0.104	1.61	0.116	2.14	0.152	2.66	0.189	3.18	0.226
4.00					1.02	0.102	1.52	0.114	2.02	0.144	2.51	0.179	3.00	0.213
5.00							1.19	0.108	1.58	0.116	1.97	0.140	2.36	0.168
6.00							0.97	0.105	1.30	0.112	1.62	0.116	1.94	0.138
7.00							0.83	0.103	1.10	0.109	1.37	0.113	1.65	0.117
8.00							0.72	0.101	0.95	0.107	1.19	0.111	1.43	0.114
10.00									0.75	0.104	0.94	0.108	1.13	0.111

Table 11.10 (contd) – $f_{cu} = 35\,\text{N/mm}^2$, $k = 0.90$.

e/h	p=0.4 N/bh	p=0.4 x/h	p=1.0 N/bh	p=1.0 x/h	p=2.0 N/bh	p=2.0 x/h	p=3.0 N/bh	p=3.0 x/h	p=4.0 N/bh	p=4.0 x/h	p=5.0 N/bh	p=5.0 x/h	p=6.0 N/bh	p=6.0 x/h
0.05	15.48	1.031	17.56	1.067	21.02	......	24.57	......	28.13	......	31.68	......	35.24	......
0.10	13.87	0.926	15.85	0.967	19.06	1.014	22.23	1.048	25.39	1.075	28.55	1.096	31.72	......
0.15	12.34	0.829	14.25	0.877	17.27	0.928	20.22	0.962	23.15	0.988	26.06	1.008	28.96	1.024
0.20	10.90	0.740	12.79	0.799	15.66	0.855	18.43	0.891	21.15	0.916	23.84	0.936	26.53	0.952
0.25	9.61	0.662	11.51	0.732	14.25	0.795	16.85	0.832	19.39	0.858	21.90	0.878	24.40	0.893
0.30	8.49	0.597	10.39	0.678	13.03	0.745	15.48	0.784	17.86	0.811	20.21	0.830	22.54	0.846
0.35	7.40	0.526	9.44	0.634	11.97	0.705	14.29	0.745	16.54	0.772	18.74	0.792	20.93	0.807
0.40	6.30	0.448	8.63	0.597	11.06	0.671	13.26	0.712	15.38	0.740	17.46	0.760	19.51	0.775
0.45	5.32	0.378	7.90	0.561	10.27	0.644	12.36	0.685	14.37	0.713	16.33	0.733	18.27	0.748
0.50	4.47	0.318	7.07	0.503	9.58	0.621	11.57	0.663	13.47	0.691	15.33	0.711	17.17	0.726
0.55	3.76	0.267	6.33	0.450	8.97	0.601	10.87	0.644	12.68	0.671	14.45	0.691	16.20	0.707
0.60	3.17	0.227	5.68	0.404	8.44	0.584	10.25	0.627	11.98	0.655	13.66	0.675	15.32	0.690
0.65	2.68	0.198	5.11	0.363	7.93	0.564	9.70	0.613	11.35	0.641	12.95	0.661	14.54	0.676
0.70	2.30	0.177	4.61	0.327	7.35	0.523	9.20	0.600	10.78	0.628	12.32	0.648	13.83	0.663
0.75	2.01	0.162	4.17	0.297	6.83	0.485	8.75	0.589	10.26	0.617	11.74	0.637	13.19	0.652
0.80	1.77	0.150	3.80	0.270	6.36	0.452	8.34	0.579	9.80	0.607	11.21	0.627	12.60	0.642
0.85	1.58	0.141	3.47	0.247	5.93	0.421	7.95	0.565	9.37	0.599	10.73	0.618	12.06	0.633
0.90	1.43	0.134	3.18	0.228	5.54	0.394	7.50	0.533	8.98	0.591	10.28	0.610	11.57	0.625
0.95	1.30	0.128	2.92	0.216	5.19	0.369	7.09	0.504	8.62	0.584	9.88	0.603	11.12	0.618
1.00	1.20	0.123	2.70	0.206	4.88	0.347	6.72	0.477	8.28	0.577	9.50	0.597	10.70	0.612
1.10	1.03	0.116	2.34	0.190	4.33	0.308	6.05	0.430	7.59	0.540	8.83	0.585	9.95	0.600
1.20	0.90	0.111	2.06	0.179	3.88	0.276	5.48	0.390	6.94	0.493	8.24	0.576	9.29	0.590
1.30	0.80	0.107	1.84	0.170	3.51	0.249	5.00	0.355	6.38	0.453	7.66	0.544	8.72	0.582
1.40	0.72	0.104	1.66	0.164	3.19	0.230	4.59	0.326	5.88	0.418	7.10	0.505	8.22	0.575
1.50	0.65	0.102	1.51	0.158	2.91	0.220	4.23	0.301	5.45	0.387	6.60	0.469	7.70	0.547
1.60			1.39	0.154	2.68	0.212	3.92	0.278	5.07	0.360	6.17	0.438	7.21	0.513
1.70			1.28	0.150	2.48	0.205	3.65	0.259	4.74	0.337	5.78	0.411	6.78	0.482
1.80			1.19	0.147	2.31	0.200	3.41	0.242	4.44	0.315	5.43	0.386	6.38	0.454
1.90			1.11	0.145	2.16	0.195	3.19	0.231	4.17	0.297	5.12	0.364	6.03	0.428
2.00			1.04	0.142	2.02	0.191	3.00	0.225	3.94	0.280	4.83	0.344	5.71	0.406
2.20			0.93	0.139	1.80	0.185	2.67	0.216	3.53	0.251	4.35	0.309	5.15	0.366
2.40			0.84	0.136	1.62	0.179	2.41	0.209	3.19	0.231	3.95	0.281	4.68	0.333
2.60			0.76	0.134	1.48	0.175	2.19	0.203	2.91	0.224	3.61	0.257	4.29	0.305
2.80			0.70	0.132	1.36	0.172	2.01	0.199	2.67	0.219	3.32	0.236	3.96	0.281
3.00			0.64	0.130	1.25	0.169	1.86	0.195	2.47	0.214	3.07	0.229	3.67	0.261
3.20			0.60	0.129	1.16	0.167	1.73	0.192	2.29	0.210	2.86	0.224	3.42	0.243
3.40			0.56	0.128	1.09	0.165	1.61	0.189	2.14	0.207	2.67	0.221	3.20	0.232
3.60			0.52	0.127	1.02	0.163	1.51	0.187	2.01	0.204	2.50	0.217	3.00	0.228
3.80			0.49	0.126	0.96	0.162	1.42	0.185	1.89	0.202	2.36	0.214	2.82	0.225
4.00			0.47	0.125	0.91	0.161	1.35	0.183	1.79	0.199	2.23	0.212	2.67	0.222
5.00			0.36	0.122	0.71	0.156	1.05	0.177	1.40	0.192	1.74	0.203	2.09	0.212
6.00			0.30	0.121	0.58	0.153	0.87	0.173	1.15	0.187	1.43	0.198	1.72	0.206
7.00			0.25	0.120	0.50	0.151	0.74	0.170	0.98	0.184	1.22	0.194	1.46	0.202
8.00			0.22	0.119	0.43	0.150	0.64	0.168	0.85	0.182	1.06	0.192	1.27	0.199
10.00			0.17	0.118	0.34	0.148	0.51	0.166	0.67	0.179	0.84	0.188	1.00	0.195

Table 11.11 Rectangular columns − f_{cu} = 35 N/mm², k = 0.85.

a/h	p=0.4 N/bh	p=0.4 x/h	p=1.0 N/bh	p=1.0 x/h	p=2.0 N/bh	p=2.0 x/h	p=3.0 N/bh	p=3.0 x/h	p=4.0 N/bh	p=4.0 x/h	p=5.0 N/bh	p=5.0 x/h	p=6.0 N/bh	p=6.0 x/h
0.05	15.42	1.022	17.44	1.050	20.80	1.086	24.19	1.008	27.69	******	31.19	******	34.69	******
0.10	13.78	0.916	15.66	0.946	18.75	0.982	21.83	0.918	24.90	1.028	27.96	1.045	31.03	1.059
0.15	12.20	0.815	13.99	0.851	16.86	0.891	19.69	0.844	22.49	0.937	25.29	0.953	28.09	0.965
0.20	10.72	0.723	12.46	0.770	15.17	0.815	17.79	0.844	20.38	0.864	22.95	0.879	25.51	0.891
0.25	9.38	0.642	11.12	0.701	13.69	0.753	16.14	0.784	18.54	0.805	20.92	0.821	23.28	0.833
0.30	8.22	0.575	9.97	0.645	12.43	0.703	14.73	0.736	16.97	0.758	19.18	0.774	21.37	0.786
0.35	7.14	0.508	9.00	0.601	11.35	0.663	13.52	0.697	15.61	0.720	17.68	0.737	19.72	0.749
0.40	6.03	0.429	8.18	0.565	10.43	0.630	12.48	0.666	14.45	0.690	16.39	0.706	18.30	0.719
0.45	5.04	0.358	7.44	0.529	9.64	0.604	11.58	0.641	13.44	0.664	15.26	0.682	17.06	0.694
0.50	4.14	0.301	6.62	0.470	8.96	0.582	10.80	0.619	12.56	0.643	14.28	0.661	15.98	0.674
0.55	3.42	0.258	5.88	0.418	8.36	0.563	10.11	0.601	11.78	0.626	13.41	0.643	15.02	0.656
0.60	2.85	0.226	5.24	0.372	7.84	0.548	9.51	0.586	11.10	0.610	12.65	0.628	14.17	0.641
0.65	2.42	0.203	4.65	0.335	7.30	0.519	8.97	0.573	10.48	0.597	11.96	0.615	13.41	0.628
0.70	2.10	0.186	4.14	0.308	6.74	0.479	8.49	0.561	9.94	0.586	11.34	0.604	12.73	0.617
0.75	1.84	0.174	3.71	0.287	6.23	0.443	8.06	0.551	9.44	0.576	10.79	0.594	12.11	0.607
0.80	1.64	0.164	3.35	0.270	5.78	0.411	7.67	0.542	9.00	0.567	10.29	0.585	11.55	0.598
0.85	1.47	0.157	3.05	0.256	5.37	0.381	7.23	0.514	8.59	0.559	9.83	0.577	11.04	0.590
0.90	1.34	0.151	2.80	0.245	5.00	0.355	6.80	0.483	8.22	0.552	9.41	0.570	10.58	0.583
0.95	1.22	0.146	2.58	0.235	4.65	0.338	6.41	0.456	7.88	0.546	9.02	0.563	10.15	0.576
1.00	1.13	0.142	2.39	0.227	4.33	0.325	6.05	0.430	7.55	0.537	8.67	0.558	9.75	0.571
1.10	0.97	0.136	2.08	0.215	3.81	0.303	5.43	0.386	6.84	0.486	8.03	0.548	9.04	0.561
1.20	0.86	0.132	1.84	0.206	3.39	0.287	4.90	0.349	6.23	0.443	7.46	0.530	8.43	0.552
1.30	0.76	0.128	1.65	0.199	3.05	0.275	4.42	0.332	5.70	0.405	6.87	0.488	7.90	0.545
1.40	0.69	0.125	1.50	0.193	2.78	0.265	4.03	0.318	5.24	0.373	6.35	0.451	7.39	0.525
1.50	0.63	0.123	1.37	0.189	2.54	0.257	3.70	0.307	4.84	0.347	5.89	0.419	6.88	0.489
1.60	0.58	0.121	1.26	0.185	2.34	0.251	3.41	0.298	4.48	0.336	5.49	0.390	6.43	0.457
1.70	0.53	0.120	1.17	0.182	2.18	0.245	3.17	0.291	4.16	0.326	5.13	0.365	6.03	0.429
1.80	0.50	0.119	1.09	0.179	2.03	0.241	2.96	0.284	3.88	0.318	4.81	0.347	5.67	0.403
1.90	0.46	0.117	1.02	0.177	1.90	0.237	2.77	0.279	3.64	0.311	4.51	0.339	5.35	0.380
2.00	0.44	0.116	0.95	0.175	1.79	0.233	2.61	0.274	3.43	0.306	4.24	0.331	5.06	0.359
2.20	0.39	0.115	0.85	0.171	1.60	0.228	2.33	0.266	3.06	0.296	3.80	0.320	4.53	0.340
2.40	0.35	0.114	0.77	0.169	1.44	0.223	2.11	0.260	2.77	0.288	3.44	0.311	4.10	0.330
2.60	0.32	0.113	0.70	0.167	1.31	0.220	1.92	0.255	2.53	0.282	3.14	0.304	3.74	0.322
2.80	0.29	0.112	0.64	0.165	1.21	0.217	1.77	0.251	2.33	0.277	2.88	0.298	3.44	0.315
3.00	0.27	0.111	0.59	0.164	1.12	0.214	1.64	0.248	2.15	0.273	2.67	0.293	3.19	0.310
3.20	0.25	0.110	0.55	0.162	1.04	0.212	1.52	0.245	2.00	0.270	2.48	0.289	2.97	0.305
3.40	0.23	0.110	0.52	0.161	0.97	0.210	1.42	0.243	1.87	0.267	2.32	0.286	2.77	0.301
3.60	0.22	0.109	0.48	0.160	0.91	0.209	1.34	0.241	1.76	0.264	2.18	0.282	2.61	0.298
3.80	0.21	0.109	0.46	0.160	0.86	0.208	1.26	0.239	1.66	0.262	2.06	0.280	2.46	0.295
4.00	0.20	0.109	0.43	0.159	0.81	0.206	1.19	0.237	1.57	0.260	1.94	0.277	2.32	0.292
5.00	0.15	0.107	0.34	0.156	0.64	0.202	0.94	0.231	1.23	0.252	1.53	0.269	1.83	0.283
6.00	0.13	0.106	0.28	0.155	0.53	0.199	0.77	0.227	1.01	0.248	1.26	0.264	1.50	0.277
7.00	0.11	0.106	0.24	0.153	0.45	0.197	0.65	0.225	0.86	0.245	1.07	0.260	1.28	0.273
8.00	0.09	0.105	0.21	0.153	0.39	0.196	0.57	0.223	0.75	0.242	0.93	0.258	1.11	0.270
10.00	0.07	0.105	0.16	0.151	0.31	0.194	0.45	0.220	0.60	0.239	0.74	0.254	0.88	0.266

Table 11.11 (contd) – $f_{cu} = 35$ N/mm², $k = 0.80$.

e/h	p=0.4		p=1.0		p=2.0		p=3.0		p=4.0		p=5.0		p=6.0	
	N/bh	x/h	N/bh	x/h	N/bh	x/h	N/bh	x/h	N/bh	x/h	N/bh	x/h	N/bh	x/h
0.05	15.38	1.015	17.34	1.034	20.62	1.060	23.92	1.080	27.22	1.097	30.54	1.111	33.98	......
0.10	13.70	0.905	15.49	0.925	18.45	0.949	21.41	0.967	24.37	0.981	27.34	0.992	30.31	1.001
0.15	12.08	0.801	13.73	0.826	16.43	0.853	19.11	0.871	21.78	0.885	24.45	0.895	27.11	0.903
0.20	10.54	0.706	12.12	0.740	14.64	0.773	17.09	0.794	19.52	0.809	21.94	0.819	24.36	0.828
0.25	9.15	0.622	10.72	0.669	13.09	0.709	15.36	0.733	17.60	0.749	19.82	0.761	22.02	0.770
0.30	7.94	0.552	9.53	0.612	11.78	0.659	13.90	0.686	15.97	0.703	18.02	0.716	20.06	0.726
0.35	6.87	0.489	8.53	0.567	10.68	0.620	12.67	0.649	14.60	0.667	16.50	0.681	18.39	0.691
0.40	5.69	0.410	7.70	0.531	9.76	0.588	11.63	0.619	13.44	0.638	15.21	0.652	16.97	0.663
0.45	4.65	0.345	6.96	0.494	8.97	0.563	10.74	0.595	12.44	0.615	14.10	0.629	15.75	0.640
0.50	3.80	0.295	6.08	0.439	8.30	0.542	9.97	0.575	11.57	0.596	13.14	0.610	14.69	0.621
0.55	3.14	0.258	5.29	0.395	7.72	0.525	9.31	0.558	10.82	0.579	12.30	0.594	13.76	0.605
0.60	2.64	0.231	4.64	0.361	7.22	0.511	8.72	0.544	10.16	0.566	11.56	0.581	12.94	0.592
0.65	2.26	0.212	4.10	0.334	6.63	0.471	8.21	0.532	9.57	0.554	10.90	0.569	12.21	0.580
0.70	1.97	0.198	3.65	0.312	6.03	0.441	7.75	0.522	9.05	0.544	10.31	0.559	11.56	0.570
0.75	1.74	0.187	3.29	0.295	5.50	0.416	7.34	0.513	8.58	0.535	9.79	0.550	10.97	0.561
0.80	1.56	0.179	2.98	0.282	5.04	0.396	6.91	0.491	8.16	0.527	9.31	0.542	10.45	0.554
0.85	1.41	0.173	2.72	0.271	4.65	0.379	6.45	0.462	7.77	0.520	8.88	0.535	9.97	0.547
0.90	1.29	0.168	2.50	0.262	4.31	0.365	6.00	0.443	7.43	0.514	8.49	0.529	9.53	0.541
0.95	1.19	0.163	2.32	0.254	4.01	0.353	5.61	0.428	7.09	0.504	8.13	0.524	9.13	0.535
1.00	1.10	0.160	2.16	0.248	3.74	0.343	5.26	0.414	6.72	0.477	7.80	0.519	8.76	0.530
1.10	0.95	0.154	1.89	0.237	3.30	0.326	4.66	0.392	5.99	0.446	7.21	0.510	8.10	0.521
1.20	0.84	0.150	1.68	0.230	2.96	0.314	4.18	0.375	5.39	0.425	6.59	0.468	7.54	0.514
1.30	0.76	0.147	1.51	0.224	2.67	0.304	3.79	0.362	4.89	0.408	5.99	0.448	7.03	0.499
1.40	0.69	0.144	1.38	0.219	2.43	0.296	3.46	0.351	4.48	0.395	5.48	0.432	6.49	0.464
1.50	0.63	0.142	1.26	0.215	2.24	0.289	3.18	0.342	4.12	0.384	5.06	0.419	5.99	0.450
1.60	0.58	0.140	1.16	0.211	2.07	0.284	2.95	0.335	3.82	0.375	4.69	0.409	5.55	0.437
1.70	0.53	0.139	1.08	0.209	1.92	0.279	2.74	0.329	3.56	0.368	4.37	0.400	5.18	0.427
1.80	0.50	0.137	1.01	0.206	1.80	0.276	2.57	0.324	3.33	0.361	4.09	0.392	4.85	0.418
1.90	0.47	0.136	0.94	0.204	1.69	0.272	2.41	0.319	3.13	0.355	3.84	0.385	4.55	0.411
2.00	0.44	0.135	0.89	0.202	1.59	0.269	2.27	0.315	2.95	0.351	3.62	0.380	4.30	0.404
2.20	0.39	0.134	0.79	0.199	1.42	0.264	2.03	0.309	2.64	0.342	3.25	0.370	3.86	0.393
2.40	0.35	0.133	0.72	0.197	1.29	0.260	1.84	0.303	2.40	0.336	2.95	0.363	3.50	0.385
2.60	0.32	0.132	0.66	0.195	1.18	0.257	1.68	0.299	2.19	0.331	2.70	0.356	3.20	0.378
2.80	0.30	0.131	0.60	0.193	1.08	0.255	1.55	0.296	2.02	0.326	2.48	0.351	2.95	0.372
3.00	0.27	0.130	0.56	0.192	1.00	0.252	1.44	0.293	1.87	0.323	2.30	0.347	2.73	0.367
3.20	0.25	0.129	0.52	0.191	0.93	0.250	1.34	0.290	1.74	0.320	2.14	0.344	2.55	0.363
3.40	0.24	0.129	0.49	0.190	0.87	0.249	1.25	0.288	1.63	0.317	2.01	0.340	2.38	0.360
3.60	0.22	0.128	0.46	0.189	0.82	0.247	1.18	0.286	1.53	0.315	1.89	0.338	2.24	0.357
3.80	0.21	0.128	0.43	0.188	0.77	0.246	1.11	0.284	1.45	0.313	1.78	0.335	2.12	0.354
4.00	0.20	0.128	0.41	0.188	0.73	0.245	1.05	0.283	1.37	0.311	1.69	0.333	2.00	0.352
5.00	0.16	0.125	0.32	0.185	0.58	0.241	0.83	0.277	1.08	0.304	1.33	0.326	1.58	0.343
6.00	0.13	0.123	0.26	0.183	0.48	0.238	0.68	0.274	0.89	0.300	1.10	0.321	1.30	0.337
7.00	0.11	0.121	0.22	0.182	0.40	0.236	0.58	0.271	0.76	0.297	0.93	0.317	1.11	0.334
8.00	0.09	0.120	0.20	0.181	0.35	0.235	0.51	0.270	0.66	0.295	0.81	0.315	0.97	0.331
10.00	0.07	0.119	0.15	0.180	0.28	0.233	0.40	0.267	0.52	0.292	0.65	0.311	0.77	0.327

Table 11.12 Rectangular columns − $f_{cu} = 40\,\text{N/mm}^2$, $k = 0.95$.

e/h	p=0.4		p=1.0		p=2.0		p=3.0		p=4.0		p=5.0		p=6.0	
	N/bh	x/h	N/bh	x/h	N/bh	x/h	N/bh	x/h	N/bh	x/h	N/bh	x/h	N/bh	x/h
0.05	17.56	1.035	19.72	1.076	23.29	......	26.89	......	30.49	......	34.09	......	37.69	......
0.10	15.77	0.933	17.86	0.979	21.20	1.034	24.47	1.074	27.72	1.104	30.99	......	34.27	......
0.15	14.06	0.836	16.11	0.892	19.31	0.951	22.39	0.992	25.42	1.022	28.44	1.046	31.44	1.066
0.20	12.46	0.749	14.53	0.815	17.60	0.880	20.51	0.922	23.36	0.953	26.18	0.977	28.98	0.996
0.25	11.02	0.672	13.12	0.750	16.09	0.821	18.85	0.864	21.54	0.895	24.18	0.919	26.80	0.938
0.30	9.77	0.607	11.90	0.695	14.77	0.771	17.40	0.816	19.93	0.847	22.42	0.871	24.88	0.890
0.35	8.42	0.523	10.84	0.650	13.62	0.730	16.12	0.776	18.52	0.808	20.87	0.831	23.20	0.850
0.40	7.16	0.446	9.94	0.613	12.62	0.695	15.00	0.742	17.28	0.775	19.51	0.798	21.71	0.817
0.45	6.04	0.376	8.96	0.557	11.74	0.667	14.02	0.714	16.19	0.747	18.30	0.770	20.39	0.789
0.50	5.07	0.315	8.02	0.499	10.98	0.642	13.16	0.690	15.22	0.723	17.23	0.746	19.21	0.765
0.55	4.26	0.265	7.18	0.446	10.30	0.622	12.39	0.670	14.36	0.702	16.28	0.726	18.16	0.744
0.60	3.59	0.223	6.43	0.400	9.70	0.603	11.70	0.652	13.59	0.685	15.42	0.708	17.22	0.726
0.65	3.06	0.190	5.78	0.359	8.97	0.558	11.09	0.637	12.89	0.669	14.65	0.693	16.37	0.711
0.70	2.63	0.164	5.21	0.324	8.32	0.517	10.53	0.624	12.27	0.656	13.95	0.679	15.59	0.697
0.75	2.29	0.143	4.71	0.293	7.72	0.480	10.03	0.612	11.70	0.644	13.31	0.667	14.89	0.685
0.80	2.02	0.126	4.29	0.267	7.18	0.447	9.53	0.593	11.18	0.633	12.73	0.656	14.25	0.674
0.85	1.80	0.113	3.92	0.244	6.70	0.417	8.98	0.558	10.70	0.624	12.19	0.647	13.66	0.664
0.90	1.61	0.105	3.60	0.224	6.26	0.389	8.47	0.527	10.27	0.615	11.70	0.638	13.11	0.656
0.95			3.32	0.206	5.86	0.365	8.01	0.498	9.86	0.607	11.25	0.630	12.61	0.648
1.00			3.07	0.191	5.50	0.342	7.58	0.472	9.43	0.596	10.83	0.623	12.15	0.640
1.10			2.67	0.166	4.89	0.304	6.83	0.425	8.57	0.533	10.08	0.610	11.31	0.628
1.20			2.35	0.146	4.38	0.272	6.18	0.385	7.83	0.487	9.36	0.582	10.59	0.617
1.30			2.10	0.130	3.95	0.246	5.64	0.351	7.19	0.447	8.64	0.537	9.95	0.608
1.40			1.89	0.117	3.60	0.224	5.17	0.321	6.63	0.412	8.01	0.498	9.31	0.579
1.50			1.71	0.111	3.30	0.205	4.76	0.296	6.14	0.382	7.45	0.463	8.69	0.540
1.60			1.57	0.106	3.04	0.189	4.41	0.274	5.71	0.355	6.95	0.432	8.13	0.506
1.70			1.45	0.103	2.82	0.175	4.11	0.255	5.34	0.332	6.51	0.405	7.64	0.475
1.80					2.62	0.163	3.84	0.239	5.00	0.311	6.12	0.380	7.19	0.447
1.90					2.45	0.152	3.60	0.224	4.70	0.292	5.76	0.358	6.79	0.422
2.00					2.30	0.143	3.39	0.211	4.43	0.276	5.45	0.339	6.43	0.400
2.20					2.05	0.127	3.03	0.188	3.98	0.247	4.90	0.305	5.80	0.361
2.40					1.84	0.116	2.73	0.170	3.60	0.224	4.45	0.276	5.27	0.328
2.60					1.67	0.112	2.49	0.155	3.28	0.204	4.07	0.253	4.83	0.300
2.80					1.53	0.109	2.28	0.142	3.02	0.188	3.74	0.233	4.45	0.277
3.00					1.41	0.106	2.11	0.131	2.79	0.174	3.47	0.216	4.13	0.257
3.20					1.31	0.104	1.96	0.122	2.60	0.161	3.23	0.201	3.85	0.239
3.40					1.22	0.102	1.83	0.116	2.43	0.151	3.02	0.188	3.60	0.224
3.60					1.15	0.101	1.71	0.114	2.28	0.142	2.83	0.176	3.38	0.210
3.80							1.61	0.112	2.14	0.133	2.67	0.166	3.19	0.198
4.00							1.52	0.110	2.02	0.126	2.52	0.157	3.01	0.187
5.00							1.19	0.105	1.58	0.112	1.98	0.123	2.37	0.147
6.00							0.97	0.102	1.30	0.109	1.62	0.114	1.94	0.121
7.00									1.10	0.106	1.37	0.111	1.65	0.114
8.00									0.95	0.104	1.19	0.109	1.43	0.112
10.00									0.75	0.102	0.94	0.106	1.13	0.109

Table 11.12 (contd) − $f_{cu} = 40\,\text{N/mm}^2$, $k = 0.90$.

e/h	p=0.4 N/bh	x/h	p=1.0 N/bh	x/h	p=2.0 N/bh	x/h	p=3.0 N/bh	x/h	p=4.0 N/bh	x/h	p=5.0 N/bh	x/h	p=6.0 N/bh	x/h
0.05	17.49	1.027	19.58	1.060	23.02	1.103	26.56	......	30.12	......	33.67	......	37.23	......
0.10	15.66	0.922	17.65	0.960	20.87	1.004	24.05	1.037	27.22	1.062	30.37	1.083	33.53	1.101
0.15	13.91	0.823	15.84	0.868	18.88	0.917	21.85	0.951	24.79	0.976	27.70	0.996	30.61	1.012
0.20	12.27	0.733	14.19	0.789	17.10	0.844	19.88	0.879	22.62	0.905	25.32	0.924	28.02	0.940
0.25	10.78	0.654	12.73	0.721	15.53	0.782	18.16	0.820	20.71	0.846	23.24	0.866	25.75	0.882
0.30	9.49	0.587	11.47	0.665	14.17	0.732	16.65	0.771	19.06	0.798	21.43	0.819	23.77	0.834
0.35	8.16	0.508	10.39	0.620	12.99	0.691	15.35	0.732	17.62	0.760	19.85	0.780	22.05	0.796
0.40	6.89	0.429	9.47	0.583	11.98	0.657	14.23	0.699	16.37	0.727	18.47	0.748	20.54	0.764
0.45	5.76	0.358	8.51	0.529	11.11	0.629	13.24	0.672	15.28	0.701	17.26	0.721	19.22	0.737
0.50	4.78	0.297	7.56	0.470	10.35	0.606	12.38	0.649	14.32	0.678	16.20	0.699	18.05	0.715
0.55	3.97	0.247	6.72	0.418	9.68	0.586	11.63	0.630	13.47	0.659	15.26	0.680	17.02	0.696
0.60	3.30	0.209	5.98	0.372	9.05	0.563	10.95	0.613	12.71	0.642	14.42	0.663	16.09	0.679
0.65	2.77	0.183	5.34	0.332	8.34	0.519	10.35	0.599	12.03	0.628	13.66	0.649	15.26	0.665
0.70	2.37	0.163	4.79	0.298	7.70	0.479	9.82	0.586	11.42	0.615	12.98	0.636	14.51	0.652
0.75	2.05	0.149	4.32	0.268	7.12	0.443	9.33	0.575	10.87	0.604	12.37	0.625	13.83	0.641
0.80	1.81	0.138	3.91	0.243	6.60	0.411	8.79	0.547	10.37	0.595	11.80	0.615	13.21	0.631
0.85	1.61	0.130	3.55	0.224	6.13	0.381	8.26	0.514	9.92	0.586	11.29	0.607	12.64	0.622
0.90	1.46	0.124	3.23	0.211	5.71	0.355	7.77	0.483	9.50	0.578	10.82	0.599	12.12	0.614
0.95	1.33	0.119	2.97	0.200	5.34	0.332	7.33	0.456	9.09	0.565	10.39	0.592	11.64	0.607
1.00	1.22	0.114	2.74	0.191	5.00	0.311	6.92	0.430	8.63	0.537	9.99	0.585	11.20	0.601
1.10	1.04	0.108	2.37	0.177	4.42	0.275	6.20	0.386	7.82	0.486	9.28	0.574	10.41	0.589
1.20	0.91	0.103	2.08	0.167	3.95	0.246	5.60	0.348	7.12	0.443	8.52	0.530	9.72	0.580
1.30	0.81	0.100	1.86	0.159	3.55	0.226	5.09	0.317	6.52	0.405	7.85	0.488	9.10	0.566
1.40			1.68	0.153	3.21	0.215	4.66	0.290	5.99	0.373	7.25	0.451	8.45	0.525
1.50			1.53	0.148	2.93	0.206	4.29	0.267	5.54	0.345	6.73	0.419	7.87	0.489
1.60			1.40	0.144	2.69	0.199	3.96	0.247	5.14	0.320	6.27	0.390	7.35	0.457
1.70			1.29	0.141	2.49	0.193	3.68	0.231	4.80	0.298	5.86	0.365	6.89	0.429
1.80			1.20	0.139	2.32	0.189	3.43	0.224	4.49	0.279	5.50	0.342	6.48	0.403
1.90			1.12	0.136	2.17	0.184	3.20	0.218	4.22	0.262	5.18	0.322	6.11	0.380
2.00			1.05	0.134	2.03	0.181	3.01	0.213	3.97	0.247	4.89	0.304	5.78	0.359
2.20			0.94	0.131	1.81	0.175	2.68	0.205	3.55	0.229	4.39	0.273	5.20	0.323
2.40			0.84	0.129	1.63	0.170	2.42	0.199	3.20	0.221	3.98	0.247	4.72	0.294
2.60			0.77	0.126	1.48	0.167	2.20	0.194	2.91	0.214	3.63	0.231	4.32	0.269
2.80			0.70	0.125	1.36	0.164	2.02	0.190	2.67	0.209	3.33	0.225	3.98	0.248
3.00			0.65	0.123	1.26	0.161	1.86	0.186	2.47	0.205	3.08	0.220	3.69	0.232
3.20			0.60	0.122	1.17	0.159	1.73	0.184	2.30	0.202	2.86	0.216	3.43	0.227
3.40			0.56	0.121	1.09	0.158	1.62	0.181	2.14	0.199	2.67	0.212	3.20	0.224
3.60			0.53	0.120	1.02	0.156	1.52	0.179	2.01	0.196	2.51	0.209	3.00	0.220
3.80			0.50	0.120	0.96	0.155	1.43	0.177	1.89	0.194	2.36	0.207	2.83	0.217
4.00			0.47	0.119	0.91	0.154	1.35	0.176	1.79	0.192	2.23	0.204	2.67	0.215
5.00			0.37	0.116	0.71	0.149	1.06	0.170	1.40	0.185	1.75	0.196	2.09	0.206
6.00			0.30	0.115	0.59	0.147	0.87	0.166	1.15	0.181	1.44	0.191	1.72	0.200
7.00			0.26	0.114	0.50	0.145	0.74	0.164	0.98	0.178	1.22	0.188	1.46	0.196
8.00			0.22	0.113	0.43	0.144	0.64	0.162	0.85	0.176	1.06	0.186	1.27	0.194
10.00			0.18	0.112	0.34	0.142	0.51	0.160	0.67	0.173	0.84	0.182	1.00	0.190

Table 11.13 Rectangular columns − $f_{cu} = 40\,\text{N/mm}^2$, $k = 0.85$.

e/h	p=0.4 N/bh	p=0.4 x/h	p=1.0 N/bh	p=1.0 x/h	p=2.0 N/bh	p=2.0 x/h	p=3.0 N/bh	p=3.0 x/h	p=4.0 N/bh	p=4.0 x/h	p=5.0 N/bh	p=5.0 x/h	p=6.0 N/bh	p=6.0 x/h
0.05	17.43	1.020	19.45	1.045	22.81	1.078	26.17	1.104	29.65	......	33.15	......	36.65	......
0.10	15.57	0.913	17.45	0.940	20.55	0.974	23.63	0.999	26.70	1.019	29.77	1.035	32.84	1.048
0.15	13.77	0.811	15.57	0.845	18.45	0.883	21.29	0.909	24.10	0.928	26.91	0.944	29.71	0.956
0.20	12.07	0.717	13.84	0.762	16.57	0.806	19.21	0.834	21.81	0.854	24.39	0.870	26.96	0.882
0.25	10.54	0.635	12.32	0.692	14.93	0.743	17.41	0.774	19.82	0.795	22.21	0.811	24.58	0.824
0.30	9.20	0.566	11.01	0.635	13.53	0.692	15.86	0.725	18.12	0.748	20.34	0.765	22.54	0.777
0.35	7.89	0.491	9.91	0.569	12.33	0.651	14.53	0.686	16.65	0.710	18.73	0.727	20.79	0.740
0.40	6.60	0.411	8.99	0.552	11.31	0.618	13.40	0.655	15.39	0.679	17.35	0.697	19.28	0.710
0.45	5.44	0.340	8.02	0.499	10.44	0.591	12.42	0.629	14.30	0.654	16.14	0.671	17.96	0.685
0.50	4.42	0.283	7.07	0.440	9.69	0.569	11.57	0.607	13.35	0.632	15.09	0.651	16.81	0.664
0.55	3.60	0.241	6.24	0.388	9.03	0.550	10.82	0.589	12.52	0.615	14.17	0.633	15.79	0.647
0.60	2.99	0.211	5.50	0.344	8.37	0.521	10.17	0.574	11.78	0.599	13.35	0.618	14.89	0.632
0.65	2.52	0.189	4.83	0.312	7.68	0.477	9.59	0.560	11.13	0.586	12.62	0.605	14.09	0.619
0.70	2.17	0.173	4.28	0.287	7.05	0.438	9.07	0.549	10.54	0.575	11.96	0.593	13.36	0.607
0.75	1.90	0.162	3.82	0.267	6.49	0.403	8.57	0.533	10.01	0.565	11.37	0.583	12.71	0.597
0.80	1.69	0.153	3.44	0.251	5.99	0.372	8.01	0.498	9.53	0.556	10.84	0.574	12.12	0.588
0.85	1.52	0.146	3.13	0.238	5.54	0.346	7.50	0.466	9.10	0.548	10.35	0.566	11.58	0.580
0.90	1.37	0.141	2.86	0.228	5.11	0.329	7.04	0.438	8.70	0.541	9.91	0.559	11.09	0.573
0.95	1.26	0.137	2.63	0.219	4.74	0.315	6.61	0.411	8.23	0.512	9.50	0.553	10.63	0.567
1.00	1.16	0.133	2.44	0.212	4.41	0.303	6.23	0.387	7.80	0.485	9.12	0.547	10.22	0.561
1.10	1.00	0.127	2.12	0.201	3.87	0.284	5.55	0.347	7.03	0.437	8.39	0.522	9.47	0.551
1.20	0.88	0.123	1.88	0.193	3.44	0.270	4.96	0.328	6.38	0.396	7.66	0.476	8.83	0.543
1.30	0.78	0.120	1.68	0.186	3.09	0.258	4.47	0.312	5.82	0.362	7.02	0.437	8.17	0.508
1.40	0.71	0.118	1.52	0.181	2.81	0.250	4.07	0.300	5.32	0.341	6.47	0.403	7.56	0.470
1.50	0.65	0.116	1.39	0.177	2.57	0.243	3.73	0.290	4.88	0.328	5.99	0.373	7.02	0.437
1.60	0.59	0.114	1.28	0.174	2.37	0.237	3.44	0.282	4.51	0.318	5.57	0.348	6.55	0.407
1.70	0.55	0.113	1.19	0.171	2.20	0.232	3.20	0.275	4.19	0.309	5.18	0.338	6.13	0.381
1.80	0.51	0.112	1.10	0.169	2.05	0.228	2.98	0.269	3.91	0.302	4.83	0.330	5.75	0.358
1.90	0.48	0.110	1.03	0.167	1.92	0.224	2.79	0.264	3.66	0.296	4.53	0.322	5.40	0.345
2.00	0.45	0.110	0.97	0.165	1.81	0.221	2.63	0.260	3.45	0.291	4.27	0.316	5.09	0.337
2.20	0.40	0.108	0.87	0.162	1.61	0.216	2.35	0.253	3.08	0.282	3.82	0.305	4.55	0.325
2.40	0.36	0.107	0.78	0.160	1.46	0.212	2.12	0.248	2.79	0.275	3.45	0.297	4.12	0.316
2.60	0.33	0.106	0.71	0.158	1.33	0.209	1.94	0.243	2.54	0.270	3.15	0.291	3.76	0.309
2.80	0.30	0.105	0.65	0.156	1.22	0.206	1.78	0.240	2.34	0.265	2.90	0.286	3.46	0.303
3.00	0.28	0.105	0.60	0.155	1.13	0.204	1.65	0.237	2.16	0.261	2.68	0.281	3.20	0.298
3.20	0.26	0.104	0.56	0.154	1.05	0.202	1.53	0.234	2.01	0.258	2.50	0.277	2.98	0.293
3.40	0.24	0.104	0.52	0.153	0.98	0.200	1.43	0.232	1.88	0.255	2.33	0.274	2.78	0.290
3.60	0.23	0.103	0.49	0.152	0.92	0.199	1.35	0.230	1.77	0.253	2.19	0.271	2.61	0.287
3.80	0.21	0.103	0.46	0.151	0.87	0.198	1.27	0.228	1.67	0.251	2.07	0.269	2.46	0.284
4.00	0.20	0.103	0.44	0.151	0.82	0.197	1.20	0.227	1.58	0.249	1.95	0.267	2.33	0.281
5.00	0.16	0.101	0.34	0.148	0.64	0.193	0.94	0.221	1.24	0.242	1.54	0.259	1.83	0.273
6.00	0.13	0.101	0.28	0.147	0.53	0.190	0.78	0.218	1.02	0.238	1.26	0.254	1.51	0.267
7.00	0.11	0.100	0.24	0.146	0.45	0.188	0.66	0.216	0.87	0.235	1.08	0.251	1.28	0.264
8.00			0.21	0.145	0.39	0.187	0.57	0.214	0.75	0.233	0.93	0.249	1.12	0.261
10.00			0.17	0.144	0.31	0.185	0.45	0.212	0.60	0.231	0.74	0.245	0.89	0.257

Table 11.13 (contd) − $f_{cu} = 40 \, \text{N/mm}^2$, $k = 0.80$.

e/h	p=0.4		p=1.0		p=2.0		p=3.0		p=4.0		p=5.0		p=6.0	
	N/bh	x/h	N/bh	x/h	N/bh	x/h	N/bh	x/h	N/bh	x/h	N/bh	x/h	N/bh	x/h
0.05	17.39	1.013	19.35	1.030	22.63	1.054	25.92	1.073	29.22	1.089	32.53	1.103	35.90	......
0.10	15.49	0.904	17.27	0.921	20.24	0.944	23.20	0.961	26.16	0.974	29.13	0.985	32.09	0.994
0.15	13.64	0.799	15.30	0.821	18.01	0.847	20.70	0.865	23.37	0.878	26.04	0.889	28.70	0.897
0.20	11.89	0.702	13.49	0.734	16.02	0.766	18.48	0.787	20.92	0.802	23.35	0.813	25.77	0.822
0.25	10.29	0.616	11.89	0.661	14.30	0.701	16.59	0.726	18.84	0.742	21.07	0.754	23.28	0.764
0.30	8.90	0.544	10.54	0.602	12.84	0.650	14.99	0.677	17.08	0.695	19.14	0.709	21.18	0.719
0.35	7.61	0.473	9.41	0.556	11.62	0.610	13.64	0.639	15.59	0.659	17.51	0.673	19.40	0.683
0.40	6.23	0.394	8.47	0.520	10.59	0.578	12.50	0.609	14.33	0.629	16.12	0.644	17.89	0.655
0.45	5.03	0.329	7.50	0.467	9.73	0.552	11.53	0.585	13.25	0.606	14.93	0.621	16.59	0.632
0.50	4.07	0.278	6.46	0.414	8.99	0.531	10.69	0.564	12.32	0.586	13.90	0.602	15.46	0.613
0.55	3.34	0.242	5.59	0.371	8.35	0.513	9.97	0.548	11.51	0.570	13.00	0.585	14.47	0.597
0.60	2.79	0.217	4.87	0.338	7.64	0.475	9.34	0.534	10.79	0.556	12.21	0.572	13.60	0.584
0.65	2.38	0.198	4.28	0.312	6.90	0.440	8.78	0.522	10.17	0.544	11.51	0.560	12.83	0.572
0.70	2.07	0.185	3.80	0.292	6.24	0.412	8.28	0.511	9.61	0.534	10.89	0.550	12.14	0.562
0.75	1.83	0.175	3.41	0.276	5.68	0.390	7.72	0.480	9.10	0.525	10.33	0.541	11.53	0.553
0.80	1.63	0.168	3.09	0.264	5.19	0.371	7.14	0.453	8.65	0.517	9.82	0.533	10.97	0.545
0.85	1.47	0.162	2.82	0.254	4.78	0.356	6.61	0.433	8.24	0.510	9.36	0.526	10.46	0.538
0.90	1.34	0.157	2.59	0.245	4.42	0.343	6.14	0.416	7.77	0.483	8.94	0.520	10.00	0.532
0.95	1.24	0.153	2.39	0.238	4.11	0.332	5.73	0.402	7.31	0.460	8.56	0.515	9.57	0.527
1.00	1.14	0.150	2.22	0.232	3.84	0.322	5.36	0.389	6.86	0.445	8.21	0.509	9.19	0.522
1.10	0.99	0.145	1.95	0.223	3.38	0.307	4.75	0.369	6.09	0.420	7.42	0.464	8.50	0.513
1.20	0.88	0.141	1.73	0.216	3.02	0.296	4.26	0.354	5.47	0.401	6.68	0.442	7.83	0.487
1.30	0.79	0.138	1.56	0.210	2.73	0.287	3.85	0.342	4.96	0.386	6.06	0.424	7.16	0.457
1.40	0.71	0.135	1.41	0.206	2.49	0.279	3.52	0.332	4.54	0.374	5.55	0.410	6.56	0.441
1.50	0.65	0.134	1.30	0.202	2.28	0.274	3.24	0.324	4.18	0.364	5.11	0.398	6.04	0.427
1.60	0.60	0.132	1.20	0.199	2.11	0.269	2.99	0.318	3.87	0.356	4.74	0.388	5.60	0.416
1.70	0.55	0.131	1.11	0.197	1.96	0.265	2.79	0.312	3.60	0.349	4.41	0.380	5.22	0.407
1.80	0.52	0.129	1.03	0.195	1.83	0.261	2.60	0.307	3.37	0.343	4.13	0.373	4.89	0.399
1.90	0.48	0.128	0.97	0.193	1.72	0.258	2.44	0.303	3.16	0.338	3.88	0.367	4.59	0.392
2.00	0.45	0.128	0.91	0.191	1.62	0.255	2.30	0.299	2.98	0.334	3.66	0.362	4.33	0.386
2.20	0.40	0.125	0.82	0.188	1.45	0.251	2.06	0.293	2.67	0.326	3.28	0.353	3.89	0.376
2.40	0.36	0.122	0.74	0.186	1.31	0.247	1.87	0.289	2.42	0.321	2.97	0.347	3.53	0.368
2.60	0.33	0.120	0.67	0.184	1.20	0.244	1.71	0.285	2.21	0.316	2.72	0.341	3.22	0.362
2.80	0.30	0.118	0.62	0.183	1.10	0.242	1.57	0.282	2.04	0.312	2.51	0.336	2.97	0.357
3.00	0.28	0.117	0.57	0.182	1.02	0.240	1.46	0.279	1.89	0.309	2.32	0.333	2.75	0.353
3.20	0.26	0.116	0.53	0.181	0.95	0.238	1.36	0.277	1.76	0.306	2.16	0.329	2.57	0.349
3.40	0.24	0.115	0.50	0.180	0.89	0.237	1.27	0.275	1.65	0.303	2.03	0.326	2.40	0.346
3.60	0.23	0.114	0.47	0.179	0.84	0.235	1.19	0.273	1.55	0.301	1.90	0.324	2.26	0.343
3.80	0.21	0.113	0.44	0.178	0.79	0.234	1.13	0.271	1.46	0.299	1.80	0.322	2.13	0.340
4.00	0.20	0.112	0.42	0.178	0.74	0.233	1.07	0.270	1.38	0.298	1.70	0.320	2.02	0.338
5.00	0.16	0.109	0.33	0.175	0.59	0.229	0.84	0.265	1.09	0.292	1.34	0.313	1.59	0.330
6.00	0.13	0.108	0.27	0.174	0.48	0.227	0.69	0.262	0.90	0.288	1.11	0.308	1.31	0.325
7.00	0.11	0.106	0.23	0.173	0.41	0.225	0.59	0.260	0.77	0.285	0.94	0.305	1.12	0.322
8.00	0.10	0.105	0.20	0.172	0.36	0.224	0.51	0.258	0.67	0.283	0.82	0.303	0.97	0.319
10.00	0.08	0.104	0.16	0.171	0.28	0.222	0.41	0.256	0.53	0.281	0.65	0.300	0.77	0.316

Table 11.14 Rectangular columns − $f_{cu} = 45\,\text{N/mm}^2$, $k = 0.95$.

e/h	p=0.4 N/bh	x/h	p=1.0 N/bh	x/h	p=2.0 N/bh	x/h	p=3.0 N/bh	x/h	p=4.0 N/bh	x/h	p=5.0 N/bh	x/h	p=6.0 N/bh	x/h
0.05	19.57	1.032	21.74	1.069	25.30	......	28.90	......	32.5	......	36.10	......	39.70	......
0.10	17.56	0.928	19.66	0.972	23.02	1.024	26.31	1.062	29.56	1.092	32.82	1.034	36.09	......
0.15	15.63	0.831	17.71	0.883	20.94	0.940	24.04	0.980	27.09	1.010	30.11	1.010	33.12	1.053
0.20	13.83	0.742	15.94	0.805	19.05	0.869	22.00	0.910	24.86	0.940	27.70	0.964	30.50	0.983
0.25	12.20	0.664	14.36	0.739	17.39	0.808	20.19	0.851	22.89	0.883	25.56	0.906	28.19	0.926
0.30	10.74	0.593	12.99	0.683	15.93	0.758	18.60	0.803	21.16	0.835	23.67	0.859	26.15	0.878
0.35	9.18	0.508	11.81	0.637	14.66	0.716	17.21	0.763	19.65	0.795	22.02	0.819	24.36	0.838
0.40	7.75	0.429	10.75	0.594	13.56	0.681	16.00	0.729	18.31	0.762	20.56	0.786	22.77	0.805
0.45	6.48	0.358	9.57	0.529	12.61	0.652	14.94	0.701	17.14	0.734	19.28	0.758	21.38	0.777
0.50	5.38	0.297	8.51	0.470	11.77	0.628	14.00	0.677	16.10	0.710	18.14	0.734	20.13	0.753
0.55	4.47	0.247	7.56	0.418	11.03	0.607	13.17	0.656	15.18	0.689	17.12	0.714	19.02	0.733
0.60	3.73	0.206	6.73	0.372	10.19	0.563	12.43	0.639	14.35	0.672	16.21	0.696	18.02	0.715
0.65	3.15	0.174	6.01	0.332	9.39	0.519	11.77	0.623	13.61	0.656	15.39	0.681	17.12	0.699
0.70	2.70	0.149	5.39	0.298	8.66	0.479	11.18	0.610	12.94	0.643	14.64	0.667	16.31	0.686
0.75	2.34	0.129	4.85	0.268	8.01	0.443	10.54	0.583	12.34	0.631	13.97	0.655	15.57	0.674
0.80	2.05	0.114	4.40	0.243	7.43	0.411	9.89	0.547	11.78	0.620	13.35	0.644	14.89	0.663
0.85	1.81	0.104	4.00	0.221	6.90	0.381	9.29	0.514	11.28	0.611	12.79	0.635	14.27	0.653
0.90			3.66	0.203	6.43	0.355	8.74	0.483	10.78	0.596	12.27	0.626	13.70	0.644
0.95			3.37	0.186	6.01	0.332	8.24	0.456	10.22	0.565	11.80	0.618	13.17	0.637
1.00			3.12	0.172	5.63	0.311	7.78	0.430	9.71	0.537	11.35	0.611	12.68	0.629
1.10			2.70	0.149	4.97	0.275	6.98	0.386	8.79	0.486	10.46	0.578	11.81	0.617
1.20			2.37	0.131	4.44	0.246	6.30	0.348	8.01	0.443	9.59	0.530	11.05	0.606
1.30			2.11	0.117	4.00	0.221	5.73	0.317	7.33	0.405	8.83	0.488	10.24	0.566
1.40			1.90	0.110	3.63	0.201	5.24	0.290	6.74	0.373	8.16	0.451	9.50	0.525
1.50			1.72	0.105	3.32	0.184	4.82	0.267	6.23	0.345	7.57	0.419	8.85	0.489
1.60			1.57	0.100	3.06	0.169	4.46	0.247	5.79	0.320	7.06	0.390	8.27	0.457
1.70					2.83	0.157	4.14	0.229	5.40	0.298	6.60	0.365	7.75	0.429
1.80					2.64	0.146	3.87	0.214	5.05	0.279	6.19	0.342	7.29	0.403
1.90					2.46	0.136	3.62	0.200	4.74	0.262	5.83	0.322	6.87	0.380
2.00					2.31	0.128	3.41	0.188	4.47	0.247	5.50	0.304	6.50	0.359
2.20					2.05	0.115	3.04	0.168	4.00	0.221	4.94	0.273	5.85	0.323
2.40					1.85	0.111	2.74	0.152	3.62	0.200	4.47	0.247	5.31	0.294
2.60					1.68	0.107	2.50	0.138	3.30	0.182	4.09	0.226	4.86	0.269
2.80					1.53	0.104	2.29	0.127	3.03	0.168	3.76	0.208	4.48	0.248
3.00					1.41	0.102	2.12	0.117	2.80	0.155	3.48	0.192	4.15	0.229
3.20					1.31	0.100	1.96	0.114	2.60	0.144	3.24	0.179	3.86	0.214
3.40							1.83	0.112	2.43	0.134	3.02	0.167	3.61	0.200
3.60							1.71	0.110	2.28	0.126	2.84	0.157	3.39	0.187
3.80							1.61	0.108	2.15	0.119	2.67	0.148	3.20	0.177
4.00							1.52	0.107	2.03	0.115	2.53	0.140	3.02	0.167
5.00							1.19	0.102	1.58	0.109	1.98	0.115	2.37	0.131
6.00									1.30	0.106	1.62	0.111	1.95	0.115
7.00									1.10	0.103	1.37	0.108	1.65	0.112
8.00									0.95	0.102	1.19	0.106	1.43	0.110
10.00											0.94	0.104	1.13	0.107

Table 11.14 (contd) – $f_{cu} = 45\,\text{N/mm}^2$, $k = 0.90$.

e/h	p=0.4 N/bh	x/h	p=1.0 N/bh	x/h	p=2.0 N/bh	x/h	p=3.0 N/bh	x/h	p=4.0 N/bh	x/h	p=5.0 N/bh	x/h	p=6.0 N/bh	x/h
0.05	19.50	1.024	21.59	1.055	25.04	1.094	28.55	******	32.10	******	35.66	******	39.21	******
0.10	17.45	0.919	19.45	0.954	22.68	0.996	25.87	1.027	29.04	1.051	32.20	1.072	35.35	1.089
0.15	15.48	0.819	17.43	0.861	20.49	0.908	23.48	0.941	26.42	0.966	29.34	0.985	32.26	1.002
0.20	13.63	0.727	15.58	0.780	18.53	0.834	21.33	0.869	24.08	0.894	26.80	0.914	29.50	0.930
0.25	11.94	0.646	13.94	0.711	16.79	0.772	19.45	0.809	22.03	0.835	24.57	0.856	27.09	0.872
0.30	10.47	0.578	12.53	0.654	15.29	0.721	17.82	0.760	20.25	0.788	22.63	0.808	24.98	0.824
0.35	8.92	0.493	11.32	0.608	14.00	0.679	16.40	0.720	18.70	0.748	20.94	0.769	23.16	0.786
0.40	7.47	0.413	10.28	0.568	12.89	0.645	15.18	0.687	17.35	0.716	19.47	0.737	21.56	0.754
0.45	6.18	0.341	9.09	0.502	11.93	0.616	14.12	0.660	16.18	0.689	18.19	0.711	20.16	0.727
0.50	5.07	0.280	8.02	0.443	11.10	0.593	13.19	0.637	15.15	0.666	17.05	0.688	18.92	0.705
0.55	4.16	0.231	7.08	0.391	10.38	0.573	12.37	0.618	14.24	0.647	16.05	0.669	17.83	0.685
0.60	3.41	0.195	6.26	0.346	9.51	0.526	11.64	0.601	13.43	0.631	15.16	0.652	16.85	0.669
0.65	2.85	0.170	5.55	0.307	8.72	0.482	11.00	0.587	12.71	0.616	14.36	0.638	15.97	0.655
0.70	2.42	0.152	4.95	0.274	8.01	0.443	10.42	0.574	12.06	0.604	13.64	0.625	15.18	0.642
0.75	2.10	0.139	4.44	0.245	7.38	0.408	9.75	0.539	11.47	0.593	12.98	0.614	14.46	0.631
0.80	1.84	0.129	3.99	0.224	6.82	0.377	9.11	0.504	10.94	0.583	12.39	0.605	13.81	0.621
0.85	1.64	0.121	3.61	0.209	6.31	0.349	8.54	0.472	10.45	0.574	11.85	0.596	13.21	0.612
0.90	1.48	0.116	3.28	0.197	5.86	0.324	8.01	0.443	9.90	0.547	11.35	0.588	12.67	0.604
0.95	1.35	0.111	3.01	0.187	5.46	0.302	7.53	0.416	9.37	0.518	10.90	0.581	12.16	0.597
1.00	1.23	0.107	2.77	0.179	5.10	0.282	7.10	0.392	8.88	0.491	10.48	0.574	11.70	0.591
1.10	1.06	0.101	2.39	0.166	4.50	0.249	6.34	0.350	8.01	0.443	9.55	0.528	10.87	0.579
1.20			2.10	0.157	3.99	0.226	5.70	0.315	7.27	0.402	8.72	0.482	10.09	0.558
1.30			1.88	0.150	3.57	0.213	5.17	0.286	6.63	0.367	8.01	0.443	9.31	0.514
1.40			1.69	0.144	3.23	0.203	4.72	0.261	6.09	0.336	7.38	0.408	8.62	0.476
1.50			1.54	0.140	2.95	0.195	4.33	0.240	5.62	0.310	6.84	0.378	8.01	0.443
1.60			1.41	0.137	2.71	0.189	3.99	0.227	5.21	0.288	6.36	0.351	7.47	0.413
1.70			1.30	0.134	2.51	0.184	3.70	0.220	4.85	0.268	5.94	0.328	6.99	0.386
1.80			1.21	0.131	2.33	0.179	3.44	0.213	4.53	0.250	5.56	0.307	6.56	0.363
1.90			1.13	0.129	2.18	0.175	3.22	0.208	4.25	0.235	5.23	0.289	6.18	0.342
2.00			1.06	0.128	2.04	0.172	3.02	0.203	3.99	0.228	4.93	0.273	5.84	0.323
2.20			0.94	0.125	1.82	0.167	2.69	0.196	3.56	0.219	4.42	0.244	5.24	0.290
2.40			0.85	0.122	1.64	0.163	2.42	0.190	3.21	0.212	3.99	0.229	4.76	0.263
2.60			0.77	0.120	1.49	0.159	2.21	0.186	2.92	0.206	3.64	0.222	4.35	0.240
2.80			0.71	0.119	1.37	0.157	2.02	0.182	2.68	0.201	3.34	0.217	3.99	0.240
3.00			0.65	0.118	1.26	0.154	1.87	0.179	2.48	0.197	3.08	0.212	3.69	0.224
3.20			0.61	0.117	1.17	0.153	1.74	0.176	2.30	0.194	2.87	0.208	3.43	0.220
3.40			0.57	0.116	1.10	0.151	1.62	0.174	2.15	0.191	2.68	0.205	3.20	0.216
3.60			0.53	0.115	1.03	0.150	1.52	0.172	2.02	0.189	2.51	0.202	3.00	0.213
3.80			0.50	0.114	0.97	0.148	1.43	0.171	1.90	0.187	2.36	0.200	2.83	0.210
4.00			0.47	0.114	0.91	0.147	1.35	0.169	1.79	0.185	2.23	0.198	2.67	0.208
5.00			0.37	0.111	0.72	0.143	1.06	0.164	1.41	0.179	1.75	0.190	2.09	0.200
6.00			0.30	0.110	0.59	0.141	0.87	0.161	1.15	0.175	1.44	0.186	1.72	0.195
7.00			0.26	0.109	0.50	0.139	0.74	0.158	0.98	0.172	1.22	0.183	1.46	0.191
8.00			0.22	0.108	0.43	0.138	0.64	0.157	0.85	0.170	1.06	0.180	1.27	0.189
10.00			0.18	0.107	0.34	0.137	0.51	0.155	0.67	0.168	0.84	0.177	1.01	0.185

Table 11.15 Rectangular columns – $f_{cu} = 45\,\text{N/mm}^2$, $k = 0.85$.

e/h	p=0.4		p=1.0		p=2.0		p=3.0		p=4.0		p=5.0		p=6.0	
	N/bh	x/h	N/bh	x/h	N/bh	x/h	N/bh	x/h	N/bh	x/h	N/bh	x/h	N/bh	x/h
0.05	19.44	1.018	21.46	1.041	24.83	1.071	28.18	1.096	31.6	******	35.10	******	38.60	******
0.10	17.36	0.910	19.25	0.936	22.35	0.967	25.43	0.991	28.50	1.010	31.57	1.026	34.64	1.040
0.15	15.34	0.808	17.14	0.840	20.05	0.876	22.89	0.901	25.71	0.920	28.52	0.935	31.32	0.948
0.20	13.43	0.713	15.22	0.755	17.97	0.798	20.63	0.826	23.24	0.846	25.83	0.862	28.40	0.874
0.25	11.68	0.629	13.51	0.683	16.17	0.734	18.66	0.765	21.10	0.786	23.50	0.803	25.88	0.816
0.30	10.16	0.558	12.05	0.625	14.62	0.682	16.98	0.716	19.26	0.739	21.49	0.756	23.71	0.769
0.35	8.64	0.477	10.81	0.578	13.30	0.641	15.54	0.676	17.68	0.700	19.78	0.718	21.85	0.732
0.40	7.16	0.396	9.78	0.540	12.18	0.607	14.30	0.644	16.33	0.669	18.30	0.687	20.24	0.701
0.45	5.83	0.325	8.57	0.474	11.22	0.580	13.24	0.618	15.16	0.644	17.02	0.662	18.85	0.676
0.50	4.68	0.269	7.50	0.415	10.40	0.557	12.32	0.596	14.14	0.622	15.90	0.641	17.63	0.655
0.55	3.78	0.227	6.56	0.363	9.65	0.534	11.52	0.578	13.24	0.604	14.91	0.623	16.55	0.638
0.60	3.11	0.198	5.71	0.323	8.79	0.486	10.82	0.563	12.46	0.589	14.04	0.608	15.60	0.623
0.65	2.62	0.177	4.99	0.292	8.01	0.443	10.19	0.549	11.76	0.576	13.27	0.595	14.75	0.610
0.70	2.25	0.162	4.40	0.269	7.33	0.405	9.58	0.530	11.13	0.564	12.57	0.584	13.99	0.598
0.75	1.96	0.152	3.92	0.250	6.71	0.371	8.91	0.492	10.57	0.554	11.95	0.574	13.30	0.588
0.80	1.74	0.144	3.53	0.235	6.16	0.344	8.29	0.459	10.06	0.546	11.38	0.565	12.68	0.579
0.85	1.56	0.137	3.20	0.224	5.64	0.325	7.74	0.428	9.54	0.527	10.87	0.557	12.11	0.572
0.90	1.41	0.133	2.92	0.214	5.20	0.309	7.24	0.400	8.98	0.497	10.39	0.550	11.59	0.564
0.95	1.29	0.129	2.69	0.206	4.82	0.296	6.79	0.375	8.47	0.468	9.96	0.544	11.11	0.558
1.00	1.19	0.125	2.49	0.200	4.48	0.285	6.38	0.353	8.01	0.443	9.51	0.525	10.67	0.552
1.10	1.02	0.120	2.16	0.190	3.92	0.268	5.62	0.328	7.19	0.397	8.60	0.475	9.89	0.542
1.20	0.90	0.116	1.91	0.182	3.49	0.255	5.01	0.310	6.50	0.359	7.82	0.433	9.07	0.502
1.30	0.80	0.114	1.71	0.176	3.13	0.245	4.52	0.296	5.88	0.337	7.16	0.396	8.34	0.461
1.40	0.72	0.111	1.55	0.172	2.84	0.237	4.11	0.284	5.36	0.323	6.58	0.364	7.70	0.425
1.50	0.66	0.109	1.41	0.168	2.60	0.230	3.77	0.275	4.92	0.312	6.06	0.343	7.13	0.394
1.60	0.61	0.108	1.30	0.165	2.40	0.225	3.47	0.268	4.54	0.303	5.60	0.332	6.64	0.367
1.70	0.56	0.107	1.20	0.162	2.22	0.220	3.22	0.262	4.22	0.295	5.21	0.323	6.19	0.347
1.80	0.52	0.106	1.12	0.160	2.07	0.217	3.01	0.257	3.93	0.288	4.86	0.315	5.78	0.338
1.90	0.49	0.105	1.05	0.158	1.94	0.213	2.82	0.252	3.69	0.283	4.56	0.308	5.43	0.330
2.00	0.46	0.104	0.99	0.157	1.82	0.211	2.65	0.248	3.47	0.278	4.29	0.302	5.11	0.323
2.20	0.41	0.103	0.88	0.154	1.63	0.206	2.37	0.242	3.10	0.270	3.84	0.293	4.57	0.312
2.40	0.37	0.101	0.79	0.152	1.47	0.202	2.14	0.237	2.80	0.264	3.47	0.285	4.13	0.304
2.60	0.33	0.101	0.72	0.150	1.34	0.199	1.95	0.233	2.56	0.259	3.17	0.280	3.77	0.297
2.80			0.66	0.149	1.23	0.197	1.79	0.230	2.35	0.254	2.91	0.275	3.47	0.292
3.00			0.61	0.148	1.14	0.195	1.66	0.227	2.18	0.251	2.69	0.271	3.21	0.287
3.20			0.57	0.147	1.06	0.193	1.54	0.224	2.03	0.248	2.51	0.267	2.99	0.283
3.40			0.53	0.146	0.99	0.192	1.44	0.222	1.89	0.246	2.34	0.264	2.79	0.280
3.60			0.50	0.145	0.93	0.190	1.35	0.221	1.78	0.243	2.20	0.262	2.62	0.277
3.80			0.47	0.144	0.88	0.189	1.28	0.219	1.68	0.242	2.07	0.259	2.47	0.274
4.00			0.44	0.144	0.83	0.188	1.21	0.218	1.58	0.240	1.96	0.257	2.34	0.272
5.00			0.35	0.141	0.65	0.185	0.95	0.213	1.25	0.234	1.54	0.250	1.84	0.264
6.00			0.29	0.140	0.54	0.182	0.78	0.210	1.03	0.230	1.27	0.246	1.51	0.259
7.00			0.24	0.139	0.45	0.181	0.66	0.208	0.87	0.227	1.08	0.243	1.29	0.255
8.00			0.21	0.138	0.40	0.179	0.58	0.206	0.76	0.225	0.94	0.241	1.12	0.253
10.00			0.17	0.137	0.31	0.178	0.46	0.204	0.60	0.223	0.74	0.238	0.89	0.249

Table 11.15 (contd) – $f_{cu} = 45\,\text{N/mm}^2$, $k = 0.80$.

e/h	p=0.4 N/bh	x/h	p=1.0 N/bh	x/h	p=2.0 N/bh	x/h	p=3.0 N/bh	x/h	p=4.0 N/bh	x/h	p=5.0 N/bh	x/h	p=6.0 N/bh	x/h
0.05	19.40	1.012	21.36	1.027	24.64	1.049	27.93	1.067	31.22	1.082	34.53	1.095	37.84	1.107
0.10	17.28	0.902	19.06	0.918	22.03	0.939	24.99	0.955	27.95	0.968	30.91	0.979	33.88	0.988
0.15	15.21	0.797	16.87	0.818	19.59	0.842	22.28	0.860	24.96	0.873	27.63	0.883	30.29	0.892
0.20	13.23	0.699	14.85	0.729	17.40	0.761	19.87	0.781	22.32	0.796	24.76	0.807	27.18	0.816
0.25	11.43	0.611	13.06	0.655	15.50	0.694	17.81	0.719	20.07	0.735	22.31	0.748	24.53	0.758
0.30	9.85	0.538	11.54	0.594	13.89	0.642	16.07	0.670	18.17	0.688	20.24	0.702	22.30	0.712
0.35	8.34	0.461	10.28	0.547	12.55	0.601	14.60	0.631	16.57	0.651	18.50	0.665	20.41	0.677
0.40	6.76	0.381	9.23	0.510	11.42	0.568	13.36	0.600	15.21	0.621	17.02	0.637	18.80	0.648
0.45	5.41	0.315	7.98	0.445	10.47	0.542	12.31	0.575	14.05	0.597	15.75	0.613	17.42	0.625
0.50	4.34	0.265	6.82	0.393	9.66	0.521	11.41	0.555	13.05	0.578	14.65	0.594	16.22	0.606
0.55	3.53	0.229	5.86	0.351	8.87	0.490	10.62	0.538	12.18	0.561	13.70	0.577	15.18	0.590
0.60	2.93	0.205	5.08	0.319	7.96	0.447	9.94	0.524	11.42	0.547	12.86	0.564	14.26	0.576
0.65	2.49	0.187	4.46	0.295	7.14	0.415	9.34	0.512	10.75	0.535	12.11	0.552	13.45	0.564
0.70	2.16	0.175	3.95	0.276	6.44	0.389	8.66	0.479	10.15	0.525	11.45	0.542	12.72	0.554
0.75	1.90	0.165	3.53	0.261	5.84	0.368	7.96	0.449	9.62	0.516	10.86	0.533	12.07	0.545
0.80	1.70	0.158	3.19	0.249	5.34	0.350	7.32	0.427	9.11	0.504	10.32	0.525	11.48	0.538
0.85	1.53	0.153	2.91	0.239	4.90	0.336	6.76	0.409	8.54	0.472	9.84	0.518	10.95	0.531
0.90	1.40	0.148	2.67	0.232	4.53	0.324	6.27	0.393	7.96	0.451	9.39	0.512	10.46	0.525
0.95	1.28	0.145	2.46	0.225	4.21	0.314	5.85	0.380	7.44	0.435	8.93	0.494	10.01	0.519
1.00	1.18	0.142	2.29	0.220	3.92	0.305	5.47	0.369	6.97	0.421	8.45	0.467	9.61	0.514
1.10	1.03	0.137	2.00	0.211	3.45	0.291	4.84	0.350	6.19	0.399	7.52	0.440	8.79	0.486
1.20	0.91	0.133	1.78	0.205	3.08	0.280	4.33	0.336	5.55	0.381	6.76	0.420	7.96	0.454
1.30	0.81	0.130	1.60	0.199	2.78	0.272	3.92	0.325	5.03	0.368	6.13	0.404	7.23	0.435
1.40	0.74	0.128	1.45	0.195	2.53	0.266	3.57	0.316	4.60	0.357	5.61	0.391	6.62	0.421
1.50	0.67	0.126	1.33	0.192	2.33	0.260	3.28	0.309	4.23	0.347	5.17	0.380	6.10	0.408
1.60	0.62	0.123	1.23	0.189	2.15	0.256	3.04	0.303	3.91	0.340	4.79	0.371	5.65	0.398
1.70	0.57	0.120	1.14	0.187	2.00	0.252	2.83	0.298	3.64	0.334	4.46	0.364	5.27	0.390
1.80	0.53	0.118	1.06	0.185	1.86	0.249	2.64	0.293	3.41	0.328	4.17	0.357	4.93	0.382
1.90	0.49	0.116	0.99	0.183	1.75	0.246	2.48	0.289	3.20	0.324	3.92	0.352	4.63	0.376
2.00	0.46	0.114	0.93	0.182	1.65	0.243	2.33	0.286	3.01	0.319	3.69	0.347	4.37	0.371
2.20	0.41	0.111	0.84	0.179	1.47	0.239	2.09	0.281	2.70	0.313	3.31	0.339	3.92	0.361
2.40	0.37	0.109	0.75	0.177	1.33	0.236	1.89	0.276	2.45	0.307	3.00	0.333	3.55	0.354
2.60	0.34	0.107	0.69	0.176	1.22	0.233	1.73	0.273	2.24	0.303	2.74	0.328	3.25	0.348
2.80	0.31	0.105	0.63	0.174	1.12	0.231	1.59	0.270	2.06	0.299	2.53	0.323	2.99	0.344
3.00	0.28	0.104	0.59	0.173	1.04	0.229	1.48	0.267	1.91	0.296	2.34	0.320	2.77	0.340
3.20	0.26	0.103	0.55	0.172	0.97	0.228	1.37	0.265	1.78	0.294	2.18	0.317	2.58	0.336
3.40	0.25	0.102	0.51	0.171	0.90	0.226	1.29	0.263	1.67	0.291	2.04	0.314	2.42	0.333
3.60	0.23	0.101	0.48	0.171	0.85	0.225	1.21	0.262	1.57	0.289	1.92	0.312	2.27	0.331
3.80	0.22	0.100	0.45	0.170	0.80	0.224	1.14	0.260	1.48	0.288	1.81	0.310	2.15	0.328
4.00			0.43	0.169	0.76	0.223	1.08	0.259	1.40	0.286	1.71	0.308	2.03	0.326
5.00			0.34	0.167	0.60	0.220	0.85	0.254	1.10	0.281	1.35	0.302	1.60	0.319
6.00			0.28	0.166	0.49	0.217	0.70	0.252	0.91	0.277	1.12	0.297	1.32	0.314
7.00			0.24	0.165	0.42	0.216	0.60	0.249	0.77	0.275	0.95	0.295	1.13	0.311
8.00			0.20	0.164	0.36	0.215	0.52	0.248	0.67	0.273	0.83	0.293	0.98	0.309
10.00			0.16	0.163	0.29	0.213	0.41	0.246	0.53	0.270	0.66	0.290	0.78	0.305

Table 11.16 Rectangular columns – $f_{cu} = 50\,\text{N/mm}^2$, $k = 0.95$.

e/h	p=0.4 N/bh	p=0.4 x/h	p=1.0 N/bh	p=1.0 x/h	p=2.0 N/bh	p=2.0 x/h	p=3.0 N/bh	p=3.0 x/h	p=4.0 N/bh	p=4.0 x/h	p=5.0 N/bh	p=5.0 x/h	p=6.0 N/bh	p=6.0 x/h
0.05	21.58	1.029	23.75	1.064	27.30	1.109	30.92	......	34.52	......	38.12	......	41.72	......
0.10	19.35	0.925	21.46	0.966	24.84	1.015	28.14	1.051	31.40	1.080	34.65	1.104	37.92	......
0.15	17.21	0.827	19.31	0.876	22.56	0.931	25.68	0.969	28.75	0.998	31.78	1.022	34.79	1.042
0.20	15.19	0.736	17.34	0.797	20.50	0.858	23.47	0.899	26.36	0.929	29.20	0.953	32.02	0.972
0.25	13.37	0.656	15.59	0.729	18.68	0.797	21.51	0.840	24.24	0.871	26.92	0.895	29.57	0.915
0.30	11.69	0.581	14.06	0.672	17.08	0.746	19.79	0.791	22.39	0.823	24.91	0.847	27.41	0.867
0.35	9.94	0.494	12.76	0.625	15.70	0.704	18.30	0.751	20.76	0.783	23.15	0.808	25.51	0.827
0.40	8.33	0.414	11.47	0.571	14.50	0.669	16.99	0.717	19.33	0.750	21.60	0.775	23.83	0.794
0.45	6.90	0.343	10.15	0.505	13.46	0.640	15.84	0.688	18.08	0.722	20.24	0.747	22.36	0.766
0.50	5.67	0.282	8.97	0.446	12.55	0.615	14.83	0.664	16.97	0.698	19.03	0.723	21.04	0.742
0.55	4.66	0.232	7.92	0.394	11.61	0.578	13.94	0.644	15.98	0.677	17.95	0.702	19.87	0.722
0.60	3.86	0.192	7.01	0.349	10.64	0.529	13.15	0.626	15.11	0.660	16.99	0.685	18.82	0.704
0.65	3.24	0.161	6.22	0.310	9.77	0.486	12.44	0.611	14.32	0.644	16.12	0.669	17.87	0.689
0.70	2.75	0.137	5.55	0.276	8.98	0.447	11.69	0.582	13.61	0.631	15.33	0.656	17.02	0.675
0.75	2.37	0.118	4.98	0.248	8.27	0.412	10.92	0.543	12.96	0.619	14.62	0.644	16.24	0.663
0.80	2.07	0.106	4.49	0.224	7.64	0.380	10.21	0.508	12.38	0.609	13.97	0.633	15.53	0.652
0.85			4.08	0.203	7.08	0.352	9.57	0.476	11.75	0.584	13.38	0.624	14.87	0.643
0.90			3.72	0.185	6.58	0.327	8.98	0.447	11.10	0.552	12.83	0.615	14.27	0.634
0.95			3.42	0.170	6.13	0.305	8.45	0.420	10.51	0.523	12.33	0.607	13.72	0.626
1.00			3.15	0.157	5.73	0.285	7.96	0.396	9.96	0.495	11.78	0.586	13.21	0.619
1.10			2.72	0.135	5.05	0.251	7.11	0.354	8.99	0.447	10.72	0.533	12.29	0.606
1.20			2.39	0.119	4.50	0.224	6.40	0.319	8.16	0.406	9.79	0.487	11.33	0.563
1.30			2.12	0.110	4.04	0.201	5.81	0.289	7.45	0.371	8.99	0.447	10.45	0.520
1.40			1.90	0.104	3.67	0.182	5.30	0.264	6.84	0.340	8.29	0.412	9.67	0.481
1.50					3.35	0.167	4.87	0.242	6.31	0.314	7.68	0.382	8.99	0.447
1.60					3.08	0.153	4.50	0.224	5.85	0.291	7.14	0.355	8.39	0.417
1.70					2.85	0.142	4.17	0.208	5.45	0.271	6.67	0.332	7.85	0.391
1.80					2.65	0.132	3.89	0.194	5.09	0.253	6.25	0.311	7.37	0.367
1.90					2.47	0.123	3.64	0.181	4.78	0.238	5.88	0.292	6.94	0.345
2.00					2.32	0.116	3.42	0.170	4.50	0.224	5.54	0.276	6.56	0.326
2.20					2.06	0.111	3.05	0.152	4.02	0.200	4.97	0.247	5.90	0.293
2.40					1.85	0.106	2.75	0.137	3.63	0.181	4.50	0.224	5.35	0.266
2.60					1.68	0.103	2.50	0.125	3.31	0.165	4.11	0.204	4.89	0.243
2.80					1.54	0.100	2.30	0.116	3.04	0.151	3.77	0.188	4.50	0.224
3.00							2.12	0.113	2.81	0.140	3.49	0.174	4.16	0.207
3.20							1.96	0.110	2.61	0.130	3.25	0.161	3.88	0.193
3.40							1.83	0.108	2.44	0.121	3.03	0.151	3.62	0.180
3.60							1.72	0.107	2.28	0.116	2.84	0.142	3.40	0.169
3.80							1.61	0.105	2.15	0.114	2.68	0.133	3.20	0.159
4.00							1.52	0.104	2.03	0.112	2.53	0.126	3.03	0.151
5.00									1.58	0.107	1.98	0.112	2.37	0.118
6.00									1.30	0.103	1.62	0.109	1.95	0.113
7.00									1.10	0.101	1.37	0.106	1.65	0.110
8.00											1.19	0.104	1.43	0.108
10.00											0.94	0.102	1.13	0.105

Table 11.16 (contd) − $f_{cu} = 50\,\text{N/mm}^2$, $k = 0.90$.

e/h	p=0.4 N/bh	p=0.4 x/h	p=1.0 N/bh	p=1.0 x/h	p=2.0 N/bh	p=2.0 x/h	p=3.0 N/bh	p=3.0 x/h	p=4.0 N/bh	p=4.0 x/h	p=5.0 N/bh	p=5.0 x/h	p=6.0 N/bh	p=6.0 x/h
0.05	21.51	1.022	23.61	1.050	27.06	1.087	30.53	......	34.09	......	37.64	......	41.20	......
0.10	19.24	0.916	21.24	0.948	24.49	0.988	27.68	1.018	30.86	1.042	34.02	1.062	37.18	1.079
0.15	17.05	0.816	19.01	0.855	22.10	0.900	25.10	0.932	28.05	0.956	30.98	0.976	33.90	0.992
0.20	14.98	0.723	16.97	0.773	19.95	0.825	22.78	0.859	25.54	0.885	28.27	0.905	30.98	0.921
0.25	13.10	0.640	15.15	0.703	18.05	0.762	20.74	0.799	23.34	0.826	25.89	0.846	28.42	0.862
0.30	11.43	0.569	13.58	0.645	16.41	0.710	18.97	0.750	21.43	0.778	23.83	0.798	26.19	0.815
0.35	9.66	0.481	12.24	0.598	15.00	0.668	17.44	0.710	19.77	0.738	22.03	0.760	24.26	0.776
0.40	8.03	0.399	10.98	0.546	13.79	0.633	16.12	0.676	18.33	0.706	20.47	0.727	22.57	0.744
0.45	6.58	0.327	9.64	0.460	12.75	0.605	14.98	0.649	17.07	0.679	19.10	0.701	21.09	0.718
0.50	5.35	0.266	8.45	0.421	11.84	0.581	13.98	0.626	15.97	0.656	17.90	0.678	19.78	0.695
0.55	4.33	0.218	7.41	0.369	10.89	0.542	13.10	0.606	15.00	0.637	16.83	0.659	18.63	0.676
0.60	3.52	0.183	6.51	0.324	9.93	0.494	12.32	0.590	14.14	0.620	15.89	0.642	17.60	0.659
0.65	2.92	0.159	5.74	0.286	9.07	0.451	11.63	0.575	13.37	0.606	15.04	0.628	16.67	0.645
0.70	2.48	0.142	5.09	0.253	8.30	0.413	10.84	0.539	12.68	0.593	14.28	0.615	15.84	0.633
0.75	2.14	0.130	4.54	0.228	7.61	0.379	10.09	0.502	12.06	0.582	13.59	0.604	15.09	0.622
0.80	1.88	0.121	4.06	0.210	7.00	0.349	9.40	0.468	11.49	0.571	12.97	0.595	14.40	0.612
0.85	1.67	0.114	3.66	0.196	6.47	0.322	8.78	0.437	10.81	0.538	12.40	0.586	13.78	0.603
0.90	1.50	0.109	3.32	0.185	5.99	0.298	8.22	0.409	10.19	0.507	11.87	0.578	13.20	0.595
0.95	1.37	0.105	3.04	0.176	5.57	0.277	7.71	0.384	9.62	0.479	11.36	0.565	12.67	0.588
1.00	1.25	0.101	2.80	0.168	5.19	0.258	7.25	0.361	9.10	0.453	10.79	0.537	12.19	0.582
1.10			2.42	0.157	4.55	0.229	6.45	0.321	8.17	0.407	9.77	0.486	11.26	0.560
1.20			2.12	0.148	4.02	0.214	5.79	0.288	7.39	0.368	8.90	0.443	10.31	0.513
1.30			1.89	0.142	3.60	0.202	5.23	0.260	6.73	0.335	8.14	0.405	9.48	0.472
1.40			1.71	0.137	3.25	0.193	4.77	0.237	6.17	0.307	7.49	0.373	8.76	0.436
1.50			1.55	0.133	2.96	0.186	4.36	0.225	5.68	0.282	6.93	0.345	8.12	0.404
1.60			1.42	0.130	2.72	0.180	4.01	0.217	5.26	0.261	6.43	0.320	7.56	0.376
1.70			1.31	0.127	2.52	0.175	3.71	0.210	4.89	0.243	6.00	0.298	7.07	0.352
1.80			1.22	0.125	2.34	0.171	3.45	0.204	4.56	0.230	5.61	0.279	6.63	0.330
1.90			1.14	0.123	2.19	0.168	3.23	0.199	4.26	0.224	5.27	0.262	6.24	0.310
2.00			1.07	0.122	2.05	0.165	3.03	0.195	4.00	0.219	4.97	0.247	5.88	0.293
2.20			0.95	0.119	1.83	0.160	2.70	0.188	3.57	0.210	4.44	0.229	5.28	0.263
2.40			0.86	0.117	1.65	0.156	2.43	0.183	3.22	0.204	4.00	0.221	4.78	0.238
2.60			0.78	0.115	1.50	0.153	2.21	0.179	2.93	0.198	3.64	0.214	4.36	0.228
2.80			0.71	0.114	1.37	0.151	2.03	0.175	2.69	0.194	3.34	0.209	4.00	0.222
3.00			0.66	0.113	1.27	0.148	1.88	0.172	2.48	0.191	3.09	0.205	3.70	0.217
3.20			0.61	0.112	1.18	0.147	1.74	0.170	2.31	0.188	2.87	0.202	3.43	0.213
3.40			0.57	0.111	1.10	0.145	1.63	0.168	2.15	0.185	2.68	0.199	3.21	0.210
3.60			0.53	0.110	1.03	0.144	1.53	0.166	2.02	0.183	2.51	0.196	3.01	0.207
3.80			0.50	0.109	0.97	0.143	1.44	0.165	1.90	0.181	2.37	0.194	2.83	0.204
4.00			0.48	0.109	0.92	0.142	1.36	0.163	1.80	0.179	2.24	0.192	2.68	0.202
5.00			0.37	0.107	0.72	0.138	1.06	0.159	1.41	0.173	1.75	0.185	2.10	0.194
6.00			0.31	0.106	0.59	0.136	0.87	0.156	1.16	0.170	1.44	0.181	1.72	0.189
7.00			0.26	0.105	0.50	0.135	0.74	0.153	0.98	0.167	1.22	0.178	1.46	0.186
8.00			0.23	0.104	0.44	0.133	0.64	0.152	0.85	0.165	1.06	0.176	1.27	0.184
10.00			0.18	0.103	0.34	0.132	0.51	0.150	0.68	0.163	0.84	0.173	1.01	0.181

Table 11.17 Rectangular columns − $f_{cu} = 50 \text{ N/mm}^2$, $k = 0.85$.

e/h	p=0.4 N/bh	p=0.4 x/h	p=1.0 N/bh	p=1.0 x/h	p=2.0 N/bh	p=2.0 x/h	p=3.0 N/bh	p=3.0 x/h	p=4.0 N/bh	p=4.0 x/h	p=5.0 N/bh	p=5.0 x/h	p=6.0 N/bh	p=6.0 x/h
0.05	21.45	1.016	23.48	1.037	26.84	1.066	30.20	1.089	33.56	1.109	37.06	******	40.56	******
0.10	19.15	0.908	21.04	0.932	24.14	0.962	27.23	0.985	30.30	1.003	33.37	1.019	36.44	1.032
0.15	16.90	0.805	18.72	0.835	21.63	0.870	24.49	0.894	27.32	0.913	30.13	0.928	32.94	0.941
0.20	14.78	0.709	16.59	0.749	19.37	0.791	22.04	0.818	24.67	0.839	27.26	0.854	29.84	0.867
0.25	12.83	0.624	14.70	0.676	17.39	0.726	19.92	0.757	22.37	0.779	24.78	0.795	27.17	0.808
0.30	11.12	0.551	13.07	0.617	15.70	0.673	18.09	0.707	20.39	0.730	22.64	0.748	24.87	0.762
0.35	9.37	0.466	11.70	0.569	14.26	0.631	16.53	0.667	18.70	0.692	20.81	0.710	22.90	0.724
0.40	7.71	0.384	10.45	0.520	13.04	0.597	15.20	0.635	17.25	0.660	19.24	0.679	21.20	0.693
0.45	6.20	0.312	9.10	0.453	11.99	0.569	14.06	0.608	16.00	0.635	17.88	0.654	19.72	0.668
0.50	4.93	0.256	7.91	0.393	11.10	0.547	13.07	0.586	14.91	0.613	16.69	0.632	18.43	0.647
0.55	3.95	0.215	6.86	0.343	10.12	0.504	12.21	0.568	13.96	0.595	15.65	0.615	17.30	0.630
0.60	3.23	0.187	5.91	0.305	9.17	0.456	11.45	0.552	13.12	0.580	14.73	0.599	16.30	0.614
0.65	2.70	0.167	5.14	0.276	8.32	0.414	10.75	0.535	12.38	0.566	13.91	0.586	15.40	0.601
0.70	2.31	0.153	4.52	0.253	7.58	0.377	9.94	0.495	11.71	0.555	13.17	0.575	14.60	0.590
0.75	2.02	0.143	4.02	0.236	6.91	0.345	9.21	0.458	11.11	0.545	12.51	0.565	13.88	0.580
0.80	1.78	0.136	3.61	0.222	6.28	0.324	8.55	0.425	10.48	0.521	11.92	0.556	13.22	0.571
0.85	1.60	0.130	3.27	0.211	5.75	0.307	7.95	0.396	9.83	0.489	11.37	0.548	12.63	0.563
0.90	1.45	0.125	2.98	0.203	5.29	0.292	7.42	0.369	9.23	0.459	10.88	0.541	12.08	0.556
0.95	1.32	0.122	2.74	0.195	4.89	0.280	6.93	0.347	8.69	0.432	10.29	0.512	11.58	0.550
1.00	1.22	0.119	2.53	0.189	4.55	0.270	6.46	0.333	8.19	0.408	9.75	0.485	11.12	0.544
1.10	1.05	0.114	2.20	0.180	3.98	0.254	5.68	0.311	7.33	0.365	8.79	0.437	10.15	0.505
1.20	0.92	0.111	1.94	0.173	3.53	0.242	5.06	0.294	6.57	0.338	7.97	0.396	9.26	0.461
1.30	0.82	0.108	1.74	0.168	3.17	0.233	4.56	0.281	5.93	0.322	7.27	0.362	8.49	0.422
1.40	0.74	0.106	1.57	0.163	2.88	0.225	4.15	0.271	5.40	0.309	6.64	0.341	7.81	0.389
1.50	0.67	0.104	1.44	0.160	2.63	0.220	3.80	0.263	4.95	0.298	6.10	0.328	7.23	0.360
1.60	0.62	0.103	1.32	0.157	2.42	0.215	3.50	0.256	4.57	0.290	5.64	0.318	6.70	0.343
1.70	0.57	0.102	1.22	0.155	2.25	0.211	3.25	0.251	4.24	0.282	5.23	0.309	6.22	0.333
1.80	0.53	0.101	1.14	0.153	2.09	0.207	3.03	0.246	3.96	0.276	4.89	0.302	5.81	0.324
1.90			1.06	0.151	1.96	0.204	2.84	0.242	3.71	0.271	4.58	0.296	5.45	0.317
2.00			1.00	0.150	1.84	0.202	2.67	0.238	3.49	0.267	4.31	0.291	5.13	0.311
2.20			0.89	0.147	1.64	0.197	2.39	0.232	3.12	0.259	3.85	0.282	4.59	0.301
2.40			0.80	0.145	1.49	0.194	2.15	0.228	2.82	0.254	3.49	0.275	4.15	0.293
2.60			0.73	0.144	1.35	0.191	1.97	0.224	2.57	0.249	3.18	0.270	3.79	0.287
2.80			0.67	0.142	1.24	0.189	1.81	0.221	2.37	0.245	2.92	0.265	3.48	0.282
3.00			0.62	0.141	1.15	0.187	1.67	0.218	2.19	0.242	2.71	0.261	3.22	0.277
3.20			0.58	0.140	1.07	0.186	1.55	0.216	2.04	0.239	2.52	0.258	3.00	0.274
3.40			0.54	0.139	1.00	0.184	1.45	0.214	1.90	0.237	2.35	0.255	2.80	0.271
3.60			0.51	0.139	0.94	0.183	1.36	0.213	1.79	0.235	2.21	0.253	2.63	0.268
3.80			0.48	0.138	0.88	0.182	1.29	0.211	1.68	0.233	2.08	0.251	2.48	0.266
4.00			0.45	0.138	0.84	0.181	1.21	0.210	1.59	0.232	1.97	0.249	2.35	0.264
5.00			0.35	0.136	0.66	0.178	0.95	0.205	1.25	0.226	1.55	0.242	1.84	0.256
6.00			0.29	0.134	0.54	0.175	0.79	0.202	1.03	0.222	1.28	0.238	1.52	0.251
7.00			0.25	0.133	0.46	0.174	0.67	0.200	0.88	0.220	1.08	0.235	1.29	0.248
8.00			0.21	0.133	0.40	0.173	0.58	0.199	0.76	0.218	0.94	0.233	1.12	0.246
10.00			0.17	0.132	0.32	0.171	0.46	0.197	0.60	0.216	0.75	0.231	0.89	0.243

Table 11.17 (contd) $-\ f_{cu} = 50\,\text{N/mm}^2$, $k = 0.80$.

e/h	p=0.4 Nbh	x/h	p=1.0 Nbh	x/h	p=2.0 Nbh	x/h	p=3.0 Nbh	x/h	p=4.0 Nbh	x/h	p=5.0 Nbh	x/h	p=6.0 Nbh	x/h
0.05	21.41	1.011	23.37	1.025	26.65	1.045	29.93	1.062	33.23	1.076	36.53	1.089	39.84	1.100
0.10	19.06	0.901	20.85	0.916	23.82	0.936	26.78	0.951	29.74	0.963	32.70	0.974	35.67	0.983
0.15	16.77	0.795	18.44	0.815	21.16	0.838	23.86	0.855	26.54	0.868	29.21	0.878	31.88	0.887
0.20	14.58	0.696	16.21	0.725	18.77	0.755	21.26	0.776	23.72	0.790	26.16	0.802	28.58	0.811
0.25	12.56	0.607	14.23	0.649	16.70	0.668	19.03	0.712	21.30	0.729	23.55	0.742	25.78	0.752
0.30	10.79	0.532	12.54	0.567	14.94	0.635	17.14	0.663	19.26	0.681	21.35	0.695	23.41	0.706
0.35	9.05	0.451	11.14	0.539	13.47	0.593	15.55	0.623	17.54	0.644	19.49	0.659	21.41	0.670
0.40	7.29	0.370	9.89	0.492	12.24	0.559	14.21	0.592	16.09	0.614	17.91	0.629	19.70	0.641
0.45	5.78	0.304	8.45	0.427	11.20	0.533	13.08	0.567	14.85	0.589	16.56	0.606	18.24	0.618
0.50	4.59	0.253	7.18	0.375	10.32	0.511	12.11	0.546	13.78	0.569	15.40	0.586	16.98	0.599
0.55	3.71	0.218	6.13	0.334	9.29	0.463	11.27	0.529	12.85	0.553	14.38	0.570	15.88	0.583
0.60	3.07	0.194	5.29	0.303	8.25	0.425	10.53	0.515	12.04	0.539	13.49	0.556	14.91	0.569
0.65	2.60	0.178	4.62	0.280	7.37	0.394	9.77	0.486	11.33	0.527	12.71	0.544	14.05	0.557
0.70	2.24	0.166	4.08	0.262	6.63	0.369	8.93	0.452	10.69	0.516	12.01	0.534	13.29	0.547
0.75	1.97	0.157	3.65	0.247	6.00	0.349	8.15	0.427	10.09	0.502	11.38	0.525	12.60	0.538
0.80	1.76	0.150	3.29	0.236	5.47	0.333	7.48	0.406	9.40	0.468	10.81	0.517	11.99	0.530
0.85	1.59	0.145	2.99	0.227	5.02	0.319	6.90	0.388	8.71	0.447	10.30	0.510	11.42	0.523
0.90	1.44	0.141	2.74	0.220	4.63	0.308	6.40	0.374	8.10	0.429	9.72	0.483	10.91	0.517
0.95	1.33	0.137	2.53	0.214	4.30	0.298	5.96	0.361	7.56	0.414	9.13	0.460	10.45	0.512
1.00	1.22	0.135	2.35	0.209	4.01	0.290	5.57	0.351	7.08	0.401	8.57	0.445	9.96	0.495
1.10	1.06	0.130	2.05	0.201	3.52	0.277	4.92	0.334	6.28	0.380	7.62	0.420	8.94	0.456
1.20	0.94	0.126	1.82	0.195	3.14	0.267	4.40	0.321	5.63	0.364	6.84	0.401	8.05	0.434
1.30	0.83	0.121	1.64	0.190	2.83	0.260	3.98	0.310	5.10	0.351	6.20	0.386	7.30	0.417
1.40	0.75	0.117	1.49	0.186	2.58	0.254	3.63	0.302	4.65	0.341	5.67	0.374	6.68	0.403
1.50	0.68	0.114	1.36	0.183	2.37	0.249	3.33	0.295	4.28	0.333	5.22	0.364	6.16	0.392
1.60	0.63	0.111	1.25	0.181	2.19	0.244	3.08	0.290	3.96	0.326	4.83	0.356	5.70	0.382
1.70	0.58	0.108	1.16	0.178	2.03	0.241	2.86	0.285	3.69	0.320	4.50	0.349	5.31	0.374
1.80	0.54	0.106	1.09	0.177	1.90	0.238	2.68	0.281	3.45	0.315	4.21	0.343	4.97	0.368
1.90	0.50	0.105	1.02	0.175	1.78	0.235	2.51	0.278	3.23	0.311	3.95	0.338	4.67	0.362
2.00	0.47	0.103	0.96	0.174	1.67	0.233	2.37	0.274	3.05	0.307	3.73	0.334	4.40	0.357
2.20	0.42	0.100	0.85	0.171	1.50	0.229	2.12	0.269	2.73	0.301	3.34	0.326	3.95	0.348
2.40			0.77	0.169	1.36	0.226	1.92	0.265	2.48	0.296	3.03	0.321	3.58	0.342
2.60			0.70	0.168	1.24	0.224	1.75	0.262	2.26	0.292	2.77	0.316	3.27	0.336
2.80			0.65	0.167	1.14	0.222	1.61	0.259	2.08	0.288	2.55	0.312	3.02	0.332
3.00			0.60	0.166	1.05	0.220	1.49	0.257	1.93	0.285	2.36	0.309	2.79	0.328
3.20			0.56	0.165	0.98	0.218	1.39	0.255	1.80	0.283	2.20	0.306	2.60	0.325
3.40			0.52	0.164	0.92	0.217	1.30	0.253	1.68	0.281	2.06	0.303	2.44	0.322
3.60			0.49	0.163	0.86	0.216	1.22	0.252	1.58	0.279	1.94	0.301	2.29	0.320
3.80			0.46	0.163	0.81	0.215	1.15	0.251	1.49	0.278	1.83	0.299	2.16	0.318
4.00			0.44	0.162	0.77	0.214	1.09	0.249	1.41	0.276	1.73	0.298	2.05	0.316
5.00			0.34	0.160	0.61	0.211	0.86	0.245	1.11	0.271	1.36	0.292	1.61	0.309
6.00			0.28	0.159	0.50	0.209	0.71	0.242	0.92	0.268	1.12	0.288	1.33	0.305
7.00			0.24	0.158	0.42	0.208	0.60	0.241	0.78	0.265	0.96	0.285	1.13	0.302
8.00			0.21	0.157	0.37	0.206	0.53	0.239	0.68	0.264	0.83	0.283	0.99	0.299
10.00			0.17	0.156	0.29	0.205	0.42	0.237	0.54	0.261	0.66	0.281	0.78	0.296

Table 11.18 Circular columns – $f_{cu} = 30\,\text{N/mm}^2$, $k = 0.90$.

e/R	p=0.4 N/R²	z/R	p=1.0 N/R²	z/R	p=2.0 N/R²	z/R	p=3.0 N/R²	z/R	p=4.0 N/R²	z/R	p=5.0 N/R²	z/R	p=6.0 N/R²	z/R
0.10	42.01	0.858	48.49	0.834	59.37	0.805	70.41	0.786	81.79	******	93.28	******	104.41	******
0.20	36.70	0.952	42.61	0.923	52.34	0.892	62.05	0.870	71.78	0.855	81.54	0.843	91.32	0.834
0.30	31.59	1.039	37.15	1.002	46.02	0.966	54.74	0.943	63.42	0.927	72.07	0.915	80.73	0.906
0.40	26.95	1.115	32.36	1.068	40.57	1.027	48.50	1.002	56.33	0.985	64.11	0.973	71.87	0.963
0.50	22.98	1.179	28.32	1.122	36.00	1.075	43.28	1.048	50.41	1.031	57.47	1.018	64.50	1.008
0.60	19.22	1.245	24.99	1.164	32.21	1.113	38.94	1.085	45.49	1.066	51.96	1.053	58.38	1.043
0.70	15.93	1.305	22.18	1.200	29.07	1.143	35.32	1.113	41.37	1.094	47.34	1.081	53.25	1.071
0.80	13.30	1.353	19.31	1.246	26.45	1.167	32.28	1.137	37.91	1.117	43.44	1.104	48.91	1.093
0.90	11.25	1.391	16.94	1.283	24.24	1.186	29.70	1.156	34.96	1.136	40.11	1.122	45.21	1.112
1.00	9.68	1.419	15.01	1.313	22.06	1.212	27.50	1.171	32.42	1.151	37.25	1.137	42.02	1.127
1.20	7.34	1.467	12.11	1.356	18.26	1.262	23.82	1.199	28.30	1.176	32.57	1.162	36.80	1.151
1.40	5.70	1.507	10.09	1.385	15.49	1.296	20.38	1.238	25.08	1.194	28.94	1.180	32.72	1.169
1.60	4.57	1.533	8.64	1.405	13.41	1.321	17.76	1.266	21.94	1.226	26.02	1.193	29.46	1.183
1.80	3.80	1.551	7.54	1.420	11.81	1.339	15.71	1.288	19.46	1.249	23.13	1.219	26.76	1.194
2.00	3.23	1.564	6.69	1.431	10.54	1.354	14.07	1.304	17.47	1.268	20.80	1.239	24.09	1.216
2.20	2.81	1.574	6.01	1.440	9.52	1.365	12.74	1.317	15.84	1.282	18.88	1.255	21.88	1.233
2.40	2.48	1.581	5.41	1.453	8.67	1.374	11.63	1.328	14.48	1.294	17.28	1.268	20.04	1.247
2.60	2.22	1.587	4.91	1.465	7.97	1.381	10.70	1.336	13.34	1.304	15.92	1.279	18.48	1.259
2.80	2.01	1.592	4.49	1.475	7.36	1.388	9.91	1.344	12.36	1.312	14.76	1.288	17.14	1.268
3.00	1.83	1.596	4.13	1.484	6.85	1.393	9.22	1.350	11.51	1.319	13.76	1.295	15.98	1.276
3.25	1.65	1.601	3.75	1.493	6.29	1.399	8.49	1.357	10.60	1.327	12.68	1.303	14.73	1.285
3.50	1.50	1.604	3.44	1.501	5.82	1.403	7.86	1.362	9.83	1.333	11.76	1.310	13.67	1.292
3.75	1.38	1.607	3.17	1.507	5.42	1.407	7.32	1.367	9.16	1.338	10.96	1.316	12.74	1.298
4.00	1.27	1.609	2.94	1.512	5.07	1.411	6.85	1.371	8.57	1.342	10.26	1.321	11.93	1.304
4.25	1.18	1.611	2.73	1.516	4.76	1.414	6.43	1.374	8.06	1.346	9.65	1.325	11.22	1.308
4.50	1.11	1.613	2.56	1.519	4.48	1.416	6.07	1.377	7.60	1.350	9.10	1.329	10.59	1.312
4.75	1.04	1.615	2.40	1.522	4.24	1.419	5.74	1.380	7.19	1.353	8.62	1.332	10.03	1.316
5.00	0.98	1.616	2.27	1.525	4.02	1.421	5.44	1.383	6.82	1.356	8.18	1.335	9.52	1.319
5.50	0.87	1.618	2.03	1.529	3.64	1.424	4.94	1.387	6.19	1.360	7.42	1.340	8.64	1.324
6.00	0.79	1.620	1.84	1.533	3.33	1.427	4.52	1.390	5.67	1.364	6.80	1.344	7.92	1.329
6.50	0.72	1.622	1.69	1.536	3.07	1.430	4.16	1.393	5.23	1.367	6.27	1.348	7.30	1.332
7.00	0.66	1.623	1.55	1.539	2.84	1.432	3.86	1.395	4.85	1.370	5.82	1.351	6.77	1.336
7.50	0.61	1.624	1.44	1.541	2.65	1.434	3.60	1.397	4.52	1.372	5.42	1.353	6.32	1.338
8.00	0.57	1.625	1.34	1.543	2.48	1.435	3.37	1.399	4.23	1.374	5.08	1.355	5.92	1.341
8.50	0.54	1.626	1.26	1.544	2.33	1.436	3.17	1.401	3.98	1.376	4.78	1.357	5.57	1.343
9.00	0.50	1.626	1.18	1.546	2.20	1.438	2.99	1.402	3.76	1.378	4.51	1.359	5.26	1.344
9.50	0.47	1.627	1.12	1.547	2.08	1.439	2.83	1.404	3.56	1.379	4.27	1.361	4.98	1.346
10.00	0.45	1.628	1.06	1.548	1.98	1.440	2.69	1.405	3.38	1.380	4.06	1.362	4.73	1.348
11.00	0.40	1.629	0.95	1.550	1.79	1.441	2.44	1.407	3.07	1.382	3.69	1.364	4.30	1.350
12.00	0.37	1.629	0.87	1.552	1.64	1.443	2.24	1.408	2.81	1.384	3.38	1.366	3.94	1.352
13.00	0.34	1.630	0.80	1.553	1.51	1.446	2.06	1.409	2.59	1.386	3.11	1.368	3.63	1.354
14.00	0.31	1.631	0.74	1.554	1.40	1.448	1.91	1.411	2.41	1.387	2.89	1.369	3.37	1.355
15.00	0.29	1.631	0.69	1.555	1.30	1.451	1.79	1.412	2.24	1.388	2.70	1.370	3.14	1.356
17.50	0.25	1.632	0.59	1.557	1.11	1.455	1.53	1.413	1.92	1.390	2.31	1.373	2.69	1.359
20.00	0.22	1.633	0.51	1.558	0.97	1.459	1.34	1.415	1.68	1.392	2.02	1.374	2.36	1.361

Table 11.18 (contd) − $f_{cu} = 30\,\text{N/mm}^2$, $k = 0.80$.

e/R	p=0.4 N/R²	p=0.4 z/R	p=1.0 N/R²	p=1.0 z/R	p=2.0 N/R²	p=2.0 z/R	p=3.0 N/R²	p=3.0 z/R	p=4.0 N/R²	p=4.0 z/R	p=5.0 N/R²	p=5.0 z/R	p=6.0 N/R²	p=6.0 z/R
0.10	41.85	0.777	48.11	0.758	58.65	0.735	69.33	0.718	80.13	0.706	91.05	0.698	102.09	*****
0.20	36.42	0.873	41.99	0.851	51.25	0.827	60.53	0.810	69.82	0.798	79.15	0.788	88.49	0.780
0.30	31.17	0.962	36.31	0.933	44.61	0.905	52.82	0.886	61.00	0.874	69.16	0.864	77.33	0.856
0.40	26.40	1.041	31.34	1.002	38.95	0.967	46.33	0.946	53.63	0.932	60.89	0.922	68.14	0.914
0.50	22.32	1.106	27.19	1.056	34.26	1.015	40.99	0.993	47.59	0.977	54.13	0.967	60.65	0.958
0.60	18.67	1.169	23.81	1.099	30.44	1.053	36.62	1.028	42.65	1.012	48.60	1.001	54.51	0.992
0.70	15.39	1.229	21.09	1.132	27.31	1.082	33.03	1.056	38.57	1.039	44.03	1.027	49.45	1.018
0.80	12.79	1.276	18.32	1.175	24.73	1.105	30.04	1.078	35.17	1.061	40.22	1.049	45.21	1.039
0.90	10.79	1.312	16.02	1.210	22.58	1.124	27.54	1.096	32.31	1.078	36.99	1.066	41.63	1.057
1.00	9.21	1.339	14.15	1.239	20.57	1.145	25.41	1.111	29.87	1.093	34.24	1.080	38.56	1.071
1.20	6.85	1.385	11.39	1.280	16.97	1.191	21.99	1.133	25.93	1.115	29.79	1.102	33.59	1.093
1.40	5.28	1.419	9.48	1.307	14.37	1.223	18.78	1.169	22.91	1.131	26.35	1.118	29.75	1.109
1.60	4.24	1.442	8.08	1.325	12.43	1.246	16.34	1.195	20.09	1.157	23.63	1.131	26.70	1.121
1.80	3.53	1.457	7.02	1.338	10.94	1.263	14.44	1.214	17.80	1.178	21.10	1.150	24.21	1.131
2.00	3.02	1.468	6.20	1.347	9.76	1.276	12.93	1.229	15.97	1.195	18.95	1.168	21.89	1.146
2.20	2.63	1.477	5.55	1.355	8.81	1.286	11.70	1.240	14.48	1.208	17.20	1.182	19.88	1.162
2.40	2.33	1.483	4.98	1.365	8.03	1.294	10.68	1.250	13.23	1.218	15.73	1.194	18.20	1.174
2.60	2.09	1.489	4.51	1.373	7.37	1.301	9.83	1.258	12.19	1.227	14.50	1.203	16.78	1.184
2.80	1.89	1.493	4.11	1.381	6.82	1.307	9.10	1.265	11.29	1.234	13.44	1.211	15.56	1.193
3.00	1.73	1.496	3.78	1.387	6.34	1.312	8.47	1.270	10.52	1.241	12.52	1.218	14.51	1.200
3.25	1.56	1.500	3.43	1.393	5.82	1.316	7.79	1.276	9.69	1.247	11.54	1.225	13.37	1.207
3.50	1.42	1.503	3.14	1.399	5.37	1.320	7.22	1.281	8.98	1.253	10.70	1.231	12.40	1.214
3.75	1.31	1.505	2.89	1.403	4.99	1.323	6.72	1.285	8.36	1.257	9.97	1.236	11.56	1.219
4.00	1.21	1.507	2.68	1.407	4.66	1.326	6.29	1.289	7.83	1.262	9.34	1.241	10.83	1.224
4.25	1.12	1.509	2.50	1.410	4.37	1.328	5.91	1.292	7.36	1.265	8.78	1.244	10.18	1.228
4.50	1.05	1.511	2.34	1.413	4.11	1.331	5.57	1.295	6.94	1.268	8.28	1.248	9.61	1.232
4.75	0.99	1.512	2.20	1.416	3.88	1.332	5.27	1.297	6.57	1.271	7.84	1.251	9.10	1.235
5.00	0.93	1.513	2.08	1.418	3.68	1.334	5.00	1.300	6.23	1.273	7.44	1.254	8.64	1.238
5.50	0.83	1.515	1.87	1.422	3.33	1.337	4.53	1.303	5.66	1.278	6.76	1.258	7.84	1.243
6.00	0.75	1.517	1.69	1.425	3.04	1.339	4.15	1.307	5.18	1.281	6.19	1.262	7.18	1.247
6.50	0.69	1.518	1.55	1.428	2.80	1.341	3.82	1.309	4.77	1.284	5.70	1.265	6.62	1.250
7.00	0.63	1.519	1.43	1.430	2.59	1.343	3.55	1.311	4.43	1.286	5.29	1.268	6.15	1.253
7.50	0.59	1.520	1.33	1.432	2.41	1.344	3.31	1.313	4.13	1.288	4.93	1.270	5.73	1.255
8.00	0.55	1.521	1.24	1.433	2.26	1.345	3.09	1.315	3.87	1.290	4.62	1.272	5.37	1.257
8.50	0.51	1.522	1.16	1.434	2.12	1.346	2.91	1.316	3.64	1.292	4.35	1.274	5.05	1.259
9.00	0.48	1.522	1.09	1.436	2.00	1.347	2.74	1.317	3.43	1.293	4.10	1.275	4.77	1.261
9.50	0.45	1.523	1.03	1.437	1.89	1.348	2.59	1.318	3.25	1.295	3.89	1.276	4.52	1.262
10.00	0.43	1.524	0.97	1.438	1.79	1.349	2.46	1.319	3.09	1.296	3.69	1.278	4.29	1.263
11.00	0.39	1.524	0.88	1.439	1.63	1.350	2.23	1.320	2.80	1.298	3.35	1.280	3.90	1.266
12.00	0.35	1.525	0.80	1.441	1.49	1.351	2.04	1.322	2.57	1.299	3.07	1.281	3.57	1.267
13.00	0.32	1.526	0.74	1.442	1.37	1.352	1.88	1.323	2.37	1.301	2.83	1.283	3.29	1.269
14.00	0.30	1.526	0.68	1.443	1.27	1.353	1.75	1.323	2.20	1.302	2.63	1.284	3.06	1.270
15.00	0.28	1.527	0.63	1.444	1.19	1.353	1.63	1.324	2.05	1.303	2.45	1.285	2.85	1.271
17.50	0.24	1.527	0.54	1.445	1.01	1.355	1.39	1.326	1.76	1.305	2.10	1.287	2.44	1.274
20.00	0.21	1.528	0.47	1.447	0.89	1.356	1.22	1.327	1.54	1.306	1.84	1.289	2.14	1.275

Table 11.19 Circular columns – $f_{cu} = 30 \, \text{N/mm}^2$, $k = 0.70$.

e/R	p=0.4 N/R²	z/R	p=1.0 N/R²	z/R	p=2.0 N/R²	z/R	p=3.0 N/R²	z/R	p=4.0 N/R²	z/R	p=5.0 N/R²	z/R	p=6.0 N/R²	z/R
0.10	41.70	0.695	47.75	0.682	57.96	0.665	68.28	0.653	78.69	0.643	89.18	0.635	99.72	0.629
0.20	36.15	0.794	41.39	0.779	50.14	0.763	58.92	0.751	67.73	0.742	76.56	0.735	85.41	0.730
0.30	30.76	0.885	35.46	0.865	43.14	0.844	50.77	0.831	58.38	0.821	65.98	0.815	73.59	0.809
0.40	25.84	0.966	30.28	0.936	37.23	0.908	44.01	0.892	50.72	0.881	57.40	0.873	64.07	0.867
0.50	21.65	1.034	26.02	0.991	32.43	0.957	38.56	0.938	44.57	0.925	50.54	0.917	56.48	0.910
0.60	18.10	1.093	22.60	1.034	28.58	0.994	34.18	0.972	39.63	0.959	45.03	0.949	50.38	0.942
0.70	14.76	1.152	19.88	1.066	25.48	1.022	30.63	0.999	35.62	0.984	40.54	0.974	45.42	0.967
0.80	12.12	1.198	17.29	1.104	22.95	1.044	27.72	1.020	32.32	1.004	36.84	0.994	41.32	0.986
0.90	10.12	1.231	15.02	1.138	20.87	1.062	25.30	1.037	29.57	1.021	33.75	1.010	37.89	1.002
1.00	8.63	1.256	13.14	1.163	19.05	1.078	23.26	1.050	27.24	1.034	31.13	1.023	34.98	1.014
1.20	6.41	1.298	10.43	1.199	15.66	1.121	20.03	1.071	23.52	1.054	26.93	1.043	30.30	1.034
1.40	4.95	1.329	8.60	1.221	13.12	1.147	17.16	1.099	20.69	1.069	23.73	1.057	26.73	1.048
1.60	4.00	1.349	7.31	1.237	11.25	1.166	14.91	1.123	18.23	1.088	21.20	1.068	23.90	1.059
1.80	3.34	1.363	6.35	1.248	9.83	1.179	13.09	1.138	16.14	1.107	19.04	1.081	21.62	1.068
2.00	2.87	1.373	5.61	1.256	8.73	1.189	11.65	1.149	14.47	1.121	17.10	1.096	19.69	1.076
2.20	2.51	1.380	5.02	1.263	7.85	1.197	10.49	1.158	13.07	1.132	15.51	1.109	17.87	1.090
2.40	2.22	1.386	4.55	1.268	7.13	1.204	9.54	1.166	11.90	1.140	14.19	1.119	16.36	1.100
2.60	2.00	1.390	4.13	1.275	6.53	1.209	8.74	1.172	10.91	1.146	13.06	1.127	15.07	1.109
2.80	1.81	1.394	3.77	1.281	6.02	1.213	8.07	1.177	10.08	1.152	12.06	1.133	13.98	1.117
3.00	1.66	1.397	3.47	1.287	5.58	1.217	7.49	1.181	9.36	1.156	11.21	1.138	13.03	1.123
3.25	1.50	1.400	3.15	1.292	5.12	1.221	6.88	1.185	8.60	1.161	10.30	1.143	11.99	1.129
3.50	1.37	1.403	2.89	1.297	4.73	1.224	6.36	1.189	7.95	1.165	9.52	1.147	11.08	1.134
3.75	1.26	1.405	2.67	1.301	4.39	1.227	5.91	1.192	7.39	1.168	8.85	1.151	10.31	1.138
4.00	1.17	1.407	2.48	1.304	4.10	1.229	5.52	1.195	6.90	1.171	8.27	1.154	9.63	1.141
4.25	1.09	1.409	2.31	1.307	3.84	1.231	5.18	1.197	6.48	1.174	7.76	1.157	9.04	1.144
4.50	1.02	1.410	2.16	1.309	3.62	1.233	4.87	1.199	6.10	1.176	7.31	1.159	8.52	1.146
4.75	0.95	1.411	2.04	1.311	3.42	1.235	4.61	1.201	5.77	1.178	6.91	1.162	8.05	1.149
5.00	0.90	1.412	1.92	1.313	3.24	1.236	4.37	1.203	5.47	1.180	6.55	1.163	7.64	1.151
5.50	0.81	1.414	1.73	1.317	2.93	1.239	3.95	1.205	4.95	1.183	5.94	1.167	6.92	1.154
6.00	0.73	1.416	1.57	1.319	2.68	1.241	3.61	1.208	4.52	1.186	5.43	1.169	6.32	1.157
6.50	0.67	1.417	1.44	1.322	2.46	1.242	3.32	1.210	4.16	1.188	5.00	1.172	5.82	1.159
7.00	0.62	1.418	1.33	1.323	2.28	1.244	3.08	1.211	3.86	1.189	4.63	1.173	5.40	1.161
7.50	0.57	1.419	1.23	1.325	2.12	1.245	2.87	1.213	3.59	1.191	4.31	1.175	5.03	1.163
8.00	0.53	1.420	1.15	1.326	1.99	1.246	2.68	1.214	3.36	1.192	4.04	1.177	4.70	1.164
8.50	0.50	1.420	1.08	1.328	1.87	1.247	2.52	1.215	3.16	1.193	3.79	1.178	4.42	1.166
9.00	0.47	1.421	1.01	1.329	1.76	1.248	2.38	1.216	2.98	1.194	3.58	1.179	4.17	1.167
9.50	0.44	1.421	0.96	1.330	1.66	1.249	2.25	1.217	2.82	1.195	3.39	1.180	3.95	1.168
10.00	0.42	1.422	0.91	1.331	1.58	1.249	2.13	1.217	2.68	1.196	3.21	1.181	3.75	1.169
11.00	0.38	1.423	0.82	1.332	1.43	1.250	1.94	1.219	2.43	1.198	2.92	1.182	3.40	1.170
12.00	0.34	1.423	0.75	1.333	1.31	1.251	1.77	1.220	2.22	1.199	2.67	1.183	3.11	1.172
13.00	0.32	1.424	0.69	1.334	1.21	1.252	1.63	1.221	2.05	1.200	2.46	1.184	2.87	1.173
14.00	0.29	1.424	0.64	1.335	1.12	1.253	1.51	1.221	1.90	1.200	2.28	1.185	2.66	1.174
15.00	0.27	1.425	0.59	1.336	1.04	1.253	1.41	1.222	1.77	1.201	2.13	1.186	2.48	1.174
17.50	0.23	1.425	0.50	1.337	0.89	1.254	1.21	1.223	1.52	1.203	1.82	1.188	2.12	1.176
20.00	0.20	1.426	0.44	1.338	0.78	1.255	1.06	1.224	1.32	1.204	1.59	1.189	1.86	1.177

Table 11.19 (contd) − $f_{cu} = 30\,\text{N/mm}^2$, $k = 0.60$.

e/R	p=0.4 N/R^2	p=0.4 z/R	p=1.0 N/R^2	p=1.0 z/R	p=2.0 N/R^2	p=2.0 z/R	p=3.0 N/R^2	p=3.0 z/R	p=4.0 N/R^2	p=4.0 z/R	p=5.0 N/R^2	p=5.0 z/R	p=6.0 N/R^2	p=6.0 z/R
0.10	41.58	0.614	47.44	0.606	57.30	0.596	67.24	0.588	77.25	0.581	87.32	0.576	97.42	0.571
0.20	35.91	0.714	40.81	0.707	49.01	0.699	57.23	0.693	65.47	0.688	73.73	0.685	82.00	0.682
0.30	30.38	0.808	34.60	0.796	41.59	0.784	48.56	0.776	55.52	0.771	62.48	0.767	69.44	0.764
0.40	25.30	0.892	29.20	0.870	35.42	0.850	41.51	0.839	47.56	0.831	53.59	0.826	59.61	0.821
0.50	20.92	0.962	24.81	0.927	30.50	0.899	35.96	0.884	41.33	0.874	46.66	0.867	51.97	0.862
0.60	17.33	1.016	21.18	0.969	26.64	0.935	31.59	0.917	36.43	0.906	41.22	0.898	45.97	0.892
0.70	14.00	1.074	18.33	1.000	23.36	0.962	28.08	0.942	32.52	0.930	36.86	0.921	41.16	0.915
0.80	11.45	1.117	15.88	1.030	20.74	0.982	25.03	0.961	29.20	0.948	33.32	0.939	37.25	0.932
0.90	9.56	1.148	13.68	1.060	18.64	0.998	22.56	0.976	26.36	0.962	30.12	0.953	33.84	0.946
1.00	8.15	1.171	11.95	1.082	16.91	1.010	20.53	0.987	24.02	0.973	27.47	0.964	30.89	0.957
1.20	6.07	1.210	9.48	1.113	13.81	1.040	17.38	1.005	20.39	0.990	23.35	0.980	26.28	0.973
1.40	4.72	1.237	7.83	1.133	11.55	1.062	14.99	1.019	17.71	1.002	20.30	0.992	22.87	0.984
1.60	3.83	1.255	6.66	1.147	9.90	1.078	12.90	1.036	15.65	1.011	17.96	1.001	20.24	0.993
1.80	3.22	1.267	5.79	1.157	8.66	1.089	11.32	1.049	13.90	1.021	16.09	1.007	18.15	1.000
2.00	2.77	1.276	5.12	1.165	7.69	1.098	10.07	1.058	12.39	1.031	14.58	1.013	16.45	1.005
2.20	2.43	1.283	4.58	1.171	6.92	1.105	9.07	1.066	11.17	1.039	13.24	1.019	15.04	1.009
2.40	2.16	1.288	4.15	1.175	6.28	1.110	8.25	1.072	10.17	1.046	12.06	1.026	13.85	1.013
2.60	1.95	1.293	3.79	1.179	5.76	1.115	7.57	1.077	9.33	1.051	11.07	1.032	12.80	1.017
2.80	1.77	1.296	3.49	1.182	5.31	1.118	6.99	1.081	8.62	1.056	10.23	1.037	11.83	1.022
3.00	1.62	1.299	3.22	1.187	4.93	1.122	6.49	1.085	8.00	1.060	9.50	1.041	10.99	1.026
3.25	1.47	1.302	2.93	1.191	4.52	1.125	5.96	1.089	7.35	1.064	8.73	1.045	10.10	1.031
3.50	1.34	1.304	2.69	1.195	4.17	1.128	5.50	1.092	6.80	1.067	8.07	1.049	9.34	1.035
3.75	1.24	1.306	2.48	1.199	3.88	1.130	5.12	1.094	6.32	1.070	7.51	1.052	8.69	1.038
4.00	1.15	1.308	2.31	1.202	3.62	1.132	4.78	1.097	5.91	1.073	7.02	1.055	8.12	1.041
4.25	1.07	1.310	2.15	1.204	3.40	1.134	4.49	1.099	5.54	1.075	6.59	1.057	7.63	1.043
4.50	1.00	1.311	2.02	1.207	3.20	1.136	4.22	1.101	5.22	1.077	6.21	1.059	7.19	1.045
4.75	0.94	1.312	1.90	1.208	3.02	1.137	3.99	1.102	4.94	1.078	5.87	1.061	6.79	1.047
5.00	0.89	1.313	1.80	1.210	2.86	1.138	3.78	1.104	4.68	1.080	5.56	1.063	6.44	1.049
5.50	0.79	1.315	1.62	1.213	2.59	1.141	3.43	1.106	4.24	1.082	5.04	1.065	5.84	1.052
6.00	0.72	1.316	1.47	1.216	2.37	1.142	3.13	1.108	3.87	1.085	4.61	1.068	5.34	1.054
6.50	0.66	1.317	1.35	1.217	2.18	1.144	2.88	1.110	3.57	1.086	4.24	1.069	4.91	1.056
7.00	0.61	1.318	1.24	1.219	2.02	1.145	2.67	1.111	3.30	1.088	3.93	1.071	4.55	1.058
7.50	0.56	1.319	1.16	1.221	1.88	1.146	2.49	1.112	3.08	1.089	3.66	1.072	4.24	1.059
8.00	0.53	1.320	1.08	1.222	1.76	1.147	2.33	1.113	2.88	1.090	3.43	1.074	3.97	1.061
8.50	0.49	1.320	1.01	1.223	1.65	1.148	2.19	1.114	2.71	1.091	3.22	1.075	3.73	1.062
9.00	0.46	1.321	0.95	1.224	1.56	1.149	2.06	1.115	2.55	1.092	3.04	1.076	3.52	1.063
9.50	0.44	1.321	0.90	1.225	1.47	1.149	1.95	1.116	2.42	1.093	2.88	1.076	3.33	1.064
10.00	0.41	1.322	0.85	1.225	1.40	1.150	1.85	1.116	2.29	1.094	2.73	1.077	3.16	1.064
11.00	0.37	1.323	0.77	1.227	1.27	1.151	1.68	1.117	2.08	1.095	2.48	1.078	2.87	1.066
12.00	0.34	1.323	0.70	1.228	1.16	1.152	1.54	1.118	1.91	1.096	2.27	1.079	2.63	1.067
13.00	0.31	1.324	0.65	1.229	1.07	1.152	1.42	1.119	1.76	1.097	2.09	1.080	2.42	1.068
14.00	0.29	1.324	0.60	1.230	0.99	1.153	1.31	1.120	1.63	1.097	1.94	1.081	2.25	1.069
15.00	0.27	1.325	0.56	1.230	0.92	1.153	1.23	1.120	1.52	1.098	1.81	1.082	2.10	1.069
17.50	0.23	1.325	0.48	1.231	0.79	1.154	1.05	1.121	1.30	1.099	1.55	1.083	1.79	1.071
20.00	0.20	1.326	0.41	1.232	0.69	1.155	0.92	1.122	1.14	1.100	1.35	1.084	1.57	1.072

Table 11.20 Circular columns – $f_{cu} = 35\,\text{N/mm}^2$, $k = 0.90$.

e/R	p=0.4 N/R²	z/R	p=1.0 N/R²	z/R	p=2.0 N/R²	z/R	p=3.0 N/R²	z/R	p=4.0 N/R²	z/R	p=5.0 N/R²	z/R	p=6.0 N/R²	z/R
0.10	48.30	0.861	54.77	0.839	65.62	0.812	76.60	0.793	87.75	0.781	99.29	******	110.68	******
0.20	42.15	0.955	48.08	0.929	57.83	0.899	67.54	0.879	77.26	0.863	87.00	0.851	96.76	0.842
0.30	36.21	1.043	41.83	1.009	50.77	0.975	59.52	0.952	68.21	0.936	76.87	0.924	85.53	0.914
0.40	30.79	1.121	36.32	1.077	44.65	1.036	52.64	1.011	60.51	0.994	68.31	0.981	76.09	0.971
0.50	26.12	1.187	31.66	1.132	39.52	1.085	46.88	1.058	54.07	1.040	61.17	1.027	68.23	1.016
0.60	21.59	1.258	27.83	1.175	35.27	1.124	42.10	1.095	48.72	1.076	55.23	1.062	61.69	1.051
0.70	17.74	1.320	24.31	1.220	31.75	1.154	38.12	1.124	44.25	1.104	50.26	1.090	56.21	1.079
0.80	14.69	1.369	21.03	1.266	28.83	1.179	34.79	1.148	40.49	1.127	46.08	1.113	51.59	1.102
0.90	12.34	1.406	18.35	1.303	26.18	1.204	31.97	1.167	37.30	1.146	42.51	1.132	47.65	1.120
1.00	10.54	1.434	16.18	1.332	23.50	1.236	29.56	1.183	34.56	1.162	39.44	1.147	44.25	1.136
1.20	7.76	1.489	12.97	1.375	19.35	1.284	25.06	1.223	30.12	1.186	34.45	1.171	38.71	1.160
1.40	5.92	1.527	10.77	1.403	16.36	1.317	21.37	1.261	26.15	1.218	30.57	1.189	34.39	1.178
1.60	4.72	1.552	9.19	1.422	14.13	1.341	18.58	1.287	22.82	1.248	26.96	1.216	30.94	1.192
1.80	3.89	1.568	8.01	1.437	12.42	1.358	16.41	1.308	20.22	1.270	23.93	1.240	27.59	1.215
2.00	3.31	1.580	7.04	1.452	11.07	1.372	14.68	1.323	18.13	1.287	21.49	1.259	24.81	1.236
2.20	2.87	1.589	6.23	1.469	9.99	1.382	13.27	1.336	16.42	1.301	19.50	1.274	22.53	1.252
2.40	2.53	1.596	5.58	1.482	9.09	1.391	12.11	1.345	15.00	1.312	17.83	1.286	20.62	1.265
2.60	2.26	1.602	5.05	1.494	8.35	1.398	11.14	1.354	13.81	1.321	16.42	1.296	19.00	1.276
2.80	2.04	1.606	4.61	1.503	7.71	1.404	10.30	1.361	12.79	1.329	15.22	1.305	17.62	1.285
3.00	1.86	1.610	4.23	1.510	7.16	1.409	9.59	1.366	11.91	1.336	14.18	1.312	16.42	1.292
3.25	1.68	1.614	3.83	1.517	6.58	1.414	8.82	1.373	10.97	1.343	13.07	1.319	15.14	1.301
3.50	1.53	1.617	3.50	1.523	6.09	1.419	8.16	1.378	10.16	1.348	12.11	1.326	14.04	1.307
3.75	1.40	1.619	3.22	1.528	5.66	1.422	7.60	1.382	9.46	1.353	11.29	1.331	13.08	1.313
4.00	1.29	1.621	2.98	1.532	5.29	1.426	7.11	1.386	8.86	1.358	10.57	1.336	12.25	1.318
4.25	1.20	1.623	2.77	1.535	4.97	1.429	6.68	1.389	8.32	1.361	9.93	1.340	11.52	1.322
4.50	1.12	1.625	2.59	1.539	4.68	1.431	6.30	1.392	7.85	1.365	9.37	1.343	10.87	1.326
4.75	1.05	1.626	2.44	1.541	4.42	1.433	5.95	1.395	7.43	1.367	8.87	1.346	10.29	1.330
5.00	0.99	1.627	2.30	1.544	4.19	1.435	5.65	1.397	7.05	1.370	8.42	1.349	9.77	1.333
5.50	0.88	1.629	2.06	1.548	3.80	1.439	5.12	1.401	6.39	1.374	7.64	1.354	8.87	1.338
6.00	0.80	1.631	1.87	1.551	3.47	1.441	4.69	1.404	5.85	1.378	6.99	1.358	8.12	1.342
6.50	0.73	1.632	1.71	1.554	3.19	1.445	4.32	1.407	5.39	1.381	6.45	1.361	7.49	1.345
7.00	0.67	1.634	1.57	1.556	2.95	1.450	4.00	1.409	5.00	1.384	5.98	1.364	6.95	1.348
7.50	0.62	1.634	1.46	1.558	2.74	1.454	3.73	1.411	4.66	1.386	5.58	1.366	6.48	1.351
8.00	0.58	1.635	1.36	1.559	2.56	1.458	3.49	1.413	4.37	1.388	5.22	1.368	6.07	1.353
8.50	0.54	1.636	1.27	1.561	2.40	1.461	3.28	1.414	4.11	1.389	4.91	1.370	5.71	1.355
9.00	0.51	1.637	1.19	1.562	2.26	1.464	3.10	1.416	3.88	1.391	4.64	1.372	5.39	1.357
9.50	0.48	1.637	1.13	1.563	2.14	1.467	2.93	1.417	3.67	1.392	4.39	1.373	5.10	1.358
10.00	0.45	1.638	1.07	1.564	2.02	1.469	2.79	1.418	3.48	1.393	4.17	1.375	4.85	1.360
11.00	0.41	1.639	0.96	1.566	1.83	1.473	2.53	1.420	3.16	1.395	3.79	1.377	4.40	1.362
12.00	0.37	1.639	0.88	1.567	1.67	1.476	2.32	1.421	2.90	1.397	3.47	1.379	4.03	1.364
13.00	0.34	1.640	0.81	1.569	1.54	1.479	2.14	1.422	2.67	1.398	3.20	1.380	3.72	1.366
14.00	0.32	1.640	0.75	1.570	1.42	1.482	1.98	1.423	2.48	1.399	2.97	1.381	3.45	1.367
15.00	0.29	1.641	0.69	1.570	1.33	1.484	1.85	1.424	2.31	1.401	2.77	1.382	3.22	1.368
17.50	0.25	1.642	0.59	1.572	1.13	1.488	1.58	1.426	1.98	1.403	2.37	1.385	2.76	1.371
20.00	0.22	1.642	0.51	1.573	0.99	1.491	1.38	1.428	1.73	1.404	2.07	1.386	2.41	1.372

Table 11.20 (contd) $-$ $f_{cu} = 35$ N/mm^2, $k = 0.80$.

eR	p=0.4 N/R²	z/R	p=1.0 N/R²	z/R	p=2.0 N/R²	z/R	p=3.0 N/R²	z/R	p=4.0 N/R²	z/R	p=5.0 N/R²	z/R	p=6.0 N/R²	z/R
0.10	48.13	0.779	54.38	0.762	64.90	0.741	75.53	0.725	86.27	0.713	97.11	0.704	108.05	0.697
0.20	41.86	0.875	47.45	0.856	56.71	0.833	65.97	0.817	75.26	0.804	84.57	0.795	93.89	0.787
0.30	35.78	0.966	40.95	0.939	49.30	0.911	57.53	0.893	65.71	0.880	73.88	0.871	82.05	0.863
0.40	30.20	1.046	35.24	1.009	42.94	0.975	50.38	0.954	57.71	0.940	64.99	0.929	72.25	0.921
0.50	25.41	1.114	30.45	1.065	37.68	1.024	44.48	1.001	51.13	0.985	57.71	0.974	64.24	0.965
0.60	21.01	1.181	26.57	1.109	33.39	1.063	39.67	1.037	45.75	1.020	51.74	1.008	57.69	0.999
0.70	17.17	1.242	23.21	1.149	29.88	1.093	35.71	1.066	41.32	1.048	46.83	1.035	52.27	1.026
0.80	14.15	1.290	20.00	1.194	27.00	1.116	32.43	1.088	37.63	1.070	42.72	1.057	47.75	1.047
0.90	11.82	1.327	17.39	1.229	24.54	1.137	29.69	1.106	34.53	1.088	39.26	1.074	43.93	1.064
1.00	10.03	1.354	15.30	1.258	21.98	1.167	27.36	1.121	31.89	1.102	36.31	1.089	40.67	1.079
1.20	7.23	1.405	12.24	1.298	18.03	1.212	23.19	1.156	27.65	1.125	31.55	1.111	35.39	1.101
1.40	5.51	1.438	10.12	1.323	15.22	1.243	19.74	1.190	24.04	1.150	27.88	1.127	31.32	1.117
1.60	4.40	1.460	8.58	1.340	13.13	1.264	17.14	1.214	20.95	1.177	24.67	1.148	28.08	1.129
1.80	3.65	1.474	7.44	1.353	11.54	1.280	15.13	1.233	18.54	1.197	21.88	1.170	25.15	1.147
2.00	3.11	1.485	6.49	1.366	10.28	1.293	13.53	1.247	16.62	1.213	19.63	1.186	22.60	1.165
2.20	2.70	1.492	5.73	1.379	9.27	1.303	12.23	1.258	15.05	1.225	17.80	1.200	20.51	1.179
2.40	2.39	1.498	5.12	1.389	8.44	1.310	11.16	1.267	13.75	1.235	16.28	1.211	18.77	1.191
2.60	2.14	1.503	4.62	1.397	7.74	1.317	10.26	1.274	12.65	1.244	14.99	1.220	17.29	1.200
2.80	1.94	1.507	4.21	1.404	7.13	1.321	9.49	1.281	11.72	1.251	13.89	1.227	16.03	1.208
3.00	1.77	1.510	3.87	1.409	6.62	1.326	8.83	1.286	10.91	1.256	12.94	1.234	14.94	1.215
3.25	1.60	1.513	3.51	1.415	6.07	1.330	8.12	1.292	10.04	1.263	11.92	1.240	13.77	1.222
3.50	1.45	1.516	3.21	1.420	5.60	1.333	7.52	1.296	9.31	1.268	11.05	1.246	12.77	1.228
3.75	1.33	1.518	2.95	1.424	5.20	1.336	7.00	1.300	8.67	1.272	10.30	1.251	11.90	1.234
4.00	1.23	1.520	2.74	1.428	4.85	1.339	6.55	1.304	8.11	1.276	9.64	1.255	11.14	1.238
4.25	1.15	1.522	2.55	1.430	4.55	1.341	6.15	1.307	7.62	1.280	9.06	1.259	10.48	1.242
4.50	1.07	1.523	2.39	1.433	4.28	1.343	5.80	1.309	7.19	1.282	8.55	1.262	9.89	1.245
4.75	1.00	1.524	2.24	1.435	4.04	1.345	5.48	1.312	6.80	1.285	8.09	1.265	9.36	1.248
5.00	0.95	1.525	2.12	1.437	3.83	1.347	5.20	1.314	6.45	1.287	7.68	1.267	8.88	1.251
5.50	0.85	1.527	1.90	1.441	3.46	1.349	4.71	1.317	5.85	1.291	6.97	1.271	8.06	1.256
6.00	0.77	1.529	1.72	1.444	3.16	1.351	4.30	1.320	5.36	1.295	6.38	1.275	7.38	1.259
6.50	0.70	1.530	1.58	1.446	2.91	1.353	3.96	1.322	4.94	1.297	5.88	1.278	6.81	1.263
7.00	0.65	1.531	1.45	1.448	2.69	1.355	3.67	1.323	4.58	1.300	5.45	1.280	6.32	1.265
7.50	0.60	1.532	1.35	1.449	2.51	1.356	3.41	1.325	4.27	1.302	5.09	1.283	5.89	1.268
8.00	0.56	1.532	1.26	1.451	2.34	1.359	3.19	1.326	4.00	1.303	4.76	1.284	5.52	1.270
8.50	0.52	1.533	1.18	1.452	2.19	1.361	3.00	1.327	3.76	1.305	4.48	1.286	5.19	1.271
9.00	0.49	1.534	1.11	1.453	2.06	1.362	2.83	1.328	3.55	1.306	4.23	1.288	4.90	1.273
9.50	0.46	1.534	1.04	1.454	1.95	1.364	2.68	1.329	3.36	1.307	4.00	1.289	4.64	1.274
10.00	0.44	1.535	0.99	1.455	1.85	1.365	2.54	1.330	3.19	1.308	3.80	1.290	4.41	1.275
11.00	0.39	1.535	0.89	1.456	1.67	1.368	2.31	1.331	2.90	1.310	3.45	1.292	4.00	1.277
12.00	0.36	1.536	0.81	1.458	1.52	1.370	2.11	1.333	2.65	1.312	3.16	1.294	3.67	1.279
13.00	0.33	1.537	0.75	1.459	1.40	1.372	1.94	1.334	2.45	1.313	2.92	1.295	3.38	1.281
14.00	0.31	1.537	0.69	1.460	1.30	1.373	1.80	1.334	2.27	1.314	2.71	1.296	3.14	1.282
15.00	0.28	1.537	0.64	1.460	1.21	1.374	1.68	1.335	2.12	1.315	2.53	1.297	2.93	1.283
17.50	0.24	1.538	0.55	1.462	1.03	1.377	1.44	1.337	1.81	1.317	2.16	1.299	2.51	1.285
20.00	0.21	1.539	0.48	1.463	0.90	1.379	1.26	1.338	1.58	1.318	1.89	1.301	2.19	1.287

Table 11.21 Circular columns – $f_{cu} = 35\,\text{N/mm}^2$, $k = 0.70$.

e/R	p=0.4 N/R²	p=0.4 z/R	p=1.0 N/R²	p=1.0 z/R	p=2.0 N/R²	p=2.0 z/R	p=3.0 N/R²	p=3.0 z/R	p=4.0 N/R²	p=4.0 z/R	p=5.0 N/R²	p=5.0 z/R	p=6.0 N/R²	p=6.0 z/R
0.10	47.98	0.697	54.02	0.685	64.20	0.670	74.48	0.658	84.85	0.648	95.29	0.640	105.79	0.634
0.20	41.60	0.795	46.83	0.782	55.58	0.767	64.35	0.755	73.14	0.747	81.96	0.740	90.79	0.735
0.30	35.36	0.888	40.07	0.869	47.78	0.849	55.42	0.836	63.03	0.826	70.64	0.819	78.25	0.814
0.40	29.63	0.971	34.14	0.941	41.15	0.914	47.97	0.898	54.70	0.887	61.40	0.879	68.08	0.872
0.50	24.70	1.040	29.21	0.999	35.75	0.964	41.94	0.945	48.00	0.932	53.99	0.923	59.95	0.915
0.60	20.42	1.104	25.27	1.043	31.42	1.002	37.10	0.980	42.61	0.966	48.04	0.956	53.42	0.948
0.70	16.49	1.165	22.08	1.078	27.94	1.031	33.19	1.007	38.25	0.992	43.21	0.981	48.11	0.973
0.80	13.43	1.211	18.93	1.122	25.12	1.054	29.99	1.029	34.66	1.012	39.22	1.001	43.73	0.993
0.90	11.14	1.245	16.29	1.155	22.79	1.072	27.34	1.046	31.67	1.029	35.90	1.017	40.07	1.008
1.00	9.44	1.270	14.19	1.180	20.41	1.099	25.10	1.060	29.14	1.042	33.08	1.030	36.96	1.021
1.20	6.80	1.318	11.20	1.215	16.62	1.139	21.30	1.088	25.13	1.063	28.58	1.050	31.99	1.041
1.40	5.20	1.348	9.21	1.237	13.86	1.164	18.09	1.119	21.91	1.082	25.16	1.065	28.19	1.056
1.60	4.18	1.367	7.81	1.252	11.86	1.182	15.60	1.139	19.07	1.106	22.35	1.079	25.19	1.067
1.80	3.48	1.379	6.77	1.263	10.35	1.195	13.66	1.154	16.87	1.124	19.81	1.099	22.70	1.077
2.00	2.97	1.389	5.96	1.272	9.18	1.204	12.15	1.165	15.02	1.137	17.77	1.114	20.38	1.093
2.20	2.59	1.396	5.27	1.283	8.25	1.212	10.93	1.173	13.54	1.146	16.10	1.125	18.49	1.106
2.40	2.30	1.401	4.71	1.291	7.49	1.218	9.93	1.180	12.31	1.153	14.66	1.133	16.91	1.116
2.60	2.06	1.405	4.26	1.298	6.85	1.223	9.10	1.186	11.29	1.160	13.45	1.140	15.58	1.125
2.80	1.87	1.409	3.89	1.304	6.32	1.227	8.40	1.190	10.42	1.165	12.42	1.146	14.40	1.131
3.00	1.71	1.411	3.58	1.309	5.86	1.231	7.80	1.194	9.68	1.169	11.54	1.150	13.38	1.136
3.25	1.55	1.414	3.25	1.314	5.37	1.235	7.15	1.199	8.89	1.174	10.60	1.155	12.29	1.141
3.50	1.41	1.417	2.97	1.318	4.96	1.238	6.61	1.202	8.21	1.178	9.80	1.159	11.37	1.145
3.75	1.30	1.419	2.74	1.322	4.60	1.240	6.14	1.205	7.64	1.181	9.11	1.163	10.57	1.149
4.00	1.20	1.420	2.54	1.325	4.30	1.243	5.74	1.208	7.13	1.184	8.51	1.166	9.88	1.152
4.25	1.12	1.422	2.37	1.327	4.03	1.245	5.38	1.210	6.69	1.186	7.99	1.169	9.27	1.155
4.50	1.04	1.423	2.22	1.330	3.79	1.246	5.07	1.212	6.30	1.188	7.52	1.171	8.73	1.157
4.75	0.98	1.424	2.09	1.332	3.58	1.248	4.79	1.214	5.96	1.190	7.11	1.173	8.26	1.160
5.00	0.92	1.425	1.97	1.333	3.39	1.249	4.54	1.215	5.65	1.192	6.74	1.175	7.83	1.161
5.50	0.83	1.427	1.77	1.336	3.07	1.252	4.11	1.218	5.11	1.195	6.11	1.178	7.09	1.165
6.00	0.75	1.428	1.61	1.339	2.80	1.254	3.75	1.220	4.67	1.197	5.58	1.180	6.48	1.167
6.50	0.69	1.429	1.47	1.341	2.58	1.255	3.45	1.222	4.30	1.199	5.14	1.183	5.97	1.170
7.00	0.63	1.430	1.36	1.342	2.39	1.256	3.20	1.223	3.98	1.201	4.76	1.184	5.53	1.172
7.50	0.59	1.431	1.26	1.344	2.22	1.258	2.98	1.225	3.71	1.202	4.43	1.186	5.15	1.173
8.00	0.55	1.432	1.18	1.345	2.08	1.259	2.79	1.226	3.47	1.204	4.15	1.187	4.82	1.175
8.50	0.51	1.432	1.10	1.346	1.95	1.260	2.62	1.227	3.26	1.205	3.90	1.188	4.53	1.176
9.00	0.48	1.433	1.04	1.347	1.84	1.260	2.47	1.228	3.08	1.206	3.68	1.190	4.27	1.177
9.50	0.45	1.433	0.98	1.348	1.74	1.261	2.33	1.229	2.91	1.207	3.48	1.190	4.04	1.178
10.00	0.43	1.434	0.93	1.349	1.65	1.262	2.22	1.229	2.76	1.207	3.30	1.191	3.84	1.179
11.00	0.39	1.434	0.84	1.350	1.50	1.263	2.01	1.231	2.51	1.209	3.00	1.193	3.48	1.180
12.00	0.35	1.435	0.76	1.351	1.37	1.264	1.84	1.232	2.29	1.210	2.74	1.194	3.19	1.182
13.00	0.32	1.436	0.70	1.352	1.26	1.264	1.69	1.232	2.11	1.211	2.53	1.195	2.94	1.183
14.00	0.30	1.436	0.65	1.353	1.17	1.265	1.57	1.233	1.96	1.212	2.34	1.196	2.73	1.183
15.00	0.28	1.436	0.61	1.354	1.09	1.266	1.47	1.234	1.83	1.212	2.19	1.196	2.54	1.184
17.50	0.24	1.437	0.52	1.355	0.93	1.267	1.25	1.235	1.56	1.214	1.87	1.198	2.17	1.186
20.00	0.21	1.438	0.45	1.356	0.81	1.267	1.09	1.236	1.37	1.215	1.63	1.199	1.90	1.187

Table 11.21 (contd) – $f_{cu} = 35\,\text{N/mm}^2$, $k = 0.60$.

e/R	p=0.4 N/R²	p=0.4 z/R	p=1.0 N/R²	p=1.0 z/R	p=2.0 N/R²	p=2.0 z/R	p=3.0 N/R²	p=3.0 z/R	p=4.0 N/R²	p=4.0 z/R	p=5.0 N/R²	p=5.0 z/R	p=6.0 N/R²	p=6.0 z/R
0.10	47.86	0.614	53.71	0.607	63.55	0.598	73.47	0.591	83.45	0.585	93.48	0.579	103.55	0.575
0.20	41.35	0.715	46.25	0.708	54.44	0.701	62.65	0.695	70.89	0.691	79.14	0.687	87.40	0.685
0.30	34.97	0.809	39.20	0.799	46.19	0.787	53.17	0.779	60.13	0.774	67.09	0.770	74.05	0.767
0.40	29.06	0.895	33.01	0.874	39.27	0.855	45.33	0.843	51.46	0.835	57.50	0.829	63.53	0.825
0.50	23.92	0.967	27.93	0.933	33.73	0.905	39.24	0.889	44.65	0.879	50.00	0.872	55.33	0.867
0.60	19.61	1.026	23.73	0.977	29.34	0.942	34.41	0.924	39.29	0.912	44.10	0.903	48.88	0.897
0.70	15.71	1.085	20.46	1.010	25.64	0.970	30.42	0.949	35.02	0.936	39.39	0.927	43.72	0.920
0.80	12.75	1.129	17.41	1.046	22.72	0.991	27.08	0.969	31.29	0.955	35.44	0.945	39.53	0.938
0.90	10.58	1.161	14.93	1.076	20.38	1.007	24.38	0.984	28.23	0.969	32.01	0.959	35.76	0.952
1.00	8.95	1.185	13.00	1.098	18.29	1.023	22.16	0.996	25.71	0.981	29.18	0.970	32.62	0.963
1.20	6.48	1.229	10.26	1.129	14.73	1.056	18.74	1.013	21.80	0.998	24.78	0.987	27.73	0.979
1.40	4.99	1.255	8.45	1.149	12.29	1.078	15.78	1.035	18.91	1.010	21.53	0.999	24.12	0.991
1.60	4.03	1.272	7.17	1.162	10.52	1.093	13.57	1.052	16.52	1.022	19.03	1.008	21.33	0.999
1.80	3.38	1.284	6.22	1.172	9.19	1.104	11.89	1.064	14.50	1.036	17.05	1.014	19.12	1.006
2.00	2.90	1.293	5.49	1.179	8.15	1.113	10.58	1.073	12.91	1.045	15.22	1.025	17.32	1.011
2.20	2.54	1.299	4.89	1.186	7.33	1.119	9.52	1.080	11.64	1.053	13.72	1.033	15.79	1.017
2.40	2.25	1.304	4.39	1.194	6.65	1.125	8.66	1.086	10.59	1.060	12.49	1.040	14.38	1.024
2.60	2.03	1.308	3.98	1.200	6.09	1.129	7.93	1.091	9.71	1.065	11.46	1.045	13.20	1.030
2.80	1.84	1.311	3.63	1.205	5.61	1.133	7.32	1.095	8.97	1.069	10.59	1.050	12.20	1.035
3.00	1.69	1.313	3.35	1.209	5.21	1.136	6.80	1.099	8.33	1.073	9.84	1.054	11.34	1.039
3.25	1.53	1.316	3.04	1.214	4.78	1.139	6.24	1.102	7.65	1.077	9.04	1.058	10.41	1.043
3.50	1.40	1.318	2.79	1.217	4.41	1.142	5.76	1.105	7.07	1.080	8.36	1.061	9.63	1.047
3.75	1.29	1.320	2.58	1.220	4.10	1.144	5.36	1.108	6.58	1.083	7.77	1.064	8.96	1.050
4.00	1.19	1.322	2.39	1.223	3.82	1.146	5.00	1.110	6.14	1.085	7.26	1.067	8.37	1.053
4.25	1.11	1.323	2.23	1.225	3.58	1.148	4.69	1.112	5.76	1.087	6.82	1.069	7.86	1.055
4.50	1.04	1.324	2.09	1.227	3.37	1.149	4.42	1.114	5.43	1.089	6.42	1.071	7.40	1.057
4.75	0.97	1.325	1.97	1.229	3.19	1.151	4.18	1.115	5.13	1.091	6.07	1.073	7.00	1.059
5.00	0.92	1.326	1.86	1.231	3.02	1.152	3.96	1.117	4.86	1.092	5.75	1.074	6.64	1.060
5.50	0.82	1.328	1.67	1.233	2.73	1.154	3.58	1.119	4.41	1.095	5.21	1.077	6.01	1.063
6.00	0.75	1.329	1.52	1.236	2.49	1.156	3.27	1.121	4.03	1.097	4.76	1.079	5.50	1.065
6.50	0.68	1.330	1.39	1.237	2.30	1.157	3.01	1.122	3.71	1.099	4.39	1.081	5.06	1.067
7.00	0.63	1.331	1.29	1.239	2.13	1.158	2.79	1.124	3.43	1.100	4.07	1.083	4.69	1.069
7.50	0.58	1.332	1.19	1.240	1.98	1.159	2.60	1.125	3.20	1.101	3.79	1.084	4.37	1.070
8.00	0.54	1.333	1.11	1.241	1.85	1.160	2.43	1.126	2.99	1.102	3.54	1.085	4.09	1.072
8.50	0.51	1.333	1.04	1.242	1.74	1.161	2.29	1.127	2.81	1.103	3.33	1.086	3.84	1.073
9.00	0.48	1.334	0.98	1.243	1.64	1.162	2.16	1.128	2.65	1.104	3.14	1.087	3.63	1.074
9.50	0.45	1.334	0.93	1.244	1.55	1.162	2.04	1.128	2.51	1.105	2.97	1.088	3.43	1.074
10.00	0.43	1.334	0.88	1.245	1.47	1.163	1.94	1.129	2.38	1.106	2.82	1.088	3.26	1.075
11.00	0.39	1.335	0.79	1.246	1.33	1.164	1.76	1.130	2.16	1.107	2.56	1.090	2.96	1.076
12.00	0.35	1.336	0.73	1.247	1.22	1.165	1.61	1.131	1.98	1.108	2.34	1.091	2.71	1.077
13.00	0.32	1.336	0.67	1.248	1.13	1.165	1.48	1.132	1.82	1.109	2.16	1.092	2.49	1.078
14.00	0.30	1.337	0.62	1.248	1.04	1.166	1.37	1.132	1.69	1.109	2.00	1.092	2.31	1.079
15.00	0.28	1.337	0.57	1.249	0.97	1.166	1.28	1.133	1.58	1.110	1.87	1.093	2.16	1.080
17.50	0.24	1.338	0.49	1.250	0.83	1.167	1.10	1.134	1.35	1.111	1.60	1.094	1.85	1.081
20.00	0.21	1.338	0.43	1.251	0.73	1.168	0.96	1.135	1.18	1.112	1.40	1.095	1.61	1.082

Table 11.22 Circular columns – $f_{cu} = 40\,\text{N/mm}^2$, $k = 0.90$.

e/R	p=0.4 N/R²	p=0.4 z/R	p=1.0 N/R²	p=1.0 z/R	p=2.0 N/R²	p=2.0 z/R	p=3.0 N/R²	p=3.0 z/R	p=4.0 N/R²	p=4.0 z/R	p=5.0 N/R²	p=5.0 z/R	p=6.0 N/R²	p=6.0 z/R
0.10	54.58	0.863	61.05	0.843	71.88	0.818	82.82	0.800	93.88	0.786	105.18	******	116.80	******
0.20	47.60	0.958	53.55	0.934	63.31	0.906	73.03	0.886	82.74	0.870	92.47	0.858	102.21	0.849
0.30	40.83	1.047	46.50	1.015	55.50	0.982	64.28	0.960	72.99	0.943	81.67	0.931	90.33	0.921
0.40	34.60	1.127	40.25	1.084	48.70	1.044	56.76	1.020	64.67	1.002	72.50	0.989	80.29	0.979
0.50	29.21	1.195	34.97	1.141	43.00	1.094	50.46	1.067	57.70	1.048	64.85	1.034	71.93	1.024
0.60	23.93	1.269	30.63	1.185	38.28	1.134	45.23	1.104	51.92	1.085	58.48	1.070	64.97	1.059
0.70	19.52	1.331	26.37	1.237	34.39	1.165	40.88	1.134	47.09	1.113	53.16	1.099	59.15	1.087
0.80	16.03	1.381	22.69	1.283	31.16	1.190	37.25	1.158	43.04	1.137	48.68	1.121	54.24	1.110
0.90	13.37	1.419	19.71	1.320	27.81	1.225	34.19	1.177	39.60	1.156	44.87	1.140	50.06	1.128
1.00	11.31	1.449	17.31	1.349	24.89	1.256	31.57	1.193	36.66	1.171	41.60	1.156	46.46	1.144
1.20	8.13	1.508	13.80	1.390	20.40	1.303	26.24	1.244	31.76	1.199	36.29	1.180	40.59	1.168
1.40	6.12	1.544	11.41	1.418	17.19	1.334	22.32	1.280	27.18	1.238	31.89	1.204	36.03	1.186
1.60	4.84	1.567	9.71	1.437	14.81	1.357	19.36	1.305	23.68	1.266	27.87	1.235	31.99	1.209
1.80	3.98	1.582	8.35	1.457	13.00	1.374	17.07	1.325	20.94	1.288	24.71	1.258	28.40	1.234
2.00	3.37	1.594	7.26	1.477	11.58	1.387	15.26	1.339	18.76	1.304	22.17	1.276	25.52	1.253
2.20	2.92	1.602	6.40	1.493	10.43	1.397	13.79	1.351	16.98	1.317	20.09	1.290	23.15	1.268
2.40	2.57	1.608	5.72	1.506	9.49	1.405	12.57	1.361	15.51	1.328	18.37	1.302	21.18	1.281
2.60	2.30	1.613	5.15	1.515	8.70	1.412	11.55	1.368	14.27	1.336	16.91	1.311	19.51	1.291
2.80	2.07	1.617	4.68	1.522	8.04	1.418	10.68	1.375	13.21	1.344	15.67	1.319	18.08	1.299
3.00	1.89	1.621	4.29	1.528	7.47	1.423	9.94	1.381	12.29	1.350	14.59	1.326	16.85	1.307
3.25	1.70	1.624	3.88	1.535	6.86	1.428	9.14	1.386	11.31	1.357	13.44	1.333	15.53	1.314
3.50	1.55	1.627	3.54	1.540	6.34	1.432	8.46	1.391	10.48	1.362	12.45	1.339	14.39	1.321
3.75	1.42	1.629	3.26	1.544	5.89	1.435	7.87	1.396	9.76	1.367	11.60	1.344	13.41	1.326
4.00	1.31	1.631	3.01	1.548	5.51	1.438	7.36	1.399	9.13	1.371	10.86	1.349	12.56	1.331
4.25	1.21	1.633	2.81	1.551	5.17	1.441	6.91	1.402	8.58	1.374	10.20	1.353	11.81	1.335
4.50	1.13	1.634	2.62	1.554	4.86	1.445	6.51	1.405	8.09	1.377	9.62	1.356	11.14	1.339
4.75	1.06	1.635	2.46	1.556	4.58	1.450	6.16	1.408	7.65	1.380	9.11	1.359	10.54	1.342
5.00	1.00	1.636	2.32	1.558	4.33	1.454	5.84	1.410	7.26	1.383	8.64	1.362	10.00	1.345
5.50	0.89	1.638	2.08	1.562	3.90	1.462	5.30	1.413	6.59	1.387	7.84	1.366	9.08	1.350
6.00	0.81	1.640	1.88	1.565	3.55	1.468	4.84	1.416	6.02	1.390	7.18	1.370	8.31	1.354
6.50	0.74	1.641	1.72	1.568	3.25	1.474	4.46	1.419	5.55	1.393	6.62	1.373	7.67	1.357
7.00	0.68	1.642	1.59	1.570	3.00	1.478	4.14	1.421	5.15	1.395	6.14	1.376	7.11	1.360
7.50	0.63	1.643	1.47	1.571	2.79	1.482	3.86	1.423	4.80	1.397	5.72	1.378	6.63	1.362
8.00	0.58	1.644	1.37	1.573	2.60	1.486	3.61	1.425	4.50	1.399	5.36	1.380	6.21	1.364
8.50	0.55	1.644	1.28	1.574	2.44	1.489	3.39	1.426	4.23	1.401	5.04	1.382	5.84	1.366
9.00	0.51	1.645	1.20	1.575	2.30	1.492	3.20	1.427	3.99	1.402	4.76	1.383	5.52	1.368
9.50	0.48	1.645	1.14	1.576	2.17	1.494	3.03	1.428	3.78	1.404	4.50	1.384	5.22	1.369
10.00	0.46	1.646	1.07	1.577	2.05	1.496	2.88	1.429	3.59	1.405	4.28	1.386	4.96	1.371
11.00	0.41	1.647	0.97	1.579	1.86	1.500	2.61	1.431	3.26	1.407	3.88	1.388	4.50	1.373
12.00	0.38	1.647	0.88	1.580	1.70	1.503	2.39	1.433	2.98	1.408	3.56	1.389	4.13	1.375
13.00	0.35	1.648	0.81	1.581	1.56	1.505	2.21	1.434	2.75	1.409	3.28	1.391	3.81	1.376
14.00	0.32	1.648	0.75	1.582	1.44	1.507	2.05	1.435	2.55	1.411	3.05	1.392	3.53	1.377
15.00	0.30	1.649	0.70	1.583	1.34	1.508	1.91	1.436	2.38	1.412	2.84	1.393	3.30	1.379
17.50	0.25	1.649	0.60	1.584	1.14	1.511	1.63	1.437	2.04	1.413	2.43	1.395	2.82	1.381
20.00	0.22	1.650	0.52	1.585	1.00	1.513	1.43	1.439	1.78	1.415	2.13	1.397	2.47	1.382

Table 11.22 (contd) − $f_{cu} = 40\,\text{N/mm}^2$, $k = 0.80$.

e/R	p=0.4 N/R²	p=0.4 z/R	p=1.0 N/R²	p=1.0 z/R	p=2.0 N/R²	p=2.0 z/R	p=3.0 N/R²	p=3.0 z/R	p=4.0 N/R²	p=4.0 z/R	p=5.0 N/R²	p=5.0 z/R	p=6.0 N/R²	p=6.0 z/R
0.10	54.41	0.781	60.66	0.765	71.15	0.745	81.75	0.730	92.45	0.718	103.23	0.709	114.10	0.702
0.20	47.31	0.877	52.90	0.859	62.17	0.838	71.43	0.822	80.70	0.810	90.00	0.800	99.31	0.793
0.30	40.38	0.968	45.59	0.944	53.98	0.917	62.23	0.899	70.43	0.886	78.60	0.876	86.77	0.869
0.40	34.00	1.050	39.12	1.015	46.92	0.982	54.41	0.961	61.77	0.946	69.08	0.936	76.35	0.927
0.50	28.48	1.120	33.69	1.073	41.07	1.032	47.95	1.009	54.65	0.993	61.26	0.981	67.82	0.972
0.60	23.33	1.191	29.29	1.118	36.30	1.072	42.68	1.045	48.83	1.028	54.86	1.016	60.84	1.006
0.70	18.92	1.253	25.24	1.165	32.41	1.102	38.35	1.074	44.04	1.056	49.59	1.043	55.08	1.033
0.80	15.47	1.302	21.62	1.210	29.23	1.126	34.78	1.097	40.06	1.078	45.20	1.064	50.27	1.054
0.90	12.81	1.339	18.72	1.245	26.13	1.156	31.79	1.116	36.72	1.096	41.50	1.082	46.21	1.072
1.00	10.69	1.370	16.40	1.273	23.32	1.186	29.27	1.131	33.88	1.111	38.35	1.097	42.75	1.086
1.20	7.57	1.422	13.04	1.313	19.05	1.229	24.34	1.175	29.32	1.133	33.28	1.119	37.16	1.108
1.40	5.71	1.454	10.70	1.337	16.03	1.259	20.66	1.208	25.04	1.169	29.28	1.138	32.85	1.124
1.60	4.54	1.474	9.05	1.353	13.80	1.280	17.91	1.231	21.79	1.195	25.55	1.165	29.24	1.141
1.80	3.75	1.488	7.71	1.372	12.11	1.296	15.78	1.249	19.26	1.214	22.63	1.186	25.94	1.163
2.00	3.18	1.498	6.67	1.387	10.78	1.307	14.10	1.262	17.24	1.229	20.29	1.202	23.29	1.181
2.20	2.77	1.505	5.87	1.399	9.70	1.317	12.74	1.273	15.60	1.240	18.39	1.215	21.12	1.194
2.40	2.44	1.510	5.24	1.408	8.80	1.323	11.61	1.281	14.24	1.250	16.80	1.225	19.32	1.205
2.60	2.19	1.515	4.72	1.416	8.05	1.329	10.67	1.289	13.10	1.258	15.47	1.234	17.79	1.214
2.80	1.98	1.518	4.30	1.422	7.42	1.333	9.86	1.295	12.13	1.265	14.33	1.241	16.49	1.222
3.00	1.80	1.521	3.94	1.427	6.88	1.337	9.17	1.300	11.29	1.270	13.34	1.247	15.36	1.229
3.25	1.63	1.524	3.57	1.433	6.30	1.341	8.44	1.305	10.39	1.276	12.29	1.254	14.15	1.236
3.50	1.48	1.527	3.27	1.437	5.81	1.345	7.81	1.309	9.62	1.281	11.39	1.259	13.12	1.241
3.75	1.36	1.529	3.01	1.441	5.40	1.348	7.27	1.313	8.96	1.285	10.61	1.264	12.23	1.246
4.00	1.26	1.530	2.78	1.444	5.04	1.350	6.79	1.316	8.38	1.289	9.93	1.268	11.45	1.250
4.25	1.17	1.532	2.59	1.447	4.72	1.352	6.37	1.319	7.88	1.292	9.33	1.271	10.76	1.254
4.50	1.09	1.533	2.43	1.449	4.44	1.354	5.99	1.321	7.43	1.295	8.80	1.274	10.15	1.257
4.75	1.02	1.534	2.28	1.451	4.19	1.356	5.66	1.323	7.03	1.297	8.33	1.277	9.61	1.260
5.00	0.96	1.535	2.15	1.453	3.96	1.359	5.36	1.325	6.67	1.300	7.90	1.279	9.12	1.263
5.50	0.86	1.537	1.93	1.456	3.56	1.364	4.85	1.327	6.05	1.303	7.17	1.283	8.28	1.267
6.00	0.78	1.538	1.75	1.459	3.24	1.368	4.43	1.330	5.53	1.307	6.56	1.287	7.58	1.271
6.50	0.71	1.539	1.60	1.461	2.97	1.372	4.08	1.332	5.10	1.309	6.05	1.290	6.99	1.274
7.00	0.66	1.540	1.47	1.462	2.74	1.374	3.78	1.334	4.73	1.311	5.61	1.292	6.48	1.276
7.50	0.61	1.541	1.37	1.464	2.54	1.377	3.52	1.335	4.41	1.313	5.23	1.294	6.04	1.279
8.00	0.57	1.542	1.27	1.465	2.37	1.379	3.29	1.336	4.13	1.315	4.90	1.296	5.66	1.280
8.50	0.53	1.542	1.19	1.466	2.22	1.381	3.09	1.337	3.88	1.316	4.61	1.297	5.33	1.282
9.00	0.50	1.543	1.12	1.467	2.09	1.383	2.92	1.338	3.66	1.317	4.35	1.299	5.03	1.284
9.50	0.47	1.543	1.06	1.468	1.97	1.384	2.76	1.339	3.46	1.318	4.12	1.300	4.76	1.285
10.00	0.44	1.544	1.00	1.469	1.87	1.385	2.62	1.340	3.28	1.319	3.91	1.301	4.52	1.286
11.00	0.40	1.544	0.91	1.470	1.69	1.388	2.37	1.341	2.98	1.320	3.55	1.303	4.11	1.288
12.00	0.36	1.545	0.83	1.471	1.54	1.389	2.17	1.342	2.73	1.322	3.25	1.304	3.76	1.290
13.00	0.34	1.545	0.76	1.472	1.42	1.391	2.00	1.343	2.51	1.323	3.00	1.306	3.47	1.291
14.00	0.31	1.546	0.70	1.473	1.31	1.392	1.86	1.344	2.33	1.323	2.78	1.307	3.22	1.292
15.00	0.29	1.546	0.65	1.474	1.22	1.393	1.73	1.345	2.17	1.324	2.60	1.308	3.00	1.293
17.50	0.25	1.547	0.56	1.475	1.04	1.396	1.48	1.346	1.86	1.326	2.22	1.310	2.57	1.295
20.00	0.21	1.547	0.48	1.476	0.91	1.397	1.29	1.347	1.62	1.327	1.94	1.311	2.25	1.297

Table 11.23 Circular columns $- f_{cu} = 40\,\text{N/mm}^2$, $k = 0.70$.

e/R	p=0.4 N/R²	z/R	p=1.0 N/R²	z/R	p=2.0 N/R²	z/R	p=3.0 N/R²	z/R	p=4.0 N/R²	z/R	p=5.0 N/R²	z/R	p=6.0 N/R²	z/R
0.10	54.26	0.698	60.30	0.687	70.45	0.673	80.71	0.662	91.04	0.653	101.44	0.645	111.90	0.639
0.20	47.04	0.796	52.28	0.785	61.02	0.770	69.78	0.759	78.56	0.751	87.37	0.744	96.19	0.739
0.30	39.95	0.890	44.69	0.872	52.41	0.853	60.06	0.840	67.69	0.831	75.30	0.824	82.91	0.818
0.40	33.40	0.974	37.97	0.946	45.06	0.920	51.91	0.903	58.68	0.892	65.39	0.883	72.09	0.877
0.50	27.74	1.046	32.39	1.005	39.05	0.971	45.31	0.951	51.41	0.938	57.43	0.928	63.41	0.921
0.60	22.71	1.113	27.91	1.051	34.23	1.010	40.00	0.987	45.57	0.972	51.04	0.962	56.45	0.954
0.70	18.19	1.175	24.07	1.093	30.37	1.040	35.72	1.015	40.84	0.999	45.84	0.987	50.78	0.979
0.80	14.70	1.222	20.46	1.136	27.24	1.063	32.23	1.037	36.96	1.020	41.57	1.008	46.12	0.999
0.90	12.12	1.256	17.51	1.169	24.40	1.088	29.33	1.054	33.74	1.037	38.01	1.024	42.22	1.015
1.00	10.07	1.287	15.21	1.194	21.72	1.116	26.91	1.068	31.02	1.050	35.00	1.037	38.92	1.028
1.20	7.15	1.334	11.95	1.228	17.51	1.154	22.42	1.106	26.70	1.071	30.21	1.058	33.64	1.048
1.40	5.43	1.363	9.79	1.250	14.57	1.179	18.92	1.134	22.88	1.099	26.56	1.073	29.62	1.063
1.60	4.34	1.381	8.28	1.265	12.45	1.196	16.24	1.153	19.88	1.123	23.21	1.096	26.46	1.074
1.80	3.60	1.393	7.10	1.279	10.86	1.208	14.21	1.167	17.46	1.138	20.54	1.114	23.47	1.093
2.00	3.07	1.402	6.16	1.292	9.62	1.217	12.63	1.178	15.53	1.149	18.39	1.128	21.06	1.108
2.20	2.67	1.408	5.43	1.303	8.63	1.225	11.36	1.186	13.99	1.158	16.57	1.138	19.09	1.120
2.40	2.37	1.413	4.85	1.311	7.83	1.231	10.32	1.193	12.72	1.166	15.08	1.145	17.42	1.129
2.60	2.12	1.417	4.38	1.317	7.16	1.235	9.45	1.198	11.66	1.172	13.83	1.152	15.98	1.136
2.80	1.93	1.420	4.00	1.323	6.60	1.239	8.72	1.203	10.76	1.177	12.77	1.157	14.76	1.142
3.00	1.76	1.423	3.67	1.327	6.12	1.243	8.09	1.206	9.99	1.181	11.86	1.162	13.71	1.146
3.25	1.59	1.425	3.33	1.332	5.61	1.246	7.42	1.210	9.17	1.185	10.89	1.166	12.60	1.151
3.50	1.45	1.428	3.05	1.336	5.18	1.249	6.85	1.214	8.48	1.189	10.07	1.170	11.65	1.156
3.75	1.33	1.430	2.81	1.339	4.81	1.252	6.37	1.217	7.88	1.192	9.36	1.174	10.83	1.159
4.00	1.23	1.431	2.60	1.342	4.48	1.254	5.95	1.219	7.36	1.195	8.74	1.177	10.12	1.162
4.25	1.15	1.432	2.43	1.344	4.20	1.256	5.58	1.221	6.90	1.197	8.21	1.179	9.50	1.165
4.50	1.07	1.434	2.27	1.346	3.95	1.258	5.25	1.223	6.50	1.199	7.73	1.181	8.94	1.167
4.75	1.00	1.435	2.14	1.348	3.73	1.259	4.96	1.225	6.14	1.201	7.30	1.183	8.45	1.169
5.00	0.95	1.436	2.02	1.349	3.54	1.261	4.70	1.226	5.82	1.203	6.92	1.185	8.01	1.171
5.50	0.85	1.437	1.81	1.352	3.20	1.263	4.25	1.229	5.27	1.205	6.27	1.188	7.26	1.174
6.00	0.77	1.438	1.64	1.354	2.92	1.265	3.88	1.231	4.81	1.208	5.73	1.190	6.63	1.177
6.50	0.70	1.439	1.50	1.356	2.69	1.266	3.57	1.233	4.43	1.210	5.27	1.193	6.11	1.179
7.00	0.65	1.440	1.39	1.358	2.49	1.267	3.31	1.234	4.10	1.211	4.89	1.194	5.66	1.181
7.50	0.60	1.441	1.29	1.359	2.32	1.269	3.08	1.235	3.82	1.213	4.55	1.196	5.27	1.182
8.00	0.56	1.442	1.20	1.360	2.16	1.270	2.88	1.237	3.58	1.214	4.26	1.197	4.93	1.184
8.50	0.52	1.442	1.12	1.361	2.03	1.271	2.71	1.238	3.36	1.215	4.00	1.198	4.64	1.185
9.00	0.49	1.443	1.06	1.362	1.91	1.273	2.55	1.238	3.17	1.216	3.77	1.199	4.37	1.186
9.50	0.46	1.443	1.00	1.363	1.80	1.274	2.42	1.239	3.00	1.217	3.57	1.200	4.14	1.187
10.00	0.44	1.443	0.95	1.364	1.71	1.275	2.29	1.240	2.85	1.217	3.39	1.201	3.93	1.188
11.00	0.40	1.444	0.85	1.365	1.55	1.277	2.08	1.241	2.58	1.219	3.08	1.202	3.56	1.189
12.00	0.36	1.445	0.78	1.366	1.41	1.279	1.90	1.242	2.36	1.220	2.81	1.203	3.26	1.190
13.00	0.33	1.445	0.72	1.367	1.30	1.280	1.75	1.243	2.18	1.221	2.59	1.204	3.01	1.191
14.00	0.31	1.445	0.66	1.367	1.20	1.281	1.63	1.243	2.02	1.221	2.41	1.205	2.79	1.192
15.00	0.29	1.446	0.62	1.368	1.12	1.282	1.52	1.244	1.88	1.222	2.24	1.206	2.60	1.193
17.50	0.24	1.446	0.53	1.369	0.96	1.284	1.30	1.245	1.61	1.223	1.92	1.207	2.22	1.195
20.00	0.21	1.447	0.46	1.370	0.83	1.285	1.13	1.246	1.41	1.224	1.68	1.208	1.94	1.196

Table 11.23 (contd) − $f_{cu} = 40\,\text{N/mm}^2$, $k = 0.60$.

e/R	p=0.4 NR²	p=0.4 z/R	p=1.0 NR²	p=1.0 z/R	p=2.0 NR²	p=2.0 z/R	p=3.0 NR²	p=3.0 z/R	p=4.0 NR²	p=4.0 z/R	p=5.0 NR²	p=5.0 z/R	p=6.0 NR²	p=6.0 z/R
0.10	54.14	0.615	59.99	0.609	69.81	0.600	79.70	0.593	89.66	0.588	99.66	0.583	109.71	0.578
0.20	46.80	0.715	51.69	0.710	59.87	0.702	68.08	0.697	76.31	0.693	84.55	0.689	92.80	0.687
0.30	39.56	0.811	43.79	0.801	50.80	0.790	57.77	0.782	64.74	0.776	71.71	0.772	78.67	0.769
0.40	32.82	0.897	36.81	0.878	43.11	0.858	49.26	0.847	55.35	0.839	61.40	0.833	67.44	0.828
0.50	26.91	0.971	31.04	0.938	36.94	0.910	42.51	0.884	47.95	0.884	53.33	0.876	58.67	0.871
0.60	21.86	1.035	26.25	0.984	31.98	0.949	37.19	0.930	42.13	0.917	46.97	0.908	51.77	0.902
0.70	17.39	1.095	22.48	1.019	27.89	0.977	32.74	0.956	37.44	0.942	41.91	0.932	46.26	0.925
0.80	14.02	1.140	18.90	1.059	24.67	0.999	29.11	0.976	33.37	0.961	37.55	0.951	41.68	0.943
0.90	11.56	1.172	16.14	1.089	22.09	1.015	26.18	0.991	30.08	0.976	33.89	0.965	37.66	0.957
1.00	9.59	1.201	14.01	1.112	19.46	1.038	23.78	1.003	27.37	0.987	30.88	0.976	34.34	0.968
1.20	6.86	1.245	11.01	1.142	15.63	1.071	19.78	1.027	23.18	1.005	26.20	0.993	29.17	0.985
1.40	5.24	1.270	9.03	1.162	13.00	1.092	16.56	1.049	19.98	1.019	22.74	1.005	25.35	0.996
1.60	4.22	1.287	7.65	1.175	11.11	1.107	14.22	1.065	17.20	1.036	20.09	1.014	22.41	1.005
1.80	3.52	1.298	6.61	1.185	9.70	1.118	12.45	1.077	15.09	1.049	17.68	1.027	20.08	1.012
2.00	3.01	1.306	5.75	1.197	8.60	1.126	11.06	1.086	13.43	1.058	15.75	1.037	18.04	1.020
2.20	2.63	1.311	5.08	1.206	7.72	1.132	9.96	1.093	12.10	1.066	14.20	1.045	16.27	1.029
2.40	2.34	1.316	4.55	1.213	7.00	1.137	9.05	1.099	11.00	1.072	12.92	1.052	14.82	1.035
2.60	2.10	1.320	4.12	1.219	6.41	1.142	8.29	1.104	10.09	1.077	11.85	1.057	13.60	1.041
2.80	1.91	1.323	3.76	1.223	5.91	1.145	7.65	1.107	9.31	1.081	10.95	1.061	12.56	1.046
3.00	1.75	1.325	3.46	1.227	5.48	1.148	7.10	1.111	8.65	1.085	10.17	1.065	11.67	1.050
3.25	1.58	1.328	3.14	1.232	5.02	1.151	6.51	1.114	7.94	1.089	9.34	1.069	10.72	1.054
3.50	1.44	1.330	2.88	1.235	4.64	1.154	6.02	1.117	7.34	1.092	8.64	1.072	9.92	1.057
3.75	1.33	1.332	2.66	1.238	4.30	1.156	5.59	1.120	6.82	1.094	8.03	1.075	9.22	1.060
4.00	1.23	1.333	2.47	1.240	4.02	1.158	5.22	1.122	6.37	1.097	7.50	1.078	8.62	1.063
4.25	1.15	1.334	2.30	1.242	3.77	1.160	4.90	1.124	5.98	1.099	7.04	1.080	8.09	1.065
4.50	1.07	1.335	2.16	1.244	3.54	1.161	4.61	1.125	5.63	1.101	6.63	1.082	7.62	1.067
4.75	1.01	1.336	2.03	1.246	3.35	1.163	4.36	1.127	5.32	1.102	6.27	1.084	7.20	1.069
5.00	0.95	1.337	1.91	1.247	3.17	1.164	4.13	1.128	5.04	1.104	5.94	1.085	6.83	1.071
5.50	0.85	1.339	1.72	1.250	2.87	1.166	3.74	1.130	4.57	1.106	5.38	1.088	6.19	1.073
6.00	0.77	1.340	1.56	1.252	2.62	1.167	3.41	1.132	4.17	1.108	4.92	1.090	5.65	1.075
6.50	0.71	1.341	1.43	1.254	2.41	1.169	3.14	1.134	3.84	1.110	4.53	1.091	5.21	1.077
7.00	0.65	1.342	1.32	1.255	2.23	1.170	2.91	1.135	3.56	1.111	4.20	1.093	4.82	1.079
7.50	0.60	1.342	1.23	1.256	2.08	1.171	2.71	1.136	3.32	1.112	3.91	1.094	4.49	1.080
8.00	0.56	1.343	1.15	1.257	1.94	1.172	2.53	1.137	3.10	1.113	3.66	1.095	4.21	1.081
8.50	0.53	1.343	1.07	1.258	1.82	1.172	2.38	1.138	2.92	1.114	3.44	1.096	3.95	1.082
9.00	0.49	1.344	1.01	1.259	1.72	1.173	2.25	1.139	2.75	1.115	3.24	1.097	3.73	1.083
9.50	0.47	1.344	0.95	1.260	1.63	1.174	2.12	1.139	2.60	1.116	3.07	1.098	3.53	1.084
10.00	0.44	1.345	0.90	1.260	1.54	1.174	2.02	1.140	2.47	1.116	2.91	1.099	3.35	1.085
11.00	0.40	1.345	0.82	1.261	1.40	1.175	1.83	1.141	2.24	1.117	2.64	1.100	3.04	1.086
12.00	0.36	1.346	0.75	1.262	1.28	1.176	1.67	1.142	2.05	1.118	2.42	1.101	2.78	1.087
13.00	0.33	1.346	0.69	1.263	1.18	1.177	1.54	1.143	1.89	1.119	2.23	1.102	2.57	1.088
14.00	0.31	1.347	0.63	1.264	1.09	1.177	1.43	1.143	1.75	1.120	2.07	1.102	2.38	1.089
15.00	0.29	1.347	0.59	1.264	1.02	1.178	1.33	1.144	1.63	1.120	1.93	1.103	2.22	1.089
17.50	0.25	1.348	0.50	1.265	0.87	1.179	1.14	1.145	1.40	1.121	1.65	1.104	1.90	1.091
20.00	0.21	1.348	0.44	1.266	0.76	1.179	1.00	1.146	1.22	1.122	1.44	1.105	1.66	1.092

Table 11.24 Circular columns − $f_{cu} = 45$ N/mm², $k = 0.90$.

e/R	p=0.4 N/R²	p=0.4 z/R	p=1.0 N/R²	p=1.0 z/R	p=2.0 N/R²	p=2.0 z/R	p=3.0 N/R²	p=3.0 z/R	p=4.0 N/R²	p=4.0 z/R	p=5.0 N/R²	p=5.0 z/R	p=6.0 N/R²	p=6.0 z/R
0.10	60.86	0.865	67.33	0.846	78.15	0.823	89.06	0.805	100.07	0.792	111.21	0.782	122.69	······
0.20	53.05	0.960	59.01	0.938	68.79	0.911	78.51	0.892	88.22	0.877	97.94	0.865	107.68	0.855
0.30	45.43	1.050	51.16	1.020	60.21	0.988	69.04	0.966	77.77	0.950	86.46	0.938	95.12	0.927
0.40	38.41	1.131	44.17	1.091	52.73	1.051	60.86	1.027	68.81	1.009	76.67	0.996	84.49	0.985
0.50	32.22	1.202	38.25	1.148	46.45	1.102	54.00	1.075	61.31	1.056	68.50	1.042	75.61	1.031
0.60	26.24	1.278	33.37	1.194	41.25	1.142	48.32	1.113	55.09	1.093	61.70	1.078	68.23	1.066
0.70	21.25	1.341	28.40	1.251	36.98	1.174	43.61	1.143	49.90	1.122	56.03	1.106	62.06	1.094
0.80	17.33	1.391	24.31	1.297	33.20	1.205	39.68	1.167	45.55	1.145	51.26	1.129	56.86	1.117
0.90	14.37	1.430	21.02	1.334	29.40	1.242	36.36	1.186	41.87	1.164	47.20	1.148	52.43	1.136
1.00	11.96	1.465	18.39	1.363	26.24	1.273	33.08	1.212	38.73	1.180	43.73	1.164	48.63	1.151
1.20	8.43	1.522	14.58	1.404	21.41	1.319	27.39	1.262	33.00	1.218	38.10	1.188	42.45	1.176
1.40	6.29	1.557	12.02	1.431	17.98	1.350	23.23	1.296	28.17	1.255	32.95	1.222	37.62	1.194
1.60	4.95	1.579	10.11	1.454	15.47	1.372	20.12	1.321	24.50	1.283	28.75	1.252	32.90	1.226
1.80	4.06	1.594	8.59	1.479	13.55	1.388	17.71	1.339	21.65	1.303	25.45	1.274	29.19	1.249
2.00	3.43	1.604	7.44	1.498	12.06	1.400	15.82	1.354	19.37	1.319	22.82	1.291	26.20	1.268
2.20	2.96	1.612	6.54	1.512	10.85	1.410	14.28	1.365	17.52	1.331	20.67	1.305	23.76	1.282
2.40	2.61	1.618	5.81	1.522	9.87	1.418	13.01	1.374	15.99	1.341	18.88	1.316	21.72	1.294
2.60	2.33	1.623	5.23	1.530	9.05	1.424	11.95	1.381	14.70	1.350	17.38	1.325	20.00	1.304
2.80	2.10	1.626	4.75	1.537	8.35	1.430	11.05	1.388	13.61	1.357	16.09	1.332	18.53	1.312
3.00	1.91	1.629	4.35	1.543	7.75	1.434	10.27	1.393	12.66	1.363	14.99	1.339	17.27	1.319
3.25	1.72	1.633	3.93	1.549	7.12	1.439	9.44	1.399	11.65	1.369	13.80	1.346	15.90	1.327
3.50	1.56	1.635	3.58	1.554	6.57	1.444	8.73	1.403	10.79	1.374	12.78	1.351	14.74	1.333
3.75	1.43	1.637	3.29	1.558	6.07	1.452	8.13	1.407	10.04	1.379	11.90	1.356	13.73	1.338
4.00	1.32	1.639	3.05	1.561	5.65	1.459	7.60	1.411	9.39	1.382	11.14	1.360	12.86	1.342
4.25	1.23	1.641	2.83	1.564	5.27	1.465	7.13	1.414	8.82	1.386	10.47	1.364	12.08	1.346
4.50	1.14	1.642	2.65	1.567	4.95	1.470	6.72	1.416	8.32	1.389	9.87	1.367	11.40	1.350
4.75	1.07	1.643	2.49	1.569	4.66	1.475	6.36	1.419	7.87	1.391	9.34	1.370	10.79	1.353
5.00	1.01	1.644	2.34	1.571	4.40	1.479	6.03	1.421	7.46	1.394	8.86	1.373	10.24	1.356
5.50	0.90	1.646	2.10	1.574	3.96	1.486	5.46	1.424	6.77	1.398	8.04	1.377	9.29	1.360
6.00	0.82	1.647	1.90	1.577	3.60	1.492	5.00	1.427	6.19	1.401	7.36	1.381	8.50	1.364
6.50	0.74	1.648	1.74	1.579	3.30	1.497	4.60	1.430	5.71	1.404	6.78	1.384	7.84	1.367
7.00	0.68	1.649	1.60	1.581	3.04	1.501	4.27	1.432	5.29	1.406	6.29	1.386	7.27	1.370
7.50	0.63	1.650	1.48	1.583	2.83	1.505	3.97	1.434	4.93	1.408	5.86	1.388	6.78	1.372
8.00	0.59	1.651	1.38	1.584	2.63	1.508	3.72	1.435	4.62	1.410	5.49	1.390	6.35	1.374
8.50	0.55	1.651	1.29	1.585	2.47	1.510	3.50	1.436	4.34	1.411	5.16	1.392	5.97	1.376
9.00	0.52	1.652	1.21	1.586	2.32	1.511	3.30	1.438	4.10	1.413	4.87	1.393	5.64	1.378
9.50	0.49	1.652	1.14	1.587	2.19	1.513	3.12	1.439	3.88	1.414	4.61	1.394	5.34	1.379
10.00	0.46	1.653	1.08	1.588	2.07	1.514	2.96	1.440	3.68	1.415	4.38	1.396	5.07	1.380
11.00	0.42	1.653	0.98	1.589	1.87	1.517	2.69	1.441	3.34	1.417	3.98	1.398	4.60	1.382
12.00	0.38	1.654	0.89	1.590	1.71	1.519	2.46	1.443	3.06	1.418	3.64	1.399	4.22	1.384
13.00	0.35	1.654	0.82	1.591	1.57	1.520	2.27	1.446	2.82	1.419	3.36	1.401	3.89	1.386
14.00	0.32	1.655	0.76	1.592	1.45	1.522	2.10	1.448	2.62	1.420	3.12	1.402	3.61	1.387
15.00	0.30	1.655	0.70	1.593	1.35	1.523	1.96	1.451	2.44	1.421	2.91	1.403	3.37	1.388
17.50	0.25	1.656	0.60	1.594	1.15	1.525	1.67	1.455	2.09	1.423	2.49	1.405	2.88	1.390
20.00	0.22	1.656	0.52	1.595	1.00	1.527	1.46	1.459	1.83	1.425	2.18	1.406	2.52	1.392

Table 11.24 (contd) − $f_{cu} = 45\,\text{N/mm}^2$, $k = 0.80$.

eR	p=0.4 N/R²	p=0.4 z/R	p=1.0 N/R²	p=1.0 z/R	p=2.0 N/R²	p=2.0 z/R	p=3.0 N/R²	p=3.0 z/R	p=4.0 N/R²	p=4.0 z/R	p=5.0 N/R²	p=5.0 z/R	p=6.0 N/R²	p=6.0 z/R
0.10	60.69	0.782	66.93	0.768	77.41	0.749	87.98	0.735	98.64	0.723	109.38	0.714	119.33	......
0.20	52.75	0.879	58.35	0.862	67.62	0.842	76.88	0.827	86.15	0.815	95.43	0.806	104.74	0.798
0.30	44.98	0.971	50.22	0.948	58.65	0.922	66.92	0.905	75.14	0.892	83.32	0.882	91.49	0.874
0.40	37.78	1.054	42.99	1.020	50.88	0.988	58.42	0.967	65.82	0.952	73.15	0.941	80.45	0.932
0.50	31.53	1.125	36.91	1.079	44.43	1.039	51.40	1.015	58.15	0.999	64.80	0.987	71.39	0.977
0.60	25.62	1.199	31.97	1.126	39.17	1.079	45.66	1.053	51.88	1.035	57.96	1.022	63.97	1.012
0.70	20.64	1.262	27.22	1.178	34.90	1.111	40.97	1.082	46.72	1.063	52.33	1.050	57.86	1.039
0.80	16.76	1.312	23.20	1.223	31.35	1.136	37.09	1.105	42.45	1.086	47.65	1.072	52.76	1.061
0.90	13.77	1.349	19.99	1.259	27.67	1.173	33.86	1.124	38.87	1.104	43.71	1.089	48.46	1.078
1.00	11.29	1.384	17.46	1.287	24.63	1.202	30.86	1.145	35.83	1.119	40.36	1.104	44.80	1.093
1.20	7.88	1.436	13.77	1.325	20.03	1.245	25.45	1.191	30.53	1.151	34.98	1.126	38.90	1.115
1.40	5.90	1.467	11.25	1.349	16.81	1.273	21.55	1.223	26.01	1.185	30.31	1.154	34.36	1.131
1.60	4.66	1.486	9.36	1.371	14.44	1.294	18.64	1.246	22.59	1.210	26.40	1.181	30.14	1.157
1.80	3.84	1.499	7.92	1.390	12.65	1.309	16.41	1.263	19.94	1.228	23.36	1.201	26.71	1.178
2.00	3.25	1.508	6.83	1.405	11.23	1.320	14.64	1.276	17.84	1.242	20.93	1.216	23.96	1.195
2.20	2.82	1.515	6.00	1.416	10.07	1.328	13.22	1.286	16.13	1.254	18.95	1.229	21.72	1.208
2.40	2.49	1.520	5.35	1.424	9.13	1.334	12.04	1.294	14.72	1.263	17.31	1.238	19.85	1.218
2.60	2.23	1.524	4.82	1.431	8.35	1.340	11.06	1.301	13.53	1.271	15.93	1.247	18.28	1.227
2.80	2.01	1.528	4.38	1.437	7.69	1.344	10.22	1.307	12.52	1.277	14.75	1.253	16.93	1.234
3.00	1.84	1.530	4.01	1.442	7.12	1.348	9.51	1.312	11.65	1.282	13.73	1.259	15.77	1.241
3.25	1.66	1.533	3.63	1.447	6.53	1.352	8.73	1.316	10.72	1.288	12.64	1.265	14.53	1.247
3.50	1.51	1.535	3.32	1.451	6.02	1.355	8.06	1.320	9.93	1.293	11.71	1.271	13.46	1.253
3.75	1.38	1.537	3.05	1.454	5.57	1.359	7.49	1.323	9.24	1.297	10.91	1.275	12.54	1.257
4.00	1.28	1.539	2.83	1.457	5.17	1.364	6.99	1.326	8.64	1.300	10.21	1.279	11.74	1.262
4.25	1.19	1.540	2.63	1.460	4.82	1.368	6.55	1.328	8.12	1.303	9.59	1.282	11.04	1.265
4.50	1.11	1.541	2.46	1.462	4.52	1.372	6.17	1.331	7.65	1.306	9.05	1.285	10.41	1.268
4.75	1.04	1.542	2.31	1.464	4.25	1.375	5.83	1.332	7.24	1.308	8.56	1.288	9.85	1.271
5.00	0.98	1.543	2.18	1.466	4.02	1.378	5.52	1.334	6.87	1.310	8.12	1.290	9.35	1.273
5.50	0.88	1.545	1.96	1.469	3.61	1.382	5.00	1.337	6.23	1.314	7.37	1.294	8.48	1.278
6.00	0.79	1.546	1.77	1.471	3.28	1.386	4.56	1.339	5.69	1.317	6.74	1.297	7.77	1.281
6.50	0.72	1.547	1.62	1.473	3.01	1.389	4.20	1.341	5.24	1.319	6.21	1.300	7.16	1.284
7.00	0.67	1.548	1.49	1.474	2.77	1.392	3.89	1.343	4.85	1.321	5.76	1.302	6.64	1.286
7.50	0.62	1.549	1.38	1.476	2.57	1.394	3.62	1.344	4.52	1.322	5.37	1.304	6.19	1.288
8.00	0.57	1.549	1.29	1.477	2.40	1.396	3.39	1.345	4.23	1.323	5.03	1.306	5.80	1.290
8.50	0.54	1.550	1.21	1.478	2.25	1.398	3.18	1.346	3.97	1.325	4.73	1.307	5.45	1.292
9.00	0.50	1.550	1.14	1.479	2.12	1.398	3.00	1.347	3.75	1.326	4.47	1.309	5.15	1.293
9.50	0.48	1.551	1.07	1.480	2.00	1.400	2.84	1.348	3.54	1.327	4.23	1.310	4.87	1.295
10.00	0.45	1.551	1.01	1.480	1.89	1.402	2.69	1.349	3.36	1.327	4.01	1.311	4.63	1.296
11.00	0.41	1.552	0.92	1.482	1.71	1.404	2.44	1.350	3.05	1.329	3.65	1.313	4.20	1.298
12.00	0.37	1.552	0.84	1.483	1.56	1.405	2.23	1.351	2.79	1.330	3.34	1.314	3.85	1.299
13.00	0.34	1.553	0.77	1.484	1.44	1.407	2.06	1.352	2.57	1.331	3.08	1.315	3.55	1.301
14.00	0.31	1.553	0.71	1.484	1.33	1.408	1.91	1.353	2.39	1.332	2.85	1.316	3.30	1.302
15.00	0.29	1.553	0.66	1.485	1.24	1.409	1.78	1.353	2.22	1.333	2.66	1.317	3.08	1.303
17.50	0.25	1.554	0.56	1.486	1.05	1.411	1.52	1.355	1.90	1.334	2.28	1.318	2.63	1.305
20.00	0.22	1.554	0.49	1.487	0.92	1.413	1.33	1.356	1.66	1.335	1.99	1.320	2.30	1.306

Table 11.25 Circular columns – $f_{cu} = 45\,\text{N/mm}^2$, $k = 0.70$.

e/R	p=0.4 N/R²	z/R	p=1.0 N/R²	z/R	p=2.0 N/R²	z/R	p=3.0 N/R²	z/R	p=4.0 N/R²	z/R	p=5.0 N/R²	z/R	p=6.0 N/R²	z/R
0.10	60.55	0.699	66.58	0.689	76.71	0.676	86.94	0.665	97.24	0.656	107.61	0.649	117.82	******
0.20	52.48	0.797	57.72	0.787	66.46	0.773	75.21	0.763	83.99	0.754	92.78	0.748	101.59	0.742
0.30	44.55	0.891	49.29	0.875	57.04	0.857	64.71	0.844	72.34	0.835	79.96	0.827	87.56	0.821
0.40	37.17	0.977	41.80	0.950	48.95	0.924	55.85	0.908	62.64	0.896	69.38	0.888	76.08	0.881
0.50	30.76	1.050	35.54	1.011	42.32	0.977	48.65	0.957	54.80	0.943	60.85	0.933	66.86	0.925
0.60	24.99	1.120	30.52	1.058	37.01	1.017	42.87	0.994	48.50	0.978	54.01	0.967	59.45	0.959
0.70	19.87	1.183	26.01	1.105	32.76	1.047	38.22	1.022	43.41	1.005	48.46	0.993	53.43	0.984
0.80	15.95	1.231	21.94	1.149	29.33	1.071	34.43	1.044	39.24	1.027	43.90	1.014	48.48	1.004
0.90	13.07	1.266	18.70	1.182	25.89	1.103	31.30	1.062	35.77	1.044	40.10	1.030	44.35	1.021
1.00	10.67	1.301	16.19	1.206	22.95	1.130	28.58	1.078	32.86	1.057	36.90	1.044	40.86	1.034
1.20	7.47	1.348	12.66	1.240	18.38	1.167	23.49	1.121	28.02	1.083	31.80	1.065	35.28	1.054
1.40	5.63	1.375	10.35	1.261	15.26	1.191	19.68	1.147	23.82	1.114	27.63	1.086	31.03	1.069
1.60	4.48	1.393	8.64	1.280	13.02	1.208	16.87	1.166	20.57	1.135	24.04	1.110	27.34	1.088
1.80	3.71	1.404	7.32	1.297	11.34	1.220	14.75	1.179	18.03	1.150	21.24	1.127	24.21	1.107
2.00	3.16	1.412	6.34	1.309	10.04	1.229	13.10	1.189	16.03	1.161	18.91	1.139	21.71	1.121
2.20	2.75	1.418	5.58	1.319	9.01	1.236	11.77	1.197	14.43	1.170	17.03	1.149	19.61	1.132
2.40	2.43	1.423	4.98	1.327	8.16	1.242	10.69	1.204	13.12	1.177	15.49	1.156	17.84	1.140
2.60	2.18	1.427	4.49	1.333	7.46	1.246	9.79	1.209	12.02	1.182	14.21	1.162	16.37	1.146
2.80	1.97	1.430	4.09	1.338	6.88	1.250	9.03	1.213	11.09	1.187	13.12	1.167	15.12	1.152
3.00	1.80	1.432	3.76	1.342	6.37	1.253	8.37	1.217	10.30	1.191	12.18	1.172	14.04	1.156
3.25	1.63	1.435	3.41	1.346	5.84	1.257	7.68	1.221	9.45	1.195	11.18	1.176	12.89	1.161
3.50	1.48	1.437	3.11	1.350	5.39	1.260	7.09	1.224	8.73	1.199	10.33	1.180	11.92	1.165
3.75	1.36	1.439	2.87	1.353	5.00	1.262	6.59	1.227	8.11	1.202	9.61	1.183	11.08	1.168
4.00	1.26	1.440	2.66	1.355	4.66	1.264	6.15	1.229	7.58	1.205	8.97	1.186	10.35	1.171
4.25	1.17	1.441	2.48	1.358	4.37	1.266	5.77	1.231	7.11	1.207	8.42	1.189	9.72	1.174
4.50	1.09	1.442	2.32	1.360	4.11	1.268	5.43	1.233	6.69	1.209	7.93	1.191	9.15	1.176
4.75	1.03	1.443	2.18	1.361	3.88	1.269	5.13	1.235	6.32	1.211	7.49	1.193	8.65	1.178
5.00	0.97	1.444	2.06	1.363	3.67	1.272	4.86	1.236	5.99	1.212	7.10	1.194	8.20	1.180
5.50	0.87	1.446	1.85	1.365	3.30	1.276	4.40	1.239	5.42	1.215	6.43	1.197	7.43	1.183
6.00	0.79	1.447	1.68	1.367	3.00	1.279	4.01	1.241	4.95	1.217	5.88	1.200	6.79	1.186
6.50	0.72	1.448	1.53	1.369	2.75	1.282	3.69	1.242	4.56	1.219	5.41	1.201	6.25	1.188
7.00	0.66	1.448	1.41	1.371	2.54	1.284	3.42	1.244	4.22	1.221	5.01	1.203	5.79	1.189
7.50	0.61	1.449	1.31	1.372	2.36	1.286	3.18	1.245	3.93	1.222	4.67	1.205	5.39	1.191
8.00	0.57	1.450	1.22	1.373	2.20	1.287	2.98	1.246	3.68	1.223	4.37	1.206	5.05	1.192
8.50	0.53	1.450	1.15	1.374	2.06	1.289	2.80	1.247	3.46	1.224	4.10	1.207	4.74	1.193
9.00	0.50	1.451	1.08	1.375	1.94	1.290	2.64	1.248	3.26	1.225	3.87	1.208	4.47	1.194
9.50	0.47	1.451	1.02	1.375	1.83	1.291	2.50	1.249	3.08	1.226	3.66	1.209	4.23	1.195
10.00	0.45	1.451	0.96	1.376	1.74	1.292	2.37	1.249	2.93	1.227	3.47	1.210	4.02	1.196
11.00	0.40	1.452	0.87	1.377	1.57	1.294	2.15	1.250	2.66	1.228	3.15	1.211	3.64	1.198
12.00	0.37	1.452	0.79	1.378	1.44	1.296	1.97	1.251	2.43	1.229	2.88	1.212	3.33	1.199
13.00	0.34	1.453	0.73	1.379	1.32	1.297	1.81	1.252	2.24	1.230	2.66	1.213	3.07	1.200
14.00	0.31	1.453	0.67	1.379	1.22	1.298	1.68	1.253	2.08	1.230	2.47	1.214	2.85	1.200
15.00	0.29	1.454	0.63	1.380	1.14	1.299	1.57	1.253	1.94	1.231	2.30	1.214	2.66	1.201
17.50	0.25	1.454	0.53	1.381	0.97	1.301	1.34	1.254	1.66	1.232	1.97	1.216	2.27	1.203
20.00	0.22	1.455	0.47	1.382	0.85	1.302	1.17	1.255	1.45	1.233	1.72	1.217	1.99	1.204

Table 11.25 (contd) − $f_{cu} = 45\,N/mm^2$, $k = 0.60$.

e/R	p=0.4		p=1.0		p=2.0		p=3.0		p=4.0		p=5.0		p=6.0	
	N/R^2	z/R	N/R^2	z/R	N/R^2	z/R	N/R^2	z/R	N/R^2	z/R	N/R^2	z/R	N/R^2	z/R
0.10	60.42	0.615	66.26	0.610	76.07	0.602	85.94	0.596	95.88	0.590	105.86	0.585	115.88	0.581
0.20	52.24	0.716	57.13	0.711	65.31	0.704	73.51	0.699	81.73	0.695	89.96	0.691	98.21	0.688
0.30	44.14	0.811	48.39	0.802	55.40	0.792	62.38	0.784	69.35	0.779	76.32	0.774	83.28	0.771
0.40	36.58	0.899	40.60	0.881	46.95	0.862	53.13	0.850	59.23	0.842	65.30	0.836	71.34	0.831
0.50	29.89	0.975	34.11	0.943	40.13	0.915	45.75	0.899	51.23	0.888	56.64	0.880	62.00	0.874
0.60	24.10	1.041	28.75	0.990	34.60	0.955	39.96	0.935	44.94	0.922	49.82	0.913	54.65	0.906
0.70	19.05	1.103	24.33	1.030	30.11	0.984	35.04	0.962	39.78	0.948	44.40	0.938	48.78	0.930
0.80	15.26	1.149	20.36	1.071	26.59	1.006	31.12	0.982	35.43	0.967	39.64	0.956	43.80	0.948
0.90	12.51	1.182	17.33	1.101	23.44	1.028	27.95	0.998	31.91	0.982	35.75	0.970	39.55	0.962
1.00	10.20	1.214	14.99	1.123	20.61	1.051	25.36	1.010	29.01	0.994	32.55	0.982	36.04	0.973
1.20	7.20	1.258	11.73	1.154	16.49	1.083	20.72	1.040	24.54	1.011	27.59	0.999	30.59	0.990
1.40	5.47	1.283	9.60	1.173	13.70	1.104	17.32	1.062	20.78	1.031	23.93	1.011	26.57	1.002
1.60	4.38	1.298	8.07	1.188	11.69	1.119	14.86	1.078	17.87	1.048	20.82	1.025	23.47	1.011
1.80	3.65	1.309	6.86	1.203	10.19	1.129	12.99	1.089	15.66	1.060	18.27	1.038	20.85	1.021
2.00	3.12	1.316	5.96	1.214	9.03	1.137	11.54	1.098	13.93	1.070	16.27	1.048	18.58	1.031
2.20	2.72	1.322	5.26	1.222	8.10	1.143	10.38	1.105	12.55	1.077	14.66	1.056	16.75	1.039
2.40	2.42	1.326	4.70	1.229	7.34	1.148	9.43	1.110	11.41	1.083	13.34	1.062	15.25	1.046
2.60	2.17	1.330	4.25	1.234	6.72	1.153	8.63	1.115	10.46	1.088	12.24	1.067	13.99	1.051
2.80	1.97	1.333	3.88	1.239	6.19	1.156	7.96	1.118	9.65	1.092	11.30	1.072	12.92	1.056
3.00	1.80	1.335	3.56	1.242	5.74	1.159	7.39	1.122	8.96	1.095	10.49	1.075	12.01	1.060
3.25	1.63	1.337	3.24	1.246	5.26	1.162	6.78	1.125	8.23	1.099	9.64	1.079	11.03	1.064
3.50	1.49	1.339	2.96	1.250	4.85	1.164	6.26	1.128	7.60	1.102	8.91	1.083	10.20	1.067
3.75	1.37	1.341	2.73	1.252	4.50	1.167	5.82	1.130	7.06	1.105	8.28	1.085	9.48	1.070
4.00	1.27	1.342	2.54	1.255	4.20	1.169	5.43	1.132	6.60	1.107	7.74	1.088	8.86	1.073
4.25	1.18	1.343	2.36	1.257	3.94	1.170	5.09	1.134	6.19	1.109	7.26	1.090	8.32	1.075
4.50	1.10	1.345	2.22	1.258	3.71	1.172	4.79	1.136	5.83	1.111	6.84	1.092	7.83	1.077
4.75	1.04	1.345	2.08	1.260	3.50	1.173	4.53	1.137	5.51	1.112	6.46	1.093	7.40	1.078
5.00	0.98	1.346	1.97	1.260	3.32	1.174	4.29	1.138	5.22	1.113	6.13	1.095	7.02	1.080
5.50	0.88	1.348	1.77	1.261	3.00	1.176	3.88	1.141	4.73	1.116	5.55	1.097	6.36	1.082
6.00	0.79	1.349	1.61	1.264	2.74	1.178	3.55	1.142	4.32	1.118	5.07	1.099	5.81	1.085
6.50	0.73	1.350	1.47	1.265	2.52	1.179	3.26	1.144	3.97	1.119	4.67	1.101	5.35	1.086
7.00	0.67	1.350	1.36	1.267	2.33	1.180	3.02	1.145	3.68	1.121	4.32	1.102	4.96	1.088
7.50	0.62	1.351	1.26	1.268	2.17	1.181	2.82	1.146	3.43	1.122	4.03	1.104	4.62	1.089
8.00	0.58	1.352	1.17	1.269	2.03	1.182	2.63	1.147	3.21	1.123	3.77	1.105	4.32	1.090
8.50	0.54	1.352	1.10	1.270	1.91	1.183	2.47	1.148	3.02	1.124	3.54	1.106	4.06	1.091
9.00	0.51	1.352	1.04	1.271	1.80	1.184	2.33	1.149	2.84	1.125	3.34	1.106	3.83	1.092
9.50	0.48	1.353	0.98	1.272	1.70	1.185	2.21	1.149	2.69	1.125	3.16	1.107	3.63	1.093
10.00	0.45	1.353	0.93	1.273	1.61	1.186	2.09	1.150	2.55	1.126	3.00	1.108	3.44	1.094
11.00	0.41	1.354	0.84	1.273	1.45	1.187	1.90	1.151	2.32	1.127	2.72	1.109	3.12	1.095
12.00	0.37	1.354	0.76	1.274	1.33	1.188	1.74	1.152	2.12	1.128	2.49	1.110	2.86	1.096
13.00	0.34	1.355	0.70	1.275	1.22	1.189	1.60	1.152	1.95	1.129	2.30	1.111	2.63	1.097
14.00	0.32	1.355	0.65	1.276	1.13	1.190	1.49	1.153	1.81	1.129	2.13	1.111	2.44	1.097
15.00	0.30	1.355	0.60	1.276	1.05	1.191	1.38	1.153	1.69	1.130	1.99	1.112	2.28	1.098
17.50	0.25	1.356	0.52	1.277	0.90	1.193	1.18	1.154	1.44	1.131	1.70	1.113	1.95	1.099
20.00	0.22	1.357	0.45	1.278	0.78	1.194	1.03	1.155	1.26	1.132	1.48	1.114	1.70	1.100

Table 11.26 Circular columns $- f_{cu} = 50\,\text{N/mm}^2$, $k = 0.90$.

e/R	p=0.4 NR²	z/R	p=1.0 NR²	z/R	p=2.0 NR²	z/R	p=3.0 NR²	z/R	p=4.0 NR²	z/R	p=5.0 NR²	z/R	p=6.0 NR²	z/R
0.10	67.15	0.866	73.62	0.849	84.43	0.827	95.31	0.810	106.28	0.797	117.35	0.786	128.57	*****
0.20	58.50	0.962	64.47	0.941	74.27	0.916	84.00	0.897	93.71	0.882	103.42	0.870	113.15	0.861
0.30	50.04	1.052	55.81	1.024	64.92	0.993	73.78	0.972	82.54	0.956	91.24	0.943	99.92	0.933
0.40	42.21	1.135	48.07	1.096	56.75	1.058	64.94	1.033	72.94	1.016	80.83	1.002	88.67	0.991
0.50	35.22	1.209	41.50	1.155	49.87	1.109	57.52	1.082	64.90	1.063	72.13	1.048	79.28	1.037
0.60	28.53	1.285	35.88	1.206	44.19	1.150	51.38	1.120	58.23	1.100	64.90	1.085	71.47	1.073
0.70	22.96	1.349	30.38	1.263	39.53	1.183	46.30	1.151	52.68	1.129	58.87	1.113	64.94	1.101
0.80	18.61	1.401	25.88	1.310	35.05	1.220	42.06	1.175	48.03	1.153	53.80	1.137	59.45	1.124
0.90	15.34	1.439	22.28	1.346	30.93	1.258	38.43	1.196	44.11	1.172	49.50	1.156	54.78	1.143
1.00	12.56	1.478	19.43	1.375	27.53	1.288	34.54	1.229	40.76	1.188	45.83	1.171	50.78	1.159
1.20	8.71	1.535	15.33	1.416	22.37	1.333	28.49	1.277	34.21	1.234	39.71	1.199	44.27	1.183
1.40	6.45	1.568	12.60	1.442	18.74	1.363	24.11	1.310	29.13	1.270	33.97	1.238	38.70	1.210
1.60	5.05	1.590	10.41	1.473	16.09	1.384	20.84	1.334	25.30	1.297	29.60	1.266	33.80	1.241
1.80	4.12	1.604	8.81	1.497	14.08	1.400	18.33	1.352	22.33	1.316	26.18	1.288	29.95	1.264
2.00	3.48	1.613	7.59	1.514	12.51	1.412	16.35	1.366	19.96	1.332	23.45	1.304	26.86	1.281
2.20	3.00	1.621	6.64	1.526	11.26	1.421	14.75	1.377	18.04	1.344	21.23	1.317	24.34	1.295
2.40	2.64	1.626	5.90	1.536	10.23	1.429	13.43	1.386	16.46	1.353	19.38	1.328	22.25	1.307
2.60	2.35	1.631	5.30	1.544	9.37	1.435	12.33	1.393	15.13	1.361	17.83	1.336	20.48	1.316
2.80	2.12	1.634	4.81	1.550	8.65	1.440	11.40	1.399	13.99	1.368	16.51	1.344	18.97	1.324
3.00	1.93	1.637	4.40	1.555	8.00	1.447	10.59	1.404	13.02	1.374	15.37	1.350	17.67	1.330
3.25	1.74	1.640	3.97	1.561	7.29	1.458	9.73	1.409	11.97	1.380	14.14	1.357	16.27	1.338
3.50	1.58	1.642	3.62	1.565	6.69	1.466	9.00	1.414	11.08	1.385	13.10	1.362	15.07	1.343
3.75	1.45	1.644	3.33	1.569	6.18	1.474	8.37	1.418	10.32	1.389	12.20	1.367	14.04	1.348
4.00	1.33	1.646	3.08	1.572	5.74	1.480	7.83	1.421	9.65	1.393	11.41	1.371	13.14	1.353
4.25	1.24	1.647	2.86	1.575	5.36	1.486	7.35	1.424	9.06	1.396	10.72	1.374	12.35	1.357
4.50	1.15	1.648	2.67	1.577	5.02	1.491	6.92	1.426	8.54	1.399	10.11	1.377	11.65	1.360
4.75	1.08	1.650	2.51	1.579	4.72	1.495	6.54	1.429	8.08	1.401	9.56	1.380	11.02	1.363
5.00	1.02	1.650	2.36	1.581	4.46	1.499	6.21	1.431	7.66	1.404	9.07	1.383	10.46	1.365
5.50	0.91	1.652	2.12	1.584	4.01	1.506	5.62	1.434	6.95	1.407	8.23	1.387	9.49	1.370
6.00	0.82	1.653	1.92	1.587	3.64	1.510	5.14	1.437	6.35	1.411	7.53	1.390	8.69	1.374
6.50	0.75	1.654	1.75	1.589	3.33	1.514	4.73	1.439	5.85	1.413	6.94	1.393	8.01	1.377
7.00	0.69	1.655	1.61	1.590	3.07	1.516	4.39	1.441	5.43	1.415	6.44	1.395	7.43	1.379
7.50	0.64	1.656	1.49	1.592	2.85	1.519	4.09	1.443	5.06	1.417	6.00	1.397	6.92	1.381
8.00	0.59	1.657	1.39	1.593	2.65	1.521	3.82	1.447	4.74	1.419	5.62	1.399	6.49	1.383
8.50	0.56	1.657	1.30	1.594	2.48	1.523	3.58	1.451	4.45	1.420	5.28	1.401	6.10	1.385
9.00	0.52	1.658	1.22	1.595	2.34	1.525	3.37	1.454	4.20	1.422	4.99	1.402	5.76	1.387
9.50	0.49	1.658	1.15	1.596	2.20	1.526	3.18	1.456	3.98	1.423	4.72	1.404	5.45	1.388
10.00	0.47	1.658	1.09	1.597	2.09	1.527	3.02	1.459	3.78	1.424	4.48	1.405	5.17	1.389
11.00	0.42	1.659	0.98	1.598	1.89	1.530	2.73	1.463	3.43	1.426	4.07	1.407	4.70	1.391
12.00	0.38	1.660	0.90	1.599	1.72	1.531	2.49	1.466	3.14	1.427	3.73	1.408	4.30	1.393
13.00	0.35	1.660	0.82	1.600	1.58	1.533	2.30	1.469	2.89	1.428	3.44	1.409	3.97	1.394
14.00	0.32	1.660	0.76	1.601	1.46	1.534	2.13	1.472	2.69	1.429	3.19	1.411	3.68	1.395
15.00	0.30	1.661	0.71	1.602	1.36	1.535	1.98	1.474	2.50	1.430	2.98	1.412	3.44	1.397
17.50	0.26	1.661	0.60	1.603	1.16	1.537	1.69	1.478	2.14	1.432	2.55	1.413	2.94	1.399
20.00	0.22	1.662	0.52	1.604	1.01	1.539	1.47	1.481	1.87	1.433	2.23	1.415	2.57	1.400

Table 11.26 (contd) − $f_{cu} = 50\,\text{N/mm}^2$, $k = 0.80$.

e/R	p=0.4 N/R²	z/R	p=1.0 N/R²	z/R	p=2.0 N/R²	z/R	p=3.0 N/R²	z/R	p=4.0 N/R²	z/R	p=5.0 N/R²	z/R	p=6.0 N/R²	z/R
0.10	66.98	0.783	73.21	0.770	83.67	0.753	94.22	0.739	104.85	0.728	115.56	0.718	126.33	0.711
0.20	58.20	0.880	63.80	0.865	73.08	0.846	82.34	0.831	91.60	0.819	100.88	0.810	110.17	0.802
0.30	49.58	0.973	54.85	0.951	63.31	0.926	71.61	0.909	79.84	0.897	88.03	0.886	96.21	0.878
0.40	41.56	1.057	46.85	1.025	54.82	0.993	62.42	0.972	69.86	0.958	77.22	0.946	84.53	0.937
0.50	34.57	1.130	40.10	1.085	47.76	1.045	54.82	1.022	61.63	1.005	68.32	0.993	74.93	0.983
0.60	27.90	1.206	34.62	1.133	42.02	1.086	48.61	1.060	54.90	1.041	61.03	1.028	67.08	1.018
0.70	22.33	1.270	29.17	1.189	37.36	1.118	43.54	1.089	49.38	1.070	55.05	1.056	60.61	1.045
0.80	17.99	1.321	24.73	1.235	33.15	1.151	39.37	1.113	44.82	1.093	50.07	1.078	55.23	1.067
0.90	14.67	1.359	21.23	1.271	29.16	1.187	35.90	1.132	40.99	1.111	45.90	1.096	50.69	1.085
1.00	11.85	1.397	18.48	1.298	25.89	1.216	32.27	1.160	37.76	1.126	42.35	1.111	46.83	1.099
1.20	8.17	1.448	14.46	1.336	20.98	1.258	26.53	1.205	31.70	1.166	36.65	1.133	40.62	1.122
1.40	6.06	1.478	11.74	1.361	17.55	1.286	22.41	1.236	26.94	1.199	31.30	1.169	35.55	1.143
1.60	4.77	1.497	9.62	1.387	15.06	1.306	19.35	1.258	23.37	1.223	27.23	1.195	31.01	1.171
1.80	3.92	1.509	8.10	1.405	13.15	1.320	17.01	1.275	20.61	1.241	24.07	1.214	27.45	1.191
2.00	3.32	1.517	6.98	1.419	11.63	1.330	15.17	1.287	18.42	1.255	21.55	1.229	24.61	1.207
2.20	2.87	1.524	6.12	1.430	10.43	1.338	13.68	1.297	16.65	1.266	19.50	1.240	22.29	1.220
2.40	2.53	1.529	5.44	1.438	9.45	1.344	12.46	1.305	15.18	1.274	17.81	1.250	20.37	1.230
2.60	2.26	1.532	4.90	1.444	8.63	1.349	11.44	1.312	13.95	1.282	16.38	1.258	18.75	1.238
2.80	2.05	1.535	4.45	1.450	7.95	1.353	10.56	1.317	12.91	1.288	15.16	1.265	17.37	1.245
3.00	1.87	1.538	4.08	1.454	7.35	1.358	9.79	1.321	12.00	1.293	14.11	1.270	16.17	1.251
3.25	1.68	1.541	3.69	1.459	6.68	1.365	8.98	1.326	11.04	1.298	12.99	1.276	14.89	1.258
3.50	1.53	1.543	3.37	1.463	6.12	1.371	8.29	1.329	10.22	1.303	12.03	1.281	13.80	1.263
3.75	1.40	1.544	3.10	1.466	5.65	1.376	7.70	1.332	9.51	1.307	11.20	1.285	12.85	1.268
4.00	1.30	1.546	2.87	1.469	5.24	1.381	7.18	1.335	8.90	1.310	10.48	1.289	12.03	1.272
4.25	1.20	1.547	2.67	1.471	4.89	1.384	6.73	1.337	8.35	1.313	9.84	1.292	11.30	1.275
4.50	1.12	1.548	2.49	1.473	4.58	1.388	6.34	1.339	7.87	1.316	9.28	1.295	10.66	1.278
4.75	1.05	1.549	2.34	1.475	4.31	1.391	5.99	1.341	7.44	1.318	8.78	1.297	10.09	1.281
5.00	0.99	1.550	2.21	1.477	4.07	1.393	5.67	1.343	7.05	1.319	8.33	1.300	9.57	1.283
5.50	0.89	1.552	1.98	1.479	3.66	1.397	5.13	1.345	6.38	1.322	7.56	1.303	8.68	1.287
6.00	0.80	1.553	1.79	1.481	3.32	1.401	4.68	1.348	5.82	1.325	6.91	1.307	7.95	1.290
6.50	0.73	1.554	1.64	1.483	3.04	1.404	4.31	1.349	5.36	1.327	6.37	1.309	7.33	1.293
7.00	0.67	1.554	1.51	1.485	2.80	1.406	3.99	1.351	4.96	1.329	5.91	1.311	6.79	1.295
7.50	0.62	1.555	1.40	1.486	2.60	1.408	3.72	1.352	4.62	1.330	5.51	1.313	6.34	1.297
8.00	0.58	1.556	1.31	1.487	2.43	1.410	3.48	1.353	4.32	1.331	5.16	1.315	5.93	1.299
8.50	0.54	1.556	1.22	1.488	2.27	1.412	3.26	1.354	4.06	1.333	4.85	1.316	5.58	1.301
9.00	0.51	1.557	1.15	1.489	2.14	1.413	3.08	1.355	3.83	1.333	4.57	1.317	5.27	1.302
9.50	0.48	1.557	1.08	1.490	2.02	1.414	2.91	1.356	3.62	1.334	4.32	1.318	4.99	1.303
10.00	0.46	1.557	1.03	1.490	1.91	1.416	2.76	1.358	3.44	1.335	4.10	1.319	4.73	1.304
11.00	0.41	1.558	0.93	1.491	1.73	1.417	2.49	1.360	3.12	1.337	3.72	1.320	4.30	1.306
12.00	0.38	1.558	0.84	1.492	1.58	1.419	2.28	1.363	2.85	1.338	3.41	1.322	3.94	1.308
13.00	0.34	1.559	0.78	1.493	1.45	1.420	2.09	1.364	2.63	1.339	3.14	1.323	3.63	1.309
14.00	0.32	1.559	0.72	1.494	1.34	1.421	1.94	1.366	2.44	1.339	2.91	1.323	3.37	1.310
15.00	0.30	1.559	0.67	1.494	1.25	1.422	1.80	1.367	2.27	1.340	2.72	1.324	3.14	1.311
17.50	0.25	1.560	0.57	1.495	1.06	1.424	1.54	1.370	1.95	1.341	2.32	1.326	2.69	1.313
20.00	0.22	1.560	0.49	1.496	0.93	1.426	1.34	1.371	1.70	1.342	2.03	1.327	2.35	1.314

Table 11.27 Circular columns − $f_{cu} = 50\,\text{N/mm}^2$, $k = 0.70$.

e/R	p=0.4 N/R²	p=0.4 z/R	p=1.0 N/R²	p=1.0 z/R	p=2.0 N/R²	p=2.0 z/R	p=3.0 N/R²	p=3.0 z/R	p=4.0 N/R²	p=4.0 z/R	p=5.0 N/R²	p=5.0 z/R	p=6.0 N/R²	p=6.0 z/R
0.10	66.83	0.700	72.85	0.691	82.97	0.678	93.18	0.668	103.46	0.660	113.80	0.653	124.20	0.646
0.20	57.93	0.798	63.16	0.788	71.90	0.775	80.65	0.765	89.41	0.757	98.20	0.751	107.00	0.745
0.30	49.14	0.893	53.90	0.877	61.67	0.860	69.35	0.847	76.99	0.838	84.61	0.831	92.22	0.825
0.40	40.94	0.979	45.62	0.954	52.83	0.929	59.77	0.912	66.59	0.901	73.35	0.892	80.07	0.885
0.50	33.76	1.054	38.67	1.016	45.58	0.982	51.98	0.962	58.17	0.948	64.26	0.938	70.29	0.930
0.60	27.24	1.127	33.09	1.064	39.76	1.023	45.72	1.000	51.41	0.984	56.96	0.972	62.43	0.964
0.70	21.52	1.191	27.92	1.116	35.12	1.054	40.69	1.028	45.95	1.011	51.05	0.999	56.06	0.989
0.80	17.16	1.240	23.39	1.159	31.20	1.082	36.60	1.051	41.48	1.033	46.20	1.020	50.82	1.010
0.90	13.92	1.277	19.87	1.192	27.34	1.116	33.23	1.069	37.79	1.050	42.17	1.037	46.46	1.026
1.00	11.23	1.312	17.15	1.217	24.09	1.143	29.95	1.092	34.68	1.064	38.77	1.050	42.77	1.040
1.20	7.78	1.359	13.35	1.251	19.23	1.179	24.47	1.133	29.15	1.097	33.38	1.071	36.89	1.060
1.40	5.82	1.386	10.85	1.273	15.93	1.202	20.42	1.159	24.71	1.127	28.60	1.099	32.37	1.076
1.60	4.61	1.403	8.91	1.296	13.57	1.218	17.49	1.177	21.23	1.146	24.85	1.123	28.19	1.101
1.80	3.81	1.414	7.53	1.312	11.81	1.230	15.28	1.190	18.59	1.161	21.82	1.138	24.94	1.119
2.00	3.24	1.422	6.50	1.324	10.45	1.239	13.55	1.200	16.52	1.171	19.42	1.149	22.28	1.132
2.20	2.81	1.427	5.72	1.333	9.36	1.246	12.18	1.207	14.86	1.180	17.48	1.158	20.07	1.141
2.40	2.49	1.432	5.09	1.346	8.48	1.251	11.05	1.214	13.50	1.186	15.90	1.166	18.26	1.149
2.60	2.23	1.435	4.59	1.346	7.75	1.256	10.12	1.219	12.37	1.192	14.57	1.172	16.75	1.155
2.80	2.02	1.438	4.18	1.351	7.14	1.260	9.33	1.223	11.41	1.197	13.45	1.177	15.46	1.161
3.00	1.84	1.440	3.83	1.354	6.62	1.263	8.65	1.226	10.59	1.201	12.49	1.181	14.36	1.165
3.25	1.66	1.443	3.47	1.359	6.06	1.266	7.93	1.230	9.72	1.205	11.46	1.185	13.19	1.170
3.50	1.51	1.445	3.18	1.362	5.59	1.269	7.32	1.233	8.98	1.208	10.59	1.189	12.19	1.174
3.75	1.39	1.446	2.92	1.365	5.16	1.273	6.80	1.236	8.34	1.211	9.85	1.192	11.33	1.177
4.00	1.29	1.448	2.71	1.367	4.80	1.277	6.35	1.238	7.79	1.214	9.20	1.195	10.59	1.180
4.25	1.20	1.449	2.52	1.369	4.48	1.280	5.95	1.240	7.30	1.216	8.63	1.197	9.93	1.182
4.50	1.12	1.450	2.36	1.371	4.20	1.283	5.60	1.242	6.88	1.218	8.12	1.199	9.35	1.185
4.75	1.05	1.451	2.22	1.373	3.95	1.285	5.29	1.244	6.50	1.219	7.68	1.201	8.84	1.186
5.00	0.99	1.451	2.09	1.374	3.73	1.288	5.01	1.245	6.16	1.221	7.28	1.203	8.38	1.188
5.50	0.88	1.453	1.88	1.376	3.36	1.291	4.53	1.247	5.57	1.224	6.59	1.205	7.59	1.191
6.00	0.80	1.454	1.70	1.378	3.05	1.294	4.14	1.249	5.09	1.226	6.02	1.208	6.93	1.194
6.50	0.73	1.455	1.56	1.380	2.80	1.297	3.81	1.251	4.68	1.227	5.54	1.210	6.38	1.196
7.00	0.67	1.455	1.44	1.381	2.58	1.299	3.53	1.252	4.34	1.229	5.13	1.211	5.91	1.197
7.50	0.62	1.456	1.33	1.382	2.40	1.301	3.28	1.254	4.04	1.230	4.78	1.213	5.51	1.199
8.00	0.58	1.457	1.24	1.383	2.24	1.302	3.07	1.255	3.78	1.231	4.47	1.214	5.15	1.200
8.50	0.54	1.457	1.16	1.384	2.10	1.304	2.89	1.256	3.55	1.232	4.20	1.215	4.84	1.201
9.00	0.51	1.458	1.09	1.385	1.97	1.305	2.72	1.256	3.35	1.233	3.96	1.216	4.57	1.202
9.50	0.48	1.458	1.03	1.386	1.86	1.306	2.57	1.257	3.17	1.234	3.75	1.217	4.32	1.203
10.00	0.46	1.458	0.98	1.386	1.76	1.307	2.44	1.258	3.01	1.235	3.56	1.217	4.10	1.204
11.00	0.41	1.459	0.88	1.387	1.60	1.309	2.21	1.259	2.73	1.236	3.23	1.219	3.72	1.205
12.00	0.38	1.459	0.81	1.388	1.46	1.310	2.03	1.260	2.49	1.237	2.95	1.220	3.41	1.206
13.00	0.35	1.460	0.74	1.389	1.34	1.311	1.87	1.260	2.30	1.238	2.72	1.221	3.14	1.207
14.00	0.32	1.460	0.69	1.389	1.24	1.312	1.73	1.261	2.13	1.238	2.52	1.221	2.91	1.208
15.00	0.30	1.460	0.64	1.390	1.15	1.313	1.61	1.262	1.99	1.239	2.35	1.222	2.71	1.209
17.50	0.25	1.461	0.54	1.391	0.98	1.315	1.38	1.263	1.70	1.240	2.01	1.223	2.32	1.210
20.00	0.22	1.461	0.47	1.392	0.86	1.316	1.21	1.264	1.49	1.241	1.76	1.224	2.03	1.211

Table 11.27 (contd) $- f_{cu} = 50\,\text{N/mm}^2$, $k = 0.60$.

e/R	p=0.4 N/R²	z/R	p=1.0 N/R²	z/R	p=2.0 N/R²	z/R	p=3.0 N/R²	z/R	p=4.0 N/R²	z/R	p=5.0 N/R²	z/R	p=6.0 N/R²	z/R
0.10	66.71	0.616	72.54	0.611	82.34	0.603	92.19	0.597	102.11	0.592	112.07	0.588	122.07	0.583
0.20	57.68	0.716	62.57	0.711	70.74	0.705	78.94	0.700	87.15	0.696	95.38	0.693	103.62	0.690
0.30	48.73	0.812	52.98	0.804	60.00	0.793	66.99	0.786	73.96	0.781	80.93	0.776	87.89	0.773
0.40	40.33	0.901	44.38	0.883	50.77	0.865	56.98	0.853	63.10	0.845	69.19	0.839	75.24	0.834
0.50	32.86	0.978	37.17	0.947	43.30	0.919	48.99	0.903	54.50	0.892	59.93	0.884	65.32	0.878
0.60	26.33	1.047	31.22	0.996	37.20	0.960	42.68	0.940	47.74	0.927	52.66	0.917	57.51	0.910
0.70	20.69	1.110	26.15	1.040	32.32	0.990	37.33	0.967	42.11	0.953	46.80	0.942	51.29	0.934
0.80	16.47	1.157	21.80	1.081	28.49	1.012	33.10	0.988	37.46	0.972	41.71	0.961	45.90	0.953
0.90	13.33	1.193	18.48	1.111	24.77	1.040	29.71	1.004	33.71	0.987	37.60	0.976	41.42	0.967
1.00	10.77	1.226	15.95	1.133	21.73	1.063	26.84	1.018	30.63	0.999	34.21	0.987	37.72	0.978
1.20	7.53	1.269	12.43	1.164	17.34	1.095	21.64	1.051	25.73	1.019	28.97	1.005	32.00	0.995
1.40	5.69	1.293	10.14	1.183	14.37	1.115	18.06	1.073	21.57	1.043	24.98	1.019	27.77	1.007
1.60	4.54	1.308	8.36	1.203	12.25	1.129	15.47	1.088	18.53	1.059	21.50	1.036	24.43	1.018
1.80	3.77	1.318	7.09	1.217	10.66	1.139	13.52	1.099	16.23	1.071	18.86	1.049	21.45	1.031
2.00	3.22	1.326	6.14	1.228	9.44	1.147	12.00	1.108	14.43	1.080	16.79	1.058	19.11	1.041
2.20	2.81	1.331	5.41	1.236	8.47	1.153	10.79	1.115	12.99	1.087	15.12	1.066	17.22	1.049
2.40	2.49	1.335	4.84	1.242	7.67	1.158	9.80	1.120	11.80	1.093	13.75	1.072	15.67	1.055
2.60	2.23	1.338	4.37	1.248	7.02	1.162	8.97	1.124	10.82	1.098	12.61	1.077	14.38	1.060
2.80	2.03	1.341	3.98	1.252	6.46	1.166	8.27	1.128	9.98	1.101	11.64	1.081	13.28	1.065
3.00	1.86	1.343	3.66	1.255	5.99	1.168	7.67	1.131	9.27	1.105	10.81	1.085	12.33	1.069
3.25	1.68	1.345	3.32	1.259	5.49	1.171	7.04	1.135	8.50	1.108	9.93	1.089	11.33	1.073
3.50	1.53	1.347	3.04	1.262	5.06	1.174	6.50	1.137	7.86	1.111	9.17	1.092	10.47	1.076
3.75	1.41	1.349	2.80	1.265	4.70	1.176	6.04	1.140	7.30	1.114	8.53	1.094	9.74	1.079
4.00	1.30	1.350	2.60	1.267	4.38	1.178	5.63	1.142	6.82	1.116	7.97	1.097	9.10	1.081
4.25	1.21	1.351	2.42	1.269	4.11	1.180	5.28	1.143	6.40	1.118	7.48	1.099	8.54	1.083
4.50	1.13	1.352	2.27	1.270	3.86	1.181	4.97	1.145	6.02	1.120	7.04	1.101	8.04	1.085
4.75	1.06	1.353	2.13	1.272	3.65	1.182	4.70	1.146	5.69	1.121	6.65	1.102	7.60	1.087
5.00	1.00	1.354	2.01	1.273	3.45	1.184	4.45	1.148	5.39	1.123	6.31	1.104	7.20	1.088
5.50	0.90	1.355	1.81	1.275	3.11	1.187	4.03	1.150	4.88	1.125	5.71	1.106	6.53	1.091
6.00	0.81	1.356	1.64	1.277	2.83	1.190	3.68	1.151	4.46	1.127	5.22	1.108	5.96	1.093
6.50	0.74	1.358	1.51	1.278	2.59	1.192	3.38	1.153	4.10	1.128	4.80	1.110	5.49	1.095
7.00	0.68	1.359	1.39	1.280	2.40	1.194	3.13	1.154	3.80	1.130	4.45	1.111	5.09	1.096
7.50	0.63	1.360	1.29	1.281	2.22	1.195	2.92	1.155	3.54	1.131	4.14	1.112	4.74	1.097
8.00	0.59	1.361	1.20	1.282	2.08	1.197	2.73	1.156	3.31	1.132	3.88	1.113	4.43	1.099
8.50	0.55	1.361	1.13	1.282	1.95	1.198	2.56	1.157	3.11	1.133	3.64	1.114	4.17	1.100
9.00	0.52	1.362	1.06	1.283	1.83	1.199	2.42	1.158	2.93	1.133	3.44	1.115	3.93	1.100
9.50	0.49	1.363	1.00	1.284	1.73	1.200	2.29	1.158	2.78	1.134	3.25	1.116	3.72	1.101
10.00	0.46	1.363	0.95	1.284	1.64	1.201	2.17	1.159	2.63	1.135	3.09	1.116	3.53	1.102
11.00	0.42	1.364	0.86	1.285	1.48	1.202	1.97	1.160	2.39	1.136	2.80	1.117	3.20	1.103
12.00	0.38	1.365	0.78	1.286	1.36	1.203	1.80	1.161	2.19	1.136	2.56	1.118	2.93	1.104
13.00	0.35	1.365	0.72	1.287	1.25	1.204	1.66	1.161	2.02	1.137	2.36	1.119	2.70	1.105
14.00	0.32	1.366	0.66	1.287	1.16	1.205	1.54	1.162	1.87	1.138	2.19	1.120	2.51	1.105
15.00	0.30	1.366	0.62	1.288	1.08	1.206	1.43	1.162	1.74	1.138	2.04	1.120	2.34	1.106
17.50	0.26	1.367	0.53	1.289	0.92	1.208	1.23	1.163	1.49	1.139	1.75	1.121	2.00	1.107
20.00	0.22	1.368	0.46	1.289	0.80	1.209	1.07	1.164	1.30	1.140	1.53	1.122	1.75	1.108

Index